Membrane Distillation

Membrane Distillation

Membranes, Hybrid Systems, and Pilot Studies

Edited by

Kang-Jia Lu
Tai-Shung Chung

CRC Press
Taylor & Francis Group
Boca Raton London New York

CRC Press is an imprint of the
Taylor & Francis Group, an informa business

CRC Press
Taylor & Francis Group
6000 Broken Sound Parkway NW, Suite 300
Boca Raton, FL 33487-2742

CRC Press is an imprint of Taylor & Francis Group, an Informa business

Printed on acid-free paper

International Standard Book Number-13: 978-0-367-25447-6 (Hardback)

Visit the Taylor & Francis Web site at
www.taylorandfrancis.com

and the CRC Press Web site at
www.crcpress.com

MIX
Paper from responsible sources
FSC™ C013985

Printed in the United Kingdom
by Henry Ling Limited

Contents

SECTION B Hybrid Systems

SECTION C Pilots & Economic Studies

Preface

Clean water, clean energy, global warming, and affordable healthcare are four major concerns globally resulting from clean water shortages, high fluctuations of oil prices, climate changes, and high costs of healthcare. Clean water and public health are highly related, while energy and sustainable prosperity are hand in hand. Among many potential solutions, advances in membrane technology are one of the most direct, effective, and feasible approaches to solve these sophisticated issues. Membrane technology is a fully integrated science and engineering which consists of materials science and engineering, chemistry and chemical engineering, separation and purification phenomena, environmental science and sustainability, statistical mechanics-based molecular simulation, process and product design.

In this book, we will focus on membrane distillation (MD) technology for water reuse and seawater desalination. MD was used to be considered as an energy intensive process for seawater desalination. However, with the advances in solar panels, nanotechnology, and membrane technology, it has been reborn as an emerging technology that draws significant attention from both academia and industries. In addition to the traditional MD-crystallizer (MD-C) systems, totally new integrated processes such as forward osmosis (FO)-MD, freeze distillation (FD)-MD, FD-MD-C, MD-solid hollow fiber cooling crystallization (MD-SHFCC), pressure retarded osmosis (PRO)-MD, and seawater reverse osmosis (SWRO)-MD-PRO have been recently proposed for a wide range of applications that totally open up new perspectives for future MD applications. Using FO as an example, it is not a low-energy process unless there is no regeneration of draw solutions. In addition, a draw solution with a higher osmotic pressure may encounter more difficulties to be regenerated using the traditional pressure-driven processes. To economically regenerate the draw solution, the FO-MD integration provides a potential solution. With the aid of solar energy and waste heat, FO-MD may be more cost-effective than those using RO or NF to recycle the draw solutions.

In the beginning of this book, we will introduce the fundamentals of MD processes, including heat and mass transfer analyses, energy evaluation, and mathematical modeling. Since the key component of MD processes is the hydrophobic membrane, several chapters after the introduction sections will focus on the engineering and molecular design of MD membranes from different aspects. After that, we shift our attention to MD integration with other processes to improve the overall water recovery and separation efficiency in order to meet the goal of zero liquid discharge (ZLD). Various types of hybrid systems, including FD-MD, MD-crystallization (MDC), PRO-MD, and FO-MD, will be discussed in this section. The last part of this book covers MD pilot studies for desalination and wastewater treatment. The first-hand knowledge of MD pilots would provide valuable guidance for the commercialization of MD technology.

Finally, we would like to express our heartfelt thanks to the contributors of each chapter for their dedication and willingness to share their knowledge. We also like to thank the funding agents in last two decades such as Singapore National Research Foundation, Public Utility Board (PUB), A*Star, BASF, and many others. Without their strong provisions, it is hard for us to accomplish this book. In addition, Dr. Lu is grateful for her husband's strong support, while Prof. Chung is thankful for his family's continuous support. We believe this book may provide valuable guidance to design useful MD systems with zero liquid discharge in the near future.

Kang-Jia Lu
Tai-Shung Chung

About the Editors

Dr. Kang-Jia Lu is a research fellow at the Department of Chemical and Biomolecular Engineering, National University of Singapore (NUS). Her Ph.D. and Bachelor (1st honor) were obtained from the same department. Her research interests include membrane fabrication, surface modification, system integration, and simulation of various membrane applications, especially membrane distillation (MD). She was the leader of an MD project of S$2,600,880. She has published 12 research articles in Journal of Membrane Science, Environmental Science & Technology and other leading research journals. She has also won the Faculty of Engineering 28th Innovation & Research Award of NUS in 2014 and Aseanian Membrane Society 11 New Member Special Award in 2018.

Prof. Tai-Shung Chung is a Provost's Chair Professor at the Department of Chemical and Biomolecular Engineering, National University of Singapore (NUS). His research focuses on polymeric membranes. After working in U.S. industries for 15 years, he joined NUS in 1995. In 2005–2008, he worked as a senior consultant for Hyflux Inc., led and built its membrane research team. He became a Fellow in the Academy of Engineering Singapore in 2012 and received IChemE (Institute of Chemical Engineers, UK) Underwood Medal for exceptional research in separations and Singapore President's Technology Award in 2015. He was a highly cited researcher in Chemical Engineering & Materials Science and Engineering by the Elsevier and Shanghai Global Ranking in 2016 and received Distinction Award in Water Reuse and Conservation from International Desalination Association (IDA) in 2016. He is also a highly cited researcher in the list of Clarivate Analytics in 2018.

Contributors

Alicia Kyoungjin An
School of Energy and Environment
City University of Hong Kong
Kowloon
Hong Kong Special Administrative Region, China

Jian Chang
Chemical and Biomolecular Engineering
University of Pennsylvania
Philadelphia, PA, USA

Yuanmiaoliang Chen
NUS Graduate School for Integrative Science and Engineering
National University of Singapore
Singapore

Zhen Lei Cheng
Chemical and Biomolecular Engineering
National University of Singapore
Singapore

Yong-jun Choi
School of Civil and Environmental Engineering
Kookmin University
Jeongneung-Dong, Seongbuk-Gu
Seoul, Republic of Korea

Tai-Shung Chung
Chemical and Biomolecular Engineering
National University of Singapore
Singapore

Bhaskar Jyoti Deka
School of Energy and Environment
City University of Hong Kong
Kowloon
Hong Kong Special Administrative Region, China

Noel Dow
Institute for Sustainable Industries and Liveable Cities
Victoria University
Melbourne, Victoria, Australia

Mikel C. Duke
Institute for Sustainable Industries and Liveable Cities
Victoria University
Melbourne, Victoria, Australia

Jiaxin Guo
School of Energy and Environment
City University of Hong Kong
Kowloon
Hong Kong Special Administrative Region, China

Gang Han
Chemical and Biomolecular Engineering
National University of Singapore
Singapore

Sangho Lee
School of Civil and Environmental Engineering
Kookmin University
Jeongneung-Dong, Seongbuk-Gu
Seoul, Republic of Korea

Kang-Jia Lu
Chemical and Biomolecular Engineering
National University of Singapore
Singapore

Lin Luo
Chemical and Biomolecular Engineering
National University of Singapore
Singapore

Liang Shen
Chemical and Biomolecular Engineering
National University of Singapore
Singapore

Panu Sukitpaneenit
Koch Membrane Systems
Wilmington, MA, USA

Peng Wang
Chemical and Biomolecular Engineering
National University of Singapore
Singapore

Yan Wang
Department of Chemical Engineering
Huazhong University of Science and Technology
Wuhan, People's Republic of China

Sui Zhang
Chemical and Biomolecular Engineering
National University of Singapore
Singapore

Dieling Zhao
Chemical and Biomolecular Engineering
National University of Singapore
Singapore

Jian Zuo
Chemical Engineering and Food Technology
Singapore Institute of Technology
Singapore

Section A

Fundamentals

1 Introduction to Membrane Distillation

Kang-Jia Lu and Tai-Shung Chung

1.1 MEMBRANE DISTILLATION AND ITS HISTORICAL DEVELOPMENT

Water is one of the most essential natural resources, yet it is depleting rapidly due to human activities. An increasing number of countries, especially in arid regions, have reached the limit at which sustainable water can be hardly delivered (United Nations, 2018). In order to mitigate water scarcity, various desalination technologies have been developed in past few decades. Among them, desalination based on membrane technologies contributes to nearly 68% of the total desalination capacity, and this figure is constantly increasing as the membrane market grows rapidly (Nakoa et al., 2016). Among various types of membrane technologies, membrane distillation (MD) has drawn increasing attention in past few decades due to its unique characteristics.

MD is a thermally driven separation process that involves the transport of water vapor across a hydrophobic membrane (Alkhudhiri et al., 2012; Alklaibi and Lior, 2005; Camacho et al., 2013; Drioli et al., 2015; Johnson and Nguyen, 2017; Khayet, 2011; Khayet and Matsuura, 2011; Lawson and Lloyd, 1997; Schneider et al., 1988; Wang and Chung, 2015). As its name suggests, the driving force of MD is similar to that of the conventional distillation process, namely, the water vapor pressure difference. Unlike the traditional distillation, MD uses hydrophobic membranes as the contact surfaces where water evaporation takes place. The concept was first brought up by Bodell in his U.S. patent entitled "Silicone rubber vapor diffusion in saline water distillation" in June 1963 (Bodell, 1963). Five years later, another patent entitled "Distillation of saline water using silicone rubber membrane" was published by Bodell as a continuation-in-part of his first MD patent (Bodel, 1968). In his patents, Bodell described an apparatus that extracted potable water from unpotable aqueous mediums through membrane-assisted distillation. The apparatus consisted of several vapor permeable, liquid impermeable tubular silicone rubber membranes positioned in a hot feed tank. Water vapor that diffused into the lumen side of the membranes was collected by an air stream and condensed externally to a salt-free fluid. This invention gave rise to the first sweeping gas membrane distillation (SGMD) configuration. Bodell also mentioned an alternative method to collect water vapor by applying vacuum on the membrane lumen side, which was regarded as the first description of vacuum membrane distillation (VMD). More detailed descriptions of various types of MD configurations will be summarized in Chapter 2.

The first paper on MD was published in 1967 by Findley (1967). He introduced an "infinite-stage flash evaporation" apparatus where water evaporated from one interface of "a nonwettable porous membrane" and condensed at the other interface of the membrane, giving rise to the first direct contact membrane distillation (DCMD) configuration. Findley listed several advantages of this technology, including the mild operating temperature and pressure, low tendency to scale, space saving, and the flexibility to carry out the heat exchange either inside or outside of the evaporation system. Findley also pointed out that the porous medium would introduce resistance to vapor diffusion and the heat conduction across the membrane would cause energy loss. In addition, Findley specified the desired properties of membranes for this application. According to him, the suitable membrane should have (1) a high resistance to conductive heat flow; (2) a sufficient but not excessive thickness; (3) a negligible permeability to liquids and nonvolatile compounds; (4) a low absorptivity of moisture; (5) a high porosity and relatively straight-through pores to reduce the mass transfer resistance; and (6) a uniform porosity. These prospective views are still providing important guidance today in developing suitable membranes for MD (Alkhudhiri et al., 2012; Camacho et al., 2013; Drioli et al., 2015a; Khayet, 2011; Suk and Matsuura, 2006; Wang and Chung, 2015). In the same year, a U.S. patent was filed by Weyl et al. (Weyl, 1967)using a polytetrafluoroethylene (PTFE) membrane to recover demineralized water from saline waters. The membrane was 3175 μm thick and had a porosity and mean pore size of 42% and 9 μm, respectively. Weyl et al. addressed another important advantage of MD, namely, 100% theoretical rejections to nonvolatile solutes if wetting did not occur. Other than PTFE, polyethylene (PE), polypropylene (PP), and polyvinyl chloride (PVC) and hydrophobic ceramic composition were also suggested. They also proposed to fabricate composite membranes that consisted of a hydrophilic support and a hydrophobic coating to achieve desirable antiwetting properties. In addition, two setup configurations were proposed to improve the heat recovery of the systems. One was a five-stage system consisting of alternating membrane and metal plate with heat supplied through the bottom plate. The heat released in the condensed water compartment of a stage was transferred to the saline water in the next stage by heat conduction through the metal plates. The other design was to have the hydrophobic flat sheet membrane coiled up into a hot cylinder to give a multistage spiral wound module.

1967 was an exciting year for MD. Another two papers were published by Henderyckx et al. from Belgium (Henderyckx, 1967; Van Haute et al., 1967). In one of the works, they established a curve relating the vapor permeability to the temperature for a given membrane in an SGMD configuration (Van Haute et al., 1967). In their second study, a diffusion doublet that consisted of a vapor permeable membrane and a plastic condensation surface that were separated by a thin layer of gas was investigated, giving rise to the air gap membrane distillation (AGMD) configuration (Henderyckx, 1967). They also stated the possibility of utilizing waste hot water and solar energy to power the system, though no further experiment was conducted. In 1969, Findley et al. published another fundamental study on mass and heat transfer relations involved in evaporation through porous membranes (Findley et al., 1969). The study showed that the major factor influencing the rate of vapor transfer was the diffusion through the stagnant gas inside the membrane pores. After that, the passion toward MD quickly faded away, which could partially be ascribed to the lack of suitable membranes and the relatively low production rate compared to other membrane-based desalination technologies such as reverse osmosis (RO).

In early 1980s, with the advances in membrane fabrication techniques, the interest in MD was recovered. Especially, the academic interest in MD was fueled by its versatility and fundamental engineering concepts embodied in MD processes. In 1969, Bob Gore accidentally applied a sudden and accelerating stretch to the heated PTFE rods of about 800% and formed a microporous structure with a porosity over 70%. This happy accident led to the creation of the expanded polytetrafluoroethylene (ePTFE) under the trademark Gore-Tex (Schneider et al., 1988). In 1982, Gore & Associated Co. proposed to apply the Gore-Tex membrane for MD in a liquid-gap MD configuration. The adopted membrane had a thickness of 25 μm, a porosity up to 80%, and a pore size between 0.2 and 0.45 μm (Gore, 1982). During the same period, Cheng et al. filed a series of U.S. patents on multilayer hydrophobic–hydrophilic composite membranes for MD (Cheng, 1981; Cheng and Wiersma, 1982, 1983a, 1983b). They claimed that the attachment of hydrophilic support layers to hydrophobic support layer could significantly enhance the water vapor flux. It was reported that a composite membrane with the maximum hydrophobic pore diameter of 0.48 μm generated a flux of 75.2 kg m^{-2} day^{-1} for a feed temperature of 62.8 °C and a distillate temperature of 56.7 °C. That was one order of magnitude higher than that reported by Weyl et al. in 1967. At the same time, several other companies such as the Swedish Development Co. and Enka AG also conducted research on MD actively. The former invented the plate-and-frame PTFE membrane for AGMD in 1983 and the latter developed trans MD tubular PP membranes for DCMD in 1984 (Carlsson, 1983; Johnson and Nguyen, 2017; Khayet and Matsuura, 2011). The breakthroughs and problems associated with these membranes and systems highlighted the need for further research on MD. Since then, MD has drawn more attention from research groups worldwide and the number of studies on this topic has been increasing exponentially. Figure 1.1 depicts the numbers of publications on MD each year since 1982 until 2017. The data were extracted from Scopus.com. There is no doubt that MD is getting increasingly popular in the last few decades.

Even though MD has been thoroughly studied in the past few decades, a full commercialization has not been achieved yet. Nevertheless, pilot studies and small-to-medium-scale commercial implementations of MD have been carried out all over the world and promising progresses have been made. Table 1.1 lists the developers and promoters of commercial MD technology. Detailed

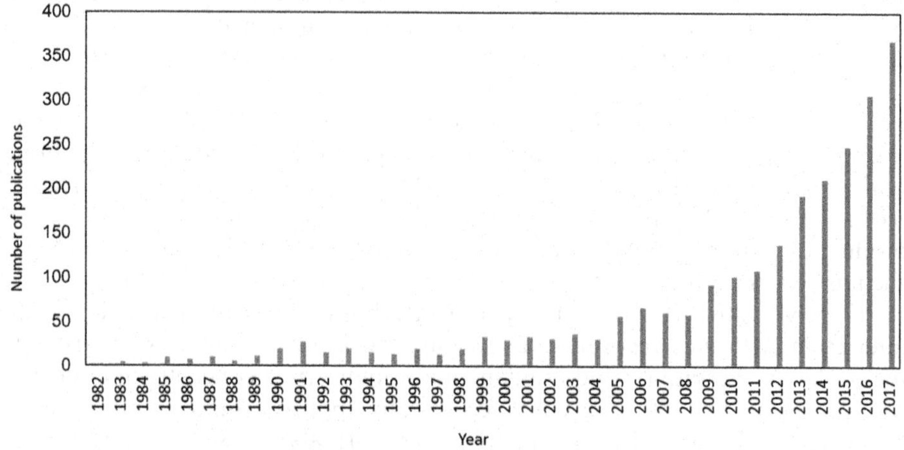

FIGURE 1.1 Numbers of publications on MD from 1983 to 2017. The data were extracted from Scopus.com on 31 July 2018.

FIGURE 1.2 Close-up of the tri-bore hollow fiber membranes commercialized by START Center and Memsift Innovation Pte Ltd. Photo taken by Lester Kok Wei Ming, Nanyang Technological University (NTU).

TABLE 1.1

Commercial MD technology developers and promoters. Copyright (2017) Elsevier

Company name	Headquarters	Type	Notes
SolarSpring GmbH	Germany	Technology developer	Spin-off company of the Fraunhofer Institute for Solar Energy Systems/provided complete MD systems
TNO	Netherlands	Technology developer	Developed Memstill technology which was licensed to Aquastill and Keppel Seghers, part of a Dutch consortium for investigation of MD as desalination alternative
Aquastill	Netherlands	Technology promoter	Holder of Memstill MD technology license
Keppel Seghers	Belgium	Technology promoter	Holder of Memstill MD technology license
Memsys GmbH	Germany	Technology developer	The developer of multistage MD systems. In 2016, NCHL acquired all the assets and IPs of Memsys.
Memsys Clearwater Pte Ltd.	Singapore	Technology promoter	
Aquaver	Netherlands	Technology developer	In 2012, Memsys and Aquaver agreed on an exclusive license agreement for worldwide applications in potable water production and produced water treatment. In 2015, Aquaver merged with Memsys.
Abengoa Water	Spain	Technology promoter	Collaborates with Aquaver for the development of an MD pilot unit for SWRO brine concentration. Also, has its own MD business section
Condorchem Envitech	Spain	Technology promoter	Utilized Memsys systems
General Electric Co. (GE)	USA	Technology promoter	GE and Memsys have entered into a partnership to jointly develop MD technology for applications related to shale gas and coal seam gas production
Scarab Development AB	Sweden	Technology developer	Technology developer for XZero and HVR Water purification

(Continued)

TABLE 1.1 (Cont.)

Company name	Headquarters	Type	Notes
Xzero AB	Sweden	Technology promoter	Production of ultrapure water for the semiconductor industry
HVR Water Purification	Sweden	Technology promoter	
KmX Corporation	Canada	Technology developer	
BlueGold Technologies	USA	Technology developer	Technology developer for Bluestill MD technology
Separation Technologies Applied Research and Translation (START) Centre and Memsift Innovations Pte Ltd.	Singapore	Technology promoter	The technology leveraged a new type of tri-bore hollow fiber membrane invented by Professor Neal Chung at the National University of Singapore (NUS)

information of these commercial attempts can be found elsewhere (Thomas et al., 2017; Zaragoza et al., 2018). In February 2019, a new MD pilot plant is being built in Singapore to reduce the amount of liquid waste disposal and to recover precious metals from the semiconductor wastewater (Nanyang Technological University, 2019). The plant is being built jointly by the Separation Technologies Applied Research and Translation (START) Center and a local water technology firm Memsift Innovations Pte Ltd. in Singapore. The commercialized technology leveraged a new type of tri-bore hollow fiber membrane invented by Chung et al. from National University of Singapore (Lu et al., 2016; Wang et al., 2014; Zuo and Chung, 2017). The novel membranes shown in Figure 1.2 possess great mechanical stabilities and provide a remarkable MD performance in a VMD mode.

1.2 ADVANTAGES, APPLICATIONS, AND CHALLENGES

As its name suggests, MD combines the properties of both membrane separation and traditional distillation. Similar to other membrane-based processes, MD modules can be made compact due to the large surface-area-to-volume ratio. In addition, the large vapor space needed in a conventional distillation unit is substituted by the pore volume of microporous membranes. This translates to the real saving of footprint and capital investment (Lawson and Lloyd, 1997). Another difference between MD and conventional distillation is the operating temperature. MD typically operates at a much lower temperature of 40 to 90 °C, which brings about at least three advantages: (1) It can be powered by low grade energy sources such as solar energy or industrial waste heat. This will significantly lower its energy cost (Henderyckx, 1967); (2) A mild temperature is often critical in some applications to avoid damaging the bioactivity or nutrients of the valuable compounds (Gunko et al., 2006; Hausmann et al., 2011); (3) Feed solutions with milder temperatures are less corrosive and the heat conduction loss to the surrounding environment is also less owing to the smaller temperature gradient.

MD is also unique compared to other conventional pressure-driven membrane separation processes such as RO, nanofiltration (NF), ultrafiltration (UF), and microfiltration (MF). These membrane processes are based on size exclusion and Donnan effect. Thus, the rejection factor toward extremely small solutes or neutral compounds may not be satisfactory. In MD processes, however, only vapor is designed to pass through the membrane. Therefore, it has a theoretically 100% rejection to all nonvolatile species regardless of their molecular sizes and charges (Drioli et al., 2015; Lu et al., 2016). In addition, the driving force of those conventional membrane separation processes is the difference between the applied hydraulic pressure and the osmotic pressure of the feed solution. As osmotic pressure is greatly affected by the feed concentration, the required hydraulic pressure will also increase accordingly if the feed concentration increases. However, the feed concentration has a much less impact on the vapor pressure, which is the driving force for MD. For example, from pure water to a synthesized seawater containing 3.5% sodium chloride, the osmotic pressure increases from 0 to 26.4 bar, whereas the vapor pressure at 70 °C only drops from 0.312 to 0.308 bar. This advantage makes MD especially suitable for the treatment of high salinity wastewater (Lu et al., 2018). In fact, MD is often integrated with other membrane processes to treat their concentrated brine solutions or to regenerate valuable draw solutes. In addition, the lower operating pressure of MD translates to the higher process safety and the less demand on mechanical properties of the MD system (Lawson and Lloyd, 1997). Furthermore, common materials for MD membranes such as PTFE, PP, and polyvinylidenedifluoride (PVDF) are chemically resistant polymers, which can preserve the integrity of membranes in harsh environments like concentrated acid.

Thanks to the above-mentioned special advantages, MD has found numbers of applications in different areas. Among them, the most studied application is desalination. Thomas et al. reported that in the first two decades since MD was invented, 67% of the publications were about desalination. Even in recent years from 2011 until 2016, 48% of the MD papers were still about desalination (Thomas et al., 2017). Similar to the academic research, industrial pilot studies also focused on desalination application. Interestingly, the majority of these pilots were coupled with either solar energy or waste energy to minimize the high energy consumption of MD (Thomas et al., 2017). Recently, another MD application – brine concentration – has drawn increasing attention. MD was integrated with other desalination processes such as RO and freeze desalination (FD) to treat their brine solutions in order to reduce the discharge volume and to improve the water recovery (Guan et al., 2012; Julian et al., 2016; Wang and Chung, 2012). This application aimed to take the advantage of unique MD ability in treating high concentration saline water. It may become an attractive and important MD application in the future by both academia and industries (Thomas et al., 2017).

When MD is integrated with salt crystallization, a zero liquid discharge desalination (ZLDD) becomes possible. Not only it converts seawater into potable water but also produce valuable salts without generating a waste stream (Creusen et al., 2013; Edwie and Chung, 2012; Guan et al., 2012; Lu et al., 2019; Quist-Jensen et al., 2016). ZLDD is a sustainable desalination technology that gains increasing attention recently. MD for wastewater treatment is another hot topic because textile, olive mills, petrochemical, oil and gas, mining, and rubber industries need to treat their effluents in order to reduce waste discharge and to recover valuable water and solutes (Calabro et al., 1991; El-Abbassi et al., 2013; Hussain et al., 2015; Khaing et al., 2010; Mokhtar et al., 2015; Sivakumar et al., 2013). Compared to other technologies such as NF, MD often shows higher rejections due to its special separation mechanism. In addition, shale gas is gaining worldwide attention recently as one of the most promising energy sources. However, it is challenging to treat its produced water owing to the presence of complex chemicals and high total dissolved solids (TDS) (Fakhru'l-Razi et al., 2009). MD is one of the emerging technologies showing promise to treat and recycle the produced water. Despite its capability in handling high salinity water with TDS, MD can also leverage the hot produced water to drive the separation process and significantly lower its energy cost (Shaffer et al., 2013).

However, even with the aforementioned advantages, MD is still in its infancy and there are very few commercially available membranes specially designed for MD. One of the major reasons is that MD membranes require a high porosity of over 75% and a large pore size of about 100–300 nm in order to achieve competitive performance (Wang and Chung, 2012, 2015); nonetheless, the high porosity and large pore size reduce the membrane strength and make the membrane vulnerable during real industrial applications. In normal operations, a hydraulic pressure must be applied to the fluid to counterbalance the pressure drop. The hydraulic pressure can be high if the process is operated at a high flow rate with a long and congested module. Moreover, membrane mechanical integrity will be further impaired due to the elevated operation temperature (Lu et al., 2017). Therefore, a technology to improve the mechanical properties of PVDF MD membranes remains a hot subject in the field.

Another challenge that sabotages the commercialization of MD is membrane wetting. Wetting occurs when the liquid solution penetrates into the membrane pores. It could lead to a reduction of flux, a drop of energy efficiency, and a loss of rejection. Wetting phenomena involve complex mechanisms. It is widely accepted that a Cassie–Baxter state where a liquid is retained on the membrane surface by a liquid–gas–solid interface is the key to avoid the wetting problem (Lu et al., 2018; Rezaei et al., 2018). However, the Cassie–Baxter state is metastable; it may transit to a wetting state in many cases (Banerjee, 2008; Koishi et al., 2009; Murakami et al., 2014). The primary cause of membrane wetting is fouling when the deposition of foreign materials alters the hydrophobicity of the membrane surface. Types of fouling include organic fouling, scaling deposition (inorganic fouling), and particulate or colloidal fouling (Rezaei et al., 2018; Warsinger et al., 2015). Wetting also occurs when the hydraulic pressure exceeds the liquid entry pressure (LEP), which is the highest hydraulic pressure applied to the contacting fluid before the liquid–gas–solid interface formed by the surface tension is destroyed (Israelachvili, 2011). Chemical degradation of membranes is regarded as another reason to cause the membrane wetting during a long-term MD process (Gryta, 2005). Because of the detrimental effects due to wetting and the difficulty to restore the membrane after wetting, how to prevent or delay it becomes a hot research field in MD.

1.3 BOOK OBJECTIVES

This book aims to provide a comprehensive summary of MD fundamentals and strategies to overcome the challenges for commercialization. The book comprises three main parts: (1) fundamentals, (2) hybrid systems, and (3) pilot and economic studies.

In the first part, an overview of MD fundamentals including its historical development and current status will be covered (Chapter 1), followed by the theoretical background and modeling of different MD configurations (Chapter 2). Next, fundamentals of membrane fabrication for MD will be discussed from both academic and industrial perspectives (Chapters 3 and 4), followed by five chapters on novel membranes for MD applications including electrospun nanofiber membranes (Chapter 5), dual-layer membranes and multibore membranes (Chapter 6), hydrophobic membranes with enhanced antiwetting properties (Chapter 7), omniphobic membranes for feeds containing low surface tension materials (Chapter 8), and membranes fabricated using green solvents (Chapter 9).

The second part of the book discusses a series of hybrid systems that contain a MD unit. Chapter 10 illustrates an integrated MD-crystallization (MD-C) system to reclaim water and recover valuable salts. Chapter 11 describes a hybrid system consisting of freeze desalination and MD (FD-MD) for harvesting pure ice and liquid water from seawater. Chapters 12 and 13 illustrate two hybrid systems, namely, pressure retarded osmosis (PRO)-MD and forward osmosis (FO)-MD systems, respectively. In both cases, MD is employed to concentrate draw solutions by removing excess water. In Chapter 14, a hybrid system consisting of freeze desalination, membrane distillation and crystallization (FD-MD-C) to achieve zero liquid discharge desalination (ZLDD) is illustrated in details.

In the last part of the book, Chapter 15 is dedicated to the energy sources and economic analysis of MD processes, which is critical to a profitable commercialization. Chapter 16 discusses the lessons learned from previous pilot studies for wastewater treatment. The valuable antecedent experience may help scale up and industrialize MD in the near future. We hope this book would provide useful insights to design better MD technologies for earth sustainability.

REFERENCES

Ahmadun, F., Pendashteh, A., Abdullah, L.C., Biak, D.R.A., Madaeni, S.S., Abidin, Z. Z. 2009. Review of technologies for oil and gas produced water treatment. J. Hazard. Mater. 170, 530–551. doi:10.1016/j.jhazmat.2009.05.044

Alkhudhiri, A., Darwish, N., Hilal, N. 2012. Membrane distillation: a comprehensive review. Desalination 287, 2–18. doi:10.1016/j.desal.2011.08.027

Alklaibi, A.M., Lior, N. 2005. Membrane-distillation desalination: status and potential. Desalination 171, 111–131. doi:10.1016/j.desal.2004.03.024

Banerjee, S., 2008. Simple derivation of Young, Wenzel and Cassie-Baxter equations and its interpretations. 1–10.

Bodel, B.R., 1968. Distillation of saline water using silicone rubber membrane. 3,361,645.

Bodell, B.R., 1963. Silicone rubber vapor diffusion in saline water distillation. 285,032.

Calabro, V., Drioli, E., Matera, F. 1991. Membrane distillation in the textile wastewater treatment. Desalination 83, 209–224. doi:10.1016/0011-9164(91)85096-D

Camacho, L.M., Dumée, L., Zhang, J., Li, J., Duke, M., Gomez, J., Gray, S. 2013. Advances in membrane distillation for water desalination and purification applications. Water 5, 94–196. doi:10.3390/w5010094

Carlsson, L. 1983. The new generation in sea water desalination SU membrane distillation system. Desalination 45, 221–222. doi:10.1016/0011-9164(83)87030-1

Cheng, D.Y., 1981. Method and apparatus for distillation. 4,265,713.

Cheng, D.Y., Wiersma, S.J., 1982. Composite membrane for a membrane distillation system. 4,316,772.

Cheng, D.Y., Wiersma, S.J., 1983a. Apparatus and method for thermal membrane distillation. 4,419,187.

Cheng, D.Y., Wiersma, S.J., 1983b. Composite membrane for a membrane distillation system. 4,419,242.

Creusen, R., van Medevoort, J., Roelands, M., van Renesse van Duivenbode, A., Hanemaaijer, J.H., van Leerdam, R. 2013. Integrated membrane distillation-crystallization: process design and cost estimations for seawater treatment and fluxes of single salt solutions. Desalination 323, 8–16. doi:10.1016/j.desal.2013.02.013

Drioli, E., Ali, A., Macedonio, F. 2015a. Membrane distillation: recent developments and perspectives. Desalination 356, 56–84. doi:10.1016/j.desal.2014.10.028

Edwie, F., Chung, T.S. 2012. Development of hollow fiber membranes for water and salt recovery from highly concentrated brine via direct contact membrane distillation and crystallization. J. Memb. Sci. 421–422, 111–123. doi:10.1016/j.memsci.2012.07.001

El-Abbassi, A., Hafidi, A., Khayet, M., García-Payo, M.C. 2013. Integrated direct contact membrane distillation for olive mill wastewater treatment. Desalination 323, 31–38. doi:10.1016/j.desal.2012.06.014

Findley, M.E. 1967. Vaporization through porous membranes. Ind. Eng. Chem. Process Des. Dev. 6, 226–230. doi:10.1021/i260022a013

Findley, M.E., Tanna, V. V., Rao, Y.B., Yeh, C.L. 1969. Mass and heat transfer relations in evaporation through porous membranes. AIChE J 15, 483–489. doi:10.1002/aic.690150404

Gore, D.W. 1982. Gore-Tex membrane distillation. In Proc. of the 10th Ann. Convention of the Water Supply Improvement Assoc. Honolulu, HI.

Gryta, M. 2005. Long-term performance of membrane distillation process. J. Memb. Sci. 265, 153–159. doi:10.1016/j.memsci.2005.04.049

Guan, G., Wang, R., Wicaksana, F., Yang, X., Fane, A.G. 2012. Analysis of membrane distillation crystallization system for high salinity brine treatment with zero discharge using Aspen flowsheet simulation. Ind. Eng. Chem. Res. 51, 13405–13413. doi:10.1021/ie3002183

Gunko, S., Verbych, S., Bryk, M., Hilal, N. 2006. Concentration of apple juice using direct contact membrane distillation. Desalination 190, 117–124. doi:10.1016/j.desal.2005.09.001

Hausmann, A., Sanciolo, P., Vasiljevic, T., Ponnampalam, E., Quispe-Chavez, N., Weeks, M., Duke, M. 2011. Direct contact membrane distillation of dairy process streams. Membranes (Basel) 1, 48–58. doi:10.3390/membranes1010048

Henderyckx, Y. 1967. Diffusion doublet research. Desalination 3, 237–242. doi:10.1016/0011-9164(67)80014-6

Hussain, A., Minier-Matar, J., Janson, A., Adham, S., 2015. Treatment of produced water from oil & gas operations by membrane distillation. In Proceedings of the 4th International Gas Processing Symposium. Elsevier B.V., pp. 285–292. doi:10.1016/B978-0-444-63461-0.50029-8

Israelachvili, J.N. 2011. Intermolecular and surface forces, 3rd ed. Elsevier, Amsterdam.

Johnson, R.A., Nguyen, M.H. 2017. Understanding membrane distillation and osmotic distillation, 1st ed. John Wiley & Sons, Inc., Hoboken, NJ.

Julian, H., Meng, S., Li, H., Ye, Y., Chen, V. 2016. Effect of operation parameters on the mass transfer and fouling in submerged vacuum membrane distillation crystallization (VMDC) for inland brine water treatment. J. Memb. Sci. 520, 679–692. doi:10.1016/j.memsci.2016.08.032

Khaing, T.H., Li, J., Li, Y., Wai, N., Wong, F.S. 2010. Feasibility study on petrochemical wastewater treatment and reuse using a novel submerged membrane distillation bioreactor. Sep. Purif. Technol. 74, 138–143. doi:10.1016/j.seppur.2010.05.016

Khayet, M. 2011. Membranes and theoretical modeling of membrane distillation: a review. Adv. Colloid Interface Sci. 164, 56–88. doi:10.1016/j.cis.2010.09.005

Khayet, M., Matsuura, T. 2011. Introduction to Membrane Distillation. In Membrane Distillation,Khayet, M., Matsuura, T. 1st ed. Elsevier B.V., Amsterdam.1-16 doi:10.1016/B978-0-444-53126-1.10001-6

Koishi, T., Yasuoka, K., Fujikawa, S., Ebisuzaki, T., Zeng, X.C. 2009. Coexistence and transition between Cassie and Wenzel state on pillared hydrophobic surface. Proc. Natl. Acad. Sci. 106, 8435–8440. doi:10.1073/pnas.0902027106

Lawson, K.W., Lloyd, D.R. 1997. Membrane distillation. J. Memb. Sci. 124, 1–25. doi:10.1016/S0376-7388(96)00236-0

Lu, K.J., Cheng, Z.L., Chang, J., Luo, L., Chung, T. 2019. Design of zero liquid discharge desalination (ZLDD) systems consisting of freeze desalination, membrane distillation, and crystallization powered by green energies. Desalination 458, 66–75. doi:10.1016/j.desal.2019.02.001

Lu, K.J., Zuo, J., Chang, J., Kuan, H.N., Chung, T.S. 2018. Omniphobic hollow-fiber membranes for vacuum membrane distillation. Environ. Sci. Technol. 52, 4472–4480. doi:10.1021/acs.est.8b00766

Lu, K.J., Zuo, J., Chung, T.S. 2016. Tri-bore PVDF hollow fibers with a super-hydrophobic coating for membrane distillation. J. Memb. Sci. 514, 165–175. doi:10.1016/j.memsci.2016.04.058

Lu, K.J., Zuo, J., Chung, T.S. 2017. Novel PVDF membranes comprising n-butylamine functionalized graphene oxide for direct contact membrane distillation. J. Memb. Sci. 539, 34–42. doi:10.1016/j.memsci.2017.05.064

Mokhtar, N.M., Lau, W.J., Ismail, A.F., Veerasamy, D. 2015. Membrane distillation technology for treatment of wastewater from rubber industry in Malaysia. Procedia CIRP 26, 792–796. doi:10.1016/j.procir.2014.07.161

Murakami, D., Jinnai, H., Takahara, A. 2014. Wetting transition from the cassie-baxter state to the wenzel state on textured polymer surfaces. Langmuir 30, 2061–2067. doi:10.1021/la4049067

Nakoa, K., Rahaoui, K., Date, A., Akbarzadeh, A. 2016. Sustainable zero liquid discharge desalination (SZLDD). Sol. Energy 135, 337–347. doi:10.1016/j.solener.2016.05.047

Nanyang Technological University. 2019. New membrane water treatment system to reduce toxic waste by over 90 percent [WWW Document]. URL https://phys.org/news/2019-02-membrane-treatment-toxic-disposal-cent.html

Quist-Jensen, C.A., Macedonio, F., Drioli, E. 2016. Membrane crystallization for salts recovery from brine – an experimental and theoretical analysis. Desalin. Water Treat. 57, 7593–7603. doi:10.1080/19443994.2015.1030110

Rezaei, M., Warsinger, D.M., Lienhard V, J.H., Duke, M.C., Matsuura, T., Samhaber, W. M. 2018. Wetting phenomena in membrane distillation: mechanisms, reversal, and prevention. Water Res 139, 329–352. doi:10.1016/j.watres.2018.03.058

Schneider, K., Hölz, W., Wollbeck, R., Ripperger, S. 1988. Membranes and modules for transmembrane distillation. J. Memb. Sci. 39, 25–42. doi:10.1016/S0376-7388(00)80992-8

Shaffer, D.L., Arias Chavez, L.H., Ben-Sasson, M., Romero-Vargas Castrillón, S., Yip, N.Y., Elimelech, M. 2013. Desalination and reuse of high-salinity shale gas produced water: drivers, technologies, and future directions. Environ. Sci. Technol. 47, 9569–9583. doi:10.1021/es401966e

Sivakumar, M., Ramezanianpour, M., O'Halloran, G. 2013. Mine water treatment using a vacuum membrane distillation system. APCBEE Procedia 5, 157–162. doi:10.1016/j.apcbee.2013.05.028

Suk, D.E., Matsuura, T. 2006. Membrane-based hybrid processes: a review. Sep. Sci. Technol. 41, 595–626. doi:10.1080/01496390600552347

Thomas, N., Mavukkandy, M.O., Loutatidou, S., Arafat, H.A. 2017. Membrane distillation research & implementation: lessons from the past five decades. Sep. Purif. Technol. 189, 108–127. doi:10.1016/j.seppur.2017.07.069

United Nations. 2018. Water scarcity [WWW Document]. URL www.unwater.org/water-facts/scarcity/; (accessed 7.6.2018).

Van Haute, A., Henderyckx, Y., 1967. The permeability of membranes to water vapor. Desalination 3, 169–173. doi:10.1016/0011-9164(67)80006-7

Wang, P., Chung, T.S. 2012. A conceptual demonstration of freeze desalination-membrane distillation (FD-MD) hybrid desalination process utilizing liquefied natural gas (LNG) cold energy. Water Res 46, 4037–4052. doi:10.1016/j.watres.2012.04.042

Wang, P., Chung, T.S. 2015. Recent advances in membrane distillation processes: membrane development, configuration design and application exploring. J. Memb. Sci. 474, 39–56. doi:10.1016/j.memsci.2014.09.016

Wang, P., Luo, L., Chung, T.S. 2014. Tri-bore ultra-filtration hollow fiber membranes with a novel triangle-shape outer geometry. J. Memb. Sci. 452, 212–218. doi:10.1016/j.memsci.2013.10.033

Warsinger, D.M., Swaminathan, J., Guillen-Burrieza, E., Arafat, H.A., Lienhard V.J.H. 2015. Scaling and fouling in membrane distillation for desalination applications: a review. Desalination 356, 294–313. doi:10.1016/j.desal.2014.06.031

Weyl, P.K., 1967. Recovery of demineralized water from saline waters. 3,340,186.

Zaragoza, G., Andrés-Mañas, J.A., Ruiz-Aguirre, A. 2018. Commercial scale membrane distillation for solar desalination. npj Clean Water 1, 1–6. doi:10.1038/s41545-018-0020-z

Zuo, J., Chung, T.S. 2017. PVDF hollow fibers with novel sandwich structure and superior wetting resistance for vacuum membrane distillation. Desalination 417, 94–101. doi:10.1016/j.desal.2017.05.022

2 Theoretical Background and Mathematical Modeling of Membrane Distillation

Kang-Jia Lu and Zhen Lei Cheng

2.1 MEMBRANE DISTILLATION MEMBRANE MATERIALS AND CONFIGURATIONS

2.1.1 MEMBRANE MATERIALS

As mentioned previously, membrane distillation (MD) membranes should be hydrophobic to resist wetting, which require materials with a low surface energy. Commonly used materials include polyethylene (PE), polypropylene (PP), polyvinylidene difluoride (PVDF), and polytetrafluoroethylene (PTFE). General properties of these materials are summarized in Table 2.1

PP and PTFE porous membranes produced via stretching processes were often used in the early MD research because of their superior hydrophobicity (Cheng and Wiersma, 1983; Schneider et al., 1988; Weyl, 1967). Today, PTFE membranes still predominate in both academic and pilot MD studies (Thomas et al., 2017). On the other hand, PVDF has gained increasing attention in the MD community because of its superior processibility. It can be dissolved in various types of solvents, including common solvents like *n*-methyl pyrrolidone (NMP), dimethylformamide (DMF), and dimethylacetamide (DMA_C) (Hou et al., 2009; Lu et al., 2017; Zhao et al., 2005) and green solvents such as triethyl phosphate (TEP) (Chang et al., 2017).

2.1.2 MEMBRANE FABRICATION TECHNOLOGIES

Polymeric MD membranes can be fabricated with different methods including melt extrusion, controlled stretching, track etching, electrospinning, and phase inversion. Among them, phase inversion is the most common method owing to its versatility and scalability (Kim et al., 2016). Recently, electrospinning has regained increasing attention as an alternative to phase inversion method for membrane fabrication due to a surging interest in nanotechnology. Figure 2.1 shows the typical morphologies of membranes produced from phase inversion and electrospinning methods.

TABLE 2.1

General properties of common polymers (Lide, 2004; Mark, 1999; Speight, 2005)

Chemical name	Structure of repeat unit	Density, ρ (g/cm^3)	Surface free energy at 293 k (mJ/m^2)	Melting temperature, T_m (K)	Glass transition temperature, T_g (K)	Thermal conductivity, λ (W/m K)
Polyvinylidene fluoride (PVDF)	$[-CF_2-CH_2-]_n$	1.75–1.78	30.3	451	238	0.13 (at 293 K)
Polytetrafluoroethylene (PTFE)	$[-CF_2-CF_2-]_n$	2.00–2.34	20	608	399 (α phase) 292 (β phase)	0.25 (at 293 K)
Polypropylene (PP)	$[-CH(CH_3)-CH_2-]_n$	0.90–0.94	30.1	444	283.7	0.12 (at 293 K)
Polyethylene (PE)	$[-CH_2-CH_2-]_n$	0.92–0.97	35.7	418	150	0.33–0.52

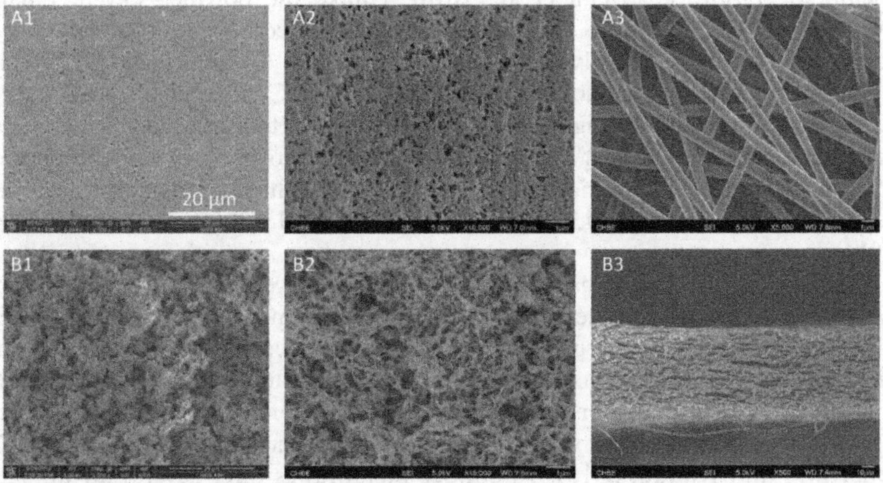

FIGURE 2.1 SEM pictures of membranes fabricated via TIPS (A1, B1) (Cui et al., 2013), copyright (2013) Elsevier, NIPS (A2, B2) (Lu et al., 2017), copyright (2017) Elsevier, and electrospinning method (A3, B3). Top and bottom rows are top and cross-sectional images, respectively.

There are four types of phase inversion methods, namely, thermally induced phase separation (TIPS), nonsolvent-induced phase separation (NIPS), vapor-induced phase separation (VIPS), and solvent evaporation. TIPS and NIPS are the most widely applied phase separation technologies. TIPS was first introduced in late 1980s, and it was used to fabricate microporous membranes (Caneba and Soong, 1985; Lloyd et al., 1990). In this technology, a polymer is first dissolved in a high-boiling, low-molecular-weight solvent at a temperature near or higher than the melting point of the polymer to form a homogeneous polymer solution. Then the dope is cast or extruded to form a desired geometry. Subsequently, the cast solution is cooled to induce phase separation because the polymer solubility in a solvent usually decreases at a low temperature (Kim et al., 2016; Lalia et al., 2013). After demixing is induced, the solvent can be removed by solvent extraction, evaporation, or freeze drying. TIPS has many unique properties such as high producibility, high porosity, narrow pore size distribution and low tendency to form defects. On the other hand, it requires high temperatures, which translates to intensive energy consumption and high demand on the thermomechanical properties of experimental apparatus (Fu et al., 2008; Xiao et al., 2015; Zhu and Zhang, 2014). NIPS, also known as the Loeb–Sourirajan process, was first proposed by Loeb and Sourirajan to develop the first high-flux RO membrane in the 1960s, which is considered as a major breakthrough in membrane technology. Since then, NIPS has been studied intensively and became the most predominant phase inversion method for membrane fabrication (Kim et al., 2016). Unlike the polymer–solvent binary system for TIPS, the NIPS system consists of three components, namely, polymer, solvent, and nonsolvent. The cast solution is quickly immersed in a nonsolvent bath to induce the phase separation. Compared to TIPS, NIPS usually does not involve high temperature spinning. It can produce asymmetric membranes with a distinct surface and substrate morphology and properties, making it possible to fabricate membranes with highly porous substrates and relatively dense surfaces (Li et al., 2008; Xiao et al., 2015; Zhu and Zhang, 2014).

In the past few years, nanotechnology has been receiving increasing attention because of its unique properties. Nanofiber membranes fabricated via electrospinning have emerged as an alternative to conventional phase inversion membranes in various applications. Electrospinning involves producing nanofibers with a diameter ranging from micrometer to nanometer and form a nonwoven nanofibrous mat under a high electric field (An et al., 2017; Bhardwaj and Kundu, 2010). Nanofiber membranes offer several advantages including: (1) high surface-to-volume ratio, porosity, and degree of interconnection; (2) tunable pore size, porosity, and thickness; and (3) controllable nanofiber composition (Tijing et al., 2014). Over the past few decades, more than 200 polymers have been used to produce nanofiber membranes via electrospinning and the number is still increasing with time (Bhardwaj and Kundu, 2010; Tijing et al., 2014). In this book, novel membranes fabricated by both phase inversion and electrospinning will be covered.

2.1.3 Membrane and Module Configurations

Similar to other membrane applications, hollow fiber and flat sheet membranes are the most common configurations of MD membranes. The former is often mounted in tubular modules while the latter is packed in plate and frame or spiral-wound modules. Both of them have been widely applied in lab and industrial scale studies.

Figure 2.2A shows the structure of the hollow fiber membrane and module. The feed flows through the lumen or shell side of the membrane, and the distillate, vacuum, air gap, or sweeping gas is applied on the other side. Comparing to the flat sheet, the hollow fiber is self-supported. The tubular configuration can achieve a very high packing density (3000 m^2/m^3) (Camacho et al., 2013). Figure 2.2B depicts the sketch of the flat sheet membrane and plate and frame module. The fabrication of flat sheet membranes and assembly of plate and frame modules are easier than those of hollow fiber membranes and modules. Therefore, the former is used extensively in academic research. However, the flat sheet membrane has a lower ratio of surface-area-to-volume than the hollow fiber (Camacho et al., 2013). The spiral-wound module is another common configuration that has been widely applied in various industrial applications. As displayed in Figure 2.2C, a spiral-wound module consists of membranes, permeate spacers, feed spacers, and a permeate tube. The purpose of spacers is to facilitate uniform flows. Since the feed flows tangentially along the length of the

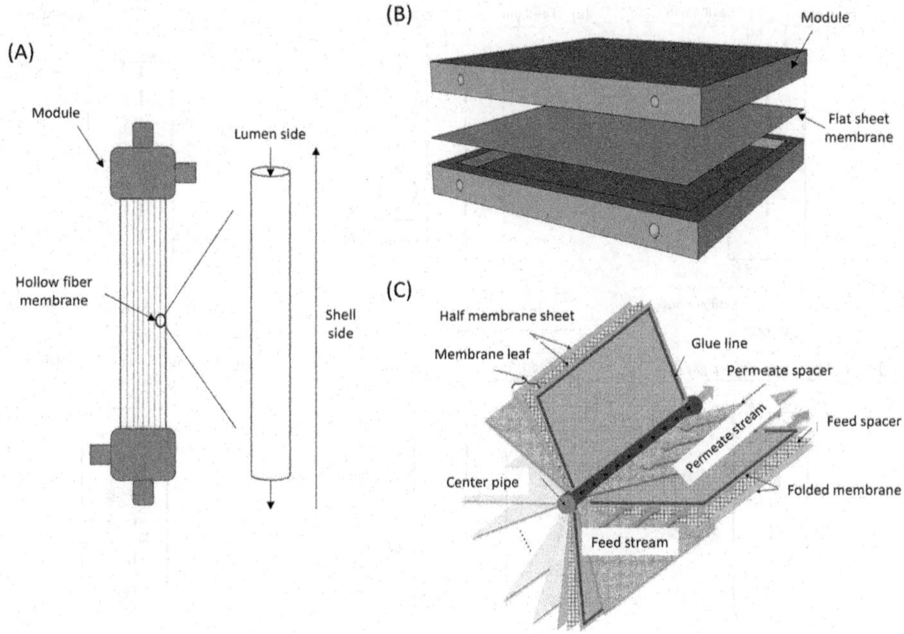

FIGURE 2.2 (A) Hollow fiber membrane and module, (B) plate and frame module, (C) spiral wound module, copyright (2017) Elsevier.

module, the permeate will transport through membranes into the permeate spacers, then flows towards the center and being collected by the permeate tube. Comparing to the plate and frame module, a spiral-wound module has a higher packing density, allows for easier cleaning and has a relatively lower cost. In addition, it provides a more robust design that prevents membrane from breakage compared with the hollow fiber configuration. Nonetheless, it suffers from more severe fouling problems and has a lower packing density than hollow fibers.

2.1.4 Different Configurations of MD

In the MD process, the hydrophobic membranes serve as the contacting surface for water evaporation to take place. Thus, the feed is always in direct contact with the membrane. Depending on the modes of contact on the permeate side, MD configurations can be categorized into four types; namely direct contact membrane distillation (DCMD), vacuum membrane distillation (VMD), air gap membrane distillation (AGMD) and sweeping gas membrane distillation (SGMD). Some variations of these four classic MD configurations have also been proposed recently. Different configurations are illustrated in Figure 2.3.

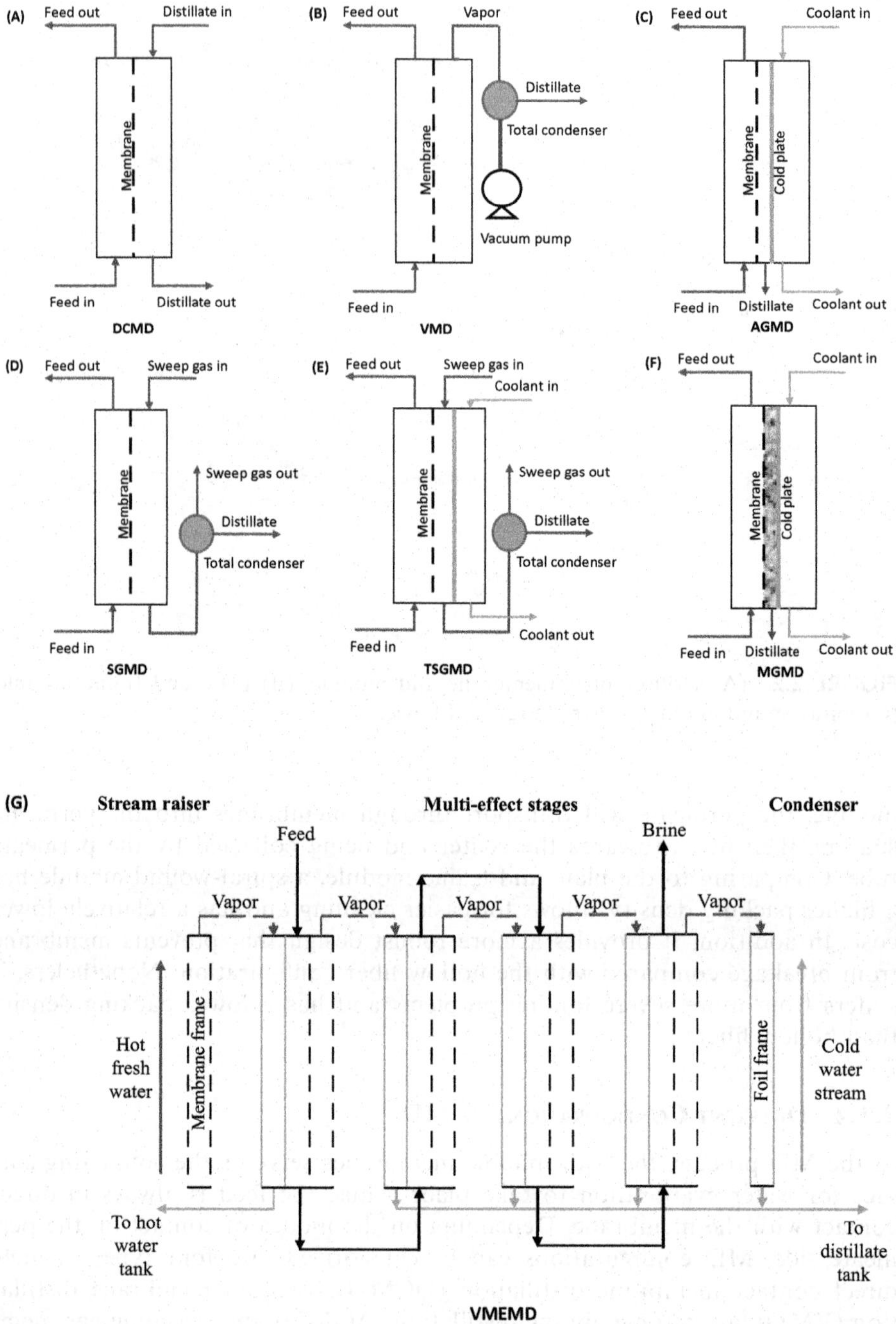

FIGURE 2.3 Schematic diagrams of (A) DCMD, (B) VMD, (C) AGMD, (D) SGMD, (E) TSGMD, (F) MGMD, (G) VMEMD.

DCMD is the simplest and most commonly applied MD configuration. In this case, the hot feed and cold distillate are in direct contact with the membrane. Water vaporizes at the membrane surface on the feed side and transports across the membrane before being condensed on the other surface of the membrane. In spite of its simplicity and short vapor transport path, DCMD has a higher conductive heat loss compared to other configurations (Khayet and Matsuura, 2011a).

If the water vapor is not condensed directly on the membrane surface but collected and condensed by means of a vacuum pump and an external chiller, it is called VMD. In such a configuration, the conductive heat loss is reduced and vapor condensation inside the membrane pores can also be avoided due to the vacuum applied. On the other hand, membranes with stronger mechanical properties and better anti-wetting abilities are required in VMD because of the higher transmembrane pressure (Sun et al., 2014; Zuo et al., 2017).

Another classic MD configuration is AGMD. In AGMD, evaporation and condensation take place on a hydrophobic membrane and a cold plate separately. A thin layer of stagnant air is interposed between the membrane and the plate. AGMD has a lower thermal loss than DCMD and no wetting phenomenon takes place on the permeate side in this configuration. Moreover, since the distillate is separated from the coolant, very pure water product can be obtained in AGMD. However, the stagnant air layer results in additional mass transfer resistance and causes extra difficulty in module design (Boutikos et al., 2017).

SGMD employs a cold sweeping gas to carry the permeate water vapor from the module to an external condenser. Similar to AGMD, wetting from the permeate side is eliminated and high quality distillate can be produced by SGMD. However, it is found that the sweeping gas gets hot rapidly along the module, causing the reduction of the local driving force for water vapor permeation across the membrane. Moreover, a pre-treatment of sweeping gas might be needed. The requirements of a sweeping gas source, a gas circulator and a vapor condenser also introduce additional complexity (Drioli et al., 2015).

A variation of SGMD, referred to as thermostatic sweeping gas membrane distillation (TSGMD), has been proposed. It aims to control the temperature of sweeping gas by contacting it with a cold wall on the permeate side, as shown in Figure 2.3E (Khayet and Matsuura, 2011a). Similarly, material gap membrane distillation (MGMD) is a variation of AGMD. Unlike AGMD where stagnant air is interposed between the membrane and the cold plate, MGMD has different materials such as polyurethane, polypropylene mesh, sand or deionized (DI) water in between, as illustrated in Figure 2.3F (Francis et al., 2013). When DI water is used, it is the same as the permeate gap membrane distillation (PGMD) (Swaminathan et al., 2016). Improved water fluxes have been reported for MGMD compared to AGMD.

Another variation termed vacuum multi-effect membrane distillation (V-MEMD) that combines VMD and multi-effect distillation (MED) has drawn significant attention recently (Boutikos et al., 2017). A VMEMD unit consists of three parts: (i) a steam raiser, (ii) four multi-effect stages, and (iii) a condenser. In the stream raiser, water vapor is generated, which passes through the membrane frames and condensed in the first effect. Similarly, hot energy can be conveyed to the feed saline water in the subsequent effects via the transfer of latent heat. Each stage comprises one foil frame and one membrane frame. Each frame has two pieces of either membranes or PP foils welded onto both sides. The space inside the foil frame acts as a condensation channel while that inside the membrane frame serves as a vapor channel. The feed saline water flows between the foil and membrane frames and gets heated by the latent heat released from the vapor condensation generated in the previous stage. At the last stage, the feed is concentrated as brine and the vapor enters the condenser (Boutikos et al., 2017). VMEMD is reported to greatly enhance the energy efficiency and has been successfully commercialized by Memsys (Zhao et al., 2013).

2.2 MATHEMATICAL MODELING OF MD

The temperature profiles of the classical four MD processes, namely DCMD, VMD, AGMD and SGMD, are drawn in Figure 2.4. In all these configurations, the flow configuration as well as the heat and mass transfer on the feed

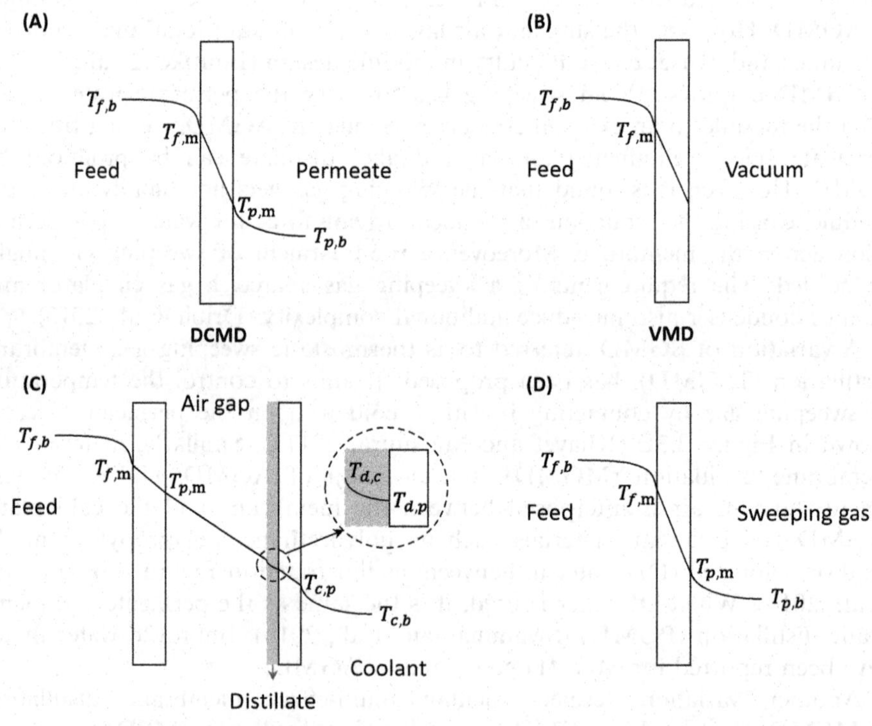

FIGURE 2.4 Schematics of temperature profiles of the most common four MD processes: (A) DCMD, (B) VMD, (C) AGMD, (D) SGMD.

side remain the same due to the fact that the feed is always in direct contact with the membrane surface. By contrast, heat and mass transfer differ on the permeate side for different configurations. In the following sections, theoretical modeling of DCMD will be discussed in detail. Based on the DCMD model, variations in the heat and mass transfer for other configurations will also be elaborated.

2.2.1 DCMD

2.2.1.1 Mass Transfer

The MD process is driven by the vapor pressure difference. It involves the transport of water vapor through stagnant air within the membrane pores. Thus, molecular diffusion plays an important role in the mass transfer. In addition, Knudsen diffusion often needs to be considered for water vapor transport in porous membranes (Pantoja et al., 2015; Shim et al., 2015). The governing quantity which provides the guideline in determining the suitable transport mechanism under given experimental conditions is the Knudsen number (Kn) (Khayet and Matsuura, 2011b).

$$Kn = \frac{\ell_w}{d_p} \tag{2.1}$$

where d_p is the membrane pore size. The mean free path of water molecules (ℓ_w) in vapor phase can be calculated using

$$\ell_w = \frac{k_B(T_m + 273.15)}{\sqrt{2}\pi P_m \sigma^2} \tag{2.2}$$

where k_B is the Boltzmann constant. T_m is the average of temperatures at the membrane surfaces, namely, $T_{f,m}$ and $T_{p,m}$. P_m is the mean pressure within the membrane pores and σ is the collision diameter of water molecules (2.65 Å). If $Kn > 10$ or $d_p < 0.1\ell_w$, the molecule–pore collisions are dominant over the molecule–molecule collisions and the Knudsen type of flow is responsible for the mass transfer across the membrane. Its corresponding mass transfer coefficient (C_K) can be expressed as (Su et al., 2010)

$$C_K = 1.064 \frac{r_p \varepsilon}{\tau \delta} \left[\frac{M_w}{R(T_m + 275.15)} \right]^{0.5} \tag{2.3}$$

where $r_p, \varepsilon, \tau,$ *and* δ are the pore size, porosity, tortuosity, and thickness of the membrane, respectively. M_w is the molecular weight of water and R is the gas constant. If $Kn < 0.01$ or $d_p > 100\ell_w$, the molecular diffusion is used to describe the mass transfer through the membrane pores due to the low solubility of air in water and the mass transfer coefficient (C_M) is computed as (Su et al., 2010)

$$C_M = \frac{1.895\times10^{-5}(T_m+273.15)^{2.072}}{P_a}\frac{\varepsilon}{\tau\delta}\frac{M_w}{R(T_m+273.15)} \quad (2.4)$$

where P_a is the atmospheric pressure. As the calculated ℓ_w is often similar to the pore size of most DCMD membranes (i.e., $0.01<Kn<10$ or $0.1l_w<d_p<100\ell_w$), a model that combines both Knudsen diffusion and molecular diffusion is commonly used to predict the water vapor flux across the membrane (J) (Schofield et al., 1987; Su et al., 2010).

$$J = \frac{1}{\left(\frac{1}{C_K}+\frac{1}{C_M}\right)}\left(P_{f,m}-P_{p,m}\right) \quad (2.5)$$

where C_K and C_M are the mass transfer coefficients in the respective flow region as defined previously. $P_{f,m}$ and $P_{p,m}$ are the vapor pressures at membrane surfaces on feed and permeate sides, respectively. Pure water vapor pressure (P_v^*) at different temperatures and vapor pressures (P_v) of aqueous salt solutions can be calculated using the following equations (Khayet and Matsuura, 2011b):

$$P_v^* = \exp\left(23.1964 - \frac{3816.44}{T+227.02}\right) \quad (2.6)$$

$$P_v = P_v^*\gamma_w(1-x) \quad (2.7)$$

$$\gamma_w = 1 - 0.5x - 10x^2 \quad (2.8)$$

where γ_w and x are the activity coefficient of water and mole fraction of NaCl, respectively. To get vapor pressures, $P_{f,m}$ and $P_{p,m}$ at the membrane surfaces, temperatures and salt concentrations of solutions at the membrane surfaces must be obtained first. They can be obtained based on the conservation of heat and concentration polarization as discussed in the following sections.

2.2.1.2 Heat Transfer

As referred to Figure 2.4A, heat transfer in the DCMD process takes place in three regions, namely, the feed layer (Q_f), the membrane (Q_m), and the permeate layer (Q_p). The heat transfer across the membrane includes both conductive (Q_c) and latent heat transfer (Q_v).

$$Q_f = A_f h_f\left(T_{f,b}-T_{f,m}\right) \quad (2.9)$$

$$Q_p = A_p h_p\left(T_{p,m}-T_{p,b}\right) \quad (2.10)$$

$$Q_m = Q_v + Q_c \quad (2.11)$$

$$Q_v = JA_f\Delta H_v \tag{2.12}$$

$$Q_c = A_f\frac{\lambda_m}{\delta}\left(T_{f,m} - T_{p,m}\right), \text{flat sheet membrane} \tag{2.13}$$

$$Q_c = A_f\frac{\lambda_m}{r_i\ln\left(\frac{r_o}{r_i}\right)}\left(T_{f,m} - T_{p,m}\right), \text{hollow fiber membrane with feed on the lumen side} \tag{2.14}$$

$$Q_c = A_f\frac{\lambda_m}{r_o\ln\left(\frac{r_o}{r_i}\right)}\left(T_{f,m} - T_{p,m}\right), \text{hollow fiber membrane with feed on the shell side} \tag{2.15}$$

$$\lambda_m = (1-\varepsilon)\times\lambda_p + \varepsilon\times\lambda_g \tag{2.16}$$

where A is the membrane area and h is the heat transfer coefficient. Subscripts f and p denote the membrane feed and permeate sides, respectively. If it is the flat sheet membrane, A_f is equal to A_p by the geometric symmetry. For hollow fiber membranes with the feed on the lumen side, A_f represents the inner surface membrane area and A_p represents the outer surface membrane area (i.e., Equation (2.14)). While these symbolic expressions reverse for hollow fiber membranes with the feed on the shell side (i.e., Equation (2.15)). $T_{f,b}$ and $T_{p,b}$ are the bulk temperatures of the feed and permeate, respectively. ΔH_v is the latent heat of water. r_o and r_i are the outer and inner radii of hollow fiber membranes, respectively. λ_m is the effective thermal conductivity of porous membranes that can be easily calculated from the porosity (ε), thermal conductivity of the membrane polymer (λ_p), and thermal conductivity of air (λ_g). Heat transfer coefficient (h) can be evaluated using Nusselt number (Nu), the thermal conductivity of the fluid (λ_w), and the hydraulic diameter (d_h).

$$h = \frac{Nu\lambda_w}{d_h} \tag{2.17}$$

Correlations for Nusselt number are available from the literature (Su et al., 2010; Taler and Taler, 2017; Wang and Chung, 2012). For readers' interest, Phattaranawik et al. give a comparison across those different sets of correlations (Phattaranawik et al., 2003). Two presentative sets of correlations governing the respective flow region are shown as follows:

$$Nu = 4.36 + \frac{0.036RePr\frac{d_h}{L}}{1+0.0011\left[RePr\left(\frac{d_h}{L}\right)\right]^{0.8}}, \text{laminar flow} \tag{2.18}$$

$$Nu = 0.036 Re^{0.8} Pr^{1/3} \left(\frac{d_h}{L}\right)^{0.055}, \text{turbulent flow} \tag{2.19}$$

where L is the channel/module length. Re and Pr are the Reynolds number and Prandtl number, respectively, and they are defined as

$$Re = \frac{d_h v \rho}{\mu} \tag{2.20}$$

$$Pr = \frac{c_p \mu}{\lambda_w} \tag{2.21}$$

where v, ρ, μ, and C_p are the solution linear velocity, density, viscosity, and heat capacity, respectively. At steady state, heat transfer across all the regions is the same. Thus, temperatures at the membrane surfaces $T_{f,m}$ and $T_{p,m}$ can be calculated by simultaneously solving Equations (2.9) to (2.15) and (2.22).

$$Q_f = Q_m = Q_p \tag{2.22}$$

The energy efficiency (*EE*) of a MD process can be quantified using the following equation:

$$EE = \frac{Q_v}{Q_v + Q_c} \tag{2.23}$$

2.2.1.3 Temperature and Concentration Polarizations

As mentioned previously, temperature and concentration polarizations must be considered in order to evaluate the driving force accurately. The temperature polarization coefficient (TPC)and concentration polarization coefficient (CPC) are given by (Martínez-Díez and Vázquez-González, 1999; Pantoja et al., 2015)

$$\text{TPC} = \frac{T_{f,m} - T_{p,m}}{T_{f,b} - T_{p,b}} \tag{2.24}$$

$$\text{CPC} = \frac{M_{f,m}}{M_{f,b}} = \exp\left(\frac{J}{\rho_w k}\right) \tag{2.25}$$

$T_{f,m}$ and $T_{p.m}$ can be obtained via the method discussed in the previous section.CPC is defined as the ratio of salt molality at the membrane surface to the bulk. It can be predicted from the membrane performance and rheological behavior of the feed solution. ρ_w is the density of the feed solution and k is the mass transfer coefficient, which can be evaluated using the correlations available in the literature (Gekas and Hallström, 1987; Martínez-Díez et al., 2000).

$$k = \frac{Sh\, D_S}{d_h} \tag{2.26}$$

$$Sh = 1.86\left(Re\, Sc\, \frac{d_h}{L}\right)^{\frac{1}{3}}, \text{laminar flow} \tag{2.27}$$

$$Sh = 0.023 Re^{0.8} Sc^{0.33}, \text{turbulent flow} \tag{2.28}$$

$$Sc = \frac{\mu}{\rho D_S} \tag{2.29}$$

where Sh and Sc are the Sherwood number and Schmidt number, respectively. D_s is the diffusion coefficient of sodium chloride in water. At elevated temperatures, D_s can be predicted by the following equation derived from the Stokes–Einstein equation (Einstein, 1956):

$$\frac{D_{S,T_1}}{D_{S,T2}} = \frac{T_1 \mu_{T2}}{T_2 \mu_{T1}} \tag{2.30}$$

2.2.1.4 Variation of Thermophysical Properties along Membrane Modules

Local heat and mass transfer as well as local temperature and concentration polarizations have been elaborated in previous sections. Above-mentioned modeling can provide a close prediction of membrane performance only when there is no significant variation of thermophysical properties along the membrane module. However, in reality such variation is almost inevitable due to the heat and mass transfer. Thus, a more comprehensive model that captures the change of fluid properties along the module is desired. As illustrated in Figure 2.5, the module is divided into N segments. Heat and mass transfer analyses are performed on every single segment by solving the local profile as shown in previous sections. The average flux is simply the algebraic mean of the flux in each segment.

Governing equations are shown as follows:

$$\frac{dm_f}{dA_f} = \frac{dm_p}{dA_f} = -J \tag{2.31}$$

$$\frac{d(m_f C_{p,f} T_{f,b})}{dA_f} = \frac{d(m_p C_{p,p} T_{p,b})}{dA_f} = -h_f(T_{f,b} - T_{f,m}) \tag{2.32}$$

$$\frac{dS_f}{dA_f} = \frac{J}{m_f} S_f \tag{2.33}$$

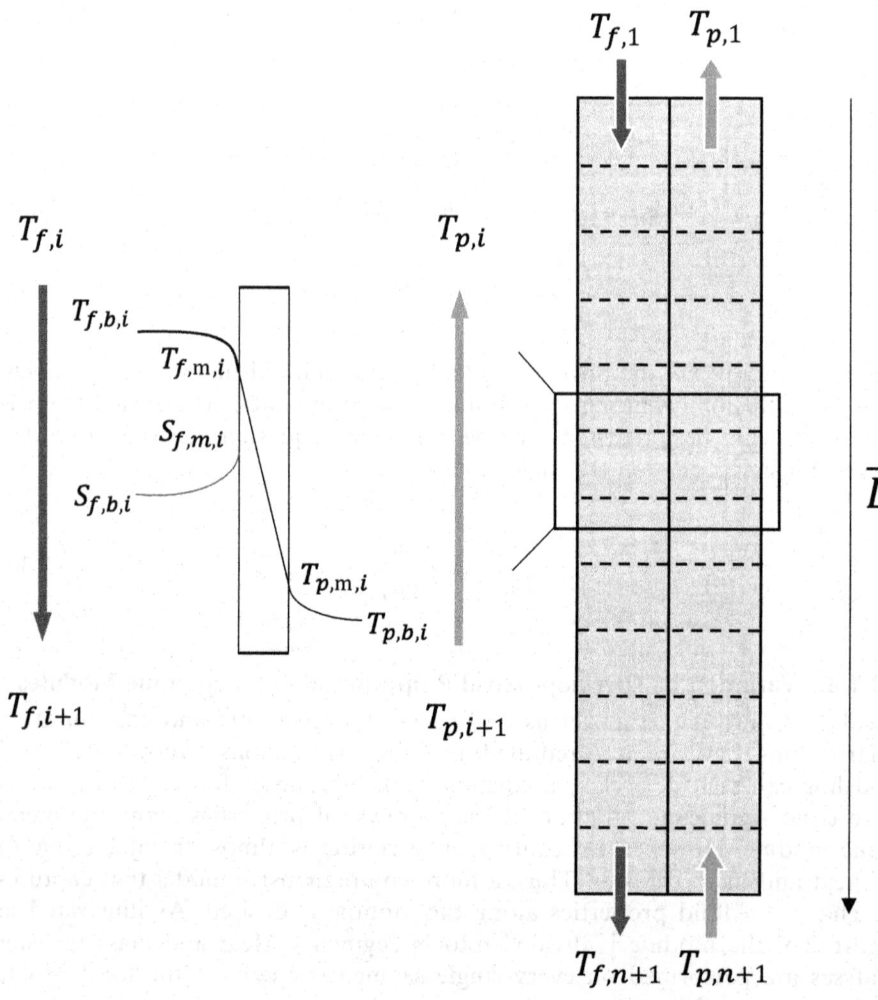

FIGURE 2.5 Modeling of the direct contact membrane distillation (DCMD) membrane module with a countercurrent flow configuration. The module is divided into N segments.

where m, C_p, T, h, and S are the mass flow rate, heat capacity, temperature, heat transfer coefficient, and salt concentration, respectively. Subscripts f, p, b, and m denote the feed, permeate, bulk, and membrane surface, respectively. J and A are the membrane flux and the membrane area, respectively. Equations (2.31)–(2.33) describing a DCMD model can be numerically solved using a MATLAB® built-in solver "bvp4c" or other equivalent solvers.

2.2.2 VMD

By applying the vacuum, it results in good heat insulation on the permeate side. Consequently, both the boundary layer (i.e., temperature polarization) on the permeate side and the conductive heat loss across the membrane could be generally neglected in VMD. The low partial pressure inside the membrane pores generated by the vacuum also greatly suppresses the molecular diffusion as compared to a DCMD process. Nonetheless, the viscous flow generated under a distinct pressure gradient across the membrane might need to be considered. On the other hand, VMD membranes should be designed with smaller pore sizes for a better mechanical stability and a greater antiwetting ability. The calculated mean free path of the transported water vapor molecules through the membrane pores is generally longer in VMD than that of DCMD because of the low pressure applied on the permeate side. As a result, most VMD studies consider $kn > 10$ or $d_p < 0.1\ell_w$ and Knudsen diffusion as the major transport mechanism. The expression of water vapor flux across the membrane can be simplified to the following equation as compared to Equation (2.5) of the DCMD process:

$$J = C_K\left(P_{f,m} - P_{p,m}\right) \tag{2.34}$$

where $P_{p,m}$ is negligible due to the applied vacuum and $P_{f,m}$ can be evaluated using Equations (2.6) to (2.8) as analogous to that of DCMD.

For cases when $kn < 0.01$ or $d_p > 100\ell_w$, the molecule–molecule collisions dominate the vapor transport in the viscous region and the mass transport coefficient of viscous flow C_v is expressed as

$$C_v = \frac{P_m\, r_p^2 \varepsilon}{8\mu_v\, \tau\delta} \frac{M_w}{R(T_m + 273.15)} \tag{2.35}$$

TABLE 2.2

A comparison of different transport mechanisms between DCMD and VMD

	DCMD		VMD	
Knudsen number (Kn) or mean free path of water molecules (ℓ_w)	**Transport mechanism**	**Most studies consider**	**Transport mechanism**	**Most studies consider**
$Kn < 0.01$ or $d_p > 100\ell_w$	Molecular diffusion		Viscous flow	
$Kn > 10$ or $d_p < 0.1\ell_w$	Knudsen diffusion		Knudsen diffusion	√
$0.01 < Kn < 10$ or $0.1\ell_w < d_p < 100\ell_w$	Molecular diffusion + Knudsen diffusion	√	Viscous flow + Knudsen diffusion	

where P_m is the mean pressure within the membrane pores and μ_v is the viscosity of the vapor. $r_p, \varepsilon, \tau,$ *and* δ are the pore size, porosity, tortuosity, and thickness of the membrane, respectively. M_w is the molecular weight of water and R is the gas constant. T_m is the average of temperatures at the membrane surfaces. When $0.01 < Kn < 10$ or $0.1\ell_w < d_p < 100\ell_w$, a combing mass transfer model of the Knudsen and viscous flow should be used to describe the transition region. A comparison between DCMD and VMD is shown in Table 2.2, indicating different transport mechanisms and the respective dominant mechanism considered in most studies.

The heat transfer equations derived previously for DCMD are generally applicable to VMD except for those concerning the permeate side (i.e., Equation (2.10)). In other words, the vacuum on the permeate side serves as a good insulator, thus the heat transfer on the permeate side is negligible.

2.2.3 AGMD

The mass transfer mechanism of AGMD greatly resembles that of DCMD, where both Knudsen diffusion and molecular diffusion are commonly considered to predict the water vapor flux across the membrane. Nonetheless, the stagnant air layer between the membrane and the cold plate imposes additional mass transfer resistance and this needs to be taken into account when formulating the respective mass transfer coefficient. To simplify the derivation steps, one could treat the stagnant air layer as an extension of the real MD membrane. In other words, by combining the real MD membrane and the air gap region as a hypothetical membrane, the AGMD process could be solved in the same way of a DCMD process where both membrane surfaces directly contact with the liquid solution. Thus, the corresponding mass transfer coefficients should be rewritten as (Attia et al., 2017):

$$C_K = 1.064 \frac{r_p \varepsilon}{\tau\delta + b} \left[\frac{M_w}{R(T_m + 273.15)} \right]^{0.5} \tag{2.36}$$

$$C_M = \frac{1.895 \times 10^{-5} (T_m + 273.15)^{2.072}}{P_a} \frac{\varepsilon}{\tau\delta + b} \frac{M_w}{R(T_m + 273.15)} \tag{2.37}$$

where all the symbols remain the same meaning as those in Equations (2.3) and (2.4) and the additional symbol b represents the air gap thickness. It can be seen that, by adding the air gap region, water vapor molecules take a longer transport path in AGMD ($\tau\delta + \tau_g b = \tau\delta + b$, where $\tau_g = 1$ as water vapor molecules could freely transport across the air gap region) than that in DCMD ($\tau\delta$) before being condensed into distillate.

The calculation of heat transfer Q_f in the feed layer of DCMD (i.e., Equation (2.9)) is applicable to that of AGMD. While the heat transfer equation across the membrane (i.e., Equation (2.11)) should be rewritten as (Attia et al., 2017):

$$Q_m = JA_f\Delta H_v + A_f\left(\frac{JC_{p,g}}{1-\exp\left(-\frac{bJC_{p,g}}{\lambda_g}\right)}\right)\left(T_{f,m} - T_{d,c}\right) \tag{2.38}$$

where J is the water vapor flux across the membrane. A_f is the membrane area on the feed side. ΔH_v is the latent heat of water. $C_{p,g}$ and λ_g are the specific heat and thermal conductivity of the gas phase, respectively. $T_{f,m}$ and $T_{d,c}$ are the membrane surface temperature on the feed side and the distillate surface temperature on the air gap side, respectively. The subsequent heat transfer across the distillate layer (Q_d), the cold plate (Q_{pl}), and the coolant region (Q_{cl}) are expressed as follows and the corresponding temperature profile is depicted in Figure 2.4C:

$$Q_d = A_p h_d\left(T_{d,c} - T_{d,p}\right) \tag{2.39}$$

$$Q_{pl} = A_p\frac{\lambda_{pl}}{l}\left(T_{d,p} - T_{c,p}\right) \tag{2.40}$$

$$Q_{cl} = A_p h_{cl}\left(T_{c,p} - T_{c,b}\right) \tag{2.41}$$

where A_p is the membrane area on the permeate side. h_d and h_{cl} are the heat transfer coefficients in the distillate layer and coolant region, respectively, which can be calculated using the correlation of Equations (2.17) to (2.19). λ_{pl} and l represent the thermal conductivity and the thickness of the cold plate, respectively. $T_{d,c}$, $T_{d,p}$, and $T_{c,p}$ are the interface temperatures among the air gap region, the distillate layer, the cold plate and the coolant region correspondingly. $T_{c,b}$ represents the coolant bulk temperature. At steady state, one could solve the temperature profile of an AGMD process by equating

$$Q_f = Q_m = Q_d = Q_{pl} = Q_{cl} \tag{2.42}$$

2.2.4 SGMD

In SGMD, air or other inert gas is used as water vapor carrier prior to condensation. Because in this process the sweeping gas rather than a liquid contacts the membrane on the permeate side, the equation describing the driving force of water vapor flux across the membrane in DCMD needs to be adjusted before it can be adopted for SGMD. More specifically, the partial pressure at membrane surface on the permeate side $\left(P_{p,m}\right)$ is modified as (Khayet and Matsuura, 2011b)

$$P_{p,m} = \frac{wP}{w + 0.622} \tag{2.43}$$

where P is the total pressure on the permeate side and w is the humidity ratio for a given moist air sample. Along the membrane module, the humidity ratio (w) grows with respect to the module inlet humidity ratio (w_i), vapor flux (J), sweeping gas mass flow rate (m_g), and the membrane area (A_f).

$$w = w_i + \frac{JA_f}{m_g} \tag{2.44}$$

As gas is present on the permeate side of the membrane, another set of correlation is required to calculate the Nusselt number (Nu) and the subsequent heat transfer coefficient (h) defined in Equation (2.17), where λ_w should be replaced by λ_g to represent the thermal conductivity of air (Khayet et al., 2003).

$$Nu = 0.206(Re \cos \alpha)^{0.63}\,\mathrm{Pr}^{0.36} \tag{2.45}$$

where Re and Pr are the Reynolds number and Prandtl number, respectively. α is the yaw angle, which varies between 0° for pure cross-flow and 90° for pure parallel flow.

2.2.5 Wetting

To mitigate temperature polarization and concentration polarization, one of the most effective methods is to enhance the heat and mass transfer by increasing the flow rates. On the other hand, it will cause a greater pressure drop along the membrane module and bring the membrane closer to wetting. Therefore, it is important to ensure that the maximum pressure drop does not exceed the liquid entry pressure (LEP) value of MD membranes.

LEP has been theoretically predicted by the Laplace–Young equation (Rácz et al., 2014).

$$\mathrm{LEP} = \frac{-2B\gamma_L \cos\theta}{r_{\max}} \tag{2.46}$$

where B is the geometry factor of the pore ($B = 1$ for cylindrical pores and $0 < B < 1$ for noncylindrical pores). θ is the effective contact angle and $r_{\max}$ is the maximum membrane pore size. γ_L is the liquid–vapor surface tension.

The pressure drop should be evaluated based on the specific flow pattern and the module geometry. The pressure drop along a duct is given by (Lee, 2010)

$$\Delta P = 4\frac{fL}{d_h} \times \frac{1}{2}\rho v^2 \tag{2.47}$$

$$f = 16/Re, \text{ laminar flow} \tag{2.48}$$

$$f = (1.58\ln(Re) - 3.28)^{-2}, \text{ turbulent flow} \quad (2.49)$$

where f is the friction factor. d_h and L are the hydraulic diameter and module length, respectively. ρ,v, and Re are the density, linear velocity, and Reynolds number, respectively. For the pressure drop on the shell side, it is given by (Lee, 2010)

$$\Delta P = \frac{1}{2} f \frac{d_s}{d_h} \rho v^2 \quad (2.50)$$

$$f = \exp(0.576 - 0.19\ln(Re)) \quad (2.51)$$

where d_s is the shell inside diameter of the membrane module.

ABBREVIATIONS

AGMD	air-gap membrane distillation
CPC	concentration polarization coefficient
DCMD	direct contact membrane distillation
DI	deionized
EE	energy efficiency
Kn	Knudsen number
MD	membrane distillation
MGMD	material gap membrane distillation
NIPS	nonsolvent-induced phase separation
Nu	Nusselt number
PE	polyethylene
PGMD	permeate gap membrane distillation
PP	polypropylene
Pr	Prandtl number
PTFE	polytetrafluoroethylene
PVDF	polyvinylidene difluoride
Re	Reynolds number
Sc	Schmidt number
SGMD	sweeping gas membrane distillation
Sh	Sherwood number
TEP	triethyl phosphate
TIPS	thermally induced phase separation
TPC	temperature polarization coefficient
TSGMD	thermostatic sweeping gas membrane distillation
VIPS	vapor-induced phase separation
VMD	vacuum membrane distillation
V-MEMD	vacuum multieffect membrane distillation

ENGLISH LETTERS

A	membrane area (m^2)
b	air gap thickness (m)
C_p	heat capacity (J/(kg K))
d	diameter (m)
d_h	hydraulic diameter (m)
d_s	module inside diameter (m)
D_s	solute diffusivity (m^2/s)
f	friction factor
h	heat transfer coefficient (W/(m^2 K))
J	membrane flux (kg m^{-2} s^{-1})
L	length of module (m)
ℓ_w	mean free path of water molecule (m)
m	mass flow rate (kg/s)
M	molality (mol/kg H_2O)
N	number of modeling segment
P	pressure (Pa)
Q	heat transfer rate (W)
Q_c	conductive heat transfer rate (W)
Q_v	latent heat transfer rate (W)
r	radius (m)
S	salt concentration (wt%)
T	temperature (°C)
v	linear velocity (m/s)
x	mole fraction of solute (mol/mol)

GREEK SYMBOLS

γ_L	liquid–vapor surface tension (N/m)
γ_w	activity coefficient of water
σ	collision diameter of water molecule (m)
δ	membrane thickness (m)
ε	porosity
λ	effective thermal conductivity (W/(m K))
θ	effective contact angle (°)
μ	viscosity of the solution (Pa s)
ρ	density of the draw solution (kg/m^3)
τ	tortuosity

SUBSCRIPTS

b	bulk solution
d	distillate (lumen) side
f	feed (shell) side

f,b	bulk of feed
f,m	membrane surface of feed side
g	gas, air
i	inner, inlet
m	membrane
o	outer, outlet
p	polymer, pore
p,m	membrane surface of permeate side
p,b	bulk of permeate
v	vapor
w, water	water

REFERENCES

An, A.K., Lee, E.J., Guo, J., Jeong, S., Lee, J.G., Ghaffour, N., 2017. Enhanced vapor transport in membrane distillation via functionalized carbon nanotubes anchored into electrospun nanofibres. Sci. Rep. 7, 1–11. doi:10.1038/srep41562.

Attia, H., Osman, M.S., Johnson, D.J., Wright, C., Hilal, N., 2017. Modelling of air gap membrane distillation and its application in heavy metals removal. Desalination 424, 27–36. doi:10.1016/j.desal.2017.09.027.

Bhardwaj, N., Kundu, S.C., 2010. Electrospinning: a fascinating fiber fabrication technique. Biotechnol. Adv. 28, 325–347. doi:10.1016/j.biotechadv.2010.01.004.

Boutikos, P., Mohamed, E.S., Mathioulakis, E., Belessiotis, V., 2017. A theoretical approach of a vacuum multi-effect membrane distillation system. Desalination 422, 25–41. doi:10.1016/j.desal.2017.08.007.

Camacho, L.M., Dumée, L., Zhang, J., Li, J., Duke, M., Gomez, J., Gray, S., 2013. Advances in membrane distillation for water desalination and purification applications. Water 5, 94–196. doi:10.3390/w5010094.

Caneba, G.T., Soong, D.S., 1985. Polymer membrane formation through the thermal-inversion process. 1. Experimental study of membrane structure formation. Macromolecules 18, 2538–2545. doi:10.1021/ma00154a031.

Chang, J., Zuo, J., Zhang, L., O'Brien, G.S., Chung, T.S., 2017. Using green solvent, triethyl phosphate (TEP), to fabricate highly porous PVDF hollow fiber membranes for membrane distillation. J. Memb. Sci. 539, 295–304. doi:10.1016/j.memsci.2017.06.002.

Cheng, D.Y., Wiersma, S.J., 1983. Composite membrane for a membrane distillation system. 4,419,242.

Cui, Z., Hassankiadeh, N.T., Lee, S.Y., Lee, J.M., Woo, K.T., Sanguineti, A., Arcella, V., Lee, Y.M., Drioli, E., 2013. Poly(vinylidene fluoride) membrane preparation with an environmental diluent via thermally induced phase separation. J. Memb. Sci. 444, 223–236. doi:10.1016/j.memsci.2013.05.031.

Drioli, E., Ali, A., Macedonio, F., 2015. Membrane distillation: recent developments and perspectives. Desalination 356, 56–84. doi:10.1016/j.desal.2014.10.028.

Einstein, A., 1956. Investigation on the theory of the Brownian movement. Dover Publications Inc, New York USA.

Francis, L., Ghaffour, N., Alsaadi, A.A., Amy, G.L., 2013. Material gap membrane distillation: a new design for water vapor flux enhancement. J. Memb. Sci. 448, 240–247. doi:10.1016/j.memsci.2013.08.013.

Fu, X.Y., Sotani, T., Matsuyama, H., 2008. Effect of membrane preparation method on the outer surface roughness of cellulose acetate butyrate hollow fiber membrane. Desalination 233, 10–18. doi:10.1016/j.desal.2007.09.022.

Gekas, V., Hallström, B., 1987. Mass transfer in the membrane concentration polarization layer under turbulent cross flow. I. Critical literature review and adaptation of existing Sherwood correlations to membrane operations. J. Memb. Sci. 30, 153–170. doi:10.1016/S0376-7388(00)83069-0.

Gu, B., Xu, X.Y., Adjiman, C.S., 2017. A predictive model for spiral wound reverse osmosis membrane modules: the effect of winding geometry and accurate geometric details. Comput. Chem. Eng. 96, 248–265. doi:10.1016/j.compchemeng.2016.07.029.

Hou, D., Wang, J., Qu, D., Luan, Z., Zhao, C., Ren, X., 2009. Preparation of hydrophobic PVDF hollow fiber membranes for desalination through membrane distillation. Water Sci. Technol. 59, 1219–1226. doi:10.2166/wst.2009.080.

Khayet, M., Godino, M.P., Mengual, J.I., 2003. Theoretical and experimental studies on desalination using the sweeping gas membrane distillation method. Desalination 157, 297–305.

Khayet, M., Matsuura, T., 2011a. Introduction to membrane distillation, 1st ed. Membrane distillation. Elsevier B.V., Amsterdam. doi:10.1016/B978-0-444-53126-1.10001-6.

Khayet, M., Matsuura, T., 2011b. Membrane distillation principles and applications, 1st ed. Elsevier B.V., Amsterdam.

Kim, J.F., Kim, J.H., Lee, Y.M., 2016. Thermally induced phase separation and electrospinning methods for emerging membrane applications: a review. AIChE J. 62, 461–490. doi:10.1002/aic.

Lalia, B.S., Kochkodan, V., Hashaikeh, R., Hilal, N., 2013. A review on membrane fabrication: structure, properties and performance relationship. Desalination 326, 77–95. doi:10.1016/j.desal.2013.06.016.

Lee, H.S., 2010. Thermal design: heat sinks, thermoelectrics, heat pipes, compact heat exchangers, and solar cells.Hoboken New Jersey. John Wiley & Sons, Inc. doi:10.1002/9780470949979.

Li, X., Wang, Y., Lu, X., Xiao, C., 2008. Morphology changes of polyvinylidene fluoride membrane under different phase separation mechanisms. J. Memb. Sci. 320, 477–482. doi:10.1016/j.memsci.2008.04.033.

Lide, D.R., 2004. CRC Handbook of Chemistry and Physics. 85th ed. Boca Raton.

Lloyd, D.R., Kinzer, K.E., Tseng, H.S., 1990. Microporous membrane formation via thermally induced phase separation. I. Solid-liquid phase separation. J. Memb. Sci. 52, 239–261. doi:10.1016/S0376-7388(00)85130-3.

Lu, K.J., Zuo, J., Chung, T.S., 2017. Novel PVDF membranes comprising n-butylamine functionalized graphene oxide for direct contact membrane distillation. J. Memb. Sci. 539, 34–42. doi:10.1016/j.memsci.2017.05.064.

Mark, J.E., (Ed.), 1999. Polymer data handbook.New York. Oxford University Press, Inc. doi:10.1021/ja907879q.

Martínez-Díez, L., Florido-Díaz, F.J., Vázquez-González, M.I., 2000. Study of polarization phenomena in membrane distillation of aqueous salt solutions. Sep. Sci. Technol. 35, 1485–1501. doi:10.1081/SS-100100237.

Martínez-Díez, L., Vázquez-González, M.I., 1999. Temperature and concentration polarization in membrane distillation of aqueous salt solutions. J. Memb. Sci. 156, 265–273. doi:10.1016/S0376-7388(98)00349-4.

Pantoja, C.E., Nariyoshi, Y.N., Seckler, M.M., 2015. Membrane distillation crystallization applied to brine desalination : a hierarchical design procedure. Ind. Eng. Chem. Res. 54, 2776–2793. doi:10.1021/ie504695p.

Phattaranawik, J., Jiraratananon, R., Fane, A.G., 2003. Heat transport and membrane distillation coefficients in direct contact membrane distillation. J. Membr. Sci. 212, 177–193.

Rácz, G., Kerker, S., Kovács, Z., Vatai, G., Ebrahimi, M., Czermak, P., 2014. Theoretical and experimental approaches of liquid entry pressure determination in membrane distillation processes. Period. Polytech. Chem. Eng. 58, 81–91. doi:10.3311/PPch.2179.

Schneider, K., Hölz, W., Wollbeck, R., Ripperger, S., 1988. Membranes and modules for transmembrane distillation. J. Memb. Sci. 39, 25–42. doi:10.1016/S0376-7388(00)80992-8.

Schofield, R.W., Fane, A.G., Fell, C.J.D., 1987. Heat and mass transfer in membrane distillation. J. Memb. Sci. 33, 299–313. doi:10.1016/S0376-7388(00)80287-2.

Shim, W.G., He, K., Gray, S., Moon, I.S., 2015. Solar energy assisted direct contact membrane distillation (DCMD) process for seawater desalination. Sep. Purif. Technol. 143, 94–104. doi:10.1016/j.seppur.2015.01.028.

Speight, J.G., 2005. Lange's handbook of chemistry. 16th Ed. CD&W Inc., Laramie, Wyoming.

Su, M., Teoh, M.M., Wang, K.Y., Su, J., Chung, T.S., 2010. Effect of inner-layer thermal conductivity on flux enhancement of dual-layer hollow fiber membranes in direct contact membrane distillation. J. Memb. Sci. 364, 278–289. doi:10.1016/j.memsci.2010.08.028.

Sun, A.C., Kosar, W., Zhang, Y., Feng, X., 2014. Vacuum membrane distillation for desalination of water using hollow fiber membranes. J. Memb. Sci. 455, 131–142. doi:10.1016/j.memsci.2013.12.055.

Swaminathan, J., Chung, H.W., Warsinger, D.M., AlMarzooqi, F.A., Arafat, H.A., Lienhard, J.H., 2016. Energy efficiency of permeate gap and novel conductive gap membrane distillation. J. Memb. Sci. 502, 171–178. doi:10.1016/j.memsci.2015.12.017.

Taler, D., Taler, J., 2017. Simple heat transfer correlations for turbulent tube flow. E3S Web Conf 13, 1–7. doi:10.1051/e3sconf/20171302008.

Thomas, N., Mavukkandy, M.O., Loutatidou, S., Arafat, H.A., 2017. Membrane distillation research & implementation: lessons from the past five decades. Sep. Purif. Technol. 189, 108–127. doi:10.1016/j.seppur.2017.07.069.

Tijing, L.D., Choi, J.S., Lee, S., Kim, S.H., Shon, H.K., 2014. Recent progress of membrane distillation using electrospun nanofibrous membrane. J. Memb. Sci. 453, 435–462. doi:10.1016/j.memsci.2013.11.022.

Wang, P., Chung, T.S., 2012. A conceptual demonstration of freeze desalination-membrane distillation (FD-MD) hybrid desalination process utilizing liquefied natural gas (LNG) cold energy. Water Res. 46, 4037–4052. doi:10.1016/j.watres.2012.04.042.

Weyl, P.K., 1967. Recovery of demineralized water from saline waters. 3,340,186.

Xiao, T., Wang, P., Yang, X., Cai, X., Lu, J., 2015. Fabrication and characterization of novel asymmetric polyvinylidene fluoride (PVDF) membranes by the nonsolvent thermally induced phase separation (NTIPS) method for membrane distillation applications. J. Memb. Sci. 489, 160–174. doi:10.1016/j.memsci.2015.03.081.

Zhao, K., Heinzl, W., Wenzel, M., Büttner, S., Bollen, F., Lange, G., Heinzl, S., Sarda, N., 2013. Experimental study of the memsys vacuum-multi-effect-membrane – distillation (VMEMD) module. Desalination. 323, 150–160.

Zhao, Z., Li, J., Yuan, X., Li, X., Zhang, Y., Sheng, J., 2005. Preparation and properties of electrospun poly(vinylidene fluoride) membranes. J. Appl. Polym. Sci. 97, 466–474. doi:10.1002/ap.21762.

Zhu, Y., Zhang, Z., 2014. PVDF hollow fiber formation via modified NIPS method: evolution elucidation of phase separation mechanism, structure and properties of membrane with coagulation strength varied. Macromol. Res. 22, 1275–1281. doi:10.1007/s13233-014-2192-9.

Zuo, J., Chung, T.S., O'Brien, G.S., Kosar, W., 2017. Hydrophobic/hydrophilic PVDF/Ultem® dual-layer hollow fiber membranes with enhanced mechanical properties for vacuum membrane distillation. J. Memb. Sci. 523, 103–110. doi:10.1016/j.memsci.2016.09.030.

3 Fundamental Principles of Membrane Distillation Membrane Fabrication

Jian Zuo
Chemical Engineering and Food Technology, Singapore Institute of Technology, 10 Dover Drive, Singapore 138683

3.1 INTRODUCTION TO MEMBRANE DISTILLATION

Membrane distillation (MD) is a thermally driven membrane separation process based on vapor–liquid equilibrium. A porous hydrophobic membrane is usually used to provide a physical and thermal barrier between the hot feed and the permeate. It retains the liquid and nonvolatile solutes in the feed side, while allowing the vapor to pass through the membrane. Thus, MD has gained much consideration in seawater desalination, where water vapor could pass through while salts are retained. Pure water can then be collected at the permeate side of the membrane. Compared to other membrane processes, MD shows unique features of: (1) operating at a moderate temperature and pressure, (2) nearly 100% rejections of nonvolatile compounds, and (3) the ability to utilize low-grade heat sources, solar or waste energy.

3.2 TYPES OF MD MEMBRANES

Membranes are the heart of MD processes. Generally, an ideal MD membrane not only needs to be hydrophobic to prevent it from wetting by the feed solution but also possesses a high permeation flux and good mechanical properties. First, membrane hydrophobicity is an essential requirement. This is because the membrane would lose its selectivity if being wetted by the feed water. Under this circumstance, water and solutes will pass through the membrane freely and separation cannot be achieved. Second, the requirements of high permeation flux and mechanical properties are vital for an efficient MD process with long-term stability.

To satisfy the first requirement, hydrophobic materials have been investigated to fabricate MD membranes, mainly including polyvinylidene fluoride (PVDF), polytetrafluoro-ethylene (PTFE), polypropylene (PP), and some other polymers, copolymers, and inorganic materials (Chandavasu et al., 2003; Teoh et al., 2011; Tomaszewska, 1996; Zuo et al., 2016; Zuo and Chung, 2016b). Figure 3.1 shows

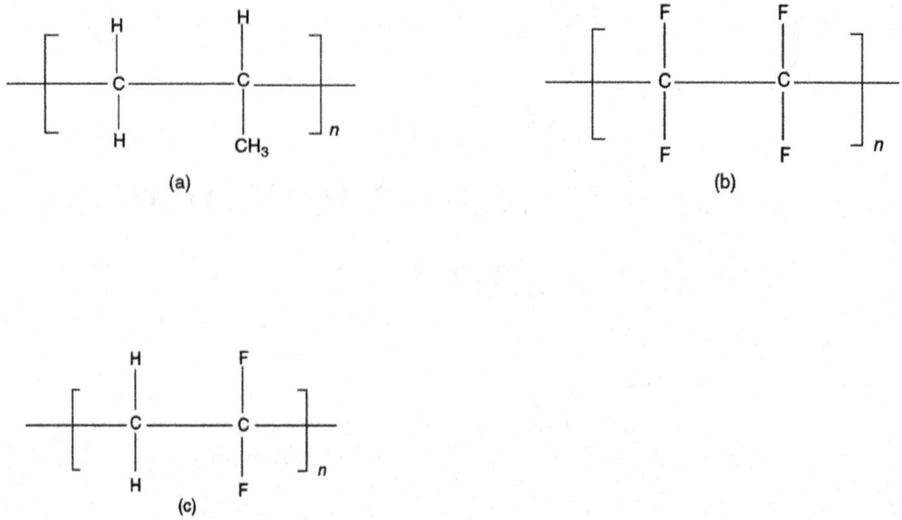

FIGURE 3.1 Molecular structure of (a) PP, (b) PTFE, and (c) PVDF.

the molecule structures of PP, PTFE, and PVDF, respectively. In this section, we will discuss about the theoretical aspects of these materials and some physical properties that are beneficial for MD membranes.

3.2.1 Polypropylene

Polypropylene, also known as polypropene, is a kind of thermoplastic polymer made up of an aliphatic backbone with a methyl group attached to every second carbon atom in the chain. Good chemical and heat stability are attained through protection of the C–C bonds by hydrophobic methyl groups in isotactic PP. Chemical stability plays an important role in MD membranes in view of the harshness and complexity of different feed solutions. Lacking of polar groups in PP contributes to its good stability in water environments. Besides, heat stability is another key factor because MD membranes are operated at elevated temperatures of ~50–80 °C. PP has a melting point of 160–166 °C, which is much higher than MD operating temperatures. In addition, its moderate melting point temperature offers a good processability of fabricating PP membranes via thermally induced phase separation method. This fabrication method will be discussed further in the next section.

Moderate tensile strength, flexibility, and Young's modulus are other requirements for MD membranes. Tensile strength as well as flexibility can enable the membranes to overcome fluctuations of transmembrane pressure happened occasionally during plant operations. In order to maintain the shape and be self-supporting, an appropriate Young's modulus is required for

MD membranes. This is particularly relevant for hollow fiber membranes which must withstand excessive shell-side pressures under the outside-in MD mode without increasing the wall thickness of hollow fibers.

Hydrophobicity is the basic requirement for MD membranes to avoid them from wetting. MD membranes with characteristics of low surface energy are highly desirable. As a result, the available energy at membrane surface could not disrupt the surface tension of an aqueous phase in contact with it. In other words, if the available surface energy is high, it could disrupt the liquid surface tension and lead to the wetting of membrane surface. On the other hand, if the available surface energy is low, there is not enough force to disrupt the liquid surface tension. The liquid could form a droplet on the surface. Thus, in order to prevent MD membranes from wetting, one must employ membrane materials with low surface energy. PP has a surface energy of 30×10^{-3} J m^{-2}, which is sufficiently lower than the surface tension of most MD feed solutions (i.e., 72.8×10^{-3} N m^{-1} for pure water). This signifies the suitability of the PP material. In addition to surface tension, another method to characterize surface hydrophobicity is the water contact angle. A general guideline is that a contact angle larger than 90° indicates a hydrophobic surface, while a contact angle smaller than 90° associates a hydrophilic surface. Figure 3.2 shows contact angles on hydrophobic and hydrophilic surfaces. PP usually shows a contact angle greater than 90°, confirming its hydrophobic property (El-Bourawi et al., 2006).

3.2.2 Polytetrafluoroethylene

Polytetrafluoroethylene is a synthetic fluoropolymer of tetrafluoroethylene. It is composed of an aliphatic backbone with fluorine atoms that occupy all positions on the carbon chain. This predominance of fluorine atoms is responsible for properties which are not encountered in any other polymeric structure. The high concentration and strength of C–F bonds provide this polymer with excellent chemical and heat stability. In addition, the symmetric structure of C–F bonds results in a nonpolar polymer overall. The lack of polarity offers the insolubility of PTFE in aqueous environments.

FIGURE 3.2 Water contact angles on (a) hydrophobic and (b) hydrophilic surfaces.

Inherently possessing a low surface energy is the most advantageous characteristic of PTFE. This is owing to the weak van der Waals forces between fluorine atoms and any other external molecules. PTFE possesses one of the lowest friction coefficients among other solid surfaces. Neither water nor water-containing substances wet PTFE. The apparent high water contact angle of PTFE membranes is consistent with its low surface tension.

3.2.3 Polyvinylidene Fluoride

Polyvinylidene fluoride is a highly nonreactive thermoplastic fluoropolymer manufactured by the polymerization of vinylidene difluoride. It is a specialty plastic applied in situations requiring high purity as well as resistance to solvents, acids, and bases. PVDF is made up of alternating CH_2 and CF_2. Due to the different electronegativity between hydrogen and fluorine atoms, the overall polymer is polar. This makes it possible to dissolve in some common organic solvents, such as N-methyl-2-pyrrolidone (NMP), N, N-dimethylacetamide (DMAc), and so on. Consequently, PVDF has a superior processability to PP and PTFE. PVDF membranes can be prepared by the simple and conventional nonsolvent-induced phase separation (NIPS) technique, while the membrane properties are highly tunable by changing the preparation conditions. Detailed analyses on membrane formation will be provided in the next section.

In addition, PVDF shares many favorable physical and mechanical properties with PTFE, owing to their structural similarity. It not only has good chemical and heat stability but also has a low thermal conductivity that reduces the conductive heat loss through membranes during MD operations. It is intrinsically hydrophobic that exhibits good wetting resistance. Thus, PVDF is the most widely studied material for fabricating MD membranes.

3.2.4 Ceramic Materials

Inorganic ceramic materials, such as alumina, titania, zirconia oxides, silicon carbide, or some glassy materials, are also potential candidates for making MD membranes. Compared to polymeric materials, ceramic materials possess excellent chemical, thermal, and mechanical stabilities. Therefore, ceramic materials have gained much consideration in membrane separation. Although current MD applications employ PP, PTFE, and PVDF widely, ceramic membranes have advantages of better mechanical strength, excellent resistance to harsh chemical environments, and good thermal stability to withstand high feed temperatures. They have been processed into both planar and tubular structures for MD applications (Kujawa et al., 2017). However, ceramic materials are hydrophilic in nature. If they are used in MD processes, surface modifications are required to change them from hydrophilic to hydrophobic one. This can be achieved by binding hydrophobic substances on their surfaces, such as fluoropolymer, organosilane molecules, or aliphatic chains. Another hurdler for ceramic materials is their high thermal conductivity, which is unfavorable because the thermal energy in the hot feed could rapidly lose to the cold permeate by conduction. To mitigate this issue, most studies used air gap membrane distillation (AGMD) or vacuum membrane distillation (VMD) to take advantage of the reduced conductivity in air and vacuum.

Ceramic membranes usually have an asymmetric structure containing a thin separation layer supported by a porous substrate. Sintered alumina (Al_2O_3), titania (TiO_2), silica (SiO_2), or zirconia (ZrO_2) particles are usually employed for the fabrication of porous ceramic membranes. Depending on different process conditions, pore sizes may vary in the range from nanometer to micrometer.

3.3 MD MEMBRANE FABRICATION

3.3.1 Fabrication of PVDF Membranes

Fabrication of PVDF membranes can be dated back to early 1980s (Munari et al., 1983; Uragami et al., 1981). Since then, significant progresses have been made to fabricate them for MD, membrane contactor, ultrafiltration and microfiltration applications by means of phase inversion, sintering, and track etching methods. Among them, the phase inversion technique is the most common one owing to its simplicity and flexibility to produce porous membranes. Therefore, this section mainly focuses on it and the effects of its operation parameters on the formation of various PVDF membranes.

Phase inversion is a process that involves the transformation of an original homogeneous polymer solution into a solid phase under a controlled manner. There are several ways to apply it, which are NIPS, TIPS, controlled evaporation of solvent, precipitation from the vapor phase, etc. Among them, NIPS and TIPS are most widely used in the fabrication of PVDF membranes. The following section will further elucidate the effects of phase inversion conditions on NIPS and TIPS processes, in terms of solvent type, polymer concentration, nonsolvents, additives, coagulation bath, spinning parameters, etc.

3.3.1.1 Nonsolvent-Induced Phase Separation

During the NIPS process, a polymer solution (polymer and solvent) is cast on a proper support and then immersed in a nonsolvent coagulation bath. Therefore, phase separation will happen because of the exchange between the solvent in the polymer solution and the nonsolvent in the coagulation bath. NIPS has been applied to produce most available PVDF membranes because PVDF can be easily dissolved in common organic solvents. However, since PVDF is semicrystalline in structure, its phase inversion process is more complicated than amorphous polymers such as polysulfone and polyethersulfone. Many researchers have been actively developing PVDF membranes through the NIPS process, working heavily on the influence of different manufacturing conditions on membrane morphology and performance. Besides, the relationship between membrane structure and its formation process is also vigorously studied (Liu et al., 2011).

3.3.1.1.1 Effect of Solvents on Membrane Morphology

Solvent is the first parameter to be considered for any NIPS process. It not only directly influences the polymer configuration in the casting or spinning solutions but also affects the subsequent phase inversion process. To ensure a high mobility of polymer chains and homogenous distribution within the solutions, a proper solvent selection is essential to form MD membranes with a desirable morphology. Solubility parameter is usually used as a guideline for the selection of solvents, where a similar solubility parameter between a polymer and a solvent may infer a good dissolution. According to Bottino et al., PVDF has a dispersive parameter of $\delta_d = 17.2$ MPa$^{1/2}$, polar parameter of $\delta_p = 12.5$ MPa$^{1/2}$, hydrogen bonding parameter of $\delta_h = 9.2$ MPa$^{1/2}$, and total solubility parameter $\delta_t = 23.2$ MPa$^{1/2}$ (Bottino et al., 1988b). Table 3.1 tabulates the solubility parameters and boiling points of common solvents for PVDF (Bottino et al., 1991; Haynes, 2010). Among these solvents, DMAc, NMP, DMF, and DMSO have been extensively studied as high boiling point strong solvents, while acetone or THF has been used as the low boiling point weak solvent in casting PVDF membranes.

TABLE 3.1
Hansen solubility parameters and boiling points of common solvents for PVDF

Solvents	Dispersion parameter δ_d (MPa$^{1/2}$)	Polar parameter δ_p (MPa$^{1/2}$)	Hydrogen bonding parameter δ_h (MPa$^{1/2}$)	Total solubility parameter δ_t (MPa$^{1/2}$)	Boiling point (°C)
N, N-Dimethylacetamide (DMAc)	16.8	11.5	10.2	22.7	165.0
N, N-Dimethylformamide (DMF)	17.4	13.7	11.3	24.8	153.0
Dimethylsulfoxide (DMSO)	18.4	16.4	10.2	26.7	189.0
Hexamethyl phosphoramide (HMPA)	18.4	8.6	11.3	23.2	232.5
N-Methyl-2-pyrrolidone (NMP)	18.0	12.3	7.2	22.9	202.0
Tetramethylurea (TMU)	16.8	8.2	11.1	21.7	176.5
Triethyl phosphate (TEP)	16.8	11.5	9.2	22.3	215.5
Trimethyl phosphate (TMP)	16.8	16.0	10.2	22.3	197.2
Acetone	–	–	–	–	56.1
Tetrahydrofuran (THF)	–	–	–	–	65.0

In the study of Bottino et al. (Bottino et al., 1991), eight different types of solvents were employed to fabricate flat sheet PVDF membranes. The results showed that the kinetic factor (e.g., the mutual diffusivity between the solvent and the nonsolvent) played a more important role than the thermodynamic properties in determining the formation mechanism and membrane structures of PVDF membranes. Yeow et al. also demonstrated that different solvents would lead to form various membrane structures (Yeow et al., 2004). Figure 3.3 shows examples of different cross-section morphologies when using TEP, NMP, DMF, and DMAc as solvents to prepare PVDF membranes. A sponge-like and almost symmetric cross section was formed by employing TEP as a solvent and water as a coagulant. Comparing to NMP, DMAc, and DMF, TEP had a weaker solvent power. This led to a slow liquid–liquid demixing during phase inversion. In addition, TEP had a weaker compatibility with water than other solvents. Thus, the mutual diffusivity between the solvent and the nonsolvent during phase inversion was retarded. Therefore, a sponge-like membrane structure was obtained. On the other hand, strong solvents led to the formation of membranes with a macrovoid structure, owing to the fast precipitation rates.

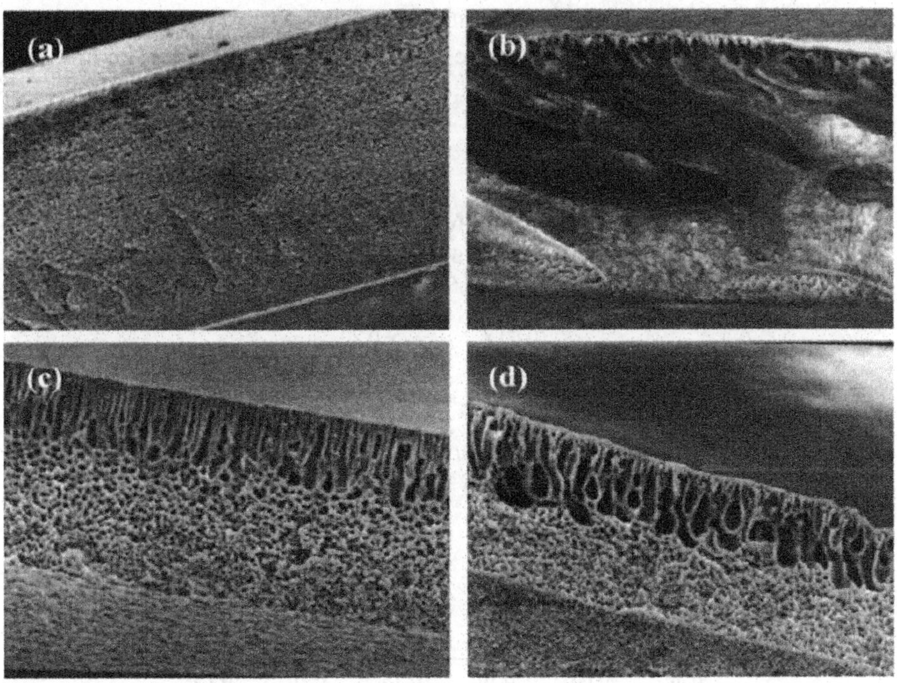

FIGURE 3.3 Images for cross sections of PVDF membranes formed by (a) TEP, (b) NMP, (c) DMF, and (d) DMAc (Yeow et al., 2004). Copyright (2004) John Wiley and Sons.

Recently, Chang et al. have also investigated the effects of TEP and NMP solvents on the formation of PVDF hollow fibers (Chang et al., 2017). Figure. 3.4 presents the ternary phase diagram of PVDF/solvent/water systems at 25 °C. In the diagram, the two curves are called binodal curves, which are the limit of two homogenous solutions. The left side of binodal curve denotes a stable homogenous phase, while the right side indicates an unstable region. The figure shows that the binodal curve of the TEP system is on the left side of the NMP system, which confirms that TEP is a weaker solvent than NMP for PVDF. As a result, a less nonsolvent is required to induce phase inversion of the TEP system. This could lead to a more porous PVDF membrane structure as polymer chains in the solution are less stretched and more spaces are left among them. Other than solvent power, Chang et al. have also studied the phase inversion kinetics. It was found that the PVDF/TEP solution had a slower phase inversion kinetics than the PVDF/NMP solution. Two reasons may be associated with this observation. First, TEP has lower affinity to water than NMP and thus water intrusion is slowed down during the phase inversion process. Second, since TEP is a weaker solvent, the polymer solution exhibits a higher viscosity, which leads to a lower diffusivity of both solvent and nonsolvent. As a result, the TEP system favors a sponge-like structure, while the NMP system errands a macrovoids morphology.

3.3.1.1.2 Effect of Polymer Concentration

When a proper solvent is selected, the next step is to determine the appropriate polymer concentration for membrane fabrication. Generally, the higher the

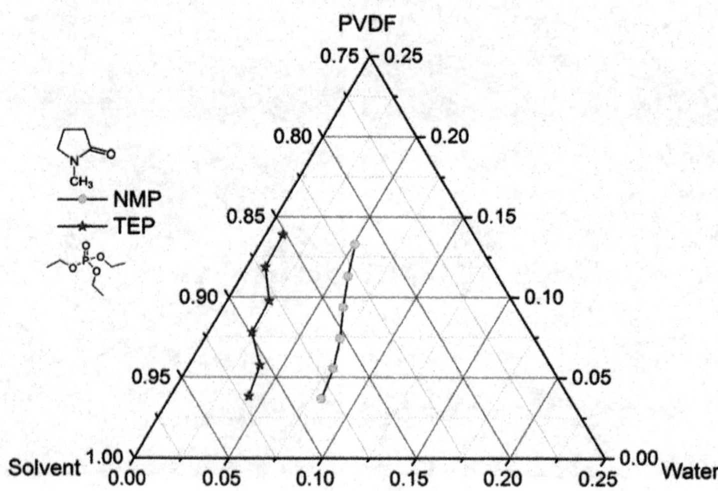

FIGURE 3.4 Phase diagram of PVDF/solvent/water system (Chang et al., 2017). Copyright (2017) Elsevier.

concentration, the higher the solution viscosity and the lower the membrane porosity. This would result in higher membrane mechanical properties, higher liquid entry pressure (LEP), but a lower membrane flux. Thus, there is a tradeoff between membrane properties and membrane performance. If the concentration is too low, the resultant membrane is easily deformed. If the concentration is too high, the membrane will lose its flux significantly. An optimum concentration or concentration range has to be determined.

Figure 3.5 shows the relationship between the shear viscosity of PVDF/NMP solutions and PVDF concentration at a shear rate of 10 s^{-1} (Sukitpaneenit and Chung, 2009). The critical concentration is about 13 wt%, which is the intersection point by extrapolating the two linear portions of the viscosity curve. Based on Chung et al., when the polymer concentration is at or above the critical concentration, significant chain entanglement may occur (Chung et al., 1997b). This favors the formation of a denser skin with minimal defects on the resultant membrane. On the other hand, when the polymer concentration is lower than the critical concentration, a porous structure consisting of macrovoids may be formed. Therefore, for the PVDF/NMP system, the concentration should be around or below its critical, because a porous membrane structure is desired. A concentration range of 10–14 wt% is usually used for the fabrication of PVDF membranes for MD applications.

3.3.1.1.3 Effect of Nonsolvents

Sukitpaneenit and Chung have explored the fundamental science of fabricating PVDF membranes and correlated membrane morphology, crystallinity, and mechanical properties with nonsolvents used in dope solutions during phase inversion (Sukitpaneenit and Chung, 2009). A series of nonsolvents (i.e., water, methanol, ethanol, and isopropanol) were employed to construct the

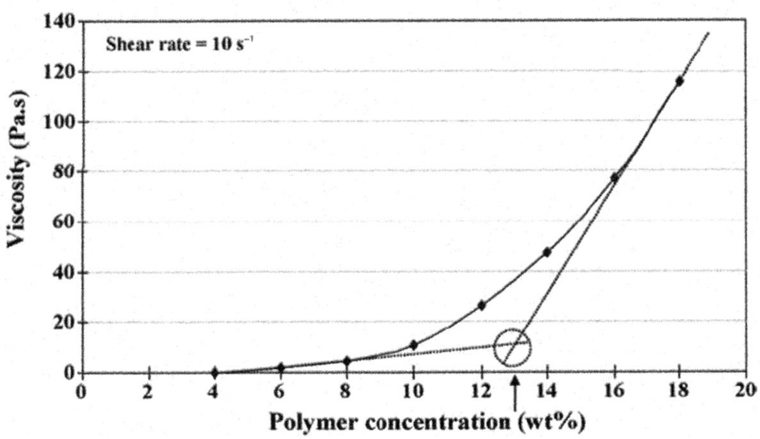

FIGURE 3.5 PVDF/NMP solution viscosity versus concentration (Sukitpaneenit and Chung, 2009). Copyright (2009) Elsevier.

polymer/solvent/nonsolvent phase diagram. Figure 3.6 shows their ternary phase diagrams at 25 °C. The binodal curve of the PVDF/NMP/water system is closer to the polymer–solvent axis. Therefore, only a small amount of water is needed to disturb its equilibrium status and trigger the polymer precipitation. According to the distance between the binodal curve and the polymer–solvent axis, the thermodynamic stability of PVDF/NMP/nonsolvent systems follows the sequence of isopropanol > ethanol > methanol > water. In other words, water is the strongest nonsolvent, while isopropanol is the weakest one. Depending on the strength of nonsolvents, phase inversion of semicrystalline PVDF would fall under the mechanism of either liquid–liquid demixing or solid–liquid demixing. The former results in an interconnected cellular morphology, while the latter produces an interconnected globule structure.

The use of a weak nonsolvent as an additive in the external coagulation bath also induces solid–liquid demixing and crystallization. The resultant membranes comprise spherulitic crystallites with a crystallite size increasing with a decrease in nonsolvent strength. The loose packing of spherulitic crystallites in membranes results in lower mechanical strength compared to those with an interconnected cellular morphology. Clearly, there is a close relationship among membrane morphology, properties, and nonsolvents used in dope solutions and coagulant baths.

3.3.1.1.4 Effect of Additives

To molecularly design PVDF membranes with desirable physiochemical properties and separation performance, one useful method is to incorporate additives into polymer solutions. An additive usually acts as a pore former, which helps create microporosity in the resultant membranes, thus enhancing the water flux. Different additives have been employed, such as the above-mentioned nonsolvents, some small molecules (i.e., lithium chloride (LiCl),

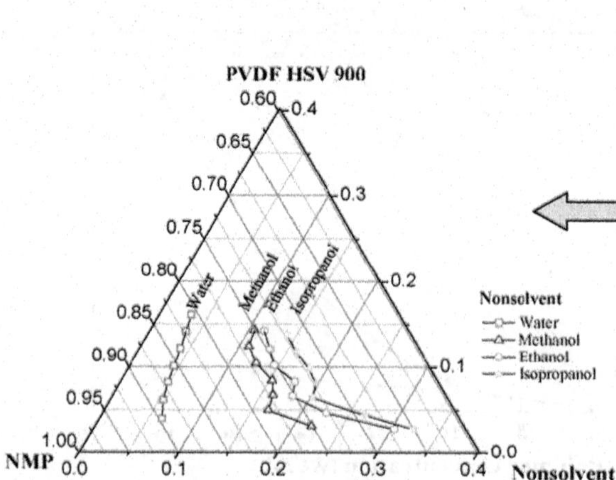

FIGURE 3.6 Phase diagram of ternary PVDF/NMP/nonsolvent systems at 25°C (Sukitpaneenit and Chung, 2009). Copyright (2009) Elsevier.

ethylene glycol (EG), ethylene diamine (EDA), graphene oxide (GO)), large molecules (i.e., polyvinylpyrrolidone (PVP) and poly(ethylene glycol) (PEG)), and so on. Other than being a pore former, some additives can also enhance membrane hydrophobicity and wetting resistance. These additives are generally hydrophobic, such as Teflon particles and hydrophobic silica. The effects of different additives on morphology and properties of PVDF membranes are discussed in this section.

Khayet and Matsuura have used water as a pore-forming additive in the polymer casting solution (Khayet and Matsuura, 2001). DMAc was used as the solvent. With the water content increased from 0 to 6 wt%, the PVDF membrane porosity was increased from 26.8% ± 3.5% to 74.9% ± 3.7%, and the MD flux was improved from 1 to 17 L/m^2 h at 25 °C. Zuo and Chung also investigated the effect of water additive in PVDF/NMP solutions (Zuo and Chung, 2016a). Figure 3.7 displays the PVDF membrane morphology casting from solutions with different water amounts. All the membranes had a macrovoid structure, while the size of macrovoid cavities increased with increasing water concentration. This was because the addition of water shifted the polymer solution closer to the binodal curve in the phase diagram, which speeded up the precipitation rate during phase inversion. Thus, a more open structure with large macrovoids was formed. The resultant membrane showed a 150% increase in MD flux when the water content was increased from 0 to 6 wt%. However, the membrane mechanical strength was largely sacrificed due to the existence of large cavities. For example, the Young's modulus decreased more than 40% when the water concentration was increased from 0 to 6 wt%.

Tomaszewska has studied the effects of LiCl additive on PVDF membrane morphology and performance (Tomaszewska, 1996). When LiCl was added into the PVDF/DMAc casting solutions, the resultant membranes showed larger cavities and pore sizes with a higher porosity. Consequently, the membrane flux was significantly increased. Similarly, Bottino et al. also observed that the incorporation of LiCl into polymer solutions resulted in more porous PVDF membranes (Bottino et al., 1988a). The enhancement effect was stronger with a higher LiCl concentration. The rationale was that LiCl had a high

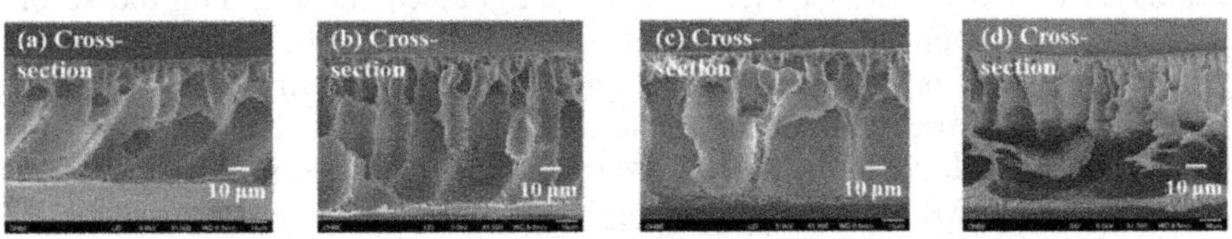

(a) No water (b) 2.4 wt% water (c) 6 wt% water (d) 8 wt% water

FIGURE 3.7 Cross-section morphology of PVDF membranes prepared from (a) no water, (b) 2.4 wt% water additive, (c) 6 wt% water additive, and (d) 8 wt% water additive (Zuo and Chung, 2016a). Copyright (2016) John Wiley and Sons.

affinity toward water, which led to a fast precipitation rate during phase inversion. Thus, large cavities were created and the resultant membranes had much higher water fluxes. Nonetheless, the major drawback of this approach was the sacrifice of membrane mechanical properties. Similar conclusions were reported by Tomaszewska (Tomaszewska, 1996). Despite the enhanced flux, the mechanical properties of membranes deteriorated rapidly when increasing the LiCl concentration from 0 to 1 wt%.

The above-mentioned issues can be overcome by using a combination of additives or by employing additives with special functions. For instance, Zuo and Chung has included LiCl and EG in the polymer dope solutions and obtained PVDF hollow fiber membranes with excellent mechanical strength (Zuo and Chung, 2017). First, the addition of LiCl and EG promoted liquid–liquid demixing because the polymer solution miscibility was reduced. When the thermodynamic miscibility was reduced, liquid–liquid phase separation took place easily during the phase inversion process. This would errand a sponge-like structure. Second, the addition of 5 wt% LiCl suppressed the formation of large-size macrovoids, while small-size macrovoids were preferred for membranes with better mechanical strength. Due to the interaction between Li^+ ions and electron-rich groups in PVDF, the solution viscosity was increased. This would delay the mutual diffusion between solvent and nonsolvent during membrane formation. Thus, the large-size macrovoids were suppressed. Both factors contributed to the high mechanical strength in the resultant membranes.

Alternatively, the employment of additives with special functions may also help to overcome the tradeoff between membrane performance and mechanical strength. Zuo and Chung have designed PVDF membranes with good MD performance without sacrificing the mechanical properties by involving water and EDA additives (Zuo and Chung, 2016a). On the one hand, both additives served as pore formers to increase membrane pore size and porosity. On the other hand, EDA acted as a cross-linking agent, which was responsible for increasing membrane mechanical properties. Figure 3.8 elucidates the two types of cross-linking mechanism of PVDF by EDA. Since EDA is a strong base, it could induce the dehydrofluorination reaction in PVDF. For the first mechanism, the dehydrofluorination reaction takes place between two polymer chains so that the two polymer chains are tightened directly. For the second type, the dehydrofluorination reaction occurs between two neighboring fluorine atoms within one polymer chain and produces unsaturated C=C double bonds for the subsequent nucleophilic addition of EDA. Then, the polymer chains are cross-linked through EDA molecules. As a result, the membrane achieved a good MD flux without compromising the mechanical properties.

Recently, graphene oxide (GO) has also been used as an additive to enhance the mechanical properties and wetting resistance of PVDF membranes. Lu et al. studied the effects of adding GO and n-butylamine modified GO (GO-NBA) in PVDF membranes (Lu et al., 2017). Since the direct incorporation of hydrophilic GO into PVDF membranes would lower their hydrophobicity and wetting resistance, surface modification of GO by means

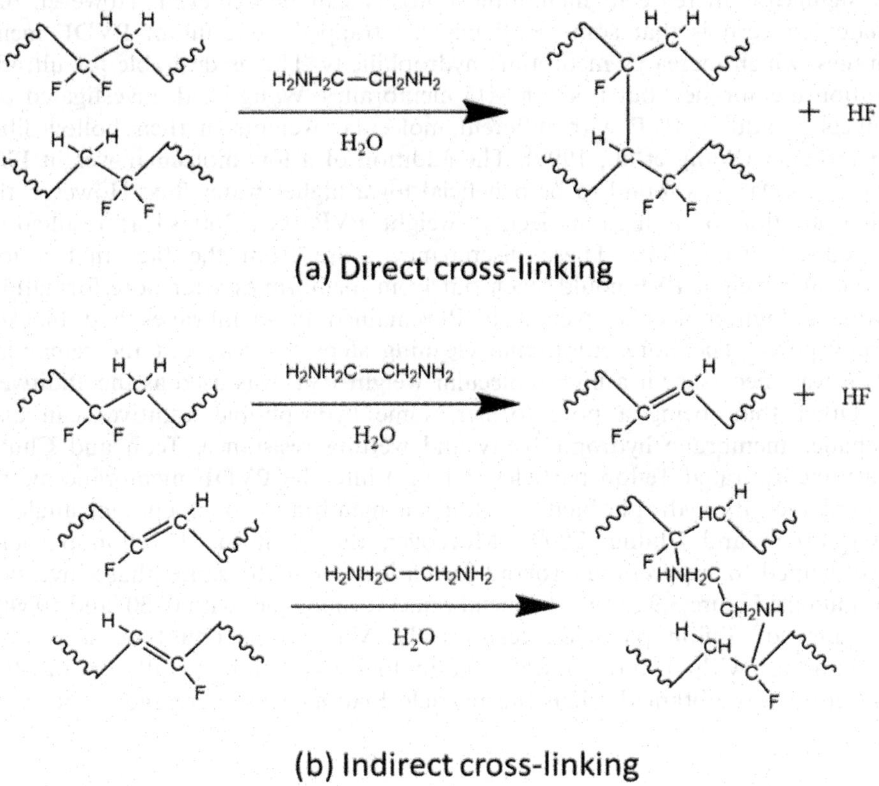

FIGURE 3.8 The cross-linking mechanisms between PVDF and EDA (Zuo and Chung, 2016a). Copyright (2016) John Wiley and Sons.

of *n*-butylamine (NBA) was carried out. It was found that the water contact angle of the pristine GO film increased from 45° to 72° for the GO-NBA film. Experimental results also confirmed that the modified GO not only improved its compatibility with the PVDF matrix but also increased membrane hydrophobicity. With a 0.5 wt% GO-NBA loading, the maximum tensile stress, maximum tensile strain, and Young's modulus of the flat sheet membranes increased by 138%, 48%, and 86%, respectively. What's more, the 0.5 wt% GO-NBA loaded hollow fiber membrane showed the best water flux of 61.9 kg/m^2 h at 80 °C under the DCMD mode.

Besides small molecule additives, large molecule additives are also commonly used in PVDF membrane fabrication. PVP is one of the most frequently studied large molecule additives to produce highly porous PVDF membranes. Since PVP is hydrophilic, it has high affinity with water. The addition of PVP in polymer solutions could enhance nonsolvent inflow during the phase inversion process, thus promoting the formation of large macrovoids.

A significant increase in membrane porosity can be achieved. However, one major concern is that some PVP may be trapped in resultant PVDF membranes, which increases membrane hydrophilicity. This is desirable for ultrafiltration membranes, but not for MD membranes. Wang et al. investigated the effects of adding PVP with different molecular weights in their hollow fiber membranes (Wang et al., 1999). The addition of a low molecular weight PVP (i.e., 10 kDa) was found to be beneficial for a higher water flux. However, the incorporation of a high molecular weight PVP (i.e., 360 kDa) resulted in a reduced water flux. These discrepancy arose from the fact that a low molecular weight PVP could leach out from membranes after pore formation, while a high molecular weight PVP remained in membranes and blocked many pores. Therefore, additional cleaning steps to wash out the remaining PVP were necessary if a high molecular weight PVP was used as the additive.

Other than being a pore former, some hydrophobic additives can also enhance membrane hydrophobicity and wetting resistance. Teoh and Chung have incorporated Teflon particles (<1 μm) into the PVDF membrane matrix to enhance its hydrophobicity, yielding a membrane water contact angle of 103° (Teoh and Chung, 2009). Moreover, the addition of Teflon particles also helped to suppress macrovoid formation under the same phase inversion conditions. Figure 3.9 shows the membrane morphologies with 0, 30, and 50 wt% loadings of Teflon particles, respectively. Macrovoids could be seen from membranes with 0 and 30 wt% Teflon loadings, while a fully sponge-like structure was obtained when the particle loading was increased to 50 wt%.

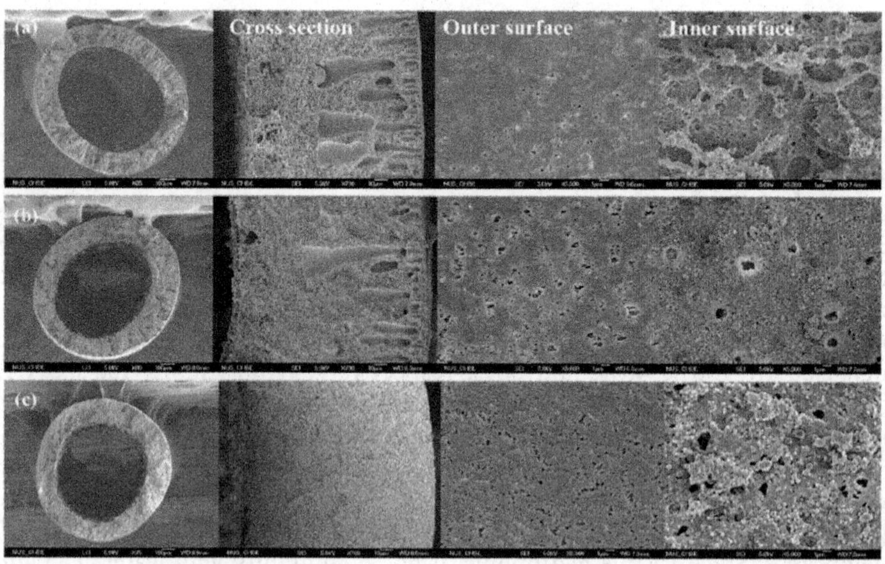

FIGURE 3.9 FESEM images of PVDF membranes with different Teflon loading (a) 0, (b) 30, and (c) 50 wt% (Teoh and Chung, 2009). Copyright (2009) Elsevier.

Undoubtedly, the addition of a large amount of hydrophobic Teflon particles might increase the dope viscosity and suppress the macrovoid formation. The reduced macrovoid structure helped enhance the mechanical properties so that the resultant hollow fiber membranes exhibited not only higher resistance to pore wetting but also stable performance in long-term tests. Similarly, Wang et al. included 3 wt% Teflon particles in PVDF/NMP dope solutions to enhance the wetting resistance (Wang et al., 2011). Their hollow fibers showed stable 200-hour DCMD desalination performance and a separation factor higher than 99.9%.

3.3.1.1.5 Effect of Coagulation Mediums

As discussed earlier, a coagulation bath is the place where the phase inversion process occurs. Thus, the medium in the coagulation bath plays a key role in governing the demixing process, namely, liquid–liquid demixing and solid–liquid demixing. For example, the use of water as a coagulant often contributes to a rapid liquid–liquid demixing process because water acts as a strong nonsolvent. As a result, it produces membranes with asymmetric structures and finger-like macrovoids. In contrast, sponge-like structures may be formed for hydrophilic glassy polymers under the same condition if a delayed liquid–liquid demixing process takes place (Widjojo et al., 2011). On the other hand, the use of a weak coagulation medium for semicrystalline polymers such as PVDF may instigate a solid–liquid demixing process and result in interconnected globule structures (Sukitpaneenit and Chung, 2009). In this section, the influences of various coagulation mediums on flat sheet and hollow fiber PVDF membranes are demonstrated.

Edwie et al. have investigated the effects of water and water/methanol coagulation mediums on PVDF membrane morphologies (Edwie et al., 2012). Figure 3.10 presents the cross sections of field-emission scanning electron microscopy (FESEM) images of membranes casted and then immersed in 100 wt% water, 60/40 wt% water/methanol, and 20/80 wt% water/methanol, respectively. Different membrane morphologies were developed by varying the coagulant compositions. Membranes casted from pure water consisted of finger-like macrovoids near the top surface and cellular structures near the bottom surface. Membranes casted from 60/40 wt% water/methanol had small and less macrovoids, while membranes casted from 20/80 wt% water/methanol totally eliminated macrovoids. These distinct structures were resulted from different precipitation rates during phase inversion, which can be explained by the concept of solubility parameters. Water and methanol have solubility parameters of 47.8 and 29.6 $MPa^{1/2}$, respectively (Hansen, 1999). Compared to the solubility parameter of PVDF (23.2 $MPa^{1/2}$), water has a larger difference from it than methanol. The larger gap in solubility parameter favors a rapid liquid–liquid demixing process (i.e., a fast precipitation rate), and hence membranes precipitated in water have asymmetric structures and finger-like macrovoids. With an increase in methanol content in the coagulation bath, delayed demixing becomes dominant and results in membranes with macrovoid-free structures. Although macrovoids structures are

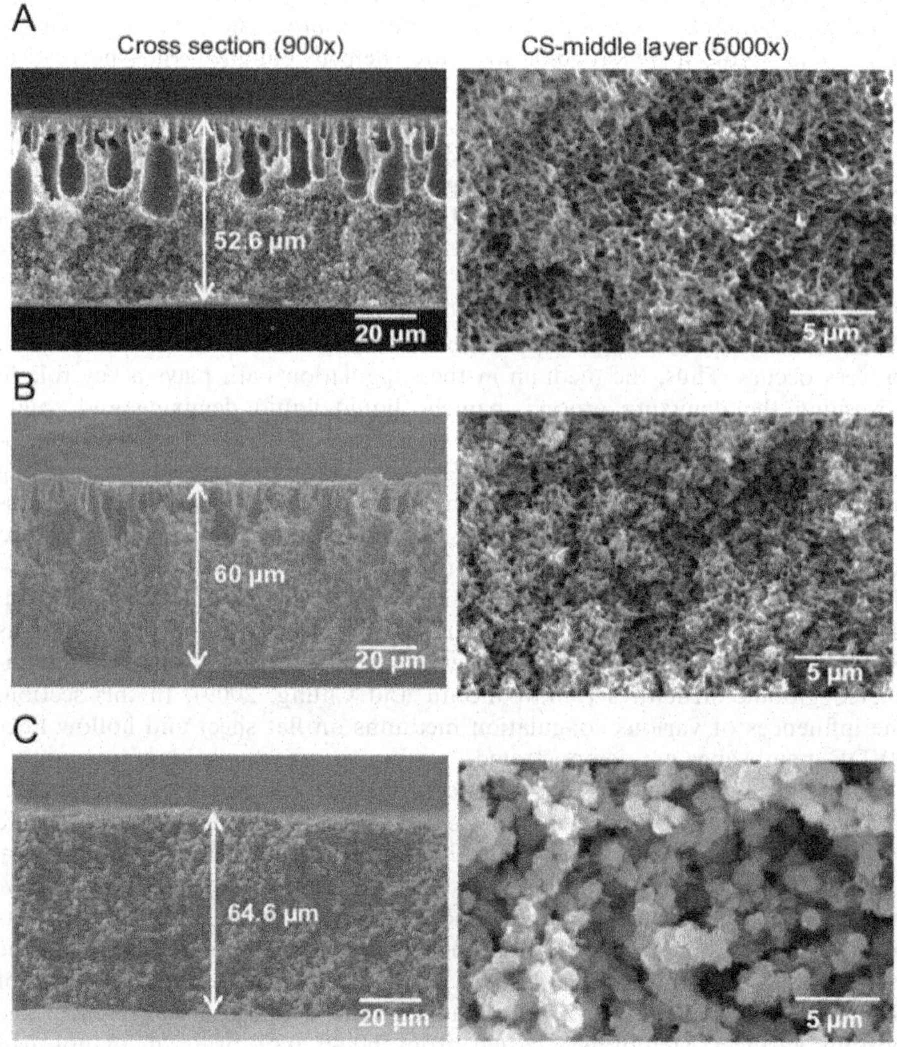

FIGURE 3.10 FESEM images of PVDF membranes casted from different coagulant (A) water, (B) 60/40 wt% water/methanol, and (C) 20/80 wt% water/methanol (Edwie et al., 2012). Copyright (2012) Elsevier.

beneficial for higher membrane fluxes, macrovoid-free morphologies are essential for better wetting resistance.

Figure 3.10 also shows another significant transformation. The membrane morphology changed from cellular to globule structures when methanol was added into the coagulant. This was associated with the solid–liquid demixing process that formed membranes with interconnected semicrystalline particles

(Lin et al., 2003; Young et al., 1999). The addition of the weak coagulant methanol could thermodynamically and kinetically decelerate the precipitation rate, and hence there was sufficient time for the crystallization process to occur.

Similar observations were also found by Lu et al. (Lu et al., 2016). In order to improve the MD flux of PVDF hollow fibers, Lu et al. enhanced membrane porosity by changing the coagulant from water to a water/isopropanol mixture. As a result, the phase inversion process changed from rapid demixing to delayed demixing. Figure 3.11 displays the relationship between membrane morphologies and isopropanol concentration. Two major changes could be observed: (1) the outer surface gradually changed from a dense structure to a porous structure, and (2) the cross-section morphology progressively shifted from an interconnected cellular structure to a globule structure. Both phenomena may be explained by the delayed demixing process in a weaker coagulation bath. When the isopropanol concentration was increased from 0 to 80 wt%, the surface porosity was increased from 5.1% to 24.4% and the surface pore size was enlarged from 128.7 to 406.8 nm. Consequently, the vapor transport resistance was reduced significantly and there was a 70% jump in water flux. However, the change in cross-section morphologies imposed higher mass transfer resistance because globule structures occupied more cross-section areas. Therefore, a tradeoff between flux and morphology would emerge at a high isopropanol concentration. The flux improvement became unnoticeable when the isopropanol concentration was increased beyond 60 wt%. Nevertheless, the use of a weaker coagulant has been demonstrated as an effective way to improve the MD flux.

In addition to control the chemistry of coagulants, coagulation temperature plays an important role in determining the activity of a coagulant, location of the binodal curve, and the crystallization process of PVDF. A higher temperature not only lowers the viscosity of the nascent membrane but also favors

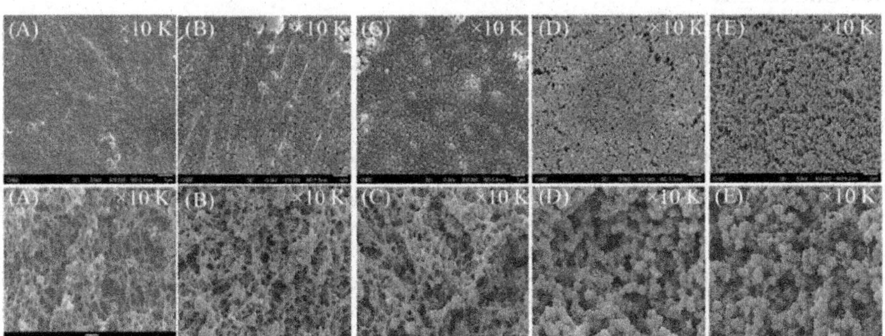

FIGURE 3.11 SEM picture of membrane outer surface and cross section using different isopropanol/water compositions in the coagulation bath: (A) 0/100, (B) 20/80, (C) 40/60, (D) 60/40, and (E) 80/20 wt% (Lu et al., 2016). Copyright (2016) Elsevier.

a faster solvent exchange rate because of the higher diffusivity. For glassy polymers, since the binodal curve shifts to the right of the ternary phase diagram at a higher temperature, a higher coagulation temperature tends to induce macrovoid-free structures (Chung et al., 1997a). However, its effects are much complicated for semicrystalline PVDF. One must take the effects of coagulant temperature on membrane viscosity, diffusivities of solvents and nonsolvents, and PVDF crystallization into consideration. Chen investigated the effects of coagulation bath temperature on PVDF membrane structures (Cheng, 1999). When the coagulant temperature was increased from 25 to 65 °C, the structure of resultant PVDF membranes changed from full globule particulates to a mixture of cellular and globule structures. The increased temperature changed the phase inversion process from solid–liquid demixing to liquid–liquid demixing. Similar phenomena were also reported by Yeow et al. (Yeow et al., 2004) and Wang et al. (Wang et al., 2008b) in their studies of the effects of coagulation bath temperature on PVDF membranes.

3.3.1.1.6 Effect of Spinning Parameters on PVDF Hollow Fiber Membranes

Hollow fiber membranes have several advantages over flat sheet membranes, such as the higher packing density and self-support property. The hollow fiber configuration has become one of the most preferred configurations in many industrial applications. However, the fabrication of hollow fiber membranes is much more complex than that of flat sheet membranes. The fabrication conditions of flat sheet membranes cannot be directly extended to produce hollow fiber membranes. Other than the aforementioned factors that control PVDF membrane morphologies, there are many other parameters during the spinning that affect the formation of hollow fibers, for instance, the dope flow rate, bore fluid flow rate, shear rate, air-gap distance, take-up speed, and so on. Chung and his coworkers have done many works to elucidate the effects of various spinning parameters on the formation of PVDF hollow fiber membranes (Bonyadi et al., 2009; Teoh and Chung, 2009; Wang et al., 2008a; Zuo and Chung, 2017).

Zuo and Chung have designed PVDF hollow fiber membranes with a unique sandwich structure by purposefully manipulating the dope formulation, additives and coagulation mediums (Zuo and Chung, 2017). The sandwich structure comprised a thin layer of small-size macrovoids in the middle, and two sponge-like porous surrounding layers. This structure resulted in hollow fibers with excellent mechanical properties and wetting resistance. However, one drawback was the relatively low MD flux. To enhance the membrane flux, spinning parameters were altered to change the porosity and/or wall thickness of the fibers. For example, the effects of bore fluid flow rate on hollow fiber morphology and performance were investigated. Figure. 3.12 compares the FESEM images of fibers spun from bore fluid flow rates of 1.5, 3.0, and 4.5 ml/min. The inner surface remained similar for all three fibers because the bore fluid chemistry was the same. The hollow fiber diameter increased with a higher bore fluid flow rate owing to the higher radial force at a higher bore fluid flow rate. Since the dope flow rate was constant, the

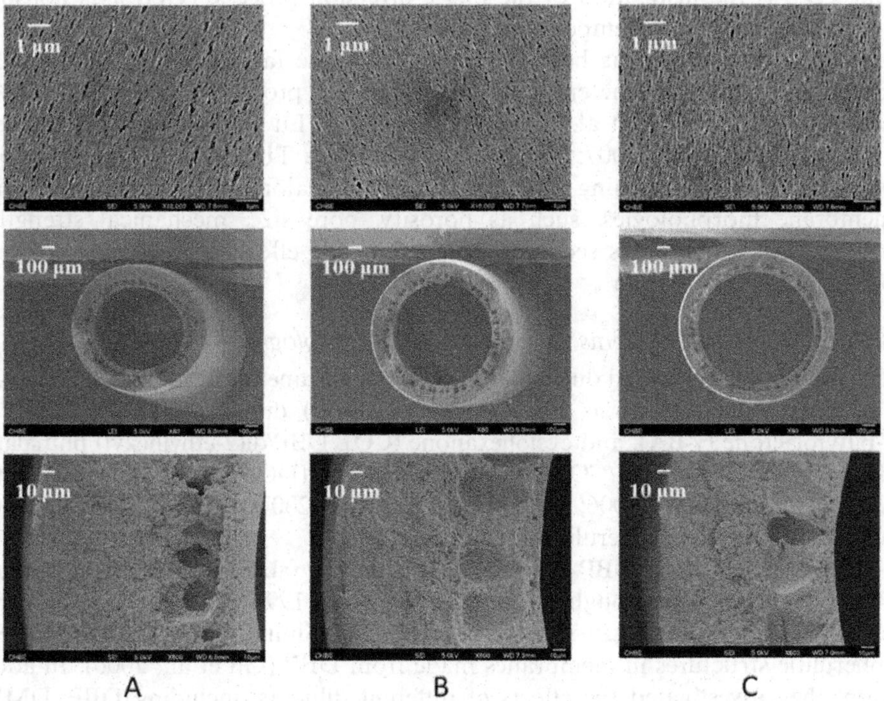

FIGURE 3.12 Inner surface and cross-section morphology of fibers spun from increasing bore fluid flow rate (A) 1.5, (B) 3.0, and (C) 4.5 ml/min (Zuo and Chung, 2017). Copyright (2017) Elsevier.

enlarged hollow fiber had a thinner wall thickness. This not only reduced the vapor transport resistance across the wall but also increased the membrane porosity. As a result, the MD flux increased 42% and 87% when bore fluid flow rates were increased to 3.0 and 4.5 ml/min, respectively.

3.3.1.2 Thermally Induced Phase Separation

In addition to immersion precipitation, another useful technique to prepare polymeric membranes via phase inversion is the TIPS. In the TIPS method, a polymer is firstly dissolved in a high boiling point diluent at a high temperature, and then cast into a desired shape. The phase inversion process is induced by cooling/quenching the as-cast solution so that polymer solidification occurs. After the formation of nascent membranes, the diluent is removed to create microporous structures. The used diluent should be a strong solvent for the polymer at a high temperature (i.e., the temperature to prepare the dope solution) and a nonsolvent at a low temperature (i.e., the quenching temperature) to induce the TIPS mechanism. The major difference between NIPS and TIPS is the instigation of the phase inversion process. NIPS uses nonsolvents while TIPS uses temperature gradient.

The TIPS method has been largely used for the fabrication of PP and PE membranes. Recently, the employment of TIPS to produce PVDF membranes has also emerged (Gu et al., 2006; Ji et al., 2007; Lu et al., 2009; Rajabzadeh et al., 2008; Su et al., 2007; Xiao et al., 2015). In a TIPS process, the selection of diluents and quenching conditions are the major factors that affect the membrane morphologies, such as porosity, pore size, mechanical strength, MD flux, etc. Thus, this section will focus on the effects of these two factors on membrane formation.

3.3.1.2.1 Effect of Diluents on Membrane Morphologies

The most commonly used diluents for PVDF membrane fabrication include single and binary diluents, such as dibutyl phthalate (DBP), dimethyl phthalate (DMP), γ-butyrolactone (γ-BA), and cyclohexanone (CO), DBP/di(2-ethylhexyl) phthalate (DEHP), CO/DBP, γ-BA/CO, DMP/dioctylsebacate (DOS), DMP/dioctyl adipate (DOA), etc. (Gu et al., 2006; Ji et al., 2007; Su et al., 2007;). Usually, these diluents would result in fuzzy spherulitic structures.

Lloyd et al. chose DBP as the diluent for the fabrication of microporous PVDF membranes through TIPS (Lloyd et al., 1990). The resultant membranes had irregular fuzzy spherulitic structures. Similarly, Gu et al. observed spherulitic structures in membranes made from DBP (Gu et al., 2006). In addition, they investigated the effects of different diluents, including DBP, DMP, DOS, DOA, and/or their mixtures on membrane structure. It was observed that PVDF membranes prepared from DMP showed the most discernable and the largest spherulitic structures compared to those made from DBP, DMP/DOS, and DMP/DOA diluents. They believed that the polymer–diluent interaction played a major role on the degree of PVDF crystallization, thus affecting the membrane morphology. Generally, the stronger the interaction, the easier the polymer phase could flow, congregate, and form the spherulitic structure. Among these dilutes, DMP had stronger interaction with PVDF, which favored the formation of discernable spherulitic structures and led to a lower membrane porosity. On the other hand, DOS or DOA had weaker interaction with PVDF, which retarded the PVDF crystallization. Thus, by using mixed diluents of DMP/DOS or DMP/DOA, the polymer–diluent interaction was reduced, which resulted in membranes with minimal spherulitic structures and high porosity.

Ji et al. also investigated the effects of mixed diluents on polymer crystallization and membrane morphology (Ji et al., 2007). A ternary system of PVDF/DBP/di(2-ehylhexyl) phthalate (DEHP) was used for membrane fabrication, where DBP was a strong solvent and DEHP was a weak solvent. When the DBP concentration was increased in dope solutions, the size of spherulites in resultant membranes was also increased. This was because of the increased compatibility between PVDF and diluents, which enhanced polymer crystallization before the liquid–liquid phase separation. Thus, by controlling the DBP/DEHP ratio, one could easily manipulate the membrane microstructure. Similarly, Li et al. observed the change of PVDF membrane structure when a weak diluent dioctyl phthalate (DOP) was added into the PVDF/DBP system (Li et al., 2008). With increasing DOP amount, membrane morphologies changed from large spherulites to bicontinuous or uniform porous structures because the cloud point was shifted to a higher temperature with the addition of DOP. Therefore, liquid–liquid phase separation became dominant with a higher DOP concentration, which favored the formation of bicontinuous porous structures.

3.3.1.2.2 Effect of Quenching Conditions

Other than diluents that affect the membrane structure, the quenching conditions also play a main part on polymer crystallization. Generally, the lower the quenching temperature, the faster the cooling rate is, and the smaller and coarser the formed crystallites are. Gu et al. quenched their membranes in ice water at 273 K and room temperature water at 303 K, respectively (Gu et al., 2006). Ice water introduced a faster cooling rate with a stronger crystallization potential. The resultant membranes comprised small spherulites. When the quenching temperature was increased to 303 K, the cooling rate was reduced and spherulites grew bigger. Similarly, Hassankiadeh et al. found that rapid cooling promoted the formation of crystal nuclei (Hassankiadeh et al., 2014). However, there was not sufficient time for the nuclei to grow. Thus, the resultant membranes possessed a larger amount of spherulites with smaller sizes, which led to denser membrane structures with a lower porosity and higher mechanical strength. On the contrary, a higher quenching temperature lowered the cooling rate and provided more time for the nuclei to grow. This resulted in larger spherulites with a higher membrane porosity and lower mechanical strength.

Besides the quenching temperature, the quenching duration could also change the structure. Figure 3.13 displays the membrane morphologies with different quenching durations (Gu et al., 2006). When the quenching time was increased from 1 to 20 min, the size of spherulites was also increased. Meanwhile, as the spherulites became larger, the diluents were squeezed out and smaller pores were formed in the spherulites, Thus, by changing the quenching conditions, membrane porosity and pore sizes can be easily controlled.

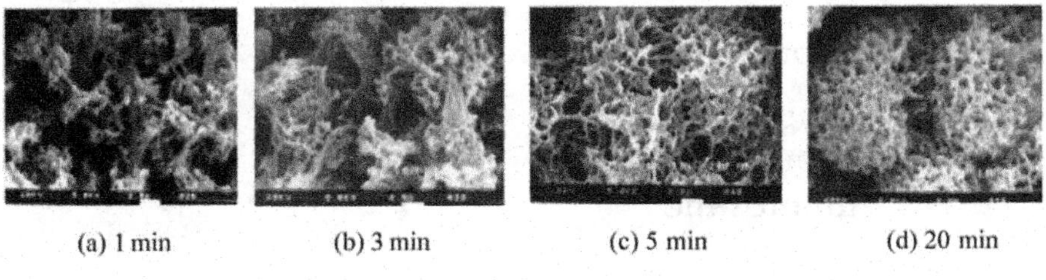

(a) 1 min (b) 3 min (c) 5 min (d) 20 min

FIGURE 3.13 SEM images of PVDF membranes as a function of quenching duration (Gu et al., 2006). Copyright (2006) Elsevier.

3.3.2 Fabrication of PTFE Membranes

PTFE possesses attractive properties such as good thermal and chemical stability, oxidation resistance, low thermal conductivity, and excellent hydrophobicity that make it suitable for MD applications. Currently, stretching and electrostatic spinning are the main techniques to fabricate PTFE membranes.

3.3.2.1 Mechanical Stretching

Stein pioneered the development of PTFE membranes via biaxial mechanical stretching (Stein, 1958). Figure 3.14(a) shows the general steps of a stretching process to fabricate PTFE membranes (Feng et al., 2018). Briefly, PTFE powder and lubricating agents were mixed together to form a paste under a certain temperature. The paste was then extruded into a flat sheet, which was further mechanically stretched under controlled conditions. During the stretching process, the lubricants were vaporized to leave behind the microporous PTFE membranes. Figure 3.14(b) shows the corresponding membrane structure at each stretching step. When the nascent flat sheet was extruded, there were no pores. At the early stage of stretching, strip cracks were formed. Then, with the uniaxial stretching, nodes were formed. Finally, with the biaxial stretching, nodes became smaller and interconnected. Porous structures of membranes were thus created. Many researchers have adopted and/or modified this process to prepare porous PTFE membranes. For instance, Huang et al. developed asymmetric porous PTFE membranes comprising a dense skin layer and a continuous porous layer via the stretching method (Huang et al., 2005). Their membranes had the maximum pore diameter in the range of 0.03–1.0 μm, a porosity and a water contact angle varying from 20% to 70% and from 120° to 135°, respectively.

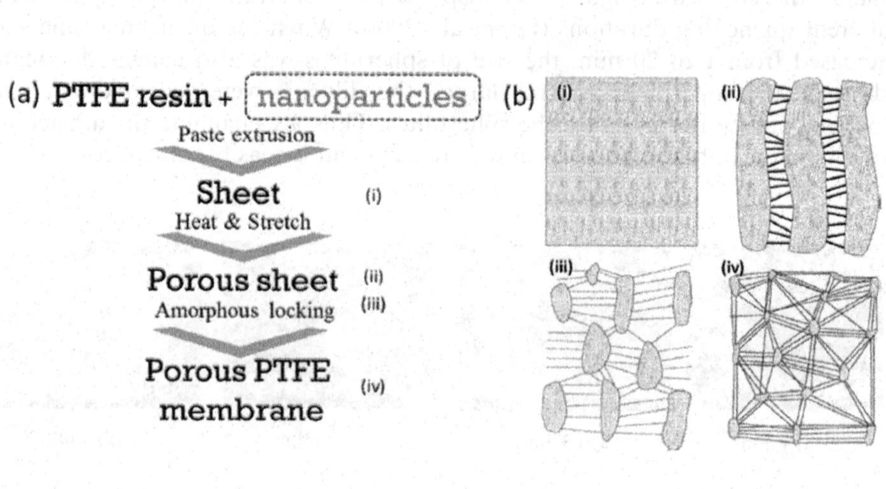

FIGURE 3.14 Preparation process of porous PTFE membranes (Feng et al., 2018). Copyright (2018) Elsevier.

3.3.2.2 Electrostatic Spinning

Electrostatic spinning is a fiber production method that utilizes a high voltage electric field. In this process, a polymer fluid (such as a polymer solution, sol-gel, particulate suspension, or polymer melt) is firstly extruded from a needle tip and then stretched under the electric field. The extruded nanofibers are collected on a substrate to form a flat sheet membrane. Since it is difficult to spin the molten PTFE due to its high viscoelasticity, PTFE emulsions are usually prepared by including additives for electrostatic spinning. For example, Xiong et al. electrospun porous PTFE membranes by using polyvinyl alcohol (PVA) and PTFE emulsions with different mass concentrations (Xiong et al., 2009). It was found that the mass ratio of PVA to PTFE had a significant influence on the nanofibers and membrane structure. Figure 3.15 shows the morphological change of electrospun nanofibers with increasing PVA content. When the mass ratio of PVA/PTFE was 10:90, a beadlike structure was obtained. When the mass ratio was increased to 30:70, nanofibers with a diameter of about 300 nm were formed. However, these nanofibers were not uniform because large nodules could be observed along the fibers. With a further increase in PVA/PTFE ratio, the fiber diameter became larger, and uniform fibers were achieved.

In addition to the PVA/PTFE ratio, polymer concentration was also critical to the nanofiber morphology. Xiong et al. found that nanofibers with beadlike structures were formed when the polymer concentration was lower than 22 wt% (Xiong et al., 2009). A higher concentration favored the formation of continuous and uniform nanofibers. After the formation of PVA/PTFE membranes, sintering treatments at 300–400 °C were carried out to completely decompose PVA and to achieve interconnected porous PTFE membranes. Dong et al. electrospun superhydrophobic nanofibrous membranes by using PVDF-PTFE blend solutions (Dong et al., 2014). The water contact angle and liquid entry pressure of resultant membranes were highly dependent on the PTFE content in the blend solutions. When the PTFE content was increased from 0 to 12 wt%, the water contact angle and liquid entry pressure were increased from 130.4° and 84 kPa to 152° and 137 kPa, respectively. The obtained superhydrophobic membranes showed a flux of 18.5 kg/m^2 and a salt rejection higher than 99.9% under VMD.

3.3.3 Fabrication of PP Membranes

PP is a thermoplastic polymer that shows good chemical resistance, heat stability, and hydrophobicity. Due to the lack of suitable solvents for PP, the conventional NIPS method cannot be used to prepare PP membranes. Instead, the TIPS method is usually employed to fabricate porous PP membranes. For instance, Lin et al. produced microporous PP membranes with bicontinuous structures and good mechanical strength via TIPS (Lin et al., 2009). The effects of polymer concentration and quenching conditions on membrane morphology and mechanical strength were studied. With an increase in

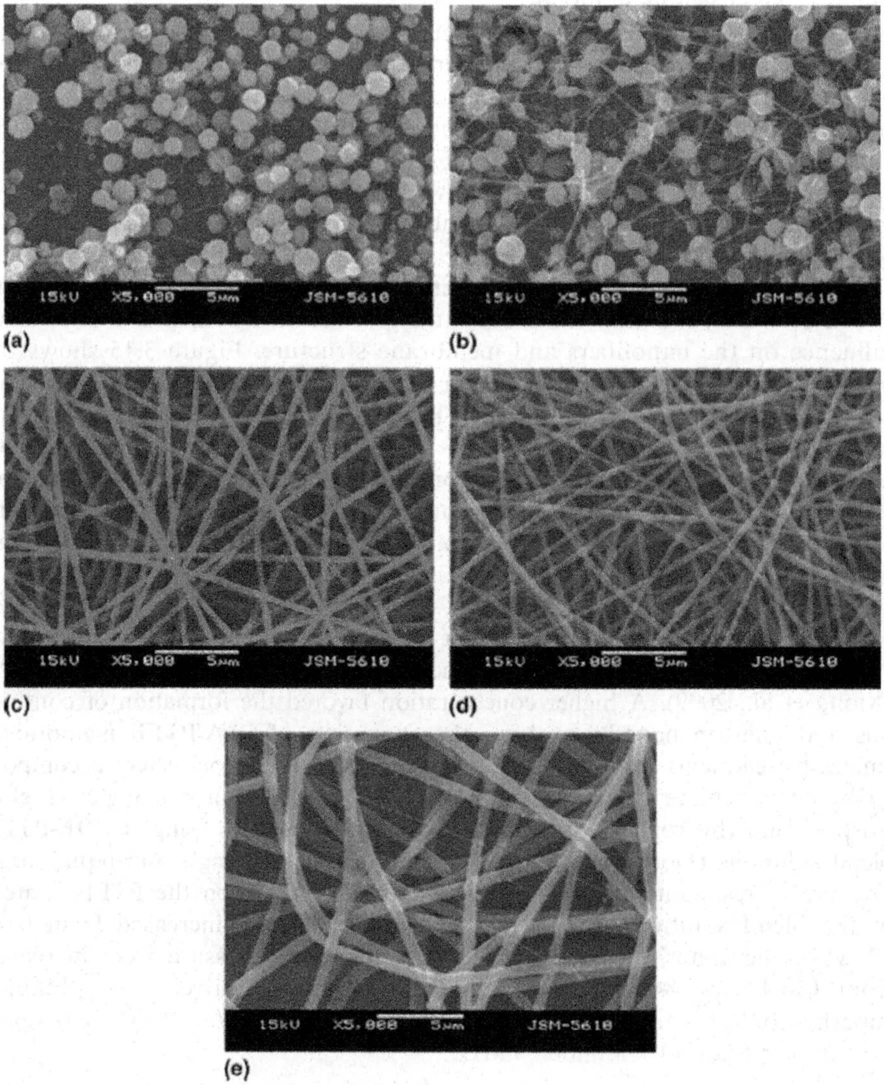

FIGURE 3.15 SEM images of electrospun fibers with different PVA/PTFE mass ratio: (a) 10:90, (b) 20:80, (c) 30:70, (d) 40:60, and (e) 50:50 (Xiong et al., 2009). Copyright (2009) Cambridge University Press

polymer concentration from 10 to 60 wt%, the membrane morphology changed from a bicontinuous structure to a cellular structure gradually. Bicontinuous structures were formed from spinodal decomposition with subsequent solidification of the polymer-rich phase. When polymer concentration was higher, nucleation and growth mechanism occurred, which led to cellular structures. Besides polymer concentration, quenching conditions also played important roles in determining membrane morphologies. With a higher quenching temperature, membranes with larger pore sizes were obtained. This was owing to the slower solidification rate of the polymer-rich phase that took a longer time for the polymer-lean phase to evolve at a higher quenching temperature. Thus, larger pore sizes were created.

3.3.4 Fabrication of Ceramic Membranes

Ceramic membranes are relatively new for MD applications. Comparing to polymeric membranes, ceramic membranes possess excellent chemical, thermal, and mechanical stability. Thus, ceramic membranes can be used in harsh environments and high temperatures without concerns of membrane deterioration. Different configurations of ceramic membranes have been fabricated for practical applications, such as flat, tubular, and hollow fiber membranes via slip casting, tape casting, pressing, and extrusion techniques. These techniques are simple and effective to convert ceramic suspensions into membrane precursors, followed by high-temperature sintering to produce the final membranes. For example, in the case of slip casting, a well-mixed ceramic suspension is transferred into a porous mold with a desired shape (e.g., flat or tubular). After the solvent is drawn into the pores of the mold by capillary force, a layer of slip particles is formed on the mold surface. Then, the membrane precursor is heated up at a high temperature to develop membranes with designed microstructures. The heat treatment process usually involves three stages, namely, presintering, thermolysis, and final sintering. The presintering stage happens at a lower temperature of up to 200 °C, which is for the removal of residual solvents and moisture. The thermolysis stage is usually conducted at 500–600 °C to decompose organic components such as binders. Lastly, the final sintering stage is to transform the ceramic particles into a strong and dense ceramic membrane.

However, ceramic membranes are hydrophilic in nature due to hydroxyl groups. If they are used in MD processes, it is crucial to change their membrane surface from hydrophilic to a hydrophobic one. Surface modifications by using low surface energy materials are the common method to produce hydrophobic ceramic membranes. For example, Zuo and Chung have prepared metal–organic framework (MOF) functionalized ceramic membranes for VMD desalination applications (Zuo and Chung, 2016b). A two-step synthesis strategy was invented to achieve hydrophobic ceramic membranes: (1) to intergrow MOF crystals on the ceramic membrane surface and (2) to introduce perfluoro molecules onto the MOF functionalized membrane surface. With the first step, the ceramic membrane surface was covered with MOF crystals, and hierarchical roughness structures were obtained. After the second step, the perfluoro molecules formed a thin hydrophobic layer, which lowered the surface energy. Therefore, a hydrophilic ceramic membrane was transformed into a hydrophobic one with a good VMD flux of 32.3 L/m^2 h at 60 °C.

3.4 CONCLUSION

In this chapter, the fundamentals of MD membrane fabrication have been reviewed. Due to the special requirements of MD membranes, selected polymer materials, such as PVDF, PTFE, PP, and ceramic materials are commonly used. For the fabrication of PVDF membranes, phase inversion methods (NIPS and TIPS) are dominating. The effects of solvents, polymer concentration, nonsolvents, additives, coagulation bath mediums, and spinning parameters on the NIPS process, and the influences of diluents and quenching conditions on the TIPS process have been discussed in detail. For the fabrication of PTFE membranes, stretching and electrostatic spinning are the major techniques. For PP membrane fabrication, the TIPS method is usually employed. Last but not least, for ceramic membrane fabrication, quite a few techniques are available. However, after the attainment of ceramic membranes, hydrophobic modifications must be done before they can be used in MD applications.

ACKNOWLEDGEMENTS

The author would gratefully thank the Singapore National Research Foundation for the funding under its Energy Innovation Research Programme for the project entitled, "Using Cold Energy from Re-gasification of Liquefied Natural Gas (LNG) for Novel Hybrid Seawater Desalination Technologies" (grant number: R-279-000-456-279).

REFERENCES

Bonyadi, S., Chung, T.S., Rajagopalan, R. 2009. A novel approach to fabricate macrovoid-free and highly permeable PVDF hollow fiber membranes for membrane distillation. AIChE Journal 55 (3), 828–833.

Bottino, A., Camera-Roda, G., Capannelli, G., Munari, S. 1991. The formation of microporous polyvinylidene difluoride membranes by phase separation. Journal of Membrane Science 57 (1), 1–20.

Bottino, A., Capannelli, G., Munari, S., Turturro, A. 1988a. High performance ultrafiltration membranes cast from LiCl doped solutions. Desalination 68 (2–3), 167–177.

Bottino, A., Capannelli, G., Munari, S., Turturro, A. 1988b. Solubility parameters of poly (vinylidene fluoride). Journal of Polymer Science B: Polymer Physics 26, 785–794.

Chandavasu, C., Xanthos, M., Sirkar, K.K., Gogos, C.G. 2003. Fabrication of microporous polymeric membranes by melt processing of immiscible blends. Journal of Membrane Science 211, 167–175.

Chang, J., Zuo, J., Zhang, L., O'Brien, G.S., Chung, T.S. 2017. Using green solvent, triethyl phosphate (TEP), to fabricate highly porous PVDF hollow fiber membranes for membrane distillation. Journal of Membrane Science 539, 295–304.

Cheng, L.P. 1999. Effect of temperature on the formation of microporous PVDF membranes by precipitation from 1-octanol/DMF/PVDF and water/DMF/PVDF systems. Macromolecules 32 (20), 6668–6674.

Chung, T.S., Kafchinski, E.R. 1997a. The effects of spinning conditions on asymmetric 6FDA/6FDAM polyimide hollow fibers for air-separation. Journal of Applied Polymer Science 65, 1555–1569.

Chung, T.S., Teoh, S.K., Hu, X.D. 1997b. Formation of ultrathin high-performance polyethersulfone hollow fiber membranes. Journal of Membrane Science 133, 161–175.

Dong, Z.Q., Ma, X.H., Xu, Z.L., You, W.T., Li, F.B. 2014. Superhydrophobic PVDF-PTFE electrospun nanofibrous membranes for desalination by vacuum membrane distillation. Desalination 347, 175–183.

Edwie, F., Teoh, M.M., Chung, T.S. 2012. Effects of addtives on dual-layer hydrophobic-hydrophilic PVDF hollow fiber membrane for membrane distillation and continuous performance. Chemical Engineering Science 68, 567–578.

El-Bourawi, M.S., Ding, Z., Ma, R. and Khayet, M. 2006. A framework for better understanding membrane distillation separation process. Journal of Membrane Science 285 (1), 4–29.

Feng, S., Zhong, Z., Wang, Y., Xing, W., Drioli, E. 2018. Progress and perspectives in PTFE membanes: Preparation, modification, and applications. Journal of Membrane Science 549, 332–349.

Gu, M., Zhang, J., Wang, X., Tao, H., Ge, L. 2006. Formation of poly(vinylidene fluoride) (PVDF) membranes via thermally induced phase separation. Desalination 192 (1–3), 160–167.

Hansen, C.M. 1999. Hansen Solubility Parameter A User's Handbook. CRC Press, New York.

Hassankiadeh, N.T., Cui, Z., Kim, J.H., Shin, D.W., Sanguineti, A., Arcella, V., Lee, Y. M., Drioli, E. 2014. PVDF hollow fiber membranes prepared from green diluent via thermally induced phase separation: Effect of PVDF molecular weight. Journal of Membrane Science 471, 237–246.

Haynes, W.M. 2010. Physical Constants of Organic Compounds, in *CRC Handbook of Chemistry and Physics*, Haynes W.M., ed., CRC Press, Boca Raton, FL, Taylor and Francis, 144–688.

Huang, J., Chou, W., Chou, D., Lai, J.Y., Lee, K.R., Wang, D.M., Ruaan, R.C., Wu, T. T., 2005. Asymmetric porous polytetrafluoroethylene membrane and process for preparing the same, US 6852223 B2.

Ji, G.L., Du, C.H., Zhu, B.K., Xu, Y.Y. 2007. Preparation of porous PVDF membrane via thermally induced phase separation with diluent mixture of DBP and DEHP. Journal of Applied Polymer Science 105 (3), 1496–1502.

Khayet, M., Matsuura, T. 2001. Preparation and characterization of polyvinylidene fluoride membranes for membrane distillation. Industrial & Engineering Chemistry Research 40, 5710–5718.

Kujawa, J., Cerneaus, S., Kujawski, W., Knozowska, K. 2017. Hydrophobic ceramic membranes for water desalination. Applied Sciences 7, 402–412.

Li, X., Xu, G., Lu, X., Xiao, C. 2008. Effects of mixed diluent compositions on poly(vinylidene fluoride) membrane morphology in a thermally induced phase-separation process. Journal of Applied Polymer Science 107 (6), 3630–3637.

Lin, D.J., Beltsios, K., Chang, C.L., Cheng, L.P. 2003. Fine structure and formation mechanism of particulate phase-inversion poly(vinylidene fluoride) membranes. Journal of Polymer Science B: Polymer Physics 41, 1578–1588.

Lin, Y.K., Chen, G., Yang, J., Wang, X.L. 2009. Formation of isotactic polypropylene membranes with bicontinuous structure and good strength via thermally induced phase separation method. Desalination 236, 8–15.

Liu, F., Hashim, N.A., Liu, Y. 2011. Progress in the production and modification of PVDF membranes. Journal of Membrane Science 375, 1–27.

Lloyd, D.R., Kinzer, K.E., Tseng, H.S. 1990. Microporous membrane formation via thermally induced phase separation. I. Solid–liquid phase separation. Journal of Membrane Science 52 (3), 239–261.

Lu, K.J., Zuo, J., Chung, T.S. 2016. Tri-bore PVDF hollow fibers with a super-hydrophobic coating for membrane distillation. Journal of Membrane Science 514, 165–175.

Lu, K.J., Zuo, J., Chung, T.S. 2017. Novel PVDF membranes comprising n-butylamine functionalized graphene oxide for direct contact membrane distillation. Journal of Membrane Science 539, 34–42.

Lu, X., Li, X. 2009. Preparation of polyvinylidene fluoride membrane via a thermally induced phase separation using a mixed diluent. Journal of Applied Polymer Science 114 (2), 1213–1219.

Munari, S., Bottino, A., Capannelli, G. 1983. Casting and performance of polyvinylidene fluoride based membranes. Journal of Membrane Science 16, 181–193.

Rajabzadeh, S., Maruyama, T., Sotani, T., Matsuyama, H. 2008. Preparation of PVDF hollow fiber membrane from a ternary polymer/solvent/nonsolvent system via thermally induced phase separation (TIPS) method. Separation and Purification Technology 63 (2), 415–423.

Stein, R.S. 1958. The x-ray diffraction, birefringence, and infrared dichroism of stretched polyethylene .2. Generalized uniaxial crystal orientation. Journal of Polymer Science 31, 327–334.

Su, Y., Chen, C., Li, Y., Li, J. 2007. Preparation of PVDF membranes via TIPS method: The effect of mixed diluents on membrane structure and mechanical property. Journal of Macromolecular Science Part A Pure and Applied Chemistry 44, 305–313.

Sukitpaneenit, P., Chung, T.S. 2009. Molecular elucidation of morphology and mechanical properties of PVDF hollow fiber membranes from aspects of phase inversion, crystallization and rheology. Journal of Membrane Science 340 (1–2), 192–205.

Teoh, M.M., Chung, T.S. 2009. Membrane distillation with hydrophobic macrovoid-free PVDF-PTFE hollow fiber membranes. Separation and Purification Technology 66, 229–236.

Teoh, M.M., Chung, T.S., Yeo, Y.S. 2011. Dual-layer PVDF/PTFE composite hollow fibers with a thin macrovoid-free selective layer for water production via membrane distillation. Chemical Engineering Journal 171, 684–691.

Tomaszewska, M. 1996. Preparation and properties of flat-sheet membranes from poly (vinylidene fluoride) for membrane distillation. Desalination 104, 1–11.

Uragami, T., Naito, Y., Sugihara, M. 1981. Studies on synthesis and permeability of special polymer membranes. Polymer Bulletin 4 (10), 617–622.

Wang, D., Li, K., Teo, W.L. 1999. Preparation and characterization of polyvinylidene fluoride (PVDF) hollow fiber membranes. Journal of Membrane Science 163 (2), 211–220.

Wang, K.Y., Chung, T.S., Gryta, M. 2008a. Hydrophobic PVDF hollow fiber membranes with narrow pore size distribution and ultra-thin skin for the fresh water production through membrane distillation. Chemical Engineering Science 63 (9), 2587–2594.

Wang, P., Teoh, M.M., Chung, T.S. 2011. Morphological architecture of dual-layer hollow fiber for membrane distillation with higher desalination performance. Water Research 45, 5489–5500.

Wang, X., Zhang, L., Sun, D., An, Q., Chen, H. 2008b. Effect of coagulation bath temperature on formation mechanism of poly(vinylidene fluoride) membrane. Journal of Applied Polymer Science 110 (3), 1656–1663.

Widjojo, N., Chung, T.S., Weber, M., Maletzko, C., Warzelhan, V. 2011. The role of sulphonated polymer and macrovoid-free structure in the support layer for thin-film composite (TFC) forward osmosis (FO) membranes. Journal of membrane science 383, 214–223.

Xiao, T., Wang, P., Yang, X., Cai, X., Lu, Ji. 2015. Fabrication and characterization of novel asymmetric polyvinylidene fluoride (PVDF) membranes by the nonsolvent thermally induced phase separation (NTIPS) method for membrane distillation applications. Journal of Membrane Science 489, 160–174.

Xiong, J., Huo, P., Ko, F.K. 2009. Fabrication of ultrafine fibrous polytetrafluoroethylene porous membranes by electrospinning. Journal of Material Research 24, 2755–2761.

Yeow, M.L., Liu, Y.T., Li, K. 2004. Morphological study of poly(vinylidene fluoride) asymmetric membranes: Effects of the solvent, additive, and dope temperature. Journal of Applied Polymer Science 92 (3), 1782–1789.

Young, T.H., Cheng, L.P., Lin, D.J., Fane, L., Chuang, W.Y. 1999. Mechanisms of PVDF membrane formation by immersion–precipitation in soft (1-octanol) and harsh (water) nonsolvents. Polymer 40, 5315–5323.

Zuo, J., Bonyadi, S., Chung, T.S. 2016. Exploring the potential of commercial polyethylene membranes for desalination by membrane distillation. Journal of Membrane Science 497, 239–247.

Zuo, J., Chung, T.S. 2016a. In-situ cross-linked PVDF membranes with enhanced mechanical durability for vacuum membrane distillation. AIChE Journal 62, 4013–4022.

Zuo, J., Chung, T.S. 2016b. Metal–organic framework-functionalized alumina membranes for vacuum membrane distillation. Water 8, 586–602.

Zuo, J., Chung, T.S. 2017. PVDF hollow fibers with novel sandwich structure and superior wetting resistance for vacuum membrane distillation. Desalination 417, 94–101.

4 Design and Fabrication of PVDF Membranes for Membrane Distillation from Industrial Perspectives

Panu Sukitpaneenit
Koch Membrane Systems, Wilmington, MA

4.1 INTRODUCTION

In recent years, membrane distillation (MD) has become a potentially promising membrane technology for desalination, water reuse, and wastewater treatment applications, especially in the fields of high-salinity waters, brackish water, and brine treatment, where separation and purification become economically and technically complex through the existing separation technologies (e.g., reverse osmosis (RO), electrodialysis (ED), as well as thermal processes including multistage flash (MSF) and multieffect distillation (MED)) (Al-Obaidani et al., 2008; Hanemaaijer et al., 2006). Through such attractive features and benefits, for a long period MD has become a targeted topic and examined by membrane researchers and scientists globally in different aspects, that is, development of membrane materials (Eykens et al., 2016a; Wang and Chung, 2015), membrane transport mechanisms (Phattaranawik and Jiraratananon, 2001; Yun et al., 2006), wetting phenomena (Rezaei et al., 2018), operation and process design (Francis et al., 2013; Lawson and Lloyd, 1996; Singh and Sirkar, 2012), and cost and energy evaluation (Al-Obaidani et al., 2008; Hogan et al., 1991). Recently, there have been positive progresses through increased interests from both academic and industry sectors on MD development and commercialization; commencing from the early 2000s, several commercially orientated R&D activities such as MD pilots and demonstration plants have been undertaken by research institutions, private and government-granted firms/startups (Drioli et al., 2015; Thomas et al., 2017). However, practical implementations on large full-scale MD plants have not yet been achieved.

One of the major challenges to the current MD technology and commercialization is the lack of effective membranes, which is the heart of the system.

To date, most membranes employed in MD studies and pilot demonstration plants are commercially available membranes made of hydrophobic polymers such as polytetrafluoroethylene (PTFE), polypropylene (PP), and polyvinylidene fluoride (PVDF) for microfiltration (MF) and ultrafiltration (UF) applications, which are designed to operate in a pressure-driven membrane filtration process. Even though those MF/UF membranes may be qualified as hydrophobic microporous membranes applicable for MD, their main characteristics and properties including morphology, thickness, porosity, pore size, and hydrophobicity are not deliberately developed and optimized for effective MD processes, which are a thermal-driven process. To overcome this shortcoming, there are extensive studies that have devoted their efforts to the development of novel MD membranes with desirable characteristics from various aspects ranging from membrane materials, fabrication techniques, membrane configuration, chemical modification and treatment methods (Drioli et al., 2015; Eykens et al., 2017; Khayet and Matsuura, 2011; Li and Sirkar, 2004; Wang and Chung, 2015). However, most studies are still in a laboratory-scale level, and particularly in some cases, the proposed preparation methods appear to be challenging for adoption and implementation in the industry because of the production cost, process control, and process complexity. Apparently, the selection of suitable membrane materials coupled with the design of fabrication methods are very crucial for MD technology development and advancement. Without the development of specially designed MD membranes based on cost-effective and scalable fabrication technology, MD is unlikely to materialize as a viable alternative desalination and water treatment technology in the market.

PVDF is one of the promising membrane materials that has prominently emerged and continued to play an important function in membrane development, not only due to its excellent chemical and physical characteristics but also owing to the ease of PVDF dissolution in common organic solvents, which allows the fabrication and production of PVDF membranes through various phase inversion processes. Among several phase inversion techniques, a nonsolvent-induced phase separation (NIPS) is the most widely used process because it provides simplicity, flexibility, and feasibility for scaling up to large-scale membrane manufacturing. Additionally, the resultant membranes with various desirable morphological structures, physicochemical properties, and configurations can be attained through manipulation and optimization of dope chemistry and fabrication parameters. For instance, commercially available PVDF membranes offered by Koch Membrane Systems have all three configurations: flat-sheet, hollow fiber, and tubular as depicted in Figure 4.1.

There have been extensive studies on the preparation and characterization of flat-sheet PVDF membranes fabricated via NIPS technique (Hou et al., 2014; Khayet and Matsuura, 2001) for MD, much lesser and limited research has focused on fabrication and characterization of PVDF hollow fiber membranes (Khayet and Matsuura, 2011). It is well known that the phase inversion process for hollow fiber membranes is more complex than that for flat-sheet membranes, and the controlling parameters of the former

FIGURE 4.1 Commercially available PVDF membranes in all three different configurations: flat-sheet, hollow fiber, and tubular, offered and manufactured by Koch Membrane Systems. Figures courtesy of Koch Membrane Systems, Inc. (www.kochmembrane.com) and used by the publisher with permission.

during membrane formation are noticeably different than those of the latter. For instance, fabricating a hollow fiber generally requires a minimum dope viscosity with an order of magnitude higher than casting a flat-sheet membrane (Baker, 2004; Chung, 2008). During the casting of flat-sheet membranes, one may neglect the moisture-induced phase inversion from air or humidity and just consider the phase inversion commencing from its top surface after being immersed in a coagulation bath. On the contrary, when hollow fiber membranes are produced via NIPS, the phase inversion starts instantly at the lumen side of the nascent fibers upon extrusion from a spinneret by the internal coagulant (bore fluid). Subsequently, the nascent fibers contact with the external coagulant, inducing phase separation at the outer surface of the hollow fiber membranes. To date, the flat-sheet membrane has been commonly employed in the plate and frame type of MD processes, while the hollow fiber configuration has recently gained momentum in various MD applications. Compared with flat-sheet membranes, the hollow fiber configuration provides numerous merits because of its inherent features and module designs including a larger packing density (i.e., membrane area per unit volume of membrane modules), which leads to higher productivity, good self-mechanical support for withstanding backwashing and vacuum, and ease of handling in module fabrication and process operation (Peng et al., 2012; Wang and Chung, 2015).

This chapter provides a comprehensive overview of the science and engineering for the formation of PVDF membranes and the fabrication technologies of designing MD membranes with desirable morphology and characteristics from industrial perspectives. In details, it covers (1) basic and fundamental understanding of MD processes, key features of practical MD membranes, and PVDF polymer as a promising membrane material; (2) fabrication and manufacturing processes of flat-sheet and hollow fiber PVDF membranes in full-scale industrial productions; (3) important parameters influencing the formation of PVDF membranes and the suppression of macrovoids; and (4) recent advancements on fabrication technology of PVDF MD membranes and emerging R&D areas from industry prospects including thermally induced phase separation (TIPS), dual-layer spinneret fabrication technology, mixed-matrix membranes, and braid-reinforced hollow fiber membranes.

A special emphasis is given to the development of PVDF MD hollow fiber membranes containing highly porous and macrovoid-free structure. The existence of macrovoids in membranes is undesirable as they are regarded as weak mechanical roots, resulting in the failure of membranes under membrane cleaning, backwashing, vacuum, and vibration operations (Peng et al., 2012; Widjojo and Chung, 2006).

4.2 FUNDAMENTALS

4.2.1 Basic Understanding of MD Processes and Key Characteristics of Practical MD Membranes

MD is a thermally driven process based on the transport of volatile vapor (for instance, water vapor) across a hydrophobic microporous membrane (Khayet and Matsuura, 2011). The hydrophobicity of the membrane retains the feed liquid directly in contact with one side of the membrane, whereas the microporous structure enables the permeation of vapor to the other side of the membrane. Because only volatile vapor is allowed to transport across it, it is important to minimize the wetting of membrane pores to prevent the feed liquid from permeating through the membrane. The driving force of MD processes is the gradient of vapor pressure across the membrane, mainly caused by the temperature difference between the streams of the heated liquid feed and cooled permeate.

As illustrated in Figure 4.2, the separation mechanism in a MD process features three major steps: (i) evaporation of volatile components in the hot feed, (ii) transportation of volatile vapor molecules through the membrane pores, and (iii) condensation of vapor molecules in the cold permeate side of the membrane. The mechanism described is based on one of the four basic MD configurations, known as direct contact membrane distillation (DCMD) whereby "the cold liquid" is applied in direct contact with the membrane at the permeate side, as shown in Figure 4.3a. The remaining configurations include air gap membrane distillation (AGMD, Figure 4.3b), sweeping gas membrane distillation (SGMD, Figure 4.3c), and vacuum membrane distillation (VMD, Figure 4.3d), where condensed vapor compounds at the permeate side are collected by means of introducing an additional air gap distance for AGMD, a cold inert gas (sweeping gas) for SGMD, and vacuum (suction) for VMD. Although DCMD is considered the most widely investigated MD configuration because of its simplicity in design and setup, the focus of the industry has been on designing VMD and AGMD modules because of their superior long-term performance. Extensive studies and reviews on fundamental principle and applications including pros and cons of all MD

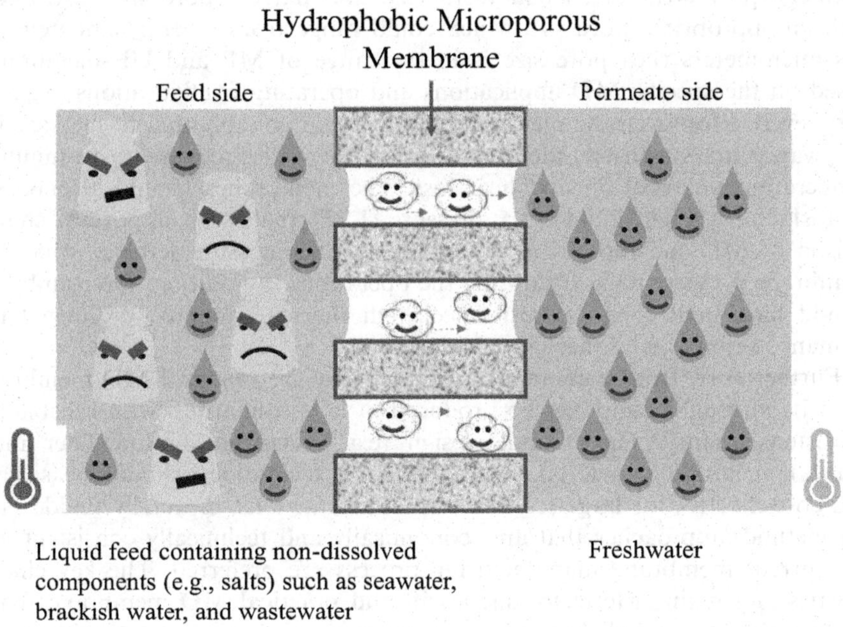

FIGURE 4.2 Basic concept and separation principle of membrane distillation (MD).

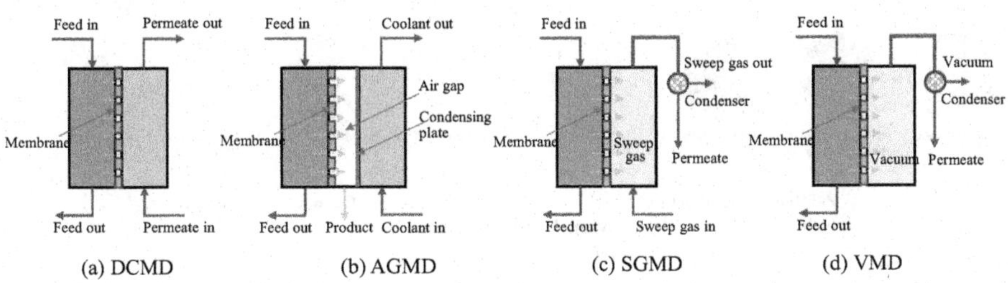

FIGURE 4.3 Basic MD configurations: (a) direct contact membrane distillation (DCMD), (b) air gap membrane distillation (AGMD), (c) sweep gas membrane distillation (SGMD), and (d) vacuum membrane distillation (VMD).

configurations could be referred to the following references (Alkhudhiri et al., 2012; Bandini et al., 1997; Khayet et al., 2012; Lawson and Lloyd, 1997; Meindersma et al., 2006; Yun et al., 2006).

Based on the principles of MD operations, the desired membranes used in MD must have the maximal vapor transportation but with minimal membrane wetting or fouling propensity. To satisfy such conditions, the preferred MD membranes should have high hydrophobicity, high surface and bulk porosities, open-cell pore structure, small tortuosity, and narrow pore size distribution with an appropriate pore size which could range from several nanometers to less micrometers (i.e., pore size is in the range of MF and UF membranes) based on the specific MD applications and operating configurations (e.g., salinity level). Moreover, the membranes must be reasonably thin to increase flux (i.e., vapor mass transfer) along with a low thermal conductivity to minimize temperature polarization and heat loss. The membranes should also possess a satisfactorily high liquid entry pressure (LEP) that is an important characteristic of MD membranes as it determines the pressure level at which the membrane pores remain dry during the operation. In addition, the membranes should have high chemical resistance, high thermal stability, prolonged performance stability, and mechanical integrity.

Furthermore, from a commercial perspective, the practical MD membranes must be affordable, and can be produced in large quantities with reliable and consistent quality. The module design (e.g., flat-sheet, hollow fiber membranes), membrane material, and fabrication technology should be scalable and cost-effective for large-scale industrial productions, especially the developing methods/approaches that are economically and technically consistent with the current membrane manufacturing process are preferred. The key characteristics and main criteria for successful and practical MD membranes based on the aforementioned discussion are summarized in Figure 4.4, whereas the overview and guidelines of membrane properties desired for MD proposed in the literature are presented in Table 4.1.

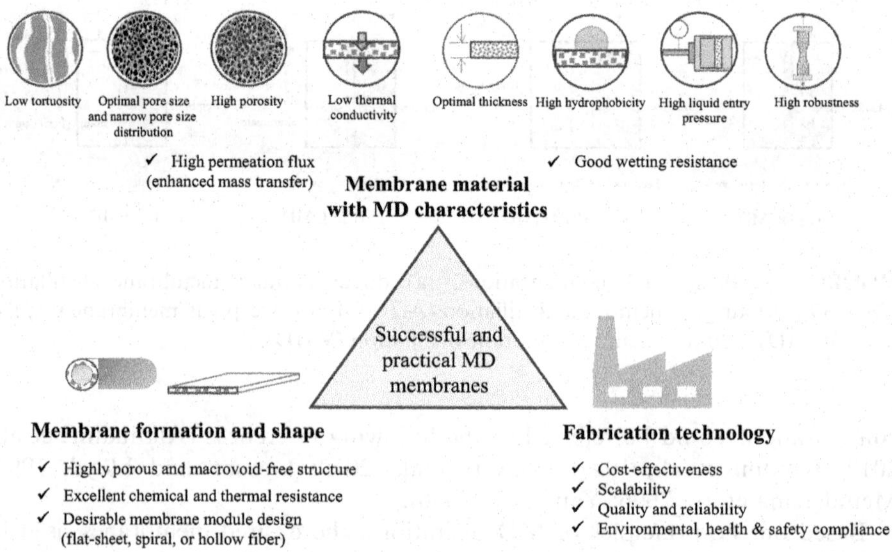

FIGURE 4.4 Key criteria for the development of successful and practical MD membranes.

TABLE 4.1

Overview of the membrane characteristics desired for MD based on a literature survey (Eykens et al., 2016a). Copyright (2016) American Chemical Society

Parameter	Affects	Range used in literature†	Reference	Recommendation	Reference
Contact angle (θ)	Wetting resistance	80°–160°		>105°, as high as possible	
Liquid entry pressure (LEP)	Wetting resistance	0.5–4.6	(Essalhi and Khayet, 2013a; García-Payo et al., 2000)	>2.5 bar	(Schneider et al., 1988)
Pore diameter (d_{av}, d_{max})	Wetting resistance, Flux and energy efficiency (for d_{av} < 0.3 μm)	0.012–1.2 μm	(Essalhi and Khayet, 2013a; Qtaishat et al., 2009; Rao et al., 2014)	0.1–1 μm	(Alkhudhiri et al., 2012; Schneider et al., 1988)
				0.3 μm	(Ali et al., 2012; Winter et al., 2013)
Porosity (ε)	Flux, energy efficiency, strength	38%–90%		>80%	(Alkhudhiri et al., 2012)
Thickness (δ)	Flux, energy efficiency, strength	10–200 μm		Low salinity: 30–60 μm	(Laganà et al., 2000; Martínez and Rodríguez-Maroto, 2008; Wu et al., 2014)
				High salinity: 2–700 μm	(Eykens et al., 2016c)
Tortuosity (τ)	Flux, energy efficiency	1.1–3.9	(Fernández-Pineda et al., 2002; Khayet et al., 2004)	As low as possible	
Thermal conductivity (κ_m)	Flux, energy efficiency	0.031–0.057 W m^{-1} K^{-1}	(García-Payo and Izquierdo-Gil, 2004)	As low as possible	
		2.5 W m^{-1} K^{-1}	(Wang et al., 2016)		
Tensile strength	Strength	3.4–57.9 MPa	(Essalhi and Khayet, 2013b; Feng et al., 2004; Liu et al., 2012)	As high as possible	

† (Basile et al., 2015; Cui et al., 2015; Eykens et al., 2016b; Khayet, 2011; Khayet and Matsuura, 2011).

4.2.2 PVDF as a Membrane Material

4.2.2.1 Properties of PVDF

PVDF is a semicrystalline polymer with the chemical composition of $-(CH_2-CF_2)_n-$. Commercially available PVDF polymers are mainly produced by the polymerization of vinylidene difluoride via emulsion or suspension processes. In recent years, PVDF has attracted significant attention as a promising membrane material in MD development due to its exceptional chemical and physical properties. The key advantages of PVDF include its high hydrophobicity, excellent chemical and thermal stability, good mechanical strength, and superior weather stability along with other desirable properties summarized in Figure 4.5. Generally, PVDF exhibits outstanding chemical resistance against several harsh and corrosive chemicals, including most organic and inorganic acids, aliphatic, cyclic, and aromatic hydrocarbons, halogens, and oxidants (Lovinger, 1982). Nevertheless, PVDF is susceptible to some strong alkali or caustic solutions with a pH value of ≥12, or to high polar amides, ethers, ketones, and esters (Hashim et al., 2011; Liu et al., 2011). An excessive caustic exposure triggers de-hydrofluorination in PVDF polymer chains, which results in polymer discoloration (darkening), embrittling, and cracking (Shinohara, 1979).

Typically, PVDF comprises crystalline phases alongside amorphous regions that, respectively, offer the thermal stability and flexibility of the membranes

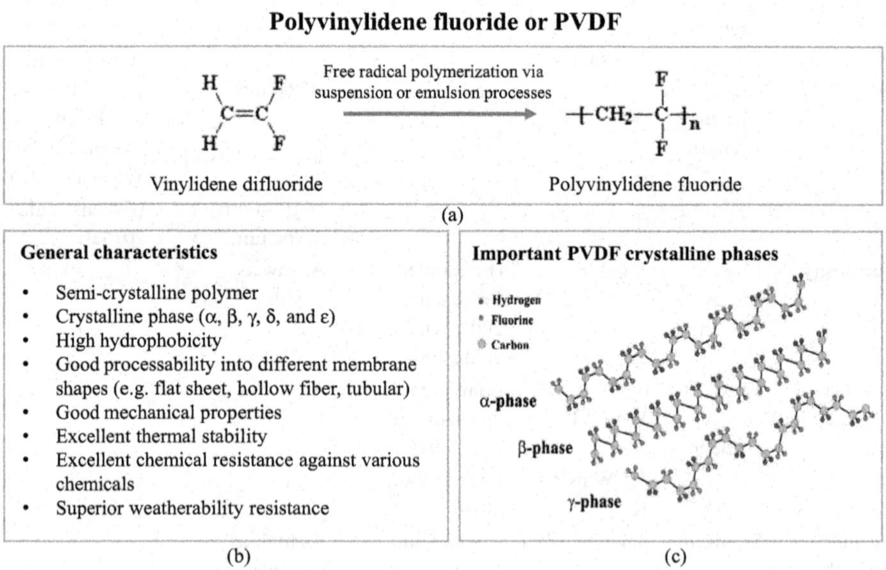

FIGURE 4.5 Basic properties and information of PVDF: (a) polymerization of PVDF and its repeating unit, (b) general characteristics, and (c) schematic representation of PVDF chain conformation for the **α**, **β**, **γ** crystalline phases of PVDF (Martins et al., 2014). Copyright (2014) Elsevier.

(Dohany and Robb, 1980). The crystalline structure in PVDF contains five potential polymorphs that include α, β, γ, δ, and ε types (Nalwa, 1995). The various types of polymorphs lead to different elastic, ferroelectric, and thermal characteristics alongside polarity and dielectric constant (Martins et al., 2014). Among them, α, β, and γ phases are the most prevalent, commonly seen and discussed phases in PVDF membrane preparation. The α is considered nonpolar whereas the β and γ are considered polar. The α phase is a kinetically favorable crystalline form, while the β phase is the most thermodynamically stable crystal form. The influence from different PVDF crystalline phases on the morphology, mechanical strength, and performance of the resulting membranes is an interesting topic investigated and discussed among membrane scientists (Cui et al., 2015; Sukitpaneenit and Chung, 2009; Zhang et al., 2008). For example, the literature suggests that the polar β and γ PVDF crystalline phases result in relatively better mechanical strength and chemical resistance than the nonpolar α phase, even though their impacts on the membrane selectivity and permeability are still unclear (Cui et al., 2015).

4.2.2.2 Phase Inversion Mechanisms of PVDF

Compared to amorphous polymers (i.e., cellulose acetate, polysulfone, polyethersulfone, polyamide, and polyimide), the phase inversion mechanisms of semicrystalline PVDF polymer are much more complex and unique. Typically, PVDF exhibits two kinds of demixing during phase inversion; they include liquid–liquid demixing and solid–liquid demixing accompanying with crystallization (Sukitpaneenit and Chung, 2009; van de Witte et al., 1996; Young et al., 1999), which subsequently control the morphological structure of the membrane. In a rapid phase inversion process, liquid–liquid demixing mainly determines the precipitation pathway and often leads to membranes containing a relatively dense skin layer, cellular microstructure, and macrovoids in the cross-sectional morphology. On the contrary, solid–liquid demixing that accompanies crystallization becomes dominant when delayed (slow) rates of phase inversion occur. In such circumstances, the resultant membranes may have a globular structure that possesses interconnected PVDF spherulitic crystals; they may be macrovoid-free but typically weak in mechanical strength.

Figure 4.6 shows a hypothetic phase diagram for semicrystalline polymers suggested by Mulder (1996). In most cases, solid–liquid demixing occurs at high polymer concentrations, whereas liquid–liquid demixing usually takes place at moderate polymer and nonsolvent concentrations. Both liquid–liquid and solid–liquid demixings coexist in the shaded region as depicted in Figure 4.6. However, the actual phase diagram varies with respect to polymer dope composition, temperature as well as solvent and nonsolvent types in the system (van de Witte et al., 1996). Figure 4.7 illustrates a typical phase diagram of PVDF/NMP/nonsolvent systems at 25 °C that was constructed experimentally based on the cloud-point measurements (Sukitpaneenit and Chung, 2009). The cloud points denote the gelation points emanated from synergistic impacts of the demixing occurring in liquid–liquid and solid–liquid polymer solutions (Cheng et al., 1999).

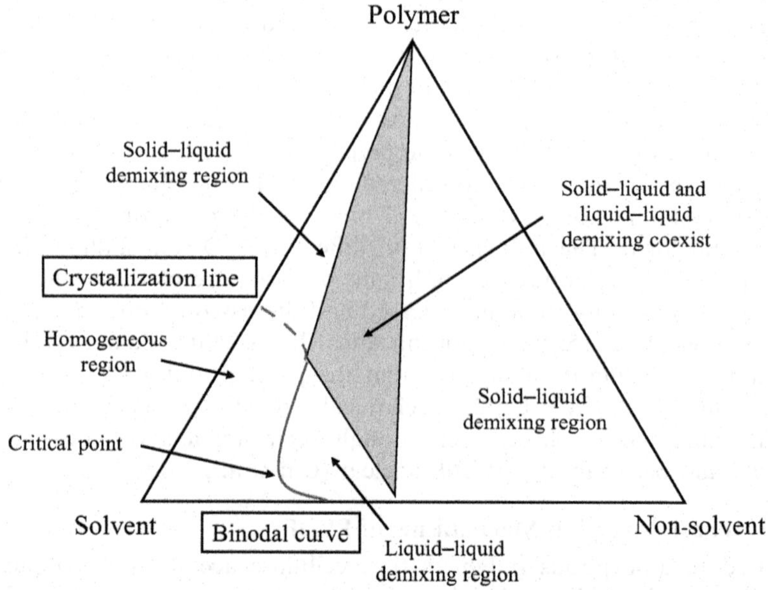

FIGURE 4.6 Hypothetic phase diagram of semicrystalline polymers (Mulder, 1996). Copyright (1996) Kluwer Academic Publishers.

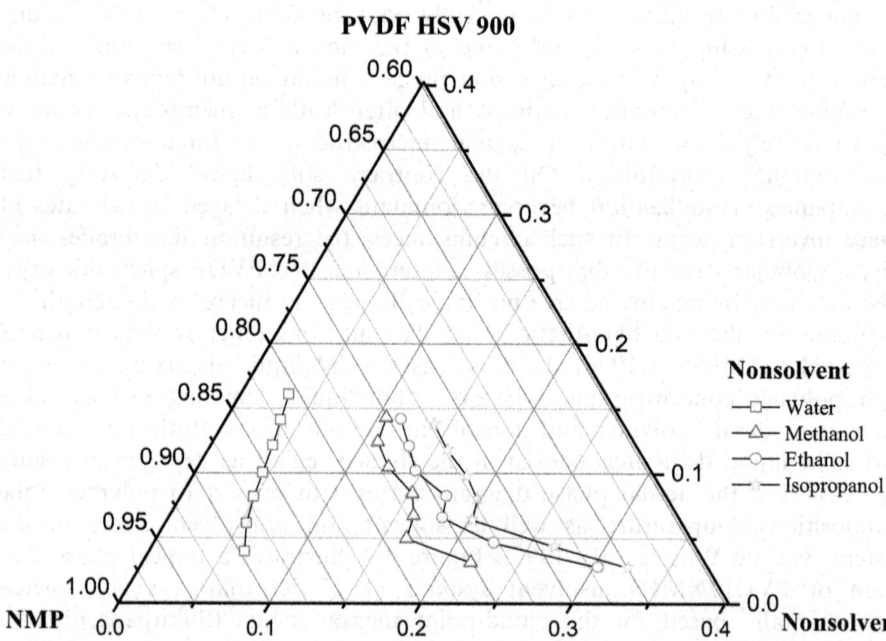

FIGURE 4.7 Phase diagram of ternary PVDF/NMP/nonsolvent systems (Sukitpaneenit and Chung, 2009). Copyright (2009) Elsevier.

As illustrated in Figure 4.7, the gelation boundary for the PVDF/NMP/water system is adjacent to the polymer–solvent axis, thus implying that only a small quantity of water is needed to induce phase inversion, as opposed to other PVDF/NMP/nonsolvent alcohol systems. This outcome signifies that the thermodynamic stability of PVDF/NMP/nonsolvent systems follows a consecutive sequence: water < methanol < ethanol < isopropanol, whereby water is a strong nonsolvent whereas alcohols are weak nonsolvents for the PVDF/NMP system. Nevertheless, it is worth noting that the phase diagram solely describes the phase inversion from a perspective of thermodynamic equilibrium among the correlated polymer, solvent, and nonsolvent components. The kinetic aspect including the rate of crystallization and the rate of solvent and nonsolvent exchange during phase inversion could also play critical roles in the actual demixing process. Indeed, it is sensible to infer that the demixing process switches from thermodynamically to kinetically favored if the duration of occurrence is reasonably short. Therefore, the overall structure of PVDF membranes is greatly reliant on two governing mechanisms, namely, crystallization rate and solvent exchange rate involved during the membrane formation.

4.3 FABRICATION AND MANUFACTURING PROCESSES OF PVDF MEMBRANES IN COMMERCIAL-SCALE PRODUCTION

4.3.1 Flat-Sheet Membrane Casting Process

The process for casting flat-sheet membranes through NIPS comprises a significant number of control parameters including dope formulation, coagulation chemistry, casting parameters such as humidity control, and take-up speed. The manufacturing process for casting flat-sheet membranes in a full-scale production, as schematically illustrated in Figure 4.8, is often undertaken as follows:

(1) Mixing the polymer dope solution (dissolving the polymer in the required solvent with or without different additives according to the dope recipe). Once the polymer dope solution is mixed homogeneously, degassing takes place to ensure the dope solution is free of bubbles.
(2) Metering the polymer solution from the feed tank to the hopper equipped with the casting blade, located above the rotating drum that is partially immersed in the coagulation tank.
(3) Casting the solution onto a moving nonwoven fabric support that is conveyed to the top surface of the rotating drum, underneath the casting blade. The nascent film is then exposed to a controlled air gap environment (depending on the dope recipe) for solvent evaporation prior to entering the coagulation bath.
(4) Immersing the nascent film into the nonsolvent coagulation bath (in most cases water) where the precipitation and solidification of the

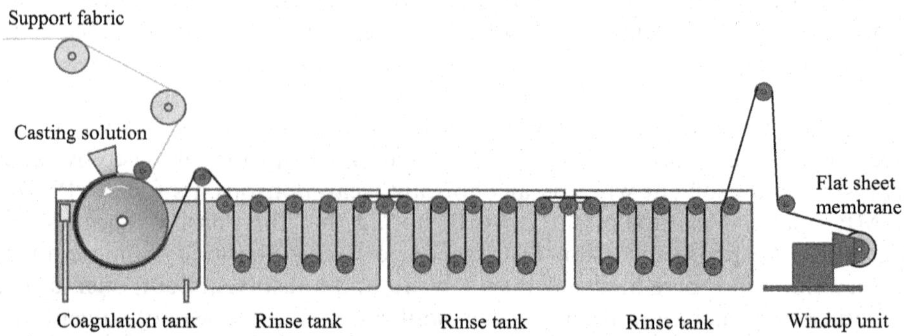

FIGURE 4.8 The manufacturing process of PVDF flat-sheet membranes in large-scale production.

solution is taking place and the membrane skin is formed on top of the fabric support.

(5) Rinsing/leaching the remainder of the solvents/additives/residues whereby the cast membranes are conveyed through several rinse tanks.
(6) Winding up the membrane rolls at the end unit of the casting process; and
(7) Applying additional post-treatments before assembling the membranes into spiral modules or plate and frame cassettes.

4.3.2 Hollow Fiber Spinning Process

A typical hollow fiber spinning line for fabricating polymeric hollow fibers via NIPS method is portrayed in Figure 4.9. Commercially available hollow fiber membranes are usually spun using at least one spinneret head that normally contains a number of nozzles, coupled with a certain air-gap distance and a reasonably moderate/high take-up speed to increase productivity. Generally, the hollow fiber spinning in industrial-scale production is undertaken as follows:

(1) Metering the polymer solution and bore fluid by different precision pumps;
(2) Extrusion of the polymer solution and bore fluid through spinnerets;
(3) Internal coagulation occurring when the dope solution contacts with the bore fluid leaving the spinnerets;
(4) Solvent evaporation/moisture-induced phase inversion from the external nascent fiber surface in the air-gap region;
(5) Fiber stretching through elongational tensions and gravity induced by the air-gap and take-up speed;
(6) Completion of phase inversion and solidification of as-spun hollow fibers in the coagulation bath;

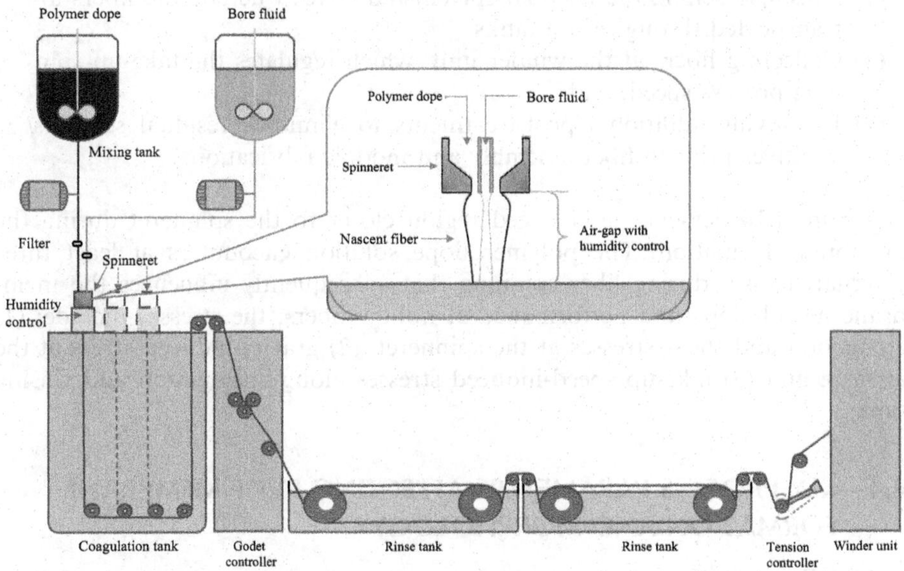

FIGURE 4.9 The manufacturing process of hollow fiber membranes in large-scale production.

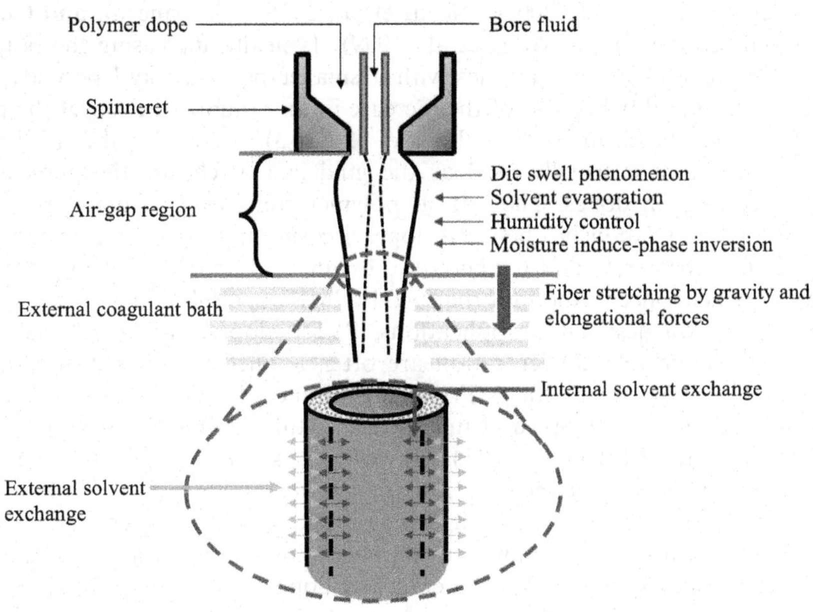

FIGURE 4.10 Schematic drawing of the enlarged regime near the spinneret and the formation of a nascent hollow fiber membrane via NIPS process.

(7) Rinsing/leaching the residual solvents/additives whereby the fibers are transported through rinse tanks;
(8) Collecting fibers at the winder unit, which regulates the take-up spinning process speed; and
(9) Employing additional post-treatments to eliminate residual solvents/additives prior to fiber assembly and module fabrication.

Figure 4.10 depicts the enlarged region closer to the spinneret during the membrane formation. The polymer dope solution encounters at least three external stresses during fiber spinning that subsequently influences the membrane morphology and performance of hollow fibers; the stresses include: (1) elongation and shear stresses at the spinneret; (2) gravity-induced stress at the air-gap; and (3) take-up speed-induced stresses along coagulation and rinsing tanks.

4.4 KEY PROCESS PARAMETERS AFFECTING PVDF MEMBRANE FORMATION AND MORPHOLOGY

4.4.1 Polymer Concentration

The polymer concentration plays a crucial role in determining membrane formation through the NIPS process. To date, the influences of polymer concentration on phase inversion mechanism and morphology of PVDF membranes have been investigated for both flat-sheet and hollow fiber configurations by many researchers (Khayet et al., 2002a; Nejati et al., 2015; Sukitpaneenit and Chung, 2009; Tomaszewska, 1996; Wang et al., 1999). Typically, increasing the polymer concentration produces membranes with a smaller pore size and porosity, and lower permeation flux because of the increase in the volume fraction of the polymer (Khayet and Matsuura, 2011; Lalia et al., 2013). A concept of critical polymer concentration is usually used as the guideline to choose the appropriate polymer content in dopes. The critical polymer concentration could be ascertained from the relationship between dope viscosity and polymer concentration at a specified shear rate and temperature. Figure 4.11 shows an example of the critical polymer concentration for the PVDF/NMP dope solution.

Depending on designated applications, PVDF dopes with a polymer content below the critical concentration are often applied to fabricate microporous PVDF hollow fibers for MF/UF membrane filtration (Cha and Yang, 2006; Khayet et al., 2002a), and membrane distillation (Bonyadi and Chung, 2007; Khayet and Matsuura, 2011). However, it is worth noting that employing a too low polymer concentration (i.e., low viscosity) may result in few undesirable consequences, that is, irregular shape formation, excessive penetration of dopes inside the nonwoven supports, low mechanical strength, and instability of fiber spinning. On the contrary, dopes containing a polymer concentration that exceeds the critical value are often selected to fabricate PVDF hollow fibers with a dense skin and a tight structure for gas separation and pervaporation applications (Matsuyama et al., 1999; Sukitpaneenit and Chung,

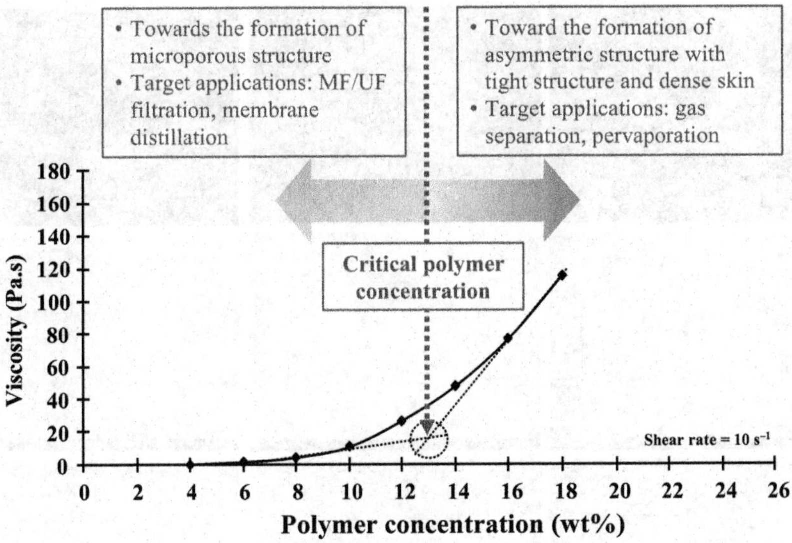

FIGURE 4.11 The critical concentration of PVDF/NMP dope solutions.

2011; Sukitpaneenit et al., 2010). The observation could be drawn due to the fact that polymer dopes possessing a higher polymer concentration typically have a greater viscosity that facilitates greater chain entanglement and efficiently decreases defects in the resultant membranes (Chung et al., 1997).

Figure 4.12 shows the external surface and cross-section morphologies of PVDF hollow fiber membranes spun from various NMP/PVDF dope concentrations of 15, 17, and 19 wt% and coagulated in a water coagulation bath. At 15 wt% PVDF, the intrusion of water is severe, leading to a huge finger-like macrovoid morphology that penetrates across the whole membrane wall. An increase in polymer content from 15 wt% to 19 wt% suppresses the water intrusion to a certain level. Specifically, it mitigates the macrovoids near the lumen side of fibers and converts it to a sponge-like structure. The phenomenon can be attributed to the higher viscoelasticity of higher concentrated polymer solutions that prohibit the immediate solvent/nonsolvent exchange during phase inversion (Smolders et al., 1992). Nevertheless, without optimizing other process parameters, a high polymer concentration may only decrease but cannot totally remove the formation of macrovoids. From a practical perspective, there is a limitation in increasing the polymer concentration in dopes because of the likelihood to induce gelation and difficulties for degassing highly viscous dopes.

4.4.2 Nonsolvent Additives in Polymer Dope Solutions

The effects of nonsolvent additives in polymer dope solutions on phase inversion, membrane morphology, separation performance, and macrovoid suppression

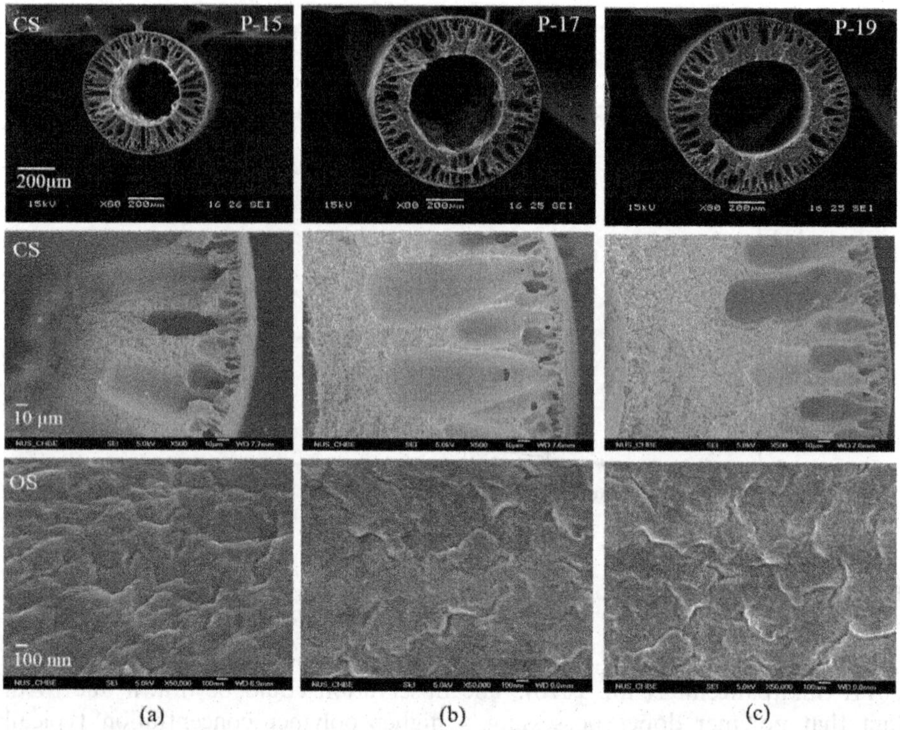

FIGURE 4.12 Cross-section and external surface morphologies of hollow fiber membranes spun from PVDF/NMP with different polymer concentrations (wt%) (a) P-15, (b) P-17, and (c) P-19 (CS, cross section; OS, outer surface) (Sukitpaneenit and Chung, 2009). Copyright (2009) Elsevier.

have been examined extensively. Typically, nonsolvent additives serve as a pore former, increase the polymer dope viscosity, adjust hydrophobicity/hydrophilicity, or accelerate the phase inversion process during membrane formation to improve membrane morphology, flux, and selectivity, as well as elimination of macrovoids. The nonsolvent additives utilized in the preparation and production of PVDF membranes could be classified into three main categories: (1) small or low molecular weight additives such as water (Sukitpaneenit and Chung, 2009; Wang et al., 2000), alcohols (Sukitpaneenit and Chung, 2009; Wang et al., 2000), acetone (Hou et al., 2014), ethylene glycol (Edwie and Chung, 2012; Khayet and Matsuura, 2001; Wang et al., 2008), glycerol (Atchariyawut et al., 2006; Naim et al., 2012), inorganic salts, that is, lithium chloride (LiCl) (Fontananova et al., 2006; Hou et al., 2014; Naim et al., 2012; Wang et al., 2000), lithium perchlorate (Yeow et al., 2005); (2) high molecular weight additives (polymeric additives) such as polyethylene glycol (PEG) (Naim et al., 2012; Tang et al., 2012; Wang et al., 2017), and polyvinylpyrrolidone (PVP) (Cha and Yang, 2006; Simone et al., 2010); and (3) other types of additives such as 1,2-ethanediol (Khayet et al., 2002b), phosphorous acid (Naim et al., 2012), and phosphoric acid (Hou et al., 2014).

Sukitpaneenit and Chung (2009) examined the effects of low molecular weight additives that include water, methanol, and ethanol in polymer dope solutions on the morphology of PVDF membranes. When water was utilized as an additive, the membrane structure featured finger-like macrovoids with a dense skin surface. Compared with membranes fabricated without additives, the water-additive membranes revealed a slight increase of macrovoids. Notably, the presence of water as an additive might suppress the polymer interaction and favors liquid–liquid demixing, resulting in the promotion of macrovoids. This finding was in agreement with the results reported by Wang et al. (2000) for a PVDF/DMAc system. When methanol and ethanol were employed as additives, the membrane morphology was substantially altered. In both cases, the membranes were composed of the packed structures of spherulitic globules in the inner substrate. There were tiny voids underneath the outer surface of fibers resulting from the solid–liquid demixing accompanying with crystallization.

Among several additives, LiCl is considered one of the most common additives for designing and producing PVDF membranes in membrane manufacturers. It has been reported that the addition of LiCl might lead to a porous structure with improved porosity and more or less macrovoids depending on the LiCl concentration (Fontananova et al., 2006; Wang et al., 2000). Fontananova et al. (2006) observed that the LiCl additive could induce delayed demixing (kinetic effect) and leads to suppress macrovoids. This can be attributed to the fact that adding a high concentration of LiCl might enhance the dope viscosity and delay the mutual diffusion between nonsolvent and solvent during the phase inversion. However, at low concentrations, LiCl might have an opposite effect and facilitate a rapid (instantaneous) phase inversion process because of the increased thermodynamic immiscibility (thermodynamic effect), thus leading to the formation of finger-like macrovoids.

PVP and PEG are well-known polymeric additives commonly used in the fabrication of highly porous PVDF membranes. Because of their hydrophilic nature and high miscibility in water, the presence of such polymers in PVDF dope solutions often enhances phase inversion via liquid–liquid demixing mechanism and facilitates the formation of big finger-like macrovoids. Wang et al. (1999) investigated the effects of PVP with different molecular weights on PVDF membrane morphology and performance. A low molecular weight PVP of 10 kDa resulted in membranes consisting of tiny pores because it leached out easily from the membrane structure, while a higher molecular weight PVP of 360 kDa seemed to remain within the membrane and obstruct the path of interconnected pores/voids. The PVDF membranes obtained from PVP with a low molecular weight had a significantly higher water flux and better solute retention than those made from PVP with a high molecular weight.

Naim et al. (2012) conducted a systematic investigation and compared the effects of phosphoric acid, methanol, polyethylene glycol (PEG-400), glycerol, and LiCl as additives in spinning dopes in fabricating microporous PVDF hollow fiber membranes for membrane contactor applications. They observed that the hydrophobicity and gas permeability of membranes decreased when nonsolvent additives were added. Nevertheless, there was an improvement in the liquid entry pressure, which was possibly attributed to the more sponge-like structures developed near the inner layer of fibers. Among the investigated additives, the addition of PEG-400 in the polymer dope led to the highest stripping flux that could have a relationship with enhanced effective surface porosity and gas permeability of the resultant membranes. Hou et al. (2009) examined PEG-1500 and LiCl as additives in PVDF/DMAc solutions for fabrication of PVDF MD hollow fiber membranes. They reported that the coupling influence of the two additives could lead to fiber membranes containing good hydrophobicity, high porosity, narrow pore size distribution, and porous sponge-like structures.

4.4.3 Coagulant Chemistry

Many studies have confirmed that the choice of coagulant chemistry and composition plays a critical role in designing the phase inversion mechanism, porosity, macrovoids, and membrane robustness in PVDF membranes (Buonomenna et al., 2007; Khayet and Matsuura, 2011; Peng et al., 2012; Sukitpaneenit and Chung, 2009). Based on the discussion in preceding sections, the phase inversion of PVDF membranes via the NIPS process involves competition between liquid–liquid demixing and solid–liquid demixing accompanying with crystallization. Generally, when a strong coagulant, for instance, water is used, liquid–liquid demixing would be dominant in the phase separation process resulting in the formation of asymmetric membranes having a relative dense and smooth surface, cellular morphology, and finger-like macrovoids. In contrast, when weaker coagulants such as alcohols, mixtures of water with various alcohols or with various solvents are employed, the liquid–liquid demixing process is delayed, and the solid–liquid demixing involving crystallization would predominantly regulate the phase separation to form typical PVDF membranes with uniform microporous structure packed with spherical particles, and less (or free) of macrovoids.

Besides coagulant chemistry, the temperature of coagulation bath constitutes another parameter influencing the formation process of PVDF membranes. Wang et al. (2009a) fabricated PVDF membranes using a PVDF/DMAc/water dope solution and investigated the influence of coagulation bath temperature on phase inversion mechanism, crystalline phase, and membrane morphology of the resultant membranes. It was reported that at low temperatures (15 and 25 °C), there was a delay in liquid–liquid demixing and gelation/crystallization of PVDF became the prevailing mechanism for membrane formation in the system. As a result, the cross section of the membranes had a globular structure comprising PVDF microcrystallites. However, when using an elevated coagulation bath temperature, 60 °C, the gelation/crystallization mechanism was suppressed by the liquid–liquid demixing because of the enhanced solvent and nonsolvent exchange rate. The resultant membranes had a cellular and bicontinuous interconnected structure.

It is important to note that the phase separation takes place from the top to bottom surfaces for as-cast flat-sheet membranes; it is different from the phase separation of hollow fiber membranes. In the latter, two coagulations occur at both the external and internal surfaces of nascent fibers. One can manipulate the external surface morphology by appropriately selecting the chemistry of the external coagulant and coagulation conditions. Likewise, the internal surface morphology can be designed and optimized by adjusting the chemistry and flow rate of the internal coagulant (i.e., bore fluid) and other coagulation conditions.

Figure 4.13 shows the impact of NMP/water bore fluid composition on the structure of PVDF hollow fiber membranes spun with an air-gap length of

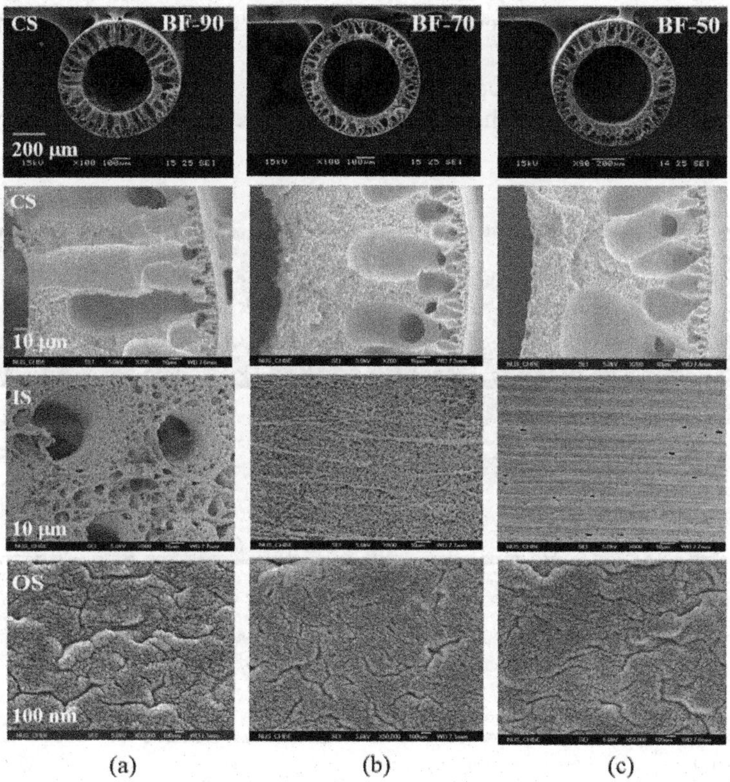

FIGURE 4.13 The membrane morphology of PVDF hollow fiber membranes spun from different NMP/water bore-fluid (BF) compositions: (a) BF-90 (90/10 wt% NMP/water), (b) BF-70 (70/30 wt% NMP/water), and (c) BF-50 (50/50 wt% NMP/water). CS, cross section; IS, inner surface; OS, outer surface (Sukitpaneenit and Chung, 2011). Copyright (2011) Elsevier.

1 cm, employing water as the external coagulant under a free fall condition (i.e., without take-up stretching) (Sukitpaneenit and Chung, 2011). When a 90/10 wt% NMP/water mixture was used as the bore fluid, the as-spun membrane had finger-like macrovoids across the entire cross section. The large holes/voids observed in the inner surface (Figure 4.13a, third row) could be directly connected through these finger-like macrovoids. The unique cross-sectional morphology might arise from the use of a high NMP content in the bore fluid, which not only induced delayed demixing leading to a soft internal skin but also allowed the rapid intrusion of bore fluid into the nascent fiber during the membrane formation. In contrast, when 70/30 and 50/50 wt% NMP/water mixtures were utilized as bore fluids, the water intrusion was suppressed to some degrees so that the finger-like structure closer to the lumen side transformed into a fully sponge-like structure. Additionally, no big holes in the internal surface could be observed for the two cases. Instead, a highly and uniform porous internal skin structure was attained for the membrane using 70/30 wt% NMP/water as the bore fluid, whereas a relatively denser skin structure was observed when 50/50 wt% NMP/water was employed as the bore fluid.

The impact of external coagulants on PVDF membrane morphology and macrovoids formation is illustrated in Figure 4.14 where the hollow fibers

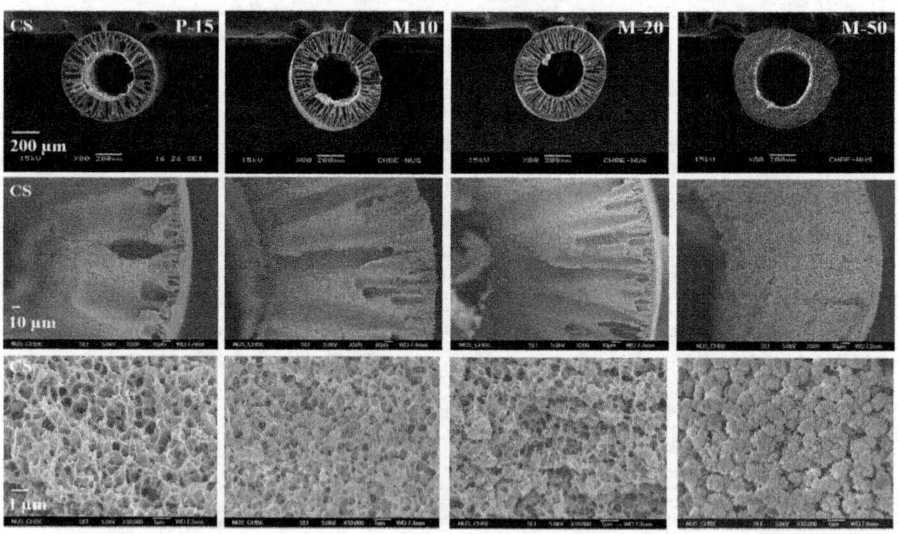

FIGURE 4.14 The cross-section morphology of hollow fiber membranes spun from 15 wt% PVDF/NMP with various compositions of water/methanol external coagulant: (a) P-15 (water), (b) M-10 (water/methanol 90/10), (c) M-20 (water/methanol 80/20), and (d) M-50 (water/methanol 50/50) (Sukitpaneenit and Chung, 2009). Copyright (2009) Elsevier.

were obtained from a 15 wt% PVDF/NMP dope solution with 0, 10, 20, and 50 wt% methanol/water mixtures as external coagulants (Sukitpaneenit and Chung, 2009). When water was utilized as the external coagulant, the resultant membrane's cross section mainly comprised large finger-like macrovoids with a small section of cellular morphology. This morphological characteristic signified that the phase inversion process was governed by instantaneous liquid–liquid demixing. With 10 and 20 wt% methanol/water mixtures as external coagulants, the size of macrovoids decreased as compared to that utilizing pure water as the external coagulant. When the methanol content was increased further to 50 wt%, a macrovoid-free structure was achieved. Moreover, the membrane structure was transferred gradually from an interconnected cellular type (P-15, water coagulant, 0 wt% methanol) to an interconnected globular transition type (10 and 20 wt% methanol/water coagulants) and eventually a globular-type structure (50 wt% methanol/water coagulant). Similar findings have been reported on the globular-type structure of PVDF (Buonomenna et al., 2007; Young et al., 1999). In summary, the considerable changes of membrane morphology can be interpreted by the coagulation strength and the amount of coagulants that initiate the delayed demixing accompanying with crystallization during phase inversion. Literally, when only water is employed as the external coagulant, the demixing process takes place rapidly and the time for inducing PVDF crystallization is inadequate. Consequently, the liquid–liquid demixing governs the phase separation. However, with the addition of methanol in the external coagulation bath, the liquid–liquid demixing process is delayed, thus the solid–liquid demixing involving crystallization becomes dominant. Therefore, through the increment of methanol content in the external coagulant, the solid–liquid demixing eventually controls the phase separation process.

It is worth noting that the incorporation of alcohols (including solvents and weak nonsolvents) in the external coagulation bath intensifies the delayed demixing, thus allowing a greater extent of PVDF crystallization. The enrichment in delayed demixing by the nonsolvents follows the sequence of IPA > ethanol > methanol > water that is in line with the phase diagram illustrated in Figure 4.7. The individual solubility parameters for PVDF, solvents, and nonsolvents are presented in Table 4.2 (Bottino et al., 1988; Hansen, 1999; Sukitpaneenit and Chung, 2009). The differences in solubility parameters among PVDF, nonsolvent water, and alcohols obey the sequence of (PVDF–water) > (PVDF–methanol) > (PVDF–ethanol) > (PVDF–IPA). A larger difference in solubility parameter indicates a faster precipitation rate and insufficient duration time for crystallization to take place. From the kinetics perspective, a greater diffusion of a nonsolvent in a solvent indicates a faster precipitation rate. The intrinsic diffusion coefficients of nonsolvents (water and alcohols) with regard to the solvent (NMP) adhere to the sequence of water > methanol > ethanol > IPA. Hence, the fundamental results from both kinetics and thermodynamics aspects clearly confirm that water, as a strong nonsolvent, could facilitate the most rapid precipitation for PVDF membranes whereas the alcohols (and other types of weak nonsolvents) could effectively

TABLE 4.2
Solubility parameters of polymers, solvent, and nonsolvents (Sukitpaneenit and Chung, 2009). Copyright (2009) Elsevier

Chemicals	Solubility parameters ($MPa^{1/2}$)			
	δ_d	δ_p	δ_h	δ_t
Water	15.60	16.00	42.30	47.80
Methanol	15.10	12.30	22.30	29.60
Ethanol	15.80	8.80	19.40	26.50
Isopropanol (IPA)	15.80	6.10	16.40	23.50
N-Methyl-2-pyrrolidone	18.00	12.30	7.21	22.90
Poly(vinylidene fluoride)	–	–	–	14.15[a]
	–	–	–	17.75[b]
	17.20	12.50	9.20	23.20[c]

Source: Hansen, C.M., (1999). Hansen solubility parameter, in *A User's Handbook*, CRC Press, NY.
δ_d, dispersive parameter; δ_p, polar parameter; δ_h, hydrogen bonding parameter; and δ_t, total solubility parameter.
[a] Simulated using material studio (synthia) based on the Van Krevelen equation
[b] Simulated using material studio (synthia) based on the Fedors equation
[c] From the reference (Bottino et al., 1988) based on solubility test

delay the demixing process, leading to the formation of spherulitic crystallites in PVDF membranes.

4.4.4 Air-Gap Distance and Take-Up Speed

In general, a high take-up speed coupled with an air-gap distance is usually applied in the industrial-scale manufacture of membranes to increase productivity and reduce production costs. An increase in take-up speed and air-gap length typically reduces the membrane thickness and permeation flux. For hollow fiber spinning, the outer and inner diameters (OD and ID) of fibers apparently become smaller at a higher take-up speed and/or air-gap.

Previous studies have demonstrated that the presence of air-gap or take-up speed may induce not only different precipitation pathways during phase inversion but also the elongation-induced polymer chain orientation and packing (Chung, 2008; Peng et al., 2012). Nevertheless, if the take-up speed or air-gap distance is very high, it might also cause defects due to the excessive elongational and gravitational stresses.

The effects of take-up speed and air-gap distance on PVDF membrane formation, morphology, and separation performance have been studied by several researchers (Khayet, 2003; Sukitpaneenit and Chung, 2011; Tang et al., 2012; Wang et al., 1999). Wang et al. (1999) designed PVDF UF hollow fiber membranes spun using various air-gap lengths of 5, 10, and 15 cm. They found that the water flux of as-spun fibers decreased with reducing air-gap distance whereas the dextran rejection only had negligible changes. Khayet (2003) systematically studied the influence of air-gap on the morphology and performance of PVDF hollow fiber membranes spun from a PVDF/DMAc/ethylene glycol (23/73/4 w/w%) dope solution using a 50 v/v% aqueous ethanol solution as the internal and external coagulants. For an air-gap distance between 1 and 80 cm, they observed that an increase in air-gap distance yielded denser and thinner sponge-like hollow fiber membrane structures because of a higher molecular orientation and polymer chain packing. They also reported that increasing air-gap length resulted in hollow fiber membranes with a thinner and denser sponge-like structure due to a greater molecular orientation and chain packing. Additionally, there was a reduction of permeation flux and an improvement of solute rejection with an increase in air-gap distance.

As demonstrated in the work of Sukitpaneenit and Chung (2011), increasing the take-up speed and air-gap length not only impeded the formation of macrovoids in PVDF membranes but also created the sponge-like structure as well. Arguably, macrovoids suppression might arise from the rapid shrinkage of fiber dimension under the elongation tension by its gravity (for the air-gap one) or the elongation stretch from the take-up unit. Theoretically, such stretching processes would induce a radial outflow (i.e., a negative normal stress) within the nascent hollow fiber to undermine the capillary intrusion of coagulants, thus removing the possible formation of macrovoids (Chung, 2008; Peng et al., 2012). In spite of the fact that macrovoid-free membranes prepared in most studies had interconnected PVDF crystallite particles (i.e., a globular structure) that were typically emanated from the solid–liquid demixing process, the PVDF hollow fiber membranes developed in their works showed an almost ideal macrovoid-free morphology with interlinked cellular structures. This morphology generally exists when the liquid–liquid demixing mechanism controls the phase inversion. It is worthwhile to note that PVDF membrane with a cellular structure resulting from liquid–liquid demixing is desired due to its superior mechanical robustness as opposed to those constructed with PVDF globule networks (Sukitpaneenit and Chung, 2009). Therefore, introducing an external elongation stress by either escalating the take-up speed or air-gap length is one of the effective means to facilitate the liquid–liquid demixing (i.e., suppress solid–liquid demixing) and to produce mechanically stable and strong PVDF membranes.

In addition, the environments in the air gap region, especially relative humidity and temperature, are important factors that greatly affect membrane morphology, quantity and size of macrovoids, and flux performance. Interestingly, while most PVDF membranes reported in the literature are made in ambient humidity, most commercially available PVDF membranes are produced under well-controlled environments with precise air-gap humidity and temperature. In most cases, employing higher relative humidity and temperature than the ambient environment in the air gap region tends to result in PVDF membranes with greater surface porosity, permeation flux, and elimination of macrovoids.

4.5 EMERGING STRATEGIES ON PVDF MEMBRANE FABRICATION TECHNOLOGY TOWARD MD PRACTICAL/INDUSTRIAL IMPLEMENTATIONS

4.5.1 Thermally Induced Phase Separation

Unlike NIPS, TIPS refers to the membrane fabrication technique where the phase separation mechanism is induced by a temperature change, the removal of thermal energy from high-temperature polymer-dope solutions. TIPS is generally used as an alternative method of membrane fabrication for semi-crystalline polymers or polymers having poor solubility in common solvents (Caneba and Soong, 1985; Lloyd et al., 1990, 1991; Matsuyama et al., 2000). Historically, the TIPS method was invented by Castro in 1980s, and then followed by studies and further developments by many investigators in the past decades (Castro, 1981). Typically, the spinning steps for hollow fibers via TIPS are as follows:

(1) Preparation of a homogeneous polymer solution by dissolving the polymer in a proper diluent at a mixing temperature above its melting temperature. The diluent should have a high boiling point. It may be either a low molecular weight solid or liquid but stable at the above-mentioned mixing temperature;
(2) Extrusion of the polymer solution and bore fluid simultaneously through a spinneret under a nitrogen gas pressure or by means of precision pumps to form hollow fibers;
(3) Immersion of the extruded fibers into a quenching tank with a choice of applying rapid or controlled cooling rates to induce phase separation and solidification;
(4) Extraction of the diluent through solvent extraction in accordance with the diluent type to completely remove the remaining diluent in the membrane structure; common solvents adopted in the extraction process are water, alcohols, and acetone.

The choice of diluents in TIPS processes plays a critical role in determining the phase separation mechanism and subsequently influences the morphology and performance of the hollow fiber membranes. Typically, diluents with good solvency at high temperatures are desirable; however, high solvency in low temperatures (for instance, room temperature) is considered unfavorable. Diluents with high boiling points such as dioctyl phthalate (DOP), dibutyl phthalate (DBP), diethyl phthalate (DEP), triacetin, γ-butyrolactone (γ-BA), and cyclohexanone (CO) are among the commonly used diluents in the fabrication of PVDF membranes via TIPS (Rajabzadeh et al., 2008; Su et al., 2007). The phase separation or demixing mechanism of TIPS greatly lies on the polymer–diluent interaction (Lloyd et al., 1990). A strong polymer–diluent interaction favors the solid–liquid demixing through crystallization, while a weak polymer–diluent interaction tends to promote the liquid–liquid demixing process.

Moreover, since TIPS is a nonequilibrium process, the influence of cooling rate on phase separation process must be considered. Several studies suggest that even though the cooling rate shows little effects on the liquid–liquid demixing mechanism, it does have substantial effects on the temperature at which the solid–liquid demixing process takes place, for instance, by increasing the cooling rate, the crystallization temperature decreases (Lloyd et al., 1990, 1991). Ji et al. (2008) investigated the impact of mixed diluents on morphology and performance of PVDF hollow fibers. Notably, an increase in the ratio of DBP in the mixed DBP and di (2-ethylhexyl) phthalate changed the as-spun fiber morphology from an interconnected sponge-like structure toward an asymmetric spherulitic (globular) bulk structure. The former structure resulted in a greater water permeability and elasticity than the latter one.

The effects of polymer concentration, extrusion temperature, quenching temperature, and take-up speed in TIPS on membrane structure and performance of microporous PVDF hollow fibers as membrane contactors for carbon dioxide (CO_2) capture applications were systematically investigated by Ghasem and co-workers (Ghasem et al., 2011, 2012a, 2012b). They found that an increase in PVDF concentration in dope solutions reduces the mean pore size and the effective surface porosity of the membrane contactors (Ghasem et al., 2012a). Besides, the fibers that were spun at a higher extrusion temperature exhibited significant improvements in mechanical properties, porosity, greater hydrophobicity, and enhanced membrane contractor efficiency (Ghasem et al., 2012b). Similar trends were also reported for fibers spun with a higher quench bath temperature. However, the quench bath temperature seemed to have no or minor effect on fiber's hydrophobicity (Ghasem et al., 2011). As reported by Li et al. (2013), the mechanical strength of TIPS PVDF hollow fibers increased with increasing take-up speed in the fiber spinning process.

To date, TIPS has drawn considerable attention and regained its momentum in designing PVDF membranes for MD applications. The feasibility and potential advantages of TIPS in developing macrovoid-free PVDF membranes with robust mechanical integrity, high porosity, and narrow pore size distribution suitable for MD have been extensively studied (Song et al., 2012). Some studies claimed that the TIPS process has merits of lesser control parameters over the NIPS process; however, the limited choice of high-boiling-point diluents with low prices and acceptable toxicity, high processing temperature, and energy consumption (e.g., precise systems for temperature controls) remains major hurdles for TIPS (Cha and Yang, 2007; Lloyd et al., 1990).

Recently, the combination of NIPS and TIPS methods, referred to as "nonsolvent thermally induced phase separation (NTIPS)," has been reported by Xiao et al. (2015) with an attempt to combine the unique merits of both techniques to fabricate PVDF membranes with desirable characteristics for MD applications. While in the TIPS process most diluents are plasticizer phthalates that are water-insoluble, the NTIPS process employs a water-soluble diluent, ε-caprolactam (CPL), thus allowing both NIPS and TIPS to take place simultaneously in the process. In their work, the developed PVDF membranes possessed an ultrathin skin layer, a short finger-like macrovoid transition layer, and a bicontinuous and cellular membrane morphology. The NTIPS membranes displayed a high overall porosity of 86%, reasonably good LEP above 3.5 bar, impressive mechanical strength, as well as superior permeation flux of 85.6 kg m^{-2} h^{-1} measured at a feed temperature of 80 °C using 3.5 wt% sodium chloride as synthetic seawater in DCMD, which met the desirable characteristics for MD applications.

4.5.2 Dual-Layer Spinneret Design in Hollow Fiber Fabrication

It is well known that spinneret design plays an essential role in determining membrane formation, morphology, and separation performance of the hollow fiber membranes. The concept to fabricate multilayer composite hollow fibers by means of single-step coextrusion of polymer solution and bore fluid via a dual-layer (or tri-orifice) spinneret has been explored widely since 1980s (He et al., 2002; Henne et al., 1979; Kuzumoto and Nitta, 1989; Ong and Chung, 2012; Pereira et al., 2003; Wang et al., 2011; Yang et al., 2009). Several advantages are associated with the configuration of dual-layer hollow fibers, including reduced costs because a relatively expensive but a small amount of functional polymers is extruded to form the thin separation layer, whereas a robust and low-cost material is selected to form the supporting layer. In fact, this spinning approach offers a higher degree of freedom for the customization and design of materials and morphology in the selective and supporting layers (Li et al., 2002; Xia et al., 2018). Moreover, from the industrial perspectives, the simultaneous coextrusion could significantly simplify the fabrication of composite hollow fiber membranes because of the elimination of the second coating in the manufacturing process. Figure 4.15 illustrates the setup of a dual-layer hollow fiber spinning line and the schematic diagram of a dual-layer spinneret with fluid channels for both bore fluid and dope solutions (Fu et al., 2013; Li et al., 2002).

The polymer–solvent coextrusion method employing a dual-layer spinneret design is another effective way to fabricate hollow fiber membranes with very high surface porosity, and controlled pore size and pore size distribution (Bonyadi and Chung, 2009; He et al., 2003; Sukitpaneenit and Chung, 2012a, 2012b). In this approach, the solvent (e.g., pure NMP) is fed at the outer-layer channel of the spinneret and extruded simultaneously with the polymer solution to induce a delayed demixing (i.e., slow phase inversion and mild solvent exchange) at their interface in the air gap region prior to entering the coagulation bath. Bonyadi and Chung (2009) employed this technique to fabricate PVDF hollow fiber membranes for MD. The use of the solvent–polymer dope coextrusion not only greatly enhanced the outer surface porosity but also eliminated the formation of macrovoids in the cross section of PVDF membranes. The developed hollow fiber membranes had a flux twice or three times higher and better energy efficiency than the typical PVDF hollow fibers for DCMD applications.

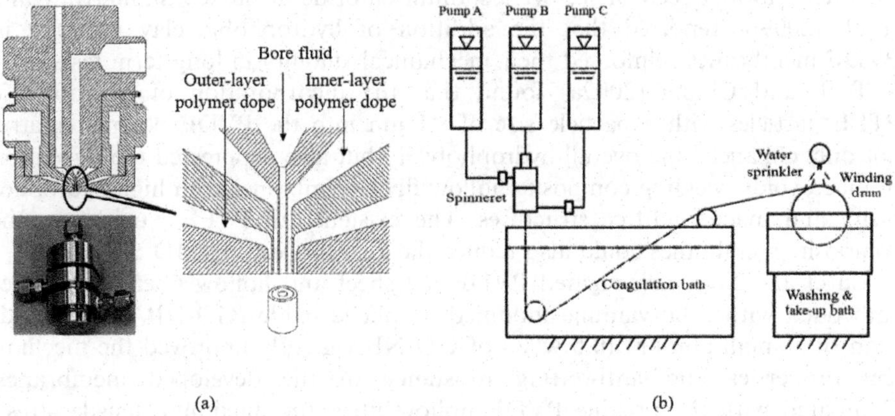

FIGURE 4.15 Schematic of (a) the dual-layer spinneret (Li et al., 2002). Copyright (2002) Elsevier; and (b) the hollow fiber spinning line setup. Pump A: inner dope solution; pump B: bore fluid; and pump C: outer dope solution (Fu et al., 2013). Copyright (2013) Elsevier.

4.5.3 PVDF Mixed-Matrix or Hybrid Membranes

The concept of fabricating mixed-matrix or hybrid membranes has gained considerable attention and significant growth in research by both academia and industry for a wide variety of membrane separation including MD applications. Typically, mixed-matrix or hybrid membranes combine the strengths of polymer materials such as good processability and low cost and inorganic/organic fillers such as diverse functionality and good mechanical strength to form membranes with desirable physicochemical properties and separation performance. Different inorganic/organic fillers have been incorporated into MD membranes. Some of them are calcium carbonate ($CaCO_3$) (Hou et al., 2012), clay (Mokhtar et al., 2014; Prince et al., 2012; Wang et al., 2009b), Teflon (PTFE) (Teoh and Chung, 2009a, 2009b), graphene oxide (GO) (Leaper et al., 2018; Lu et al., 2017; Woo et al., 2016), and silica (SiO_2) (Li et al., 2015). The resultant PVDF membranes have shown more desirable hydrophobicity, pore size and porosity, mechanical strength, wetting/fouling properties, and MD performance than the pristine ones.

The fabrication of PVDF/$CaCO_3$ mixed-matrix hollow fiber membranes for MD applications was explored by Hou et al. (2012). The $CaCO_3$ nanoparticles were first treated with octadecyl dihydrogen phosphate to enhance their hydrophobicity. The incorporation of $CaCO_3$ into PVDF led to the resultant membranes with a narrower pore size distribution, greater membrane porosity, and hydrophobicity. Compared with the pristine PVDF membrane, the hybrid membranes showed significant enhancements in mechanical properties, satisfying MD performance stability in 30-day continuous DCMD experiments using a model seawater feed of 3.5 wt% sodium chloride at 80 °C. Similarly, Wang et al. (2009b) reported that the addition of hydrophobic clay particles in PVDF membranes reinforced their mechanical stability in long-term tests.

Teoh and Chung (2009a) found that the incorporation of hydrophobic PTFE particles with a particle size of <1 μm into the PVDF polymer matrix not only enhanced the overall hydrophobicity but also suppressed the formation of macrovoids, yielding composite hollow fiber membranes with high hydrophobicity and macrovoid-free structures. The existence of PTFE particles in the composite membranes could also reduce the heat loss during MD operations.

Lu et al. (2017) investigated PVDF flat-sheet and hollow fiber membranes embedded with n-butylamine modified graphene oxide (GO-NBA). The addition of a small amount (0.5 wt%) of GO-NBA greatly improved the mechanical properties and antiwetting resistance of the developed membranes. Compared with the pristine PVDF hollow fiber, the maximum tensile stress, maximum tensile strain, and Young's modulus of the resultant hollow fiber membranes increased by 32%, 63%, and 71%, respectively, while the liquid entry pressure (LEP) and burst pressure jumped by 67% and 15%, respectively. In addition, the PVDF-modified GO composite membranes showed a reasonably high flux of 61.9 kg m^{-2} h^{-1} and a salt rejection of 99.9% using model seawater as the feed at 80 °C under DCMD tests.

Another promising approach is to fabricate multilayer mixed-matrix or hybrid membranes through the dual-layer hollow fiber spinning process. Thanks to the benefits of dual-layer spinneret technology as discussed in Section 5.2, the production cost and process complexity of mixed-matrix hollow fiber membranes could be reduced significantly by solely incorporating the fillers into the functional layer. Besides, the hydrophobicity and hydrophilicity of the outer and inner layers can be optimized through the addition of different fillers at the respective layers. Bonyadi and Chung (2007) reported the fabrication of the first hydrophobic–hydrophilic dual-layer PVDF hollow fiber for DMCD applications. They embedded the outer layer with hydrophobic Cloisite 15A particles to enhance its hydrophobicity and incorporated hydrophilic Cloisite NA+ particles and polyacrylonitrile into the inner layer to improve its hydrophilicity. As a result, the dual-layer hollow fiber membranes showed a DMCD flux of 55 kg m^{-2} h^{-1} under a feed temperature of 90 °C. Recently, Zuo et al. (2017) adopted the same coextrusion technique to fabricate the hydrophobic/hydrophilic PVDF/Ultem dual-layer hollow fiber membranes for VMD. The resultant membranes displayed an encouraging VMD flux of 45.8 kg m^{-2} h^{-1} at 70 °C, good wetting resistance, and enhanced mechanical properties compared to the conventional singe layer MD hollow fibers.

4.5.4 PVDF Braid-Reinforced Hollow Fiber Membranes

The concept of fabricating braid-reinforced hollow fiber membranes may present a new opportunity to design and fabricate high-flux MD membranes with exceptional mechanical properties exceeding conventional singer-layer and dual-layer hollow fiber configurations. Figure 4.16 illustrates the spinning

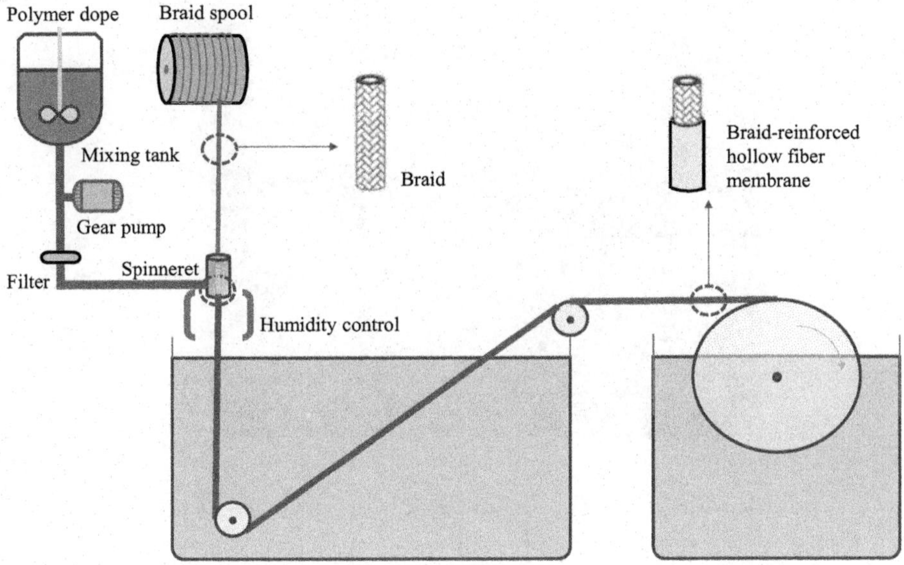

FIGURE 4.16 Schematic drawing of a braid-reinforced hollow fiber membrane fabrication process.

concept and apparatus. The braid material and polymer spinning dope are fed into the inner and outer channel of the spinneret, respectively. After that, the braid and polymer dope meet at the tip of the spinneret and pass through an air gap region before entering the coagulation and rinse baths. Finally, the as-spun braid-reinforced membrane is collected on the take-up unit.

Hao et al. (2016) reported the preparation and characterization of microporous PVDF composite hollow fiber membranes using the concept of braid reinforcement. The as-prepared membranes consisted of an outer-coated PVDF/graphene separation layer and an inner PET hollow tubular braids. The resultant membranes exhibited an improved hydrophobicity, high water LEP value, high performance consistency, and durability. Liu et al. (2009) investigated the effect of PET braids as the reinforcement material on membrane morphology and properties of PVDF membranes. The results revealed that the PET reinforcement layer had a negligible influence on pore size and porosity of the resultant membranes; however, a remarkable improvement on mechanical strength of fibers was observed.

Interestingly, even though there have been limited research from academia that reported the preparation and application of braid-reinforced hollow fiber membranes, it has in recent years been considered as one of the most popular PVDF hollow fiber membrane configurations in commercial UF and membrane bioreactor (MBR) markets. For illustration, Figure 4.17 shows the commercially available braid-reinforced PVDF hollow fiber membrane, PURON

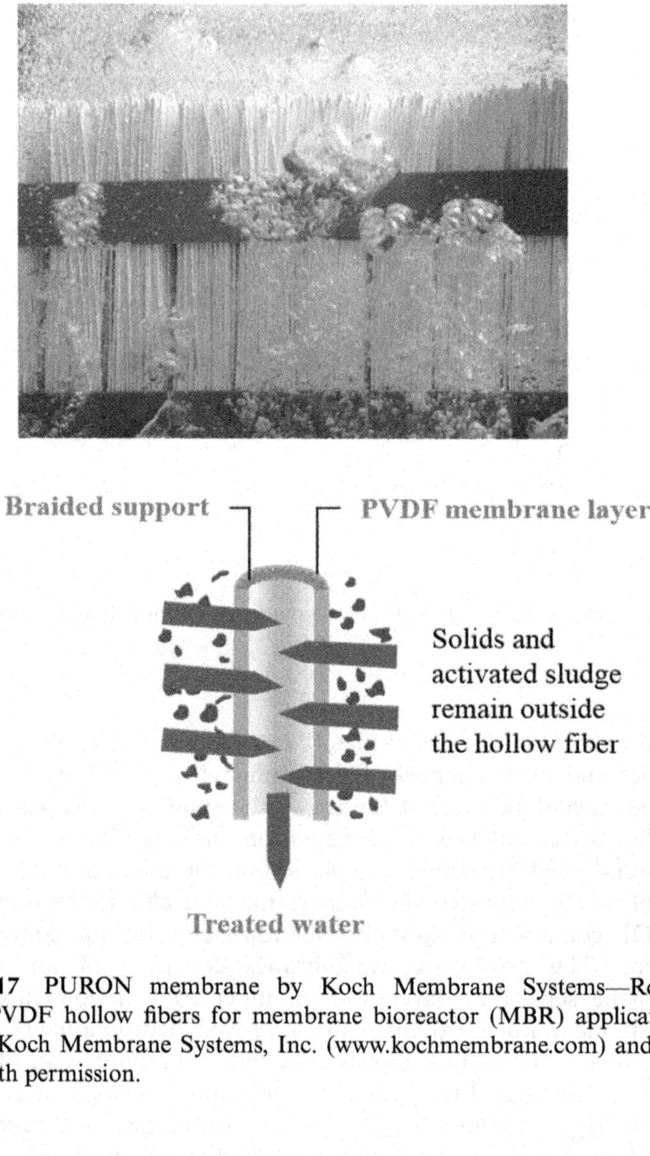

FIGURE 4.17 PURON membrane by Koch Membrane Systems—Robust braid-reinforced PVDF hollow fibers for membrane bioreactor (MBR) applications. Figure courtesy of Koch Membrane Systems, Inc. (www.kochmembrane.com) and used by the publisher with permission.

membrane, offered by Koch Membrane Systems for MBR applications. The membrane displays exceptional properties including high flux characteristic, uniform pore size, macrovoid-free morphology, and excellent long-term stability. From the industrial R&D perspective, this fabrication technique offers a great opportunity in the design of mechanically robust PVDF hollow fiber membranes with desirable characteristics such as a thin wall thickness, high porosity, open-cell pore structure, and macrovoid-free morphology for MD applications. Furthermore, the successful implementation of this fabrication technology, based on its reliability and scalability in a large-scale membrane production, is another attractive point in developing novel and practical MD membranes, which may expedite and merit the MD industrial development and commercialization.

4.6 CONCLUSION AND PERSPECTIVES

Design and engineering of PVDF membranes with desirable morphology, properties, and separation performance is of paramount importance to bring the MD technology to industrialization and commercialization. The development of such membranes via innovative fabrication techniques/strategies should exhibit cost-effective and reliable properties, as well as easy implementation in large-scale manufacturing environments. In this chapter, an overview on the state-of-the-art and recent advances in design and fabrication of PVDF membranes for MD is reviewed. Basic understanding of MD processes, fundamental principles of PVDF membrane formation and production as well as the development of robust, highly porous, and macrovoid-free morphology were elaborated. Recent progresses on fabrication methods and strategies such as TIPS, dual-layer coextrusion technology, mixed-matrix membranes, and braid-reinforced hollow fiber PVDF membranes that could be attractive from the industrial R&D perspectives were also highlighted. It goes without saying that PVDF would continue serving as the promising material in the development of novel and high-performance MD membranes. Another crucial aspect from the industrial prospect to the future success of MD is the fabrication process feasibility and economics. A better alignment between industry and academia regarding MD research themes on membrane materials and fabrication technologies is necessary to establish the advance and breakthrough of MD membranes toward industrial and commercialization.

REFERENCES

Ali, M.I., Summers, E.K., Arafat, H.A., Lienhard V.J.H., 2012. Effects of membrane properties on water production cost in small scale membrane distillation systems. Desalination 306, 60–71.

Alkhudhiri, A., Darwish, N., Hilal, N., 2012. Membrane distillation: a comprehensive review. Desalination 287, 2–18.

Al-Obaidani, S., Curcio, E., Macedonio, F., Profio, G.D., Al-Hinai, H., Drioli, E., 2008. Potential of membrane distillation in seawater desalination: thermal efficiency, sensitivity study and cost estimation. J. Membr. Sci. 323, 85–98.

Atchariyawut, S., Feng, C., Wang, R., Jiraratananon, R., Liang, D.T., 2006. Effect of membrane structure on mass-transfer in the membrane gas-liquid contacting process using microporous PVDF hollow fibers. J. Membr. Sci. 285, 272–281.

Baker, R.W., 2004. Membrane Technology and Applications, 2nd ed. John Wiley & Sons Ltd, Chichester, UK.

Bandini, S., Saavedra, A., Sarti, G.C., 1997. Vacuum membrane distillation: experiments and modeling. AIChE J. 43, 398–408.

Basile, A., Figoli, A., Khayet, M., 2015. Pervaporation, Vapour Permeation and Membrane Distillation: Principles and Applications, 1st ed. Woodhead Publishing Ltd, Oxford, UK.

Bonyadi, S., Chung, T.S., 2007. Flux enhancement in membrane distillation by fabrication of dual layer hydrophilic-hydrophobic hollow fiber membranes. J. Membr. Sci. 306, 134–146.

Bonyadi, S., Chung, T.S., 2009. Highly porous and macrovoid-free PVDF hollow fiber membranes for membrane distillation by a solvent-dope solution co-extrusion approach. J. Membr. Sci. 331, 66–74.

Bottino, A., Capannelli, G., Munari, S., Turturro, A., 1988. Solubility parameters of poly(vinylidene fluoride). J. Polym. Sci. B: Polym. Phys. 26, 785–794.

Buonomenna, M.G., Macchi, P., Davoli, M., Drioli, E., 2007. Poly(vinylidene fluoride) membranes by phase inversion: the role the casting and coagulation conditions play in their morphology, crystalline structure and properties. Eur. Polym. J. 43, 1557–1572.

Caneba, G.T., Soong, D.S., 1985. Polymer membrane formation through the thermal-inversion process. 1. Experimental study of membrane structure formation. Macromolecules. 18, 2538–2545.

Castro, A.J., 1981. Methods for making microporous products. US4247498A.

Cha, B.J., Yang, J.M., 2006. Effect of high-temperature spinning and PVP additive on the properties of PVDF hollow fiber membranes for microfiltration. Macromol. Res. 14, 596–602.

Cha, B.J., Yang, J.M., 2007. Preparation of poly(vinylidene fluoride) hollow fiber membranes for microfiltration using modified TIPS process. J. Membr. Sci. 291, 191–198.

Cheng, L.P., Lin, D.J., Shih, C.H., Dwan, A.H., Gryte, C.C., 1999. PVDF membrane formation by diffusion-induced phase separation-morphology prediction based on phase behavior and mass transfer modeling. J. Polym. Sci. B: Polym. Phys. 37, 2079–2092.

Chung, T.S., 2008. Fabrication of hollow fiber membranes by phase inversion. In Li, N., et al. (Ed.), Advanced Membrane Technology and Applications. John Wiley & Sons Inc., Hoboken, NJ, pp. 821–841.

Chung, T.S., Teoh, S.K., Hu, X.D., 1997. Formation of ultrathin high-performance polyethersulfone hollow fiber membranes. J. Membr. Sci. 133, 161–175.

Cui, Z., Hassankiadeh, N.T., Zhuang, Y., Drioli, E., Lee, Y.M., 2015. Crystalline polymorphism in poly(vinylidenefluoride) membranes. Prog. Polym. Sci. 51, 94–126.

Dohany, J.E., Robb, L.E., 1980. Polyvinylidene fluoride. In Kirk-Othmer (Ed.), Encyclopedia of Chemical Technology, Vol. 11. 3rd ed. Wiley, New York, NY, pp. 64–74.

Drioli, E., Ali, A., Macedonio, F., 2015. Membrane distillation: recent developments and perspectives. Desalination 356, 56–84.

Edwie, F., Chung, T.S., 2012. Development of hollow fiber membranes for water and salt recovery from highly concentrated brine via direct contact membrane distillation and crystallization. J. Membr. Sci. 421–422, 111–123.

Essalhi, M., Khayet, M., 2013a. Self-sustained webs of polyvinylidene fluoride electrospun nanofibers at different electrospinning times: 2. Theoretical analysis, polarization effects and thermal efficiency. J. Membr. Sci. 433, 180–191.

Essalhi, M., Khayet, M., 2013b. Self-sustained webs of polyvinylidene fluoride electrospun nanofibers at different electrospinning times: 1. Desalination by direct contact membrane distillation. J. Membr. Sci. 433, 167–179.

Eykens, L., De Sitter, K., Dotremont, C., Pinoy, L., Van der Bruggen, B., 2016a. How to optimize the membrane properties for membrane distillation: a review. Ind. Eng. Chem. Res. 55, 9333–9343.

Eykens, L., De Sitter, K., Dotremont, C., Pinoy, L., Van der Bruggen, B., 2016b. Characterization and performance evaluation of commercially available hydrophobic membranes for direct contact membrane distillation. Desalination 392, 63–73.

Eykens, L., De Sitter, K., Dotremont, C., Pinoy, L., Van der Bruggen, B., 2017. Membrane synthesis for membrane distillation: a review. Sep. Purif. Technol. 182, 36–51.

Eykens, L., Hitsov, I., De Sitter, K., Dotremont, C., Pinoy, L., Nopens, I., Van der Bruggen, B., 2016c. Influence of membrane thickness and process conditions on direct contact membrane distillation at different salinities. J. Membr. Sci. 498, 353–364.

Feng, C., Shi, B., Li, G., Wu, Y., 2004. Preliminary research on microporous membrane from F2.4 for membrane distillation. Sep. Purif. Technol. 39, 221–228.

Fernández-Pineda, C., Izquierdo-Gil, M.A., García-Payo, M.C., 2002. Gas permeation and direct contact membrane distillation experiments and their analysis using different models. J. Membr. Sci. 198, 33–49.

Fontananova, E., Jansen, J.C., Cristiano, A., Curcio, E., Drioli, E., 2006. Effect of additives in the casting solution on the formation of PVDF membranes. Desalination 192, 190–197.

Francis, L., Ghaffour, N., Alsaadi, A.A., Amy, G.L., 2013. Material gap membrane distillation: a new design for water flux enhancement. J. Membr. Sci. 448, 240–247.

Fu, F.J., Zhang, S., Sun, S.P., Wang, K.Y., Chung, T.S., 2013. POSS-containing delamination-free dual-layer hollow fiber membranes for forward osmosis and osmotic power generation. J. Membr. Sci. 443, 144–155.

García-Payo, M., Izquierdo-Gil, M., 2004. Thermal resistance technique for measuring the thermal conductivity of thin microporous membranes. J. Phys. D: Appl. Phys. 37, 3008–3016.

García-Payo, M., Izquierdo-Gil, M., Fernández-Pineda, C., 2000. Wetting study of hydrophobic membranes via liquid entry pressure measurements with aqueous alcohol solutions. J. Colloid Interface Sci. 230, 420–431.

Ghasem, N., Al-Marzouqi, M., Duaidar, A., 2011. Effect of quenching temperature on the performance of poly(vinylidene fluoride) microporous hollow fiber membranes fabricated via thermally induced phase separation technique on the removal of CO_2 from CO_2-gas mixture. Int. J. Greenh. Gas Con. 5, 1550–1558.

Ghasem, N., Al-Marzouqi, M., Duidar, A., 2012a. Effect of PVDF concentration on the morphology and performance of hollow fiber membrane employed as gas–liquid membrane contactor for CO_2 absorption. Sep. Purif. Technol. 98, 174–185.

Ghasem, N., Al-Marzouqi, M., Rahim, N.A., 2012b. Effect of polymer extrusion temperature on poly(vinylidene fluoride) hollow fiber membranes: properties and performance used as gas-liquid membrane contactor for CO_2 absorption. Sep. Purif. Technol. 99, 91–103.

Hanemaaijer, J.H., Medevoort, J.V., Jansen, A.E., Dotremont, C., Sonsbeek, E.V., 2006. Memstill membrane distillation – a future desalination technology. Desalination 199, 175–176.

Hansen, C.M., 1999. Hansen Solubility Parameters: A User's Handbook, CRC Press, Boca Raton, FL.

Hao, J., Xiao, C., Zhang, T., Zhao, J., Fan, Z., Chen, Li., 2016. Preparation and performance of PET-braid-reinforced poly(vinylidene fluoride)/Graphene hollow fiber membranes. Ind. Eng. Chem. Res. 55, 2174–2182.

Hashim, N.A., Liu, Y., Li, K., 2011. Stability of PVDF hollow fibre membranes in sodium hydroxide aqueous solution. Chem. Eng. Sci. 66, 1565–1575.

He, T., Mulder, M. H. V., Strathmann, H., Wessling, M., 2002. Preparation of composite hollow fiber membranes: co-extrusion of hydrophilic coating onto porous hydrophobic support structures. J. Membr. Sci. 207, 143–156.

He, T., Mulder, M.H.V., Wessling, M., 2003. Preparation of porous hollow fiber membranes with a triple-orifice spinneret. J. Appl. Polym. Sci. 87, 2151–2157.

Henne, W., Dunweg, G., Schmitz, W., Pohle, R., Lawitzki, F., 1979. Method of producing dialyzing membrane. US 4164437A.

Hogan, P.A., Sudjito, Fane, A.G., Morrison, G.L., 1991. Desalination by solar heated membrane distillation. Desalination 81, 81–90.

Hou, D., Fan, H., Jiang, Q., Wang, J., Zhang, X., 2014. Preparation and characterization of PVDF flat-sheet membranes for direct contact membrane distillation. Sep. Purif. Technol. 135, 211–222.

Hou, D., Wang, J., Qu, D., Luan, Z., Ren, X., 2009. Fabrication and characterization of hydrophobic PVDF hollow fiber membranes for desalination through direct contact membrane distillation. Sep. Purif. Technol. 69, 78–86.

Hou, D., Wang, J., Sun, X., Ji, Z., Luan, Z., 2012. Preparation and properties of PVDF composite hollow fiber membranes for desalination through direct contact membrane distillation. J. Membr. Sci. 405–406, 185–200.

Ji, G.L., Zhu, L.P., Zhu, B.K., Zhang, C.F., Xu, Y.Y., 2008. Structure formation and characterization of PVDF hollow fiber membrane prepared via TIPS with diluent mixture. J. Membr. Sci. 319, 264–270.

Khayet, M., 2003. The effects of air gap length on the internal and external morphology of hollow fiber membranes. Chem. Eng. Sci. 58, 3091–3104.

Khayet, M., 2011. Membranes and theoretical modeling of membrane distillation: a review. Adv. Colloid Interface Sci. 164, 56–88.

Khayet, M., Cojocaru, C., Baroudi, A., 2012. Modeling and optimization of sweeping gas membrane distillation. Desalination 287, 159–166.

Khayet, M., Feng, C.Y., Khulbe, K.C., Matsuura, T., 2002a. Preparation and characterization of polyvinylidene fluoride hollow fiber membranes for ultrafiltration. Polymer 43, 3879–3890.

Khayet, M., Feng, C.Y., Khulbe, K.C., Matsuura, T., 2002b. Study on the effect of a non-solvent additive on the morphology and performance of ultrafiltration hollow-fiber membranes. Desalination 148, 321–327.

Khayet, M., Khulbe, K., Matsuura, T., 2004. Characterization of membranes for membrane distillation by atomic force microscopy and estimation of their water vapor transfer coefficients in vacuum membrane distillation process. J. Membr. Sci. 238, 199–211.

Khayet, M., Matsuura, T., 2001. Preparation and characterization of polyvinylidene fluoride membranes for membrane distillation. Ind. Eng. Chem. Res. 40, 5710–5718.

Khayet, M., Matsuura, T., 2011. Membrane Distillation: Principles and Applications. Elsevier B.V., Amsterdam, The Netherlands.

Kuzumoto, E., Nitta, K., 1989. Production of permselective compound hollow yarn membranes. JP01015104.

Laganà, F., Barbieri, G., Drioli, E., 2000. Direct contact membrane distillation: modelling and concentration experiments. J. Membr. Sci. 166, 1–11.

Lalia, B.S., Kochkodan, V., Hashaikeh, R., Hilal, N., 2013. A review on membrane fabrication: structure, properties and performance relationship. Desalination 326, 77–95.

Lawson, K.W., Lloyd, D.R., 1996. Membrane distillation. I. Module design and performance evaluation using vacuum membrane distillation. J. Membr. Sci. 120, 111–121.

Lawson, K.W., Lloyd, D.R., 1997. Membrane distillation. J. Membr. Sci. 124, 1–25.

Leaper, S., Abdel-Karim, A., Faki, B., Luque-Alled, J.M., Alberto, M., Vijayaraghavan, A., Holmes, S.M., Szekely, G., Badawy, M.I., Shokri, N., Gorgojo, P., 2018. Flux-enhanced PVDF mixed matrix membranes incorporating APTS-functionalized graphene oxide for membrane distillation. J. Membr. Sci. 554, 309–323.

Li, B., Sirkar, K.K., 2004. Novel membrane and device for direct contact membrane distillation-based desalination process. Ind. Eng. Chem. Res. 43, 5300–5309.

Li, D.F., Chung, T.S., Wang, R., Liu, Y., 2002. Fabrication of fluoropolyimide/polyethersulfone (PES) dual-layer asymmetric hollow fiber membranes for gas separation. J. Membr. Sci. 198, 211–223.

Li, X., Liu, H., Xiao, C., Ma, S., Zhao, X., 2013. Effect of take-up speed on polyvinylidene fluoride hollow fiber membrane in a thermally induced phase separation process. J. App. Polym. Sci. 128, 1054–1060.

Li, X., Yu, X., Cheng, C., Deng, L., Wang, M., Wang, X., 2015. Electrospun superhydrophobic organic/inorganic composite nanofibrous membranes for membrane distillation. ACS Appl. Mater. Interfaces 7, 21919–21930.

Liu, F., Hashim, N.A., Liu, Y., Moghareh Abed, M.R., Li, K., 2011. Progress in the production and modification of PVDF membranes. J. Membr. Sci. 375, 1–27.

Liu, F., Tao, M.M., Xue, L.X., 2012. PVDF membranes with interconnected pores prepared via a Nat-ips process. Desalination 298, 99–105.

Liu, J., Li, P., Li, Y., Xie, L., Wang, S., Wang, Z., 2009. Preparation of PET threads reinforced PVDF hollow fiber membrane. Desalination 249, 453–457.

Lloyd, D.R., Kim, S.S., Kinzer, K.E., 1991. Microporous membrane formation via thermally-induced phase separation. II. Liquid-liquid phase separation. J. Membr. Sci. 64, 1–11.

Lloyd, D.R., Kinzer, K.E., Tseng, H.S., 1990. Microporous membrane formation via thermally induced phase separation. I. Solid–liquid phase separation. J. Membr. Sci. 52, 239–261.

Lovinger, A.J., 1982. Poly(vinylidene fluoride). In: Bassett, D.C. (Ed.), Development in Crystalline Polymers, Vol. 1. Applied Science, London, pp. 195–261.

Lu, K.J., Zuo, J., Chung, T.S., 2017. Novel PVDF membranes comprising n-butylamine functionalized graphene oxide for direct contact membrane distillation. J. Membr. Sci. 539, 34–42.

Martínez, L., Rodríguez-Maroto, J.M., 2008. Membrane thickness reduction effects on direct contact membrane distillation performance. J. Membr. Sci. 312, 143–156.

Martins, P., Lopes, A.C., Lanceros-Mendez, S., 2014. Electroactive phases of poly(vinylidene fluoride): determination, processing and applications. Prog. Polym. Sci. 39, 683–706.

Matsuyama, H., Teramoto, M., Nakatani, R., Maki, T., 1999. Membrane formation via phase separation induced by penetration of nonsolvent from vapor phase. II. Membrane morphology. J. Appl. Polym. Sci. 74, 171–178.

Matsuyama, H., Yuasa, M., Kitamura, Y., Teramoto, M., Lloyd, D.R., 2000. Structure control of anisotropic and asymmetric polypropylene membrane prepared by thermally induced phase separation. J. Membr. Sci. 179, 91–100.

Meindersma, G.W., Guijt, C.M., Haan, A.B., 2006. Desalination and water recycling by air gap membrane distillation. Desalination 187, 291–301.

Mokhtar, N.M., Lau, W.J., Ismail, A.F., Ng, B.C., 2014. Physicochemical study of polyvinylidene fluoride-cloisite15A® composite membranes for membrane distillation application. RSC Adv 4, 63367–63379.

Mulder, M., 1996. Basic Principles of Membrane Technology. Kluwer Academic Publishers, Dordrecht, The Netherlands.

Naim, R., Ismail, A.F., Mansourizadeh, A., 2012. Effect of non-solvent additives on the structure and performance of PVDF hollow fiber membrane contactor for CO2 stripping. J. Membr. Sci. 423–424, 503–513.

Nalwa, H.S., 1995. Ferroelectric Polymers: Chemistry, Physics, and Applications. CRC Press, Boca Raton.

Nejati, S., Boo, C., Osuji, C.O., Elimelech, M., 2015. Engineering flat sheet microporous PVDF films for membrane distillation. J. Membr. Sci. 492, 355–363.

Ong, Y.K., Chung, T.S., 2012. High performance dual-layer hollow fiber fabricated via novel immiscibility induced phase separation (I2PS) process for dehydration of ethanol. J. Membr. Sci. 421–422, 271–282.

Peng, N., Widjojo, N., Sukitpaneenit, P., Teoh, M.M., Lipscomb, G.G., Chung, T.S., Lai, J.Y., 2012. Evolution of polymeric hollow fibers as sustainable technologies: past, present, and future. Prog. Polym. Sci. 37, 1401–1424.

Pereira, C.C., Nobrega, R., Peinemann, K. V., Borges, C.P., 2003. Hollow fiber membranes obtained by simultaneous spinning of two polymer solutions: a morphological study. J. Membr. Sci. 226, 35–50.

Phattaranawik, J., Jiraratananon, R., 2001. Direct contact membrane distillation: effect of mass transfer on heat transfer. J. Membr. Sci. 188, 137–143.

Prince, J.A., Singh, G., Rana, D., Matsuura, T., Anbharasi, V., Shanmugasundaram, T. S., 2012. Preparation and characterization of highly hydrophobic poly(vinylidene fluoride)-clay nanocomposite nanofiber membranes (PVDF-clay NNMs) for desalination using direct contact membrane distillation. J. Membr. Sci. 397–398, 80–86.

Qtaishat, M., Khayet, M., Matsuura, T., 2009. Guidelines for preparation of higher flux hydrophobic/hydrophilic composite membranes for membrane distillation. J. Membr. Sci. 329, 193–200.

Rajabzadeh, S., Maruyama, T., Sotani, T., Matsuyama, H., 2008. Preparation of PVDF hollow fiber membrane from a ternary polymer/solvent/nonsolvent system via thermally induced phase separation (TIPS) method. Sep. Purif. Technol. 63, 415–423.

Rao, G., Hiibel, S.R., Childress, A.E., 2014. Simplified flux prediction in direct-contact membrane distillation using a membrane structural parameter. Desalination 351, 151–162.

Rezaei, M., Warsinger, D.M., Lienhard, V.J.H., Duke, M.C., Matsuura, T., Samhaber, W. M., 2018. Wetting phenomena in membrane distillation: mechanisms, reversal, and prevention. Water Res 139, 329–352.

Schneider, K., Holz, W., Wollbeck, R., 1988. Membranes and modules for transmembrane distillation. J. Membr. Sci. 39, 25–42.

Shinohara, H., 1979. Fluorination of polyhydrofluoroethylenes. II. Formation of perfluoroalkyl carboxylic acids on the surface region of poly(vinylidene fluoride) film by oxyfluorination, fluorination, and hydrolysis. J. Polym. Sci. A: Polym. Chem. 17, 1543–1556.

Simone, S., Figoli, A., Criscuoli, A., Carnevale, M.C., Rosselli, A., Drioli, E., 2010. Preparation of hollow fibre membranes from PVDF/PVP blends and their applications in VMD. J. Membr. Sci. 364, 219–232.

Singh, D., Sirkar, K.K., 2012. Desalination of brine and produced water by direct contact membrane distillation at high temperatures and pressures. J. Membr. Sci. 389, 380–388.

Smolders, C.A., Reuvers, A.J., Boom, R.M., Wienk, I.M., 1992. Microstructures in phase inversion membranes. Part 1. Formation of macrovoids. J. Membr. Sci. 73, 259–275.

Song, Z., Xing, M., Zhang, J., Li, B., Wang, S., 2012. Determination of phase diagram of a ternary PVDF/γ-BL/DOP system in TIPS process and its application in preparing hollow fiber membranes for membrane distillation. Sep. Purif. Technol. 90, 221–230.

Su, Y., Chen, C., Li, Y., Li, J., 2007. Preparation of PVDF membranes via TIPS method: the effect of mixed diluents on membrane structure and mechanical property. J. Macromol. Sci. A. 44, 305–313.

Sukitpaneenit, P., Chung, T.S., 2009. Molecular elucidation of morphology and mechanical properties of PVDF hollow fiber membranes from aspects of phase inversion, crystallization and rheology. J. Membr. Sci. 340, 192–205.

Sukitpaneenit P., Chung T.S., 2011. Molecular design of the morphology and pore size of PVDF hollow fiber membranes for ethanol-water separation employing the modified pore-flow concept. J. Membr. Sci. 374, 67–82.
Sukitpaneenit, P., Chung, T.S., 2012a. High performance thin-film composite forward osmosis hollow fiber membranes with macrovoid-free and highly porous structure for sustainable water production. Environ. Sci. Technol. 46, 7358–7365.
Sukitpaneenit, P., Chung, T.S., 2012b. PVDF/nanosilica dual-layer hollow fibers with enhanced selectivity and flux as novel membranes for ethanol recovery. Ind. Eng. Chem. Res. 51, 978–993.
Sukitpaneenit P., Chung, T.S., Jiang, L.Y., 2010. Modified pore-flow model for pervaporation mass transport in PVDF hollow fiber membranes for ethanol-water separation. J. Membr. Sci. 362, 393–406.
Tang, Y., Li, N., Liu, A., Ding, S., Yi, C., Liu, H., 2012. Effect of spinning conditions on the structure and performance of hydrophobic PVDF hollow fiber membranes for membrane distillation. Desalination 287, 326–339.
Teoh, M.M., Chung, T.S., 2009a. Membrane distillation with hydrophobic macrovoid-free PVDF–PTFE hollow fiber membranes. Sep. Purif. Technol. 66, 229–236.
Teoh, M.M., Chung, T.S., 2009b. Micelle-like macrovoids in mixed matrix PVDF-PTFE hollow fiber membranes. J. Membr. Sci. 338, 5–10.
Thomas, N., Mavukkandy, M.O., Loutatidou, S., Arafat, H.A., 2017. Membrane distillation research & implementation: lessons from the past five decades. Sep. Purif. Technol. 189, 108–127.
Tomaszewska, M., 1996. Preparation and properties of flat-sheet membranes from poly (vinylidene fluoride) for membrane distillation. Desalination 104, 1–11.
van de Witte, P., Dijkstra, P.J., van den Berg, J.W.A., Feijen, J., 1996. Phase separation processes in polymer solutions in relation to membrane formation. J. Membr. Sci. 117, 1–31.
Wang, D., Li, K., Teo, W.K., 1999. Preparation and characterization of polyvinylidene fluoride (PVDF) hollow fiber membranes. J. Membr. Sci. 163, 211–220.
Wang, D., Li, K., Teo, W.K., 2000. Porous PVDF asymmetric hollow fiber membranes prepared with the use of small molecular additives. J. Membr. Sci. 178, 13–23.
Wang, J.W., Li, L., Zhang, J.W., Xu, X., Chen, C.S., 2016. β-Sialon ceramic hollow fiber membranes with high strength and low thermal conductivity for membrane distillation. J. Eur. Ceram. Soc. 36, 59–65.
Wang, K.Y., Chung, T.S., Gryta, M., 2008. Hydrophobic PVDF hollow fiber membranes with narrow pore size distribution and ultra-thin skin for the fresh water production through membrane distillation. Chem. Eng. Sci. 63, 2587–2594.
Wang, K.Y., Foo, S.W., Chung, T.S., 2009b. Mixed matrix PVDF hollow fiber membranes with nanoscale pores for desalination through direct contact membrane distillation. Ind. Eng. Chem. Res. 48, 4474–4483.
Wang, L.Y., Yong, W.F., Yu, L.E., Chung, T.S., 2017. Design of high efficiency PVDF-PEG hollow fibers for air filtration of ultrafine particles. J. Membr. Sci. 535, 342–349.
Wang, P., Chung, T.S., 2015. Recent advances in membrane distillation processes: membrane development, configuration design and application exploring. J. Membr. Sci. 474, 39–56.
Wang, P., Teoh, M.M., Chung, T.S., 2011. Morphological architecture of dual-layer hollow fiber for membrane distillation with higher desalination performance. Water Res. 45, 5489–5500.
Wang, X., Zhang, L., Sun, D., An, Q., Chen, H., 2009a. Formation mechanism and crystallization of poly(vinylidene fluoride) membrane via immersion precipitation method. Desalination 236, 170–178.

Widjojo, N., Chung, T.S., 2006. Thickness and air gap dependence of macrovoid evolution in phase inversion asymmetric hollow fiber membranes. Ind. Eng. Chem. Res. 45, 7618–7626.

Winter, D., Koschikowski, J., Düver, D., Hertel, P., Beuscher, U., 2013. Evaluation of MD process performance: effect of backing structures and membrane properties under different operating conditions. Desalination 323, 120–133.

Woo, Y.C., Kim, Y., Shim, W.G., Tijing, L.D., Yao, M., Nghiem, L.D., Choi, J.S., Kim, S.H., Shon, H.K., 2016. Graphene/PVDF flat-sheet membrane for the treatment of RO brine from coal seam gas produced water by air gap membrane distillation. J. Membr. Sci. 513, 74–84.

Wu, H.Y., Wang, R., Field, R.W., 2014. Direct contact membrane distillation: an experimental and analytical investigation of the effect of membrane thickness upon transmembrane flux. J. Membr. Sci. 470, 257–265.

Xia, Q.C., Liu, M.L., Cao, X.L., Wang, Y., Xing, W., Sun, S.P., 2018. Structure design and applications of dual-layer polymeric membranes. J. Membr. Sci. 562, 85–111.

Xiao, T., Wang, P., Yang, X., Cai, X., Lu, J., 2015. Fabrication and characterization of novel asymmetric polyvinylidene fluoride (PVDF)membranes by the nonsolvent thermally induced phase separation (NTIPS) method for membrane distillation applications. J. Membr. Sci. 489, 160–174.

Yang, Q., Wang, K. Y., Chung, T. S., 2009. Dual-layer hollow fibers with enhanced flux as novel forward osmosis membranes for water production. Environ. Sci. Technol. 43, 2800–2805.

Yeow, M.L., Liu, Y., Li, K., 2005. Preparation of porous PVDF hollow fibre membrane via a phase inversion method using lithium perchlorate ($LiClO_4$) as an additive. J. Membr. Sci. 268, 16–22.

Young, T.H., Cheng, L.P., Lin, D.J., Fane, L., Chuang, W.Y., 1999. Mechanisms of PVDF membrane formation by immersion-precipitation in soft (1-octanol) and harsh (water) nonsolvents. Polymer 40, 5315–5323.

Yun, Y., Ma, R., Zhang, W., Fane, A.G., Li, J., 2006. Direct contact membrane distillation mechanism for high concentration NaCl solutions. Desalination 188, 251–262.

Zhang, M., Zhang, A.Q., Zhu, B.K., Du, C.H., Xu, Y.Y., 2008. Polymorphism in porous poly(vinylidene fluoride) membranes formed via immersion precipitation process. J. Membr. Sci. 319, 169–175.

Zuo, J., Chung, T.S., O'Brien, G.S., Kosar, W., 2017. Hydrophobic/hydrophilic PVDF/Ultem® dual layer hollow fiber membranes with enhanced mechanical properties for vacuum membrane distillation. J. Membr. Sci. 523, 103–110.

5 Electrospun Nanofiber Membranes for Membrane Distillation

Jiaxin Guo, Bhaskar Jyoti Deka, and Alicia Kyoungjin An

5.1 OVERVIEW OF ELECTROSPINNING TECHNOLOGY

5.1.1 Development of Electrospinning in Early 1900s

Electrospinning, or electrostatic spinning, is a versatile and reliable technique for forming polymer fibers with diameters ranging from micro-to-nanometers via the electrostatic repulsion between surface charges under a high electric field. The idea behind electrospinning was first conceived in 1882 when Rayleigh discovered that when the electrostatic force of droplets exceeded their surface tension they would break and split into a series of charged droplets (Rayleigh, 1882). The electrospinning process was first patented in 1902 as an apparatus for electrically dispersing fluids by J.F. Colley (U.S. patent 692,631). A series of patents followed which improved and modified the electrospinning apparatus, leading to the first commercial production of textile yarn by Anton Formhals in 1934 (U.S. patent 1,975,504).

However, the mechanisms behind electrospinning were still not clearly understood until Taylor's work in the 1960s. Taylor revealed the theoretical underpinning of electrospinning by mathematically modeling the conical shape formed by fluid droplets under the influence of an electric field, which is now known as the Taylor cone, shown in Figure 5.1. Electrospinning slowly gained greater academic attention in the decades since Taylor's pioneering work. Another breakthrough in electrospinning technique was made in the 1990s, when Reneker's research team tested a series of polymer materials and elaborated on the mechanism of one-dimensional (1D) fiber formation during the electrospinning process (Doshi and Reneker, 1993).

From this basic understanding of the stretching and splitting of droplets under an electric field to form a single polymer nanofiber, electrospinning techniques have advanced into a core technology possessing diverse and broad applications. In particular, rapid development of nanotechnology in recent years has led to a boost in the use of electrospinning technology for the fabrication of nanoscale fibers, including inorganic composited hybrid nanofibers, aligned/patterned nanofibers, and electrosprayed thin fibers. At

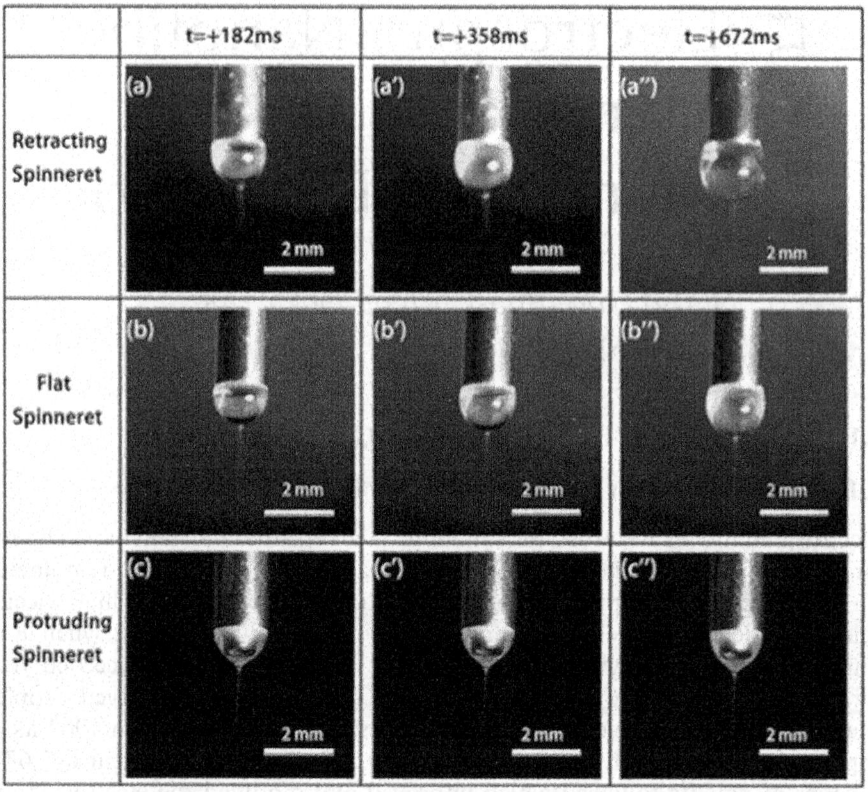

FIGURE 5.1 Formation of the Taylor cone under different spinneret structures and operation time. Reprinted with permission from Li et al. (2018). Copyright (2018) American Chemical Society.

the time of writing, almost 25,000 scientific papers have been published on electrospinning, as evidenced in Figure 5.2.

5.1.2 Electrostatic Spinning Device

Typically, an electrospinning device largely consists of three parts as illustrated in Figure 5.3: a voltage supplier, a syringe (or spinneret), and a collector.

The function of the high voltage power supply, generally providing between 1 and 50 kV, is to create a robust electric field between the electrode and the syringe, so that the charged polymer can be completely stretched. A drop of polymer exits the syringe and starts to form a Taylor cone under the influence of the electric charge. This charged polymer then stretches as it travels toward the collector. Figure 5.4 illustrates the different schematics of the membrane fabrication processes depending on the type of the syringe nozzle (single, coaxial, and dual).

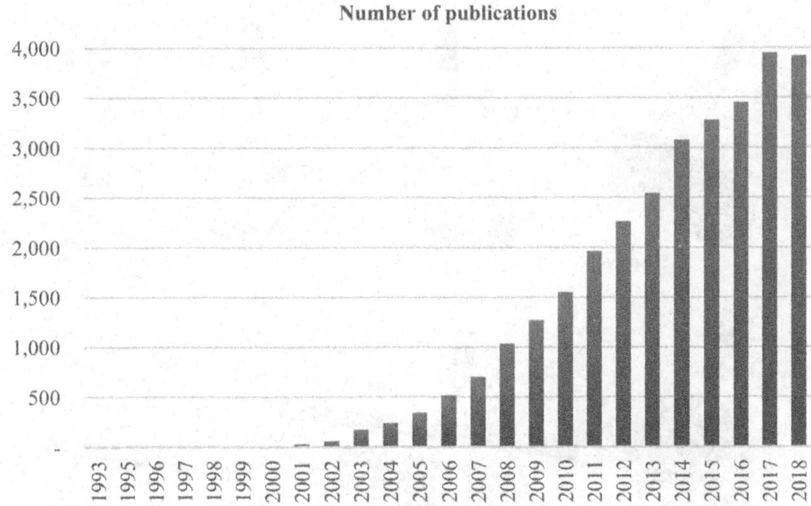

FIGURE 5.2 Publications searched with keywords of 'electrospinning' or 'electrospun' (source: web of science, as at 11 February 2019).

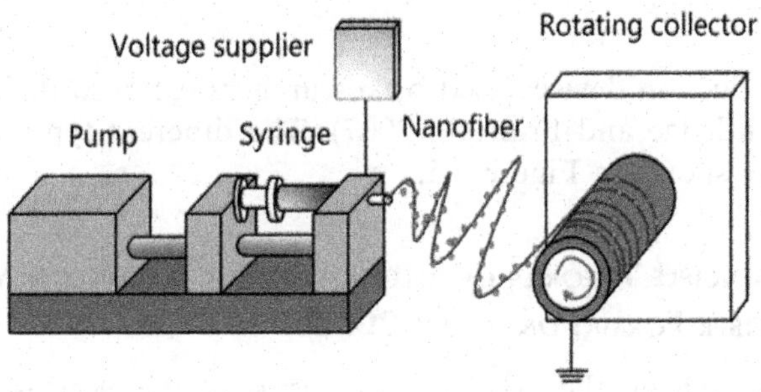

FIGURE 5.3 Single needle electrospinning device. Reprinted with permission from (Wu et al., 2017). Copyright (2017) Elsevier.

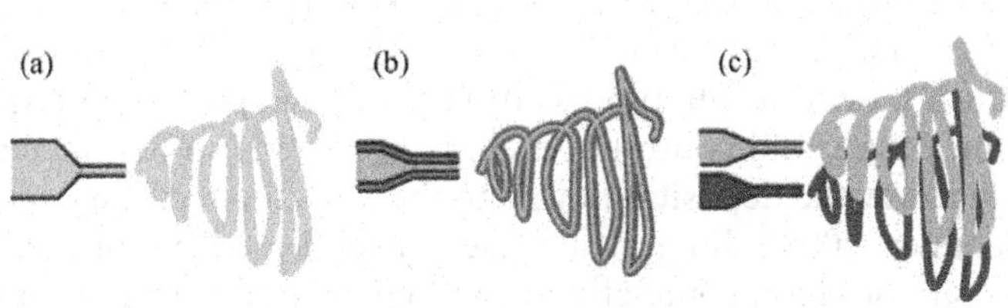

FIGURE 5.4 Different electrospinning collecting schemes. Reprinted with permission from Lee et al. (2017a). Copyright (2017) Elsevier.

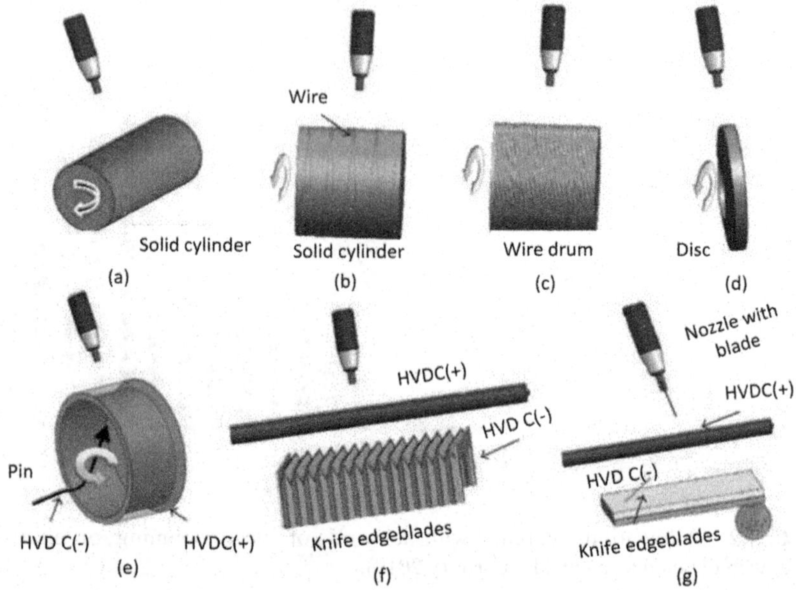

FIGURE 5.5 Different electrospinning collecting schemes (Sahay et al., 2011).

The fiber collection devices used by researchers can also differ (Grafahrend et al., 2010; Rutledge and Fridrikh, 2007). The different types of fiber collection devices are shown in Figure 5.5.

5.1.3 The Principle Theories of Electrospinning Technology for Nanofiber Formation in the 2000s

Theoretical research on the driving mechanisms of the electrospinning process has significantly advanced over the last 20 years. Several experimental and in-depth theoretical studies have been conducted on electrospinning techniques as one of the main preparation methods for nanofiber materials. Chronakis et al. demonstrated the fabrication of electrospun nanofibers and the changes occurring in the jet during the electrospinning process (Chronakis, 2005). Schiffman et al. examined the changes in the morphology, orientation, and deposition of nanofibers during the drawing process to redesign the process in terms of the jet path, receiving device, and control of melting properties so that increased control can be wielded over the deposition of nanofibers on the receiving devices (Schiffman and Schauer, 2008). Jung et al. researched the effects of surface tension, solution viscosity, solution conductivity, and polymer glass transition temperature on fiber shape and size and found that the solution viscosity has the greatest influence (Jung et al., 2005).

Additionally, Reznik extensively studied the evolution of the Taylor cone and the subsequent formation of a fluid jet (Reznik et al., 2004), and Yarin provided in-depth theoretical intuitions on the effects of surface forces on changing the polymer fluids into micro or nanosize of jets that were collected as nanofibers (Reneker and Yarin, 2008). During electrospinning, the droplets usually have a certain static voltage while they remain in the electric field. When a desired voltage is applied to the surface of a polymer, it causes the surface to generate a current (Lukáš et al., 2009). The combination of the forces created by the electric field and the surface tension of the polymer droplet produces an outward force in the opposite direction of this surface tension, enabling the formation of a hemispherical droplet. When these two forces become equal, the droplet becomes suspended at the end of the spinneret. Increasing the strength of the electric field stretches the hemispherical Taylor cone (Yarin et al., 2001), and when the force of the electric field surpasses some critical voltage values, the electric rejection force overpowers the surface tension of the droplets and causes them to elongate into a jet of polymeric nanofibers. As the polymeric nanofibers elongate, the solvents initially used to dissolve them start to evaporate.

When the jet moves from the end of the nozzle to the collector, an acceleration phenomenon occurs, drawing the jet of charged polymeric nanofibers into the electric field to form nonwoven nanofibers on the collector as shown in Figure 5.6.

5.1.4 Important Factors for Electrospinning Nanofibers

Polymer solution characteristics and electrospinning operational parameters are the two broad technological variables responsible for determining the properties of the fibers fabricated. The characteristics of the polymer solutions that affect the properties of the fibers include the solution concentration and viscosity, electrical conductivity, relative molecular mass and

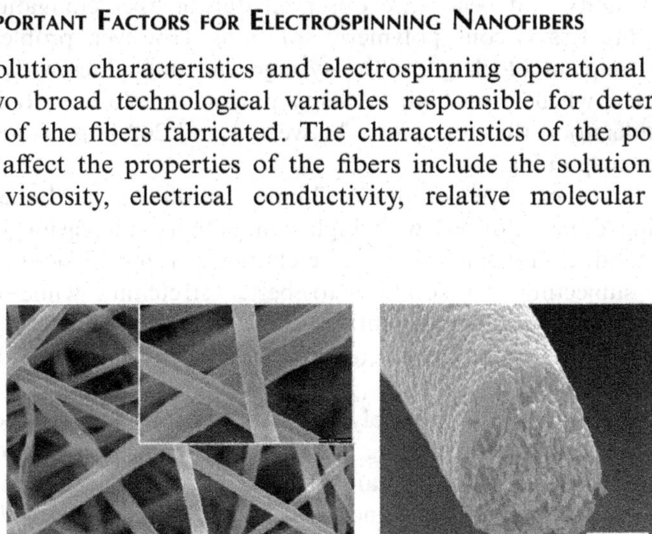

FIGURE 5.6 FESEM images of (left, reprinted with permission from Lee et al. (2016), Copyright (2016) Elsevier). 20% polyvinylidene fluoride-*co*-hexafluoropropylene nanofibers and cross-sectional FESEM images of polystyrene nanofibers (right, reprinted with permission from Wu et al. (2017), Copyright (2017) Elsevier).

structure, and surface tension. Parameters affecting the operational conditions are mainly the strength of the electrostatic field, the spacing between the capillary tube head and the receiving plate, the velocity of the polymer fluid flow, the motion of the receiving plate, and the temperature and humidity of the electrospinning environment (Christopherson et al., 2009; Ding et al., 2002). The following provides a summarized explanation of how the polymer solution's characteristics and the electrospinning parameters can affect the electrospinning process.

(a) Solution concentration: Higher concentrations generally translate to higher viscosity and surface tension as well as a lower splitting ability of the droplets as their surface tension increases after leaving the nozzle. Generally, when other conditions are constant, an increase in polymer concentration will lead to a larger fiber diameter.
(b) Surface tension: The selection of solvents plays a critical role in determining the surface tension. Generally, smoother nanofibers can be fabricated from low surface tension solutions under low applied voltages, while more beads are formed when high surface tension polymeric solutions are used. The solution's surface tension can easily be tailored to make the desired fiber diameter by manipulating the ratios of the solvents.
(c) Viscosity of dope solutions: Viscosity depends upon the polymer's molecular weight and has a substantial influence on fiber morphology. Discontinuity and beads are observed during fiber elongation when employing less viscous polymeric solutions. However, problems such as the frequent blockage of the syringe and difficulties with rotation can occur when using high viscosity polymeric solutions. It is generally understood that a viscosity between 1 and 200 poise is acceptable for electrospinning.
(d) Conductivity of dope solutions: Thinner nanofibers can be fabricated by using dope solutions with high conductivity. Increasing the solution's conductivity can enhance the charge carrying capacity of nanofibers, subjecting the solution to better stretching while traveling through the electric field. Solutions with lower conductivity can result in fibers with insufficient elongation and the formation of beads in the electric field.
(e) Molecular weight of the polymer: As previously explained, other important properties including the viscosity, surface tension, conductivity, and dielectric strength are affected by the molecular weight of the selected polymer. An increase in molecular weight leads to smoother and thicker fiber formation, whereas beads will be formed with low molecular weight polymers.
(f) Fluid flow rate: With a fixed spinneret diameter, the average velocity of the jet is directly proportional to the fiber diameter.
(g) Electric field strength: By increasing the strength of the electrical field, the charge density of the polymer fluid will increase, causing

electrostatic repulsion, and the higher intensity will cause greater acceleration of the jet. These two factors can cause the jet and fiber formation to have greater tensile strength, resulting in a higher tensile strain rate, which is beneficial for making finer fibers.

(h) Syringe nozzle tip and collector distance: As droplets are ejected through the end of the capillary tube, the solvent will immediately start to evaporate upon exposure to air. This causes the polymer to concentrate and solidify into fibers which can be collected by the receiver at the other end. Therefore, the fiber diameter decreases when there is a larger distance between the syringe nozzle and the collector. Suitable distances will provide a sufficient time for the fiber to stretch and dry completely.

(i) Diameter of the nozzle: Different nozzle sizes can be used to control the diameter of the nanofibers. Using a nozzle with a shorter inner diameter in the syringe will reduce the diameter of the electrospun nanofibers. The diameter of the nozzle also determines the amount of polymer solution that can be ejected. If the polymer solution is low in viscosity, it can easily flow out of the nozzle. By contrast, it will be difficult to extrude a high viscosity polymeric solution from a narrow nozzle. Therefore, a suitable nozzle should be selected based on the properties of the dope solution.

(j) Type/shape of the collector: Different types of collectors can lead to different nanofiber arrangements. When a flat type collector is utilized, the resulting nanofibers exhibit random irregularity; when a rotating disk collector is employed, the nanofibers become arranged in parallel. In other words, different electrospinner configurations will generate different fibers.

(k) Ambient parameters: The ambient humidity and temperature during electrospinning operation will affect the nanofiber structure and morphology. In general, higher temperatures will produce thinner fibers while a rise in humidity causes a more porous fiber structure, as the moisture present in air becomes trapped in the relatively hot fibers to form pores. Moreover, a lower evaporation rate under low humidity can cause clogging issues inside the spinneret, because higher water molecule content in the electrospinning chamber can affect the elongation of nanofibers by molecular polarization (Medeiros et al., 2008). Therefore, humidity must be considered when fabricating membranes by electrospinning.

5.1.5 Applications of Electrospun Nanofibers in Water Purification

Today, electrospinning devices and electrospun materials are rapidly undergoing industrial upscaling and commercialization thanks to their easy scale-up potential, high porosity, 3D nanoscaffold structure, and thermal insulation properties of electrospun nanofibers. Evidently, the academic, scientific, and industrial interests in electrospun nanofibers have been continuously growing.

So far, electrospun nanofibers have been applied to diverse area such as air and water filtration (21–24%), tissue engineering and composite reinforcement (15.5–44%), battery (9.7%), drug delivery (7%), smart textile and protective clothing (6.5%), catalysis (6.5%), sensor (5%), wound dressing (6.3%), and so on (Persano et al., 2013; Thavasi et al., 2008).

One of the areas in which the application of electrospun nanofibers has found success is water purification. Electrospun nanofibers' properties make them highly suitable as absorbents and filter materials, and the filtration efficiency of electrospun nanofiber membranes (ENMs) has been reported to be much higher than conventional filtration materials, leading to their recent uses as commercial filters and membranes. Moreover, nanofiber filters not only have increased filtration performance but also have offered a wider range of adaptability toward environmental applications, as their larger surface-area-to-volume ratio enables a high level of contaminant removal from water.

The development and application of electrospun nanofibers over the past ten years has yielded substantial impact on related fields and industries. In particular, electrospun nanofibers have revealed great potential in energy and water treatment applications. The following sections highlight the importance of electrospun membranes in resolving critical shortages of water through the introduction of its application in membrane distillation (MD) processes, to produce alternative sources of water through seawater distillation, wastewater treatment and reclamation, etc.

5.2 ELECTROSPUN NANOFIBER MEMBRANES FOR MEMBRANE DISTILLATION APPLICATIONS

5.2.1 Introduction of Membrane Distillation and Electrospun Membranes

Fresh water, the supply of energy, and a clean environment have been identified as key elements for the sustainable development of any nation. The conversion of seawater or wastewater into fresh water through desalination and waste reuse is a good example of how these three elements are closely associated. Membrane distillation is an emerging technique of rising importance for desalination and water reuse applications which utilizes the vapor pressure across a hydrophobic microporous membrane. MD only allows vapor molecules to transport through the membrane while rejecting the feed's inorganic ions, nonvolatile compounds, and macromolecules, thus guaranteeing a high quality of condensed permeate water (theoretically 99.99% salt rejection).

As a thermally driven process, MD operations are minimally affected by the feed salinity, which is crucial when processing seawater as well as when treating different types of high-concentrate wastewater feeds. In addition, MD possesses the potential to be as energy-efficient as, or even more so than, reverse osmosis technology when waste heat or renewable energy can be integrated, as MD operates at a low temperature (40–80 °C). Albeit, MD didn't receive much attention until the 1980s when commercial membranes (Goretex,

PTFE, Gore & Associated Co.) became available (Khayet, 2011). Since then, MD has gained significant regard from not only the industry but also the academia as evidenced by the growth in the number of published papers. As such, MD is now broadly explored for the treatment of wastewater, oil–water separation, as well as for recovering valuable resource from wastewater, seawater, or brine as shown in Table 5.1.

Researchers have identified the optimal membrane characteristics for MD, namely, high liquid entry pressure (LEP) (hydrophobicity), high permeability (high porosity and high surface-to-volume ratio), less fouling, high chemical stability, and high thermostability. Consequently, researches have been focusing on finding the best membrane morphology and structure which will allow for the satisfaction of these characteristics.

Of the various methods used to improve flux and mitigate wetting/fouling issues in the MD process, *electrospinning* has attracted attention for its effectiveness in achieving greater hydrophobicity, high porosity, and thermal stability (Baji et al., 2010; Tijing et al., 2014). Multiple studies have also noted the suitability of electrospinning over other methods for membrane fabrication.

TABLE 5.1
MD applications in water and wastewater treatment

Feed type	Membrane	Ref.
Apple juice	PVDF	(Gunko et al., 2006)
Seawater and NaCl	PTFE	(Hsu et al., 2002)
Olive mill wastewaters	PTFE/PVDF	(El-Abbassi et al., 2009)
Orange juice	PVDF	(Calabrb and Drioli, 1994)
Acid & base solution	PTFE	(Kimura et al., 1987)
Acetone, ethanol, isopropanol	PTFE	(Bandini and Sarti, 1999)
Dye wastewater	PP	(Mozia et al., 2010)
Dye wastewater	PP	(Qu et al., 2014)
Dye wastewater	PVDF-Cloisite 15A	(Mokhtar et al., 2015b)
Chinese medicine	PTFE	(Ding et al., 2008)
Pharmaceutical waste	CNT+PVDF	(Gethard et al., 2012)
Saline dairy effluent	PTFE+PP	(Kezia et al., 2015)
Rubber industry waste	PVDF-Cloisite 15A	(Mokhtar et al., 2015a)
lithium bromide solution	PTFE	(Sudoh et al., 1997)
Aqueous alcohol solutions	PVDF	(García-Payo et al., 2000)
Concentrated salts	PVDF	(Chan et al., 2005)
Agricultural wastes	PP	(Zarebska et al., 2014)
Synthetic wastewater	PTFE	(El-Bourawi et al., 2007)
Human urine	PTFE	(Qu et al., 2013)
Synthetic wastewater	PTFE	(Xie et al., 2009)
Wastewater of soda	PVDF	(Wu et al., 2015)

For instance, Shang et al. prepared hydrophobic surfaces using silica-based nanomaterials using four different techniques: a sol-gel process, dip-coating, template-based capillary pore structures, and electrospinning. It revealed that the electrospun nonwoven mats exhibited the greatest roughness, thus achieving the highest contact angle (CA) among the four methods (Shang et al., 2005).

To date, polytetrafluoroethylene (PTFE), polyvinylidene fluoride (PVDF), and polypropylene (PP) currently dominate in MD markets due to their inherent hydrophobic nature. Meanwhile, PVDF is most commonly used by researchers for fabricating membranes via electrospinning (Wang and Chung, 2015). Essalhi and Khayet reported that a longer electrospinning time led to increases in both membrane thickness and liquid entry pressure, and the membranes spun from PVDF with a concentration of 25 wt% had the highest direct contact MD performance (Essalhi and Khayet, 2014, 2013). Lalia et al. experimented the effects of various parameters such as PVDF copolymer concentration, additives, and humidity as well as heat-press on membrane properties (Lalia et al., 2013). Liao et al. not only improved membrane integrity and enhanced water flux by the application of heat-press post-treatment to nanofiber PVDF membranes (Liao et al., 2013b) but also achieved higher hydrophobicity by applying surface modifications via incorporation of silver nanoparticles (Liao et al., 2013a) and silica-blended PVDF dopes (Liao et al., 2014b). However, weak mechanical strength of PVDF requires further address for wider MD applications.

5.2.2 Nanofiber Membranes with Enhanced Antifouling and Antiwetting for Membrane Distillation

Despite the increased attention on MD and the developments in MD technology, the industrial scaling-up of the MD process is still limited by its low production capacity and lack of long-term performance data due to the fouling and wetting of membranes during MD operations (Mathioulakis et al., 2007). Thus, in recent years, improvements in MD membranes have mainly been focused on developing methods which enhance flux and reduce specific energy consumption. Studies have found that incorporating carbon nanotubes (CNTs) into polymer membranes improved the mechanical properties of membranes (Escobar and Van der Bruggen, 2015) and promoted the transport of water vapor and gas across membranes (Fujiwara et al., 2001; Gethard et al., 2012), thus enhancing flux. However, most studies used nanoparticles mainly to improve surface hydrophobicity; there was a lack of data on their effects on other factors such as mechanical strength, improved permeability, and resistance for fouling.

Over the last decade, numerous studies have focused on achieving superhydrophobicity in membranes through electrospinning to avert the inevitable problems of membrane wetting and fouling during MD processes (Abdal-hay et al., 2014; Arslan et al., 2016; Asmatulu et al., 2011; Darmanin and

Guittard, 2014; Gleason et al., 2007; Guo et al., 2010; Hu et al., 2016; Li et al., 2014b; Liao et al., 2014a; Lim et al., 2007; Liu et al., 2014; Lu et al., 2008; Muthiah et al., 2010; Park et al., 2010; Patel et al., 2013; Rawal, 2012; Sun et al., 2008; Tang et al., 2009; Wang et al., 2011). They aimed to design the model structure similar to the rough lotus leaf, which relies on trapped air pockets within the membrane structure to impart water repelling capabilities (Quéré, 2008). However, these air pockets can be destroyed by factors such as humidity, temperature, and pressure, which, in turn, can damage the overall architectural structure of the membranes (Bocquet and Lauga, 2011; Daniel et al., 2013; Quéré, 2008). Moreover, these surfaces cannot efficiently repel liquids possessing low surface tensions, making them unsuitable for oil-contaminated water (Bocquet and Lauga, 2011).

To overcome these drawbacks, different studies have been conducted on fabricating microscopic reentrant structures that possess low surface energies to create robust omniphobic surfaces (Lee et al., 2016). Recent studies have successfully fabricated membranes with superior surface omniphobicity (Boo et al., 2016; Guo et al., 2010; Huang et al., 2013; Kim et al., 2012; Lee et al., 2016; Lin et al., 2014; Tuteja et al., 2008, 2007; Yao et al., 2013). Tuteja et al. developed surfaces with reentrant textures that provide strong metastable composite solid–liquid–air interfaces even when treating extremely low surface tension liquids such as pentane ($\gamma/v = 15.7\ \mathrm{mN\ m^{-1}}$) (Tuteja et al., 2008). Elimelech et al. modified a hydrophilic glass fiber membrane utilizing silica nanoparticles, followed by surface fluorination and polymer coating, to fabricate an omniphobic microporous membrane (Lin et al., 2014).

Typically, there are two approaches to overcome the wetting issue in MD applications: enhancing membrane hydrophobicity and reducing the membrane pore size (Gryta and Barancewicz, 2010). Improvements in surface hydrophobicity can have a significant impact on antiwetting properties, since the contact angle is considered to be a quantitative indicator of surface wettability. In terms of MD applications, the water contact angle ought to be as high as possible to ensure superhydrophobic behaviors, that is, the lotus effect, which can bolster the antifouling function of membranes and increases their resistance to pore wetting. Electrospun nanofiber membranes have been reported to have higher hydrophobicity compared to phase inversion membranes.

5.2.2.1 Surface Modifications of Membranes

Modifications of the membrane surface to impart enhanced surface roughness can be achieved by both chemical and physical methods or by reducing the membrane's surface energy to imbue superhydrophobic characteristics. Lee et al. have done significant research on surface modification by changing the chemical composition. They developed a smart approach for modifying commercial PVDF membranes through a surface coating of poly(dimethylsiloxane) (PDMS) microspheres to enhance their hydrophobicity (Figure 5.7) (Lee et al., 2017b).

Liao et al. demonstrated two surface modification methods (integral and surface modification method) using silver nanoparticles and 1-dodecanethiol

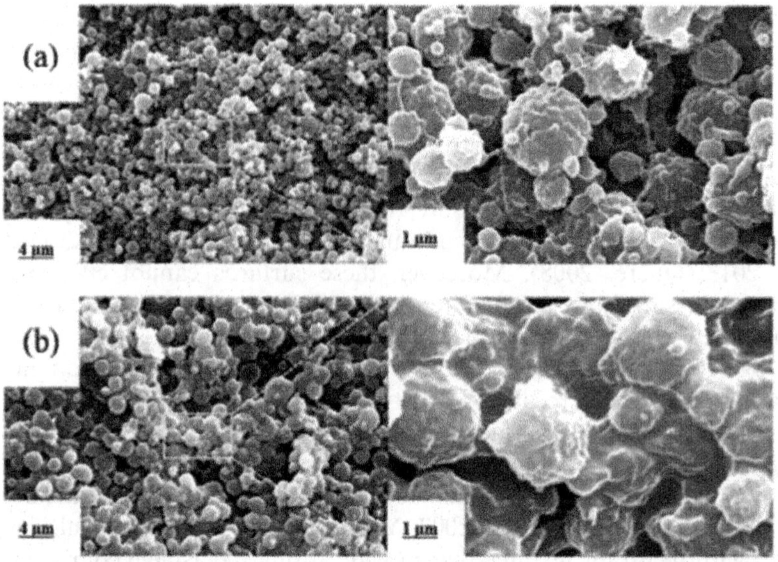

FIGURE 5.7 SEM images of the electrospun membrane coated with the microspheres using PDMS. Reprinted with permission from Lee et al. (2017b). Copyright (2017) American Chemical Society.

on PVDF electrospun nanofiber membranes (Liao et al., 2013a). The SEM images of the PVDF membranes before and after modification are shown in Figure 5.8. After the surface modification by dip-coating, the membrane's contact angle was enhanced from 138 ± 1° to 158 ± 3°, and the LEP value was increased by more than double (0.86 bar), improving the membrane's overall antiwetting ability. In contrast, the integral modification method didn't block the pores significantly and resulted in a much higher flux (31.6 LMH) than the surface modification method (5.4 LMH), making it more suitable for membrane distillation applications.

5.2.2.2 Nanocomposite Electrospun Membranes

Recently, a substantial amount of research has been introduced on fabricating electrospun membranes with nanoparticles to achieve higher functionality (Amarjargal et al., 2012). Lee et al. used TiO_2 nanoparticles with polyvinylidene fluoride-*co*-hexafluoropropylene (PVDF-HFP) to fabricate superhydrophobic membranes, as shown in Figure 5.9, which demonstrated high performance for a long-term MD operation (Lee et al., 2016).

Liao et al. also developed a 3-dimensional (3D) superhydrophobic surface by electrospinning using PVDF with silica nanoparticles (Liao et al., 2014a). The SEM images of the fabricated membranes are shown in Figure 5.10. The surface roughness was enhanced after adding silica nanoparticles, and the

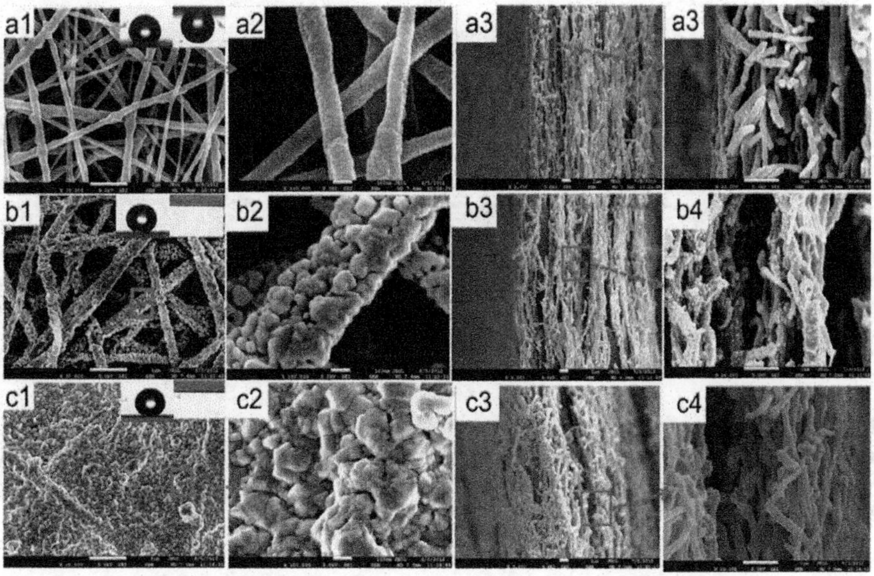

FIGURE 5.8 The surface and cross-section FESEM images of PVDF membranes (a), integrally modified PVDF-silver membrane (b), and surface-modified PVDF-silver membrane (c). Reprinted with permission from Liao et al. (2013a). Copyright (2013) Elsevier.

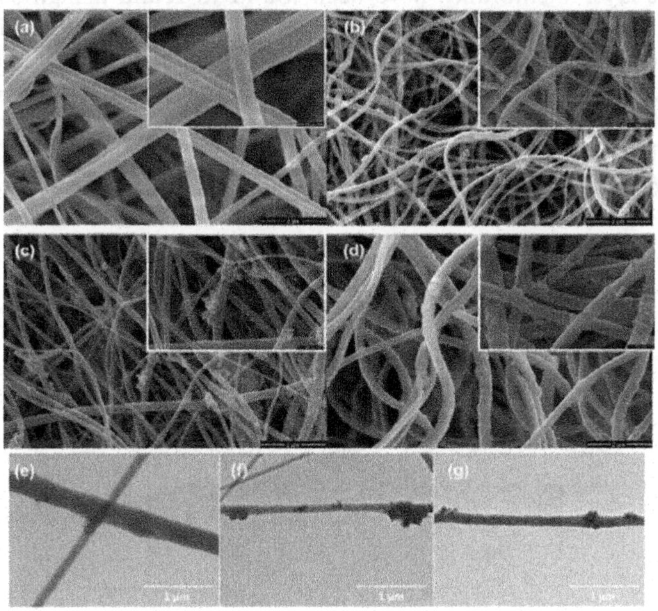

FIGURE 5.9 FESEM images of TiO_2 composited electrospun membranes. Reprinted with permission from Lee et al. (2016). Copyright (2016) Elsevier.

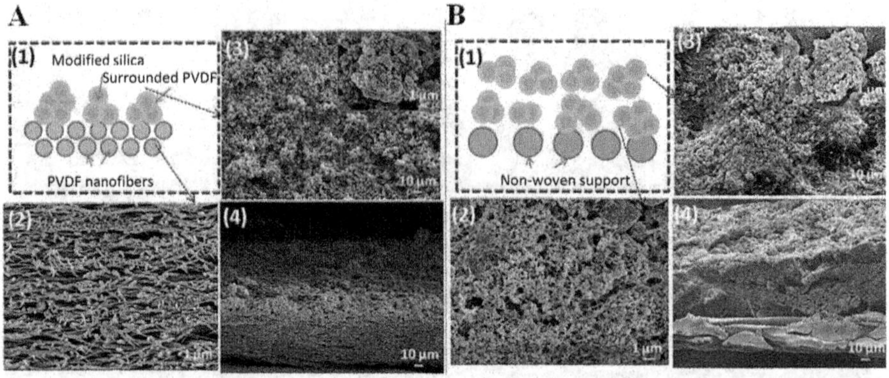

FIGURE 5.10 The surface and cross-section SEM images of 3%PVDF-6%SiO_2 membranes with a PVDF nanofiber support (A), and 5%PVDF-10%SiO_2 membranes with a nonwoven support (B). Reprinted with permission from Liao et al. (2014a). Copyright (2014) American Chemical Society.

surface energy was reduced by the surface modification using a hydrophobic reagent, FS10. The superhydrophobic surface of the 3%PVDF-6%SiO_2 membrane had a water contact angle of 152.4 ± 0.2° and a slide angle below 30°.

5.2.2.3 Multilayer Electrospun Membranes

Novel three-layer membranes have been found to retain a high salt rejection (>99.95%) and stable flux over a long-term testing period (40 hours), demonstrating good overall performance. The SEM images of the three-layer structure with a PET support layer, PVDF-casted middle layer, and electrospun PVDF nanofiber top layer are shown in Figure 5.11. By contrast, the two-layer membranes showed a decline in their salt rejection rate after just 10 hours, which was indicative of pore wetting. Each of the layers in the three-

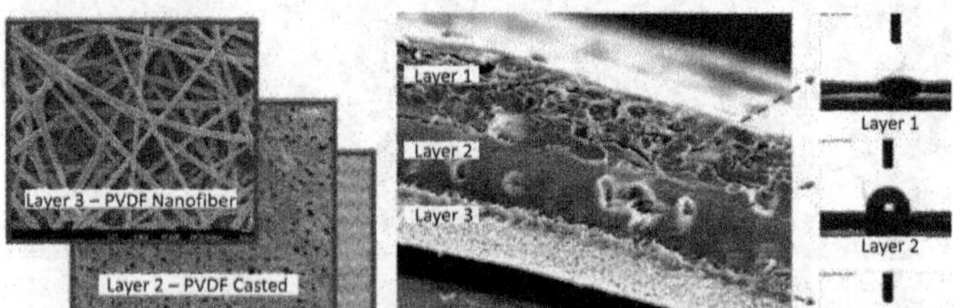

FIGURE 5.11 The surface (a) and cross-section (b) SEM images of the three-layer membrane and water contact angle information (c). Reprinted with permission from Prince et al. (2013). Copyright (2013) Elsevier.

layer membrane had a specific function. The selective nanofiber layer was highly hydrophobic and porous, contributing to lowering the overall heat loss. The small pore size of the intermediate cast film layer enhanced the membrane's LEP, while the PET support layer assisted with the fast transfer of water vapor. The three layers came together in a structure that had a foreground air-gap membrane distillation (AGMD) pattern, which contributed to a higher flux and good desalination performance (Prince et al., 2013).

5.2.2.4 Dual-Biomimetic Nanopapillose Electrospun Membranes

Hierarchical structures found in water-repelling plants, such as the lotus leaf or pitcher plant, have inspired researchers in their attempts to fabricate superhydrophobic membrane surfaces. Dual-biomimetic membranes were also born from this inspiration, mimicking the structure found in nature as shown Figure 5.12. Li et al. studied the fabrication of dual-biomimetic hierarchical polystyrene (PS) superhydrophobic membranes by a one-step electrospinning process where different PS percentages from 10% to 30% were incorporated (Li et al., 2014a). A stronger electric force was achieved by applying a small dope solution injection rate (0.3 mL h^{-1}) with a high voltage (30 kV) to produce the rough and porous nanofibers shown in Figure 5.12 from the unstable electrospinning jet, which enhanced the water contact angle to 150.2 ± 1.2°.

5.2.2.5 Multipolymer Electrospun Membranes

Poly(tetrafluoroethylene) is one of the ideal materials for hydrophobic membrane fabrication due to its great thermal stability, high hydrophobicity, and

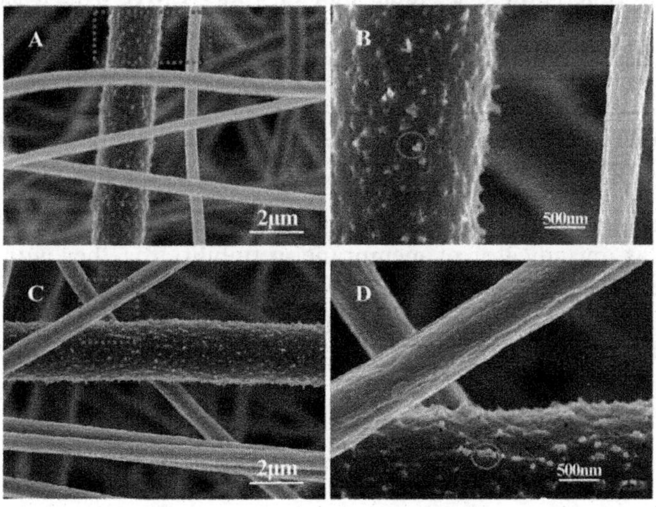

FIGURE 5.12 FESEM images of the 25% PS electrospun membrane with nanopapillose and nonporous morphology. Reprinted with permission from Li et al. (2014a). Copyright (2014) American Chemical Society.

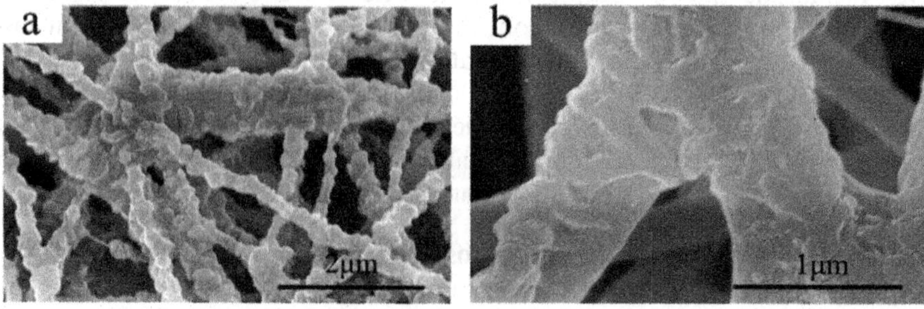

FIGURE 5.13 The SEM images of the ultrafine PTFE/PVA(2:1) and 10% PAN co-electrospun membrane (a: X20000, b: X40000). Reprinted with permission from Huang et al. (2017b). Copyright (2017) Elsevier.

sufficient mechanical strength. However, the difficulty in melting or dissolving PTFE into solutions has been an impediment of fabricating PTFE membranes. Recently, multipolymer electrospinning methods have been adopted to disperse PTFE particles in other polymer dope solutions, which has enabled the creation of PTFE membranes. Huang et al. reported on a fabrication process of poly(vinyl alcohol) (PVA)/PTFE membranes via electrospinning which combined a sintering operation and PAN co-electrospinning to enhance the mechanical strength (Huang et al., 2017b). During the sintering process under 400 °C, the PVA started to decompose while the PTFE aggregated together to become a rough fiber shown in Figure 5.13. The PVA/PTFE membrane showed a stable flux of 14.53 LMH with a salt rejection rate of 99.8% in a vacuum membrane distillation (VMD) using 3.5 wt% NaCl and10 wt% NaOH as feeds.

5.3 CHARACTERIZATIONS OF ELECTROSPUN NANOFIBER MEMBRANES FOR MEMBRANE DISTILLATION

To prevent wetting, membranes should possess a high contact angle (°) and LEP (bar) as an indicator of membrane wettability together with membrane surface free energy (mN m^{-1}). Moreover, the important parameters that determine the permeate flux (J) are porosity, average pore size, and membrane thickness, and their relationship can be expressed as follows (Lawson and Lloyd, 1997):

$$J \propto \frac{\langle \gamma^{\alpha} \rangle \varepsilon}{\delta \tau}$$

where $\langle \gamma^{\alpha} \rangle$is the mean pore size, ε is the membrane's porosity, δ is membrane thickness, and τ is the membrane's tortuosity factor.

5.3.1 Hydrophobicity

Membranes suitable for MD processes possess low thermal conductivity and low mass transfer resistance which help to minimize the heat loss between the feed and permeate sides. High chemical tolerance, especially for acids and bases, and stability at extreme temperatures are also necessary properties for all MD membranes. More than anything, however, hydrophobicity is a vital condition for MD as hydrophobicity impedes liquids from penetrating into membrane pores and allows only water vapor to transport.

The wettability of a membrane surface is a function of the chemical composition and architecture of the membrane and is assessed in terms of CA. In general, MD membranes should be of a hydrophobic or superhydrophobic nature (CA > 150°) to prevent pore wetting. CA can be analyzed by observing the shape of the liquid drop using the contact angle measuring system. This system drops a 5-mL droplet of liquid using a syringe equipped with a thin needle and captures the images of the water drop through a video-digitizer board. CA is then calculated by a geometrical method (sessile drop) based on the drop diameter and the height of the apex (An et al., 2016).

A recent study comparing the hydrophobicity between commercial and electrospun MD membranes in terms of CA confirmed that the CA values of electrospun MD membranes were higher than those of commercial membranes using water and different color dyes. Figure 5.14 shows the order of CA values as follows: electrospun poly(dimethylsiloxane) (E-PDMS) > electrospun polyvinylidene fluoride-*co*-hexafluoropropylene (E-PVDF-HFP) > commercial PVDF membrane (Durapore Membrane Filters, Merck Millipore Ltd.).

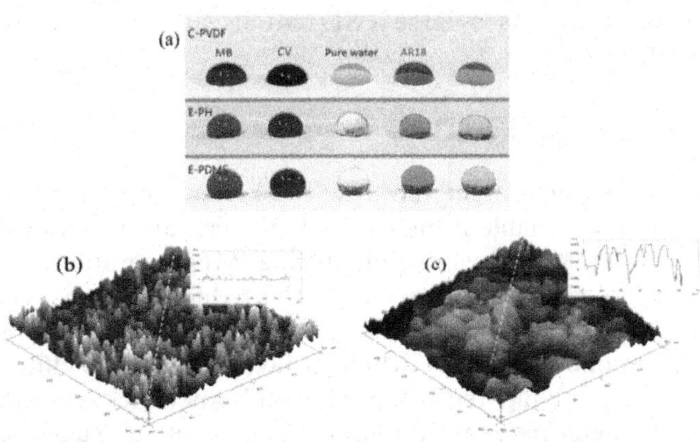

FIGURE 5.14 (a) Water and dye droplets on the surfaces of C-PVDF, E-PVDF-HFP, and E-PDMS and membranes' roughness of (b) C-PVDF and (c) E-PDMS. Reprinted with permission from An et al. (2017). Copyright (2017) Elsevier.

5.3.2 Porosity

Generally, a membrane possessing a higher porosity will have a higher flux and a lower thermal conduction than other membranes. The Smolder–Franken equation (Khayet and Matsuura, 2001) can be used to determine the porosity (ε).

$$\varepsilon = 1 - \frac{\rho_m}{\rho_{pol}} \tag{5.1}$$

where ρ_m is the density of the membrane, and ρ_{pol} is the density of the polymer material. Membrane porosity generally varies from 30% to 85% (El-Bourawi et al., 2006) in MD membranes.

5.3.3 Pore Size and Pore Size Distribution

The diameter of pores found in MD membranes is usually between 0.1 and 1 μm. Two factors need to be considered in determining the appropriate membrane pore size: the pore size must be (1) sufficiently large enough to obtain the desired flux and (2) small enough to prevent wetting.

Several studies have already illustrated the importance of pore size distribution (PSD) for an efficient MD flux (Imdakm and Matsuura, 2005; Khayet et al., 2004; Phattaranawik et al., 2003). However, while PSD is noted as having a leading role in determining MD flux, only a small number of research studies have been carried out on PSD with regards to MD (Woods et al., 2011), and Khayet et al. found that the average pore size rather than PSD can be used to determine a vapor transfer coefficient (Khayet et al., 2004). Generally, it is recommended that the PSD should be minimized as flow mechanisms vary depending on pore size. A large PSD will potentially cause pore wetting, allowing salts to migrate unexpectedly through the membrane as a large PSD entails an uneven LEP across the membrane surface.

5.3.4 Membrane Thickness

In MD applications, the larger the flux, the better the performance. In general, thin membranes enable a high flux; however, thin membranes are more inclined to suffer from pore wetting during MD operations. Thick membranes will lead to a declined flux due to the reduction in mass transfer and additional resistance; at the same time, an increase in membrane thickness can restrict heat conduction loss, which is favorable for the efficiency of MD operations. After studying the effect of membrane thickness and pore size distribution on membrane morphology, Laganà et al. suggested that the optimal membrane thickness for MD is between 30 and 60 μm (Laganà et al., 2000).

5.3.5 Surface Energy and Liquid Entry Pressure

Membrane wetting is a significant challenge which must be addressed before upscaling MD from the lab scale to industrial applications, as long-term MD operations have been continuously reported to suffer from this issue (Gryta, 2005). LEP is the pressure at which a fluid will begin to penetrate into the membrane pores and cause wetting as the fluid migrates through the membrane. Overall, the LEP of a specific membrane is affected by the surface energy of the material, the surface tension of the solution, and the pore size of the membrane. The LEP of a membrane is generally calculated using the following Young–Laplace equation (Alkhudhiri et al., 2012):

$$LEP = \frac{-4B_g \sigma \cos\theta}{d_{max}} \tag{5.2}$$

where B_g is the pore geometric factor, σ is the surface tension of the solution, θ is the angle between the membrane surface and the liquid, and d_{max} is the diameter of the largest pore size. Following the equation, it can be seen that a higher LEP leads to a better MD performance.

In earlier studies, LEP was understood based on the Young–Laplace equation and was described by the capillary model that assumed cylindrical pores with a constant radius of curvature (Washburn, 1921). However, since the capillary model could not accurately describe noncircular pores and tortuous paths or interstitial spaces between the stretch fibrils found in certain types of MD membranes (Guillen-Burrieza et al., 2015), modifications of the capillary model had been suggested (Franken et al., 1987; Purcell, 1949). For membranes with CA greater than 90°, the Purcell model was found to accurately recreate experimentally measured values for LEP, albeit with some limitations. Recently, Servi et al. suggested a new LEP model that modified the Purcell model to successfully describe the observed LEP values over a wide range of hydrophobicity (CA ranging from 63° to 129°) (Servi et al., 2016). Nevertheless, a definitive model for predicting LEP is still lacking, resulting in the preference of experimental measurements over theoretical modeling (Guillen-Burrieza et al., 2015).

5.4 THEORETICAL ANALYSIS OF MD MEMBRANES

5.4.1 Mass Transfer in MD

The following expression determines the required transmembrane vapor flux necessary to drive the mass transfer across the MD membrane (Martínez and Rodríguez-Maroto, 2007; Schofield et al., 1987):

$$J = \frac{C_m}{p_{fm} - p_{pm}} \tag{5.3}$$

where J is the mass transfer flux, C_m is the mass transfer coefficient, and p_{fm} and p_{pm} are the sectional vapor pressures at the vapor–liquid interfaces for the feed and permeate, respectively (Martínez and Rodríguez-Maroto, 2007). Larger temperature differences can be expected to result in higher driving forces which will enhance the permeate flux, as feed temperature and vapor pressure share an exponential relationship (Lawson and Lloyd, 1997).

5.4.2 Heat Transfer in MD

The heat transfer in MD depends on the transport resistance. Heat transfer resistance, when no fouling is present, involves membrane resistance and the resistance of the fluid dynamic boundary layers on the feed side and the permeate side. The presence of a fouling layer causes the boundary layers to provide further thermal resistance in the heat transfer. Figure 5.15 shows the overall mass and heat transfer process.

5.4.3 Temperature Coefficient

Mass and heat transfers occur simultaneously during MD. The temperatures of the boundary layers on the feed side and the permeate side differ from the overall temperature as a result of temperature polarization, which becomes the driving force of the MD process. The changes in this driving force are usually evaluated by determining the temperature coefficient (TPC) as shown in the following equation (Schofield et al., 1987):

$$\text{TPC} = \frac{t_{fm} - t_{pm}}{t_f - t_p} \tag{5.4}$$

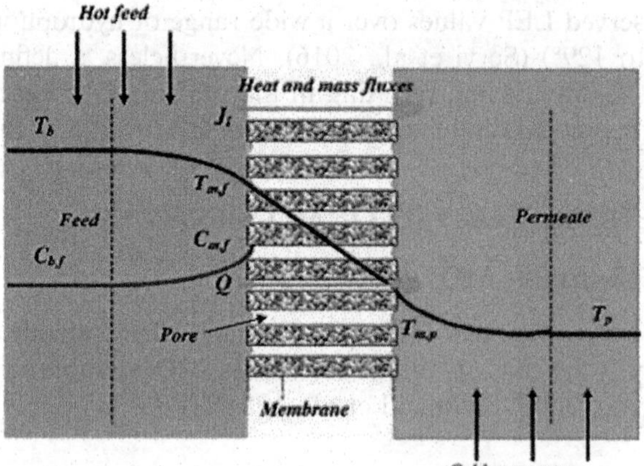

FIGURE 5.15 Demonstration of mass and heat transfer in the MD process. Reprinted with permission from Khayet (2011). Copyright (2011) Elsevier.

where TPC represents the thermal efficiency of a MD system. TPC values close to 1 (one) are indicative of good thermal efficiency, while values close to 0 (zero) indicate that the MD system has poor efficiency. However, TPC is not a direct coefficient of the driving forces in MD, and systems with the same TPC value may not have the same value of driving force (Martínez-Díez and Vázquez-González, 1999).

5.5 RECENT APPLICATIONS OF ENMS FOR MEMBRANE DISTILLATION

ENMs are now widely used in air purification, water and wastewater treatment, and desalination that show their industrial and operational capability to produce reliable permeate quality (Wang et al., 2016). It is expected that technological advances in ENMs, coupled with the use of industrial exhaust streams or waste heat, will allow MD to overcome the limitations that have blocked its industrial applications, such as high capital and operation costs, issues of membrane wetting and fouling, wastewater treatment, and resource recovery (Shannon et al., 2008). The following sections will further discuss various applications of ENMs.

5.5.1 Desalination

The application of ENMs in desalination has triggered a significant amount of research interest in recent years, and as a result, notable developments in membrane fabrication methods have overcome the basic limitations of MD. Yan et al. demonstrated the fabrication of a thin layer of CNTs interlaced over PVDF ENMs using a spraying method as shown in Figure 5.16. The

FIGURE 5.16 The cross-sectional SEM image of CNT deposition over PVDF. Reprinted with permission from Yan et al. (2018). Copyright (2018) Elsevier.

ENMs show superhydrophobic characteristics due the low surface energy of CNTs and a significant water flux of 28.4 kg m^{-2} h^{-1} under a steady VMD performance for more than 26 h (Yan et al., 2018).

In another study, optimal 5 wt% graphene-loaded ENMs were incorporated on a PVDF-HFP-supporting membrane (G5PH) fabricated by electrospinning (Shim et al., 2016). The fabricated G5PH membrane possessed superhydrophobicity (>162°), high porosity (>88%), and high LEP (>186 kPa) compared to the commercial PVDF (pore size 0.22 µm) membrane. Due to its favorable properties, the G5PH membrane achieved a very high salt rejection of 100% and a flux of 22.9 LMH when treating a 3.5 wt% NaCl feed during an AGMD operation.

Essalhi and Khayet demonstrated the desalination of real RO brine in an AGMD operation using lab-made ENMs (Essalhi and Khayet, 2014). Of the different concentrations (15–30 wt%) of PVDF in N,N-dimethyl acetamide and acetone used to fabricate ENMs, the 22.5 wt% PVDF membranes exhibited a superior bead-free fibrous structure and an enhanced deionized (DI) water flux of 12.15×10^{-3} kg m^{-2} s^{-1} and 10.8×10^{-3} kg m^{-2} s^{-1} when treating a 30 g L^{-1} NaCl aqueous feed solution. Woo et al. fabricated superhydrophobic and omniphobic PVDF membranes by electrospinning and CF_4 plasma treatment, which achieved 15.28 LMH in flux with 100% salt rejection even when treating a sodium dodecyl sulfate (SDS) brine feed that had a low surface tension of 0.7 mM (Woo et al., 2017).

Hydrophilic polysulfone (PSF) polymers have been combined with hydrophobic PVDF to fabricate dual-layer ENMs (DL-ENMs) with lower degrees of crystallinity, enthalpy of melting, and crystallization (Khayet et al., 2018). The open structure of PSF increased the interfiber space and void volume fraction of the DL-ENMs to enable very high fluxes of 53.6, 49.9, and 47.7 kg m^{-2} h^{-1} with 99.9% salt rejection in a DCMD operation at feed and permeate temperatures of 80 °C and 20 °C, respectively.

Electrospun PVDF-HFP nanofiber mats modified through a deep infiltration of the PVDF dope solution followed by phase inversion, shown in Figure 5.17, were found to have superior pore size, porosity, morphology, and DCMD performance when compared to a simple E-PVDF-HFP membrane as well as an E-PVDF-HFP membrane modified through a shallow infiltration of the PVDF dope solution followed by phase inversion (Shaulsky et al., 2017).

5.5.2 Wastewater Treatment

Conventional separation technologies are unable to deal with the increasingly complex compositions of organic and inorganic compounds in wastewaters as a result of growing industrialization. For instance, oil–water separation has recently become a major issue due to industrial oily wastewater and other oil-related water pollution. Lee et al. presented an omniphobic ENM for treating oily wastewater, prepared by creating a supporting layer of PVDF-HFP and cationic surfactant (benzyltriethylammonium) which was grafted with negatively charged silica nanoparticles (SiNP) to form multilayer reentrant

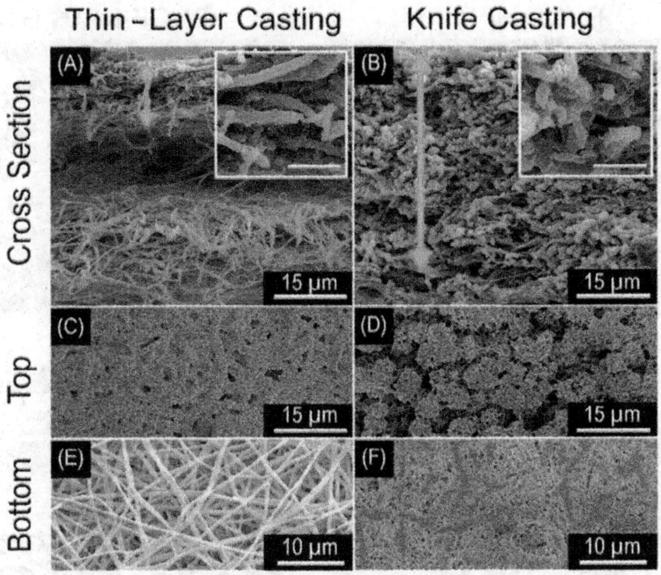

FIGURE 5.17 The SEM images of PVDF coating by thin layer and knife casting over an E-PVDF-HFP membrane showing cross-sectional (A) and (B), top surface (C) and (D), and bottom surface (E) and (F) perspectives of the E-PVDF-HFP membrane. The penetration depth is represented by arrows. Reprinted with permission from Shaulsky et al. (2017). Copyright (2017) Elsevier.

structures via dip-coating method (see Figure 5.18), then fluorinated by (heptadecafluoro-1,1,2,2-tetrahydrodecyl)trichlorosilane (FDTS) (Lee et al., 2016). The omniphobic membrane achieved superhydrophobicity (CA~150°) and retained low surface tension liquids such as mineral oil, decane, and ethanol, showing a stable MD flux of around 12.28 LMH with 100% salt rejection using a feed solution of 1 M NaCl with 0.3 mM SDS over 8 h of MD operation. This flux was comparable to the PVDF-HFP membrane but the latter experienced serious wetting and poor salt rejection even after 2 h of operation.

Liang et al. recently explained the potential application of Janus ENMs (J-ENMs) fabricated by using hydrophilic polyacrylonitrile (PAN) and hydrophobic polystyrene (PS) nanofibers to treat oil–water emulsions (Liang et al., 2019). The J-ENMs had a 1.7-time higher flux than the single-layer membrane as the PS nanofibers prevented the PAN layer from becoming clogged with oil droplets.

Wang et al. illustrated the application of hydrophobic–oleophilic ENMs fabricated with cellulose nanocrystals (CNC) and PVDF for treating water–toluene emulsions (Wang et al., 2019). The ENM consisting of 4 wt% CNC in PVDF showed a higher water CA (Figure 5.19) and an emulsion flux of 5842 LMH (which was four times higher than the commercial PVDF membrane)

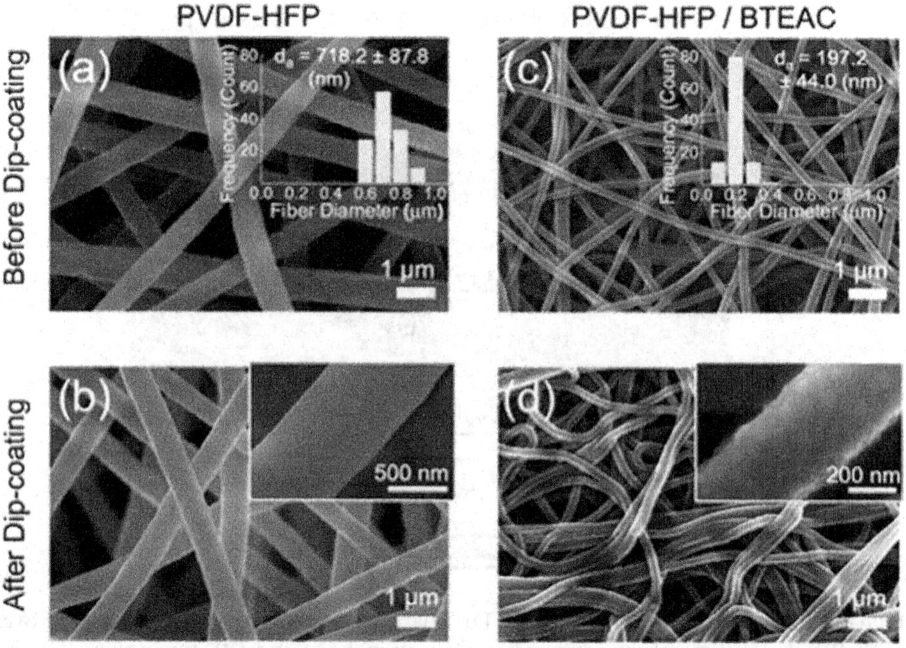

FIGURE 5.18 Role of BTEAC for SiNPs grafting over PVDF-HFP nanofibers. The SEM images present the PVDF-HFP membrane (a) before and (b) after immersion in the SiNP solution; PVDF-HFP /BTEAC nanofibers (c) before and (d) after immersion in the SiNPs solution. Reprinted with permission from Lee et al. (2016). Copyright (2016) American Chemical Society.

with 97% water–toluene separation efficiency. Moreover, the ENM maintained the same flux despite being reused ten times and showed significant separation ability for viscous lubricants and oil as well.

Colored wastewater is another type of wastewater that has received growing concern as a result of rapid growth in industrialization and the consumption of textile goods. A recent study investigated the removal of colored wastewater using clay-based ENMs (Hosseini et al., 2019) made from mixtures of montmorillonite (Mt) chitosan (0, 1.0, 2.0, and 3.0 mass%) and PVA nanocomposites to obtain the optimum pore size, flux, mechanical strength, and porosity. Among the fabricated ENMs, a 2.0 mass% combination showed the best performance in rejecting Basic Blue 41 (BB41) colored wastewater (10 mg L^{-1} of dye concentration at pH 7) with a high flux of 1765 LMH at 0.4 bar.

5.5.3 Other Applications

MD technology has also been applied to the removal of metals from water sources and wastewater. The presence of heavy metals in water presents a severe risk to human health as well as to other living species; however existing

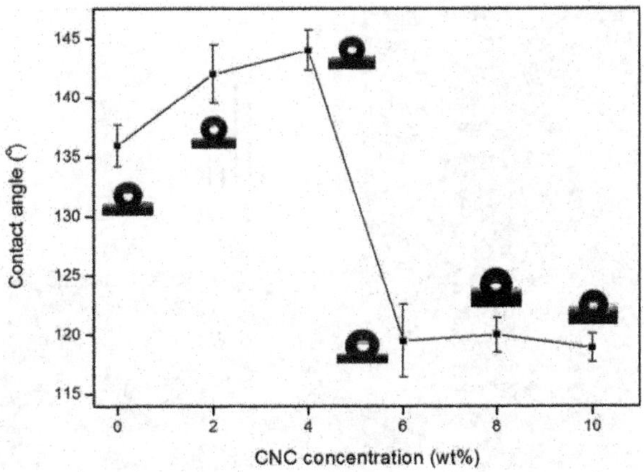

FIGURE 5.19 Water CA of PVDF/CNC nanofiber membranes. Reprinted with permission from Wang et al. (2019). Copyright (2019) Elsevier.

technologies have been unable to fully eliminate this risk. Surface-modified ENMs have shown great potential to treat water containing heavy metal ions. In recent research conducted by Chen et al., ENMs were used to adsorb and remove copper (II) from wastewater (Chen et al., 2018). The novel ENMs used in this study were fabricated by intercalating ethylenediaminetetraacetic acid (EDTA) with layered double hydroxides (LDH), then incorporating with polyacrylonitrile (PAN) to absorb heavy metal ions. The ENMs achieved 95% copper removal from real industrial wastewater collected from Zijin Mining Group Co., Ltd., Shanghang, Fujian Province, China. Recently, Attia et al. fabricated lab-made ENMs comprised of 11 wt% PVDF and 20 wt% of functionalized alumina (Al_2O_3) which achieved a heavy metal rejection of 99.36%, a high flux, a water CA of 150°, and a LEP of 186 kPa as shown in Figure 5.20 (Attia et al., 2017).

ENMs can also be used to absorb and filter semimetals and minerals. Wang et al. demonstrated the application of ENMs fabricated using poly(amic acid) (PAA) and grafted by hyperbranched polyols for the absorption of boron (Wang et al., 2017). The PAA membrane, with its uniform nanofibers with a narrow diameter distribution, achieved a boron uptake of 5.68 mmol g^{-1}, absorbing 0.82 mmol g^{-1} of boron in 15 min from a 5 mg L^{-1} solution and showed a regeneration efficiency of 93.9% even after being reused ten times.

In another recent study, ENMs made of PVDF-polyvinyl pyrrolidone (PVP) and PVDF-PVP-MnO_2 have been shown to remove bisphenol A (BPA) from drinking water (Mohd et al., 2018). Although both types of advanced ENMs possessed superior features such as high CA (>99°), porosity (>97%), flux (>355 LMH), and thickness (>570 μm), the PVDF-PVP-MnO_2 membrane was found to be the better candidate due to its reusability and higher hydrogen bond formation among the BPA molecules, PVP, and MnO_2 nanoparticles.

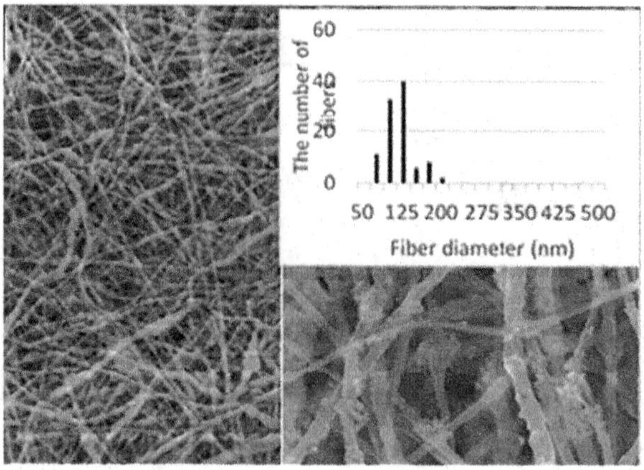

FIGURE 5.20 SEM images of the PVDF-Al_2O_3 electrospun nanofiber membrane. Reprinted with permission from Attia et al. (2017). Copyright (2017) Elsevier.

ENMs are also gaining attention for applications in food processing, especially for removing certain unfavorable compounds. For instance, grapefruit juice, which is rich in vitamin C and phenonic compounds, is a healthier antioxidant drink, but it also has a somewhat bitter taste due to the presence of naringin and limonin (Puri and Banerjee, 2000). Cellulose acetate ENMs have shown effectiveness in debittering grapefruit juice by removing the naringin (22.72%) and limonin (60.71%) (Huang et al., 2017a).

5.6 CONCLUSION

The simplicity and versatility of the electrospinning technique has paved the way for the fabrication of novel nanofibers whose properties can be easily fitted to meet specific technological needs with the potential for industrial scale-up. The growing interest in electrospun nanofibers among the academia and the industry has led to its broad applications in diverse fields that require filtration, material reinforcement, and protection. In particular, electrospun nanofiber membranes have gained significant attention as one of the key solutions for upgrading and upscaling MD processes for treating water and wastewater, due to their comparatively high flux, thermal resistance, and improved antifouling and antiwetting properties.

REFERENCES

Abdal-hay, A., Mousa, H.M., Khan, A., Vanegas, P., Lim, J.H., 2014. TiO_2 nanorods coated onto nylon 6 nanofibers using hydrothermal treatment with improved mechanical properties. Colloids Surfaces A Physicochem. Eng. Asp. 457, 275–281. doi:10.1016/j.colsurfa.2014.05.058

Alkhudhiri, A., Darwish, N., Hilal, N., 2012. Membrane distillation: A comprehensive review. Desalination. 287, 2–18. doi:10.1016/j.desal.2011.08.027

Amarjargal, A., Tijing, L.D., Ruelo, M.T.G., Lee, D.H., Kim, C.S., 2012. Facile synthesis and immobilization of Ag–TiO_2 nanoparticles on electrospun PU nanofibers by polyol technique and simple immersion. Mater. Chem. Phys. 135, 277–281. doi:10.1016/J.MATCHEMPHYS.2012.05.078

An, A.K., Guo, J., Jeong, S., Lee, E.J., Tabatabai, S.A.A., Leiknes, T.O., 2016. High flux and antifouling properties of negatively charged membrane for dyeing wastewater treatment by membrane distillation. Water Res. 103, 362–371. doi:10.1016/j.watres.2016.07.060

An, A.K., Guo, J., Lee, E.-J., Jeong, S., Zhao, Y., Wang, Z., Leiknes, T., 2017. PDMS/PVDF hybrid electrospun membrane with superhydrophobic property and drop impact dynamics for dyeing wastewater treatment using membrane distillation. J. Memb. Sci. 525, 57–67. doi:10.1016/j.memsci.2016.10.028

Arslan, O., Aytac, Z., Uyar, T., 2016. Superhydrophobic, hybrid, electrospun cellulose acetate nanofibrous mats for oil/water separation by tailored surface modification. ACS Appl. Mater. Interfaces. 8, 19747–19754. doi:10.1021/acsami.6b05429

Asmatulu, R., Ceylan, M., Nuraje, N., 2011. Study of superhydrophobic electrospun nanocomposite fibers for energy systems. Langmuir. 27, 504–507. doi:10.1021/la103661c

Attia, H., Alexander, S., Wright, C.J., Hilal, N., 2017. Superhydrophobic electrospun membrane for heavy metals removal by air gap membrane distillation (AGMD). Desalination 420, 318–329. doi:10.1016/j.desal.2017.07.022

Baji, A., Mai, Y.W., Wong, S.C., Abtahi, M., Chen, P., 2010. Electrospinning of polymer nanofibers: Effects on oriented morphology, structures and tensile properties. Compos. Sci. Technol. 70, 703–718. doi:10.1016/j.compscitech.2010.01.010

Bandini, S., Sarti, G.C., 1999. Heat and mass transport resistances in vacuum membrane distillation per drop. Aiche J. 45, 1422–1433. doi:10.1002/aic.690450707

Bocquet, L., Lauga, E., 2011. A smooth future? Nat. Mater. 10, 334–337. doi:10.1038/nmat2994

Boo, C., Lee, J., Elimelech, M., 2016. Omniphobic Polyvinylidene Fluoride (PVDF) membrane for desalination of shale gas produced water by membrane distillation. Environ. Sci. Technol. 50, 12275–12282. doi:10.1021/acs.est.6b03882

Calabrb, V., Drioli, E., 1994. Theoretical and experimental study on membrane distillation in the concentration of orange juice. Ind. Eng. Chem. Res. 33, 1803–1808. doi:10.1021/ie00031a020

Chan, M.T., Fane, A.G., Matheickal, J.T., Sheikholeslami, R., 2005. Membrane distillation crystallization of concentrated salts – Flux and crystal formation. J. Memb. Sci. 257, 144–155. doi:10.1016/j.memsci.2004.09.051

Chen, H., Lin, J., Zhang, N., Chen, L., Zhong, S., Wang, Y., Zhang, W., Ling, Q., 2018. Preparation of MgAl-EDTA-LDH based electrospun nanofiber membrane and its adsorption properties of copper(II) from wastewater. J. Hazard. Mater. 345, 1–9. doi:10.1016/j.jhazmat.2017.11.002

Christopherson, G.T., Song, H., Mao, H.Q., 2009. The influence of fiber diameter of electrospun substrates on neural stem cell differentiation and proliferation. Biomaterials. 30, 556–564. doi:10.1016/j.biomaterials.2008.10.004

Chronakis, I.S., 2005. Novel nanocomposites and nanoceramics based on polymer nanofibers using electrospinning process – A review. J. Mater. Process. Technol. 167, 283–293. doi:10.1016/j.jmatprotec.2005.06.053

Daniel, D., Mankin, M.N., Belisle, R. A., Wong, T.S., Aizenberg, J., 2013. Lubricant-infused micro/nano-structured surfaces with tunable dynamic omniphobicity at high temperatures. Appl. Phys. Lett. 102. doi:10.1063/1.4810907

Darmanin, T., Guittard, F., 2014. Homogeneous growth of conducting polymer nanofibers by electrodeposition for superhydrophobic and superoleophilic stainless steel meshes. RSC Adv. 4, 50401–50405. doi:10.1039/C4RA08938C

Ding, B., Kim, H.Y., Lee, S.C., Lee, D.R., Choi, K.J., 2002. Preparation and characterization of nanoscaled polyvinyl alcohol fibers via electrospinning. Fibers Polym. 3, 73–79. doi:10.1007/BF02875403

Ding, Z., Liu, L., Yu, J., Ma, R., Yang, Z., 2008. Concentrating the extract of traditional Chinese medicine by direct contact membrane distillation. J. Memb. Sci. 310, 539–549. doi:10.1016/j.memsci.2007.11.036

Doshi, J., Reneker, D.H., 1993. Electrospinning process and applications of electrospun fibers. Conf. Rec.1993 IEEE Ind. Appl. Conf. Twenty-Eighth IAS Annu. Meet. 35, 151–160. doi:10.1109/IAS.1993.299067

El-Abbassi, A., Hafidi, A., García-Payo, M.C., Khayet, M., 2009. Concentration of olive mill wastewater by membrane distillation for polyphenols recovery. Desalination. 245, 670–674. doi:10.1016/j.desal.2009.02.035

El-Bourawi, M.S., Ding, Z., Ma, R., Khayet, M., 2006. A framework for better understanding membrane distillation separation process. J. Memb. Sci. 285, 4–29. doi:10.1016/j.memsci.2006.08.002

El-Bourawi, M.S., Khayet, M., Ma, R., Ding, Z., Li, Z., Zhang, X., 2007. Application of vacuum membrane distillation for ammonia removal. J. Memb. Sci. 301, 200–209. doi:10.1016/j.memsci.2007.06.021

Escobar, I.C., Van der Bruggen, B., 2015. Microfiltration and ultrafiltration membrane science and technology. J. Appl. Polym. Sci. 132, n/a–n/a. doi:10.1002/ap.42002

Essalhi, M., Khayet, M., 2013. Self-sustained webs of polyvinylidene fluoride electrospun nanofibers at different electrospinning times: 1. Desalination by direct contact membrane distillation. J. Memb. Sci. 433, 167–179. doi:10.1016/j.memsci.2013.11.056

Essalhi, M., Khayet, M., 2014. Self-sustained webs of polyvinylidene fluoride electrospun nano-fibers: Effects of polymer concentration and desalination by direct contact membrane distillation. J. Memb. Sci. 454, 133–143. doi:10.1016/j.memsci.2013.11.056

Franken, A.C.M., Nolten, J.A.M., Mulder, M.H.V., Bargeman, D., Smolders, C.A., 1987. Wetting criteria for the applicability of membrane distillation. J. Memb. Sci. 33, 315–328. doi:10.1016/S0376-7388(00)80288-4

Fujiwara, A., Ishii, K., Suematsu, H., Kataura, H., Maniwa, Y., Suzuki, S., Achiba, Y., 2001. Gas adsorption in the inside and outside of single-walled carbon nanotubes. Chem. Phys. Lett. 336, 205–211. doi:10.1016/S0009-2614(01)00111-7

García-Payo, M.C., Izquierdo-Gil, M. A., Fernández-Pineda, C., 2000. Air gap membrane distillation of aqueous alcohol solutions. J. Memb. Sci. 169, 61–80. doi:10.1016/S0376-7388(99)00326-9

Gethard, K., Sae-Khow, O., Mitra, S., 2012. Carbon nanotube enhanced membrane distillation for simultaneous generation of pure water and concentrating pharmaceutical waste. Sep. Purif. Technol. 90, 239–245. doi:10.1016/j.seppur.2012.02.042

Gleason, K.K., Rutledge, G.C., Gupta, M., Ma, M., Mao, Y., 2007. Superhydrophobic fibers produced by electrospinning and chemical vapor deposition. U.S. Pat. Appl. Publ. 226.US 7.651,760 B2

Grafahrend, D., Jungbecker, P., Seide, G., Leonards, H., Gries, T., Moller, M., Klee, D., 2010. Development and optimization of an electrospraying device for the continuous collection of nano- and microparticles~!2009-12-01~!2010-01-29~!2010-04-14~! Open Chem. Biomed. Methods J. 3, 1–9. doi:10.2174/1875038901003010001

Gryta, M., 2005. Long-term performance of membrane distillation process. J. Memb. Sci. 265, 153–159. doi:10.1016/J.MEMSCI.2005.04.049

Gryta, M., Barancewicz, M., 2010. Influence of morphology of PVDF capillary membranes on the performance of direct contact membrane distillation. J. Memb. Sci. 358, 158–167. doi:10.1016/j.memsci.2010.04.044

Guillen-Burrieza, E., Servi, A., Lalia, B.S., Arafat, H.A., 2015. Membrane structure and surface morphology impact on the wetting of MD membranes. J. Memb. Sci. 483, 94–103. doi:10.1016/j.memsci.2015.02.024

Gunko, S., Verbych, S., Bryk, M., Hilal, N., 2006. Concentration of apple juice using direct contact membrane distillation. Desalination. 190, 117–124. doi:10.1016/j.desal.2005.09.001

Guo, M., Ding, B., Li, X., Wang, X., Yu, J., Wang, M., 2010. Amphiphobic nanofibrous silica mats with flexible and high-heat-resistant properties. J. Phys. Chem. C. 114, 916–921. doi:10.1021/jp909672r

Hosseini, S.A., Vossoughi, M., Mahmoodi, N.M., Sadrzadeh, M., 2019. Clay-based electrospun nanofibrous membranes for colored wastewater treatment. Appl. Clay Sci. 168, 77–86. doi:10.1016/j.clay.2018.11.003

Hsu, S.T., Cheng, K.T., Chiou, J.S., 2002. Seawater desalination by direct contact membrane distillation. Desalination. 143, 279–287. doi:10.1016/S0011-9164(02)00266-7

Hu, L., Zhang, S., Zhang, Y., Li, B., 2016. Supplemental material Flexible nanofiber-based membrane with highly superhydrophobic pinning property. J. Colloid Interface Sci. 472, 3–5. doi:10.1016/j.jcis.2016.03.056

Huang, W., Zhan, Y., Shi, X., Chen, J., Deng, H., Du, Y., 2017a. Controllable immobilization of naringinase on electrospun cellulose acetate nanofibers and their application to juice debittering. Int. J. Biol. Macromol. 98, 630–636. doi:10.1016/j.ijbiomac.2017.02.018

Huang, X., Chrisman, J.D., Zacharia, N.S., 2013. Omniphobic slippery coatings based on lubricant-infused porous polyelectrolyte multilayers. ACS Macro Lett. 2, 826–829. doi:10.1021/mz400387w

Huang, Y., Huang, Q., Liu, H., Zhang, C., You, Y., Li, N., Xiao, C., 2017b. Preparation, characterization, and applications of electrospun ultrafine fibrous PTFE porous membranes. J. Memb. Sci. 523, 317–326. doi:10.1016/j.memsci.2016.10.019

Imdakm, A.O., Matsuura, T., 2005. Simulation of heat and mass transfer in direct contact membrane distillation (MD): The effect of membrane physical properties. J. Memb. Sci. 262, 117–128. doi:10.1016/J.MEMSCI.2005.05.026

Jung, Y., Soo, Y., Park, C., 2005. Micropatterning of semicrystalline poly (vinylidene fluoride) (PVDF) solutions. Eur. Poly. J. 41, 1002–1012. doi:10.1016/j.eurpolymj.2004.11.022

Kezia, K., Lee, J., Weeks, M., Kentish, S., 2015. Direct contact membrane distillation for the concentration of saline dairy effluent. Water Res. 81, 167–177. doi:10.1016/j.watres.2015.05.042

Khayet, M., 2011. Membranes and theoretical modeling of membrane distillation: A review. Adv. Colloid Interface Sci. 164, 56–88. doi:10.1016/j.cis.2010.09.005

Khayet, M., García-Payo, M.C., García-Fernández, L., Contreras-Martínez, J., 2018. Dual-layered electrospun nanofibrous membranes for membrane distillation. Desalination. 426, 174–184. doi:10.1016/j.desal.2017.10.036

Khayet, M., Matsuura, T., 2001. Preparation and characterization of polyvinylidene fluoride membranes for membrane distillation. Ind. Eng. Chem. Res. 40, 5710–5718. doi:10.1021/ie010553y

Khayet, M., Velázquez, A., Mengual, J.I., 2004. Modelling mass transport through a porous partition: Effect of pore size distribution. J. Non-Equilibrium Thermodyn. 29, 279–299. doi:10.1515/JNETDY.2004.055

Kim, P., Wong, T.-S., Alvarenga, J., Kreder, M.J., Adorno-martinez, W.E., Aizenberg, J., Al, K.I.M.E.T., 2012. Liquid-infused nanostructured surfaces with extreme anti-ice and anti-frost performance. ACS Nano. 6, 6569–6577. doi:10.1021/nn302310q

Kimura, S., Nakao, S.-I., Shimatani, S.-I., 1987. Transport phenomena in membrane distillation. J. Memb. Sci. 33. doi:10.1016/S0376-7388(00)80286-0

Laganà, F., Barbieri, G., Drioli, E., 2000. Direct contact membrane distillation: Modelling and concentration experiments. J. Memb. Sci. 166, 1–11. doi:10.1016/S0376-7388(99)00234-3

Lalia, B.S., Guillen-Burrieza, E., Arafat, H.A., Hashaikeh, R., 2013. Fabrication and characterization of polyvinylidenefluoride-co-hexafluoropropylene (PVDF-HFP) electrospun membranes for direct contact membrane distillation. J. Memb. Sci. 428, 104–115. doi:10.1016/j.memsci.2012.10.061

Lawson, K.W., Lloyd, D.R., 1997. Membrane distillation. J. Mem. Sci. 124, 1–25. doi:10.1007/s00216-011-4733-9

Lee, E.-J., An, A.K., Hadi, P., Lee, S., Woo, Y.C., Shon, H.K., 2017a. Advanced multi-nozzle electrospun functionalized titanium dioxide/polyvinylidene fluoride-co-hexafluoropropylene (TiO_2/PVDF-HFP) composite membranes for direct contact membrane distillation. J. Memb. Sci. 524, 712–720. doi:10.1016/j.memsci.2016.11.069

Lee, E.-J., An, A.K., He, T., Woo, Y.C., Shon, H.K., 2016. Electrospun nanofiber membranes incorporating fluorosilane-coated TiO_2 nanocomposite for direct contact membrane distillation. J. Memb. Sci. 520, 145–154. doi:10.1016/j.memsci.2016.07.019

Lee, E.-J., Deka, B.J., Guo, J., Woo, Y.C., Shon, H.K., An, A.K., 2017b. Engineering the re-entrant hierarchy and surface energy of PDMS-PVDF membrane for membrane distillation using a facile and benign microsphere coating. Environ. Sci. Technol. 51, 10117–10126. doi:10.1021/acs.est.7b01108

Lee, J., Boo, C., Ryu, W.H., Taylor, A.D., Elimelech, M., 2016b. Development of omniphobic desalination membranes using a charged electrospun nanofiber scaffold. ACS Appl. Mater. Interfaces. 8, 11154–11161. doi:10.1021/acsami.6b02419

Li, M., Zheng, Y., Xin, B., Xu, Y., 2018. Roles of coaxial spinneret in taylor cone and morphology of core − shell fibers. Ind. Eng. Chem. Res. 57, 17310–17317. doi:10.1021/acs.iecr.8b04341

Li, X., Wang, C., Yang, Y., Wang, X., Zhu, M., Hsiao, B.S., 2014a. Dual-biomimetic superhydrophobic electrospun polystyrene nano fi brous membranes for membrane distillation. ACS Appl. Mater. Interfaces. doi:10.1021/am4048128

Li, X., Yu, X., Cheng, C., Deng, L., Wang, M., Wang, X., 2014b. Electrospun superhydrophobic organic/inorganic composite nano fi brous membranes for membrane distillation. Appl. Mater. Interfaces. 7, 21919–21930. doi:10.1021/acsami.5b06509

Liang, Y., Kim, S., Kallem, P., Choi, H., 2019. Capillary effect in Janus electrospun nanofiber membrane for oil/water emulsion separation. Chemosphere. 221, 479–485. doi:10.1016/j.chemosphere.2019.01.048

Liao, Y., Loh, C.-H.H., Wang, R., Fane, A.G., Fane, T., 2014a. Electrospun superhydrophobic membranes with unique structures for membrane distillation. ACS Appl. Mater. Interfaces. 6, 16035–16048. doi:10.1021/am503968n

Liao, Y., Wang, R., Fane, A.G., 2013a. Engineering superhydrophobic surface on poly (vinylidene fluoride) nanofiber membranes for direct contact membrane distillation. J. Memb. Sci. 440, 77–87. doi:10.1016/j.memsci.2013.04.006

Liao, Y., Wang, R., Fane, A.G., 2014b. Fabrication of bioinspired composite nanofiber membranes with robust superhydrophobicity for direct contact membrane distillation. Environ. Sci. Technol. 48, 6335–6341. doi:10.1021/es405795s

Liao, Y., Wang, R., Tian, M., Qiu, C., Fane, A.G., 2013b. Fabrication of polyvinylidene fluoride (PVDF) nanofiber membranes by electro-spinning for direct contact membrane distillation. J. Memb. Sci. 425–426, 30–39. doi:10.1016/j.memsci.2012.09.023

Lim, J.M., Yi, G.R., Moon, J.H., Heo, C.J., Yang, S.M., 2007. Superhydrophobic films of electrospun fibers with multiple-scale surface morphology. Langmuir. 23, 7981–7989. doi:10.1021/la700392w

Lin, S., Nejati, S., Boo, C., Hu, Y., Osuji, C.O., Elimelech, M., 2014. Omniphobic membrane for robust membrane distillation. Environ. Sci. Technol. Lett. 1, 443–447. doi:10.1021/ez500267p

Liu, H., Cao, C.-Y., Wei, F.-F., Huang, P.-P., Sun, Y.-B., Jiang, L., Song, W.-G., 2014. Flexible macroporous carbon nanofiber film with high oil adsorption capacity. J. Mater. Chem. A. 2, 3557. doi:10.1039/c3ta14468b

Lu, X., Zhou, J., Zhao, Y., Qiu, Y., Li, J., 2008. Room temperature ionic liquid based polystyrene nanofibers with superhydrophobicity and conductivity produced by electrospinning. Chem. Mater. 20, 3420–3424. doi:10.1021/cm800045h

Lukáš, D., Sarkar, A., Martinová, L., Vodsed'álková, K., Lubasová, D., Chaloupek, J., Pokorný, P., Mikeš, P., Chvojka, J., Komárek, M., 2009. Physical principles of electrospinning (electrospinning as a nano-scale technology of the twenty-first century). Text. Prog. 41, 59–140. doi:10.1080/00405160902904641

Martínez, L., Rodríguez-Maroto, J.M., 2007. On transport resistances in direct contact membrane distillation. J. Memb. Sci. 295, 28–39. doi:10.1016/J.MEMSCI.2007.02.029

Martínez-Díez, L., Vázquez-González, M., 1999. Temperature and concentration polarization in membrane distillation of aqueous salt solutions. J. Memb. Sci. 156, 265–273. doi:10.1016/S0376-7388(98)00349-4

Mathioulakis, E., Belessiotis, V., Delyannis, E., 2007. Desalination by using alternative energy: Review and state-of-the-art. EuroMed 2006 Conf. Desalin. Strateg. South Mediterr. Ctries. 203, 346–365. doi:10.1016/j.desal.2006.03.531

Medeiros, E.S., Mattoso, L.H.C., Offeman, R.D., Wood, D.F., Orts, W.J., 2008. Effect of relative humidity on the morphology of electrospun polymer fibers. Can. J. Chem. 86, 590–599. doi:10.1139/v08-029

Mohd, A., Wen, C., Sathishkumar, P., Rahim, A., Yuso, M., 2018. A reusable electrospun PVDF-PVP-MnO2 nanocomposite membrane for bisphenol A removal from drinking water. J. Environ. Chem. Eng. 6, 5801–5811. doi:10.1016/j.jece.2018.08.073

Mokhtar, N.M., Lau, W.J., Ismail, A.F., Veerasamy, D., 2015a. Membrane distillation technology for treatment of wastewater from rubber industry in Malaysia. Procedia CIRP. 26, 792–796. doi:10.1016/j.procir.2014.07.161

Mokhtar, N.M., Lau, W.J., Ismail, A.F., Youravong, W., Khongnakorn, W., Lertwittayanon, K., 2015b. Performance evaluation of novel PVDF-Cloisite 15A hollow fiber composite membranes for treatment of effluents containing dyes and salts using membrane distillation. RSC Adv. 5, 38011–38020. doi:10.1039/C5RA00182J

Mozia, S., Morawski, A.W., Toyoda, M., Tsumura, T., 2010. Integration of photocatalysis and membrane distillation for removal of mono- and poly-azo dyes from water. Desalination. 250, 666–672. doi:10.1016/j.desal.2009.06.075

Muthiah, P., Hsu, S.H., Sigmund, W., 2010. Coaxially electrospun PVDF-teflon AF and teflon AF-PVDF core-sheath nanofiber mats with superhydrophobic properties. Langmuir. 26, 12483–12487. doi:10.1021/la100748g

Park, S.H., Lee, S.M., Lim, H.S., Han, J.T., Lee, D.R., Shin, H.S., Jeong, Y., Kim, J., Cho, J.H., 2010. Robust superhydrophobic mats based on electrospun crystalline nanofibers combined with a silane precursor. ACS Appl. Mater. Interfaces. 2, 658–662. doi:10.1021/am100005x

Patel, S.U., Patel, S.U., Chase, G.G., 2013. Electrospun superhydrophobic poly(vinylidene fluoride- co -hexafluoropropylene) fibrous membranes for the separation of dispersed water from ultralow sulfur diesel. Energy and Fuels. 27, 2458–2464. doi:10.1021/ef400248c

Persano, L., Camposeo, A., Tekmen, C., Pisignano, D., 2013. Industrial upscaling of electrospinning and applications of polymer nanofibers. A Rev. Macromol. Mater. Eng. 298, 504–520. doi:10.1002/mame.201200290

Phattaranawik, J., Jiraratananon, R., Fane, A.G., 2003. Effect of pore size distribution and air flux on mass transport in direct contact membrane distillation. J. Memb. Sci. 215, 75–85. doi:10.1016/S0376-7388(02)00603-8

Prince, J. A., Anbharasi, V., Shanmugasundaram, T.S., Singh, G., 2013. Preparation and characterization of novel triple layer hydrophilic–hydrophobic composite membrane for desalination using air gap membrane distillation. Sep. Purif. Technol. 118, 598–603. doi:10.1016/j.seppur.2013.08.006

Purcell, W.R., 1949. Capillary pressures – Their measurement using mercury and the calculation of permeability therefrom. J. Pet. Technol. 1, 39–48. doi:10.2118/949039-G

Puri, M., Banerjee, U.C., 2000. Production, purification, and characterization of the debittering enzyme naringinase. Biotechnol. Adv. 18, 207–217.

Qu, D., Qiang, Z., Xiao, S., Liu, Q., Lei, Y., Zhou, T., 2014. Degradation of reactive black 5 in a submerged photocatalytic membrane distillation reactor with microwave electrodeless lamps as light source. Sep. Purif. Technol. 122, 54–59. doi:10.1016/j.seppur.2013.11.004

Qu, D., Sun, D., Wang, H., Yun, Y., 2013. Experimental study of ammonia removal from water by modified direct contact membrane distillation. Desalination. 326, 135–140. doi:10.1016/j.desal.2013.07.021

Quéré, D., 2008. Wetting and roughness. Annu. Rev. Mater. Res. 38, 71–99. doi:10.1146/annurev.matsci.38.060407.132434

Rawal, A., 2012. Design parameters for a robust superhydrophobic electrospun nonwoven mat. Langmuir. 28, 3285–3289. doi:10.1021/la204535s

Rayleigh, L., 1882. XX. On the equilibrium of liquid conducting masses charged with electricity. London Edinburgh Dublin Philos. Mag. J. Sci. 14, 184–186. doi:10.1080/14786448208628425

Reneker, D.H., Yarin, A.L., 2008. Electrospinning jets and polymer nanofibers. Polymer (Guildf). 49, 2387–2425. doi:10.1016/j.polymer.2008.02.002

Reznik, S.N., Yarin, A.L., Theron, A., Zussman, E., 2004. Transient and steady shapes of droplets attached to a surface in a strong electric field. J. Fluid Mech. 516, 349–377. doi:10.1017/S0022112004000679

Rutledge, G.C., Fridrikh, S. V., 2007. Formation of fibers by electrospinning. Adv. Drug Deliv. Rev. 59, 1384–1391. doi:10.1016/j.addr.2007.04.020

Sahay, R., Thavasi, V., Ramakrishna, S., 2011. Design modifications in electrospinning setup for advanced applications. J. Nanomater. 2011, 1–17. doi:10.1155/2011/317673

Schiffman, J.D., Schauer, C.L., 2008. A review: Electrospinning of biopolymer nanofibers and their applications. Polym. Rev. 48, 317–352. doi:10.1080/15583720802022182

Schofield, R.W., Fane, A.G., Fell, C.J.D., 1987. Heat and mass transfer in membrane distillation. J. Memb. Sci. 33, 299–313. doi:10.1016/S0376-7388(00)80287-2

Servi, A.T., Kharraz, J., Klee, D., Notarangelo, K., Eyob, B., Guillen-Burrieza, E., Liu, A., Arafat, H.A., Gleason, K.K., 2016. A systematic study of the impact of hydrophobicity on the wetting of MD membranes. J. Memb. Sci. 520, 850–859. doi:10.1016/j.memsci.2016.08.021

Shang, H.M., Wang, Y., Takahashi, K., Cao, G.Z., Li, D., Xia, Y.N., 2005. Nanostructured superhydrophobic surfaces. J. Mater. Sci. 40, 3587–3591. doi:10.1007/s10853-005-2892-9

Shannon, M.A., Bohn, P.W., Elimelech, M., Georgiadis, J.G., Marĩas, B.J., Mayes, A.M., 2008. Science and technology for water purification in the coming decades. Nature. 452, 301–310. doi:10.1038/nature06599

Shaulsky, E., Nejati, S., Boo, C., Perreault, F., 2017. Post-fabrication modification of electrospun nanofiber mats with polymer coating for membrane distillation applications. J. Memb. Sci. 530, 158–165. doi:10.1016/j.memsci.2017.02.025

Shim, W.-G., Woo, Y.C., Drioli, E., He, T., Shon, H.K., Choi, J.-S., Kim, S.-H., Tijing, L.D., 2016. Water desalination using graphene-enhanced electrospun nanofiber membrane via air gap membrane distillation. J. Memb. Sci. 520, 99–110. doi:10.1016/j.memsci.2016.07.049

Sudoh, M., Takuwa, K., Iizuka, H., Nagamatsuya, K., 1997. Effects of thermal and concentration boundary layers on vapor permeation in membrane distillation of aqueous lithium bromide solution. J. Memb. Sci. 131, 1–7. doi:10.1016/S0376-7388(97)00109-9

Sun, C., Wang, N., Zhou, S., Hu, X., Zhou, S., Chen, P., 2008. Preparation of self-supporting hierarchical nanostructured anatase/rutile composite TiO(2) film. Chem. Commun. (Camb). 3293–3295. doi:10.1039/b805072d

Tang, H., Wang, H., He, J., 2009. Superhydrophobic titania membranes of different adhesive forces fabricated by electrospinning. J. Phys. Chem. C. 113, 14220–14224. doi:10.1021/jp904221f

Thavasi, V., Singh, G., Ramakrishna, S., 2008. Electrospun nanofibers in energy and environmental applications. Energy Environ. Sci. 1, 205. doi:10.1039/b809074m

Tijing, L.D., Choi, J.S., Lee, S., Kim, S.H., Shon, H.K., 2014. Recent progress of membrane distillation using electrospun nanofibrous membrane. J. Memb. Sci. 453, 435–462. doi:10.1016/j.memsci.2013.11.022

Tuteja, A., Choi, W., Ma, M., Mabry, J.M., Mazzella, S.A., Rutledge, G.C., McKinley, G.H., Cohen, R.E., 2007. Designing superoleophobic surfaces. Science. 318, 1618–1622. doi:10.1126/science.1148326

Tuteja, A., Choi, W., Mabry, J.M., McKinley, G.H., Cohen, R.E., 2008. Robust omniphobic surfaces. Proc. Natl. Acad. Sci. U. S. A. 105, 18200–18205. doi:10.1073/pnas.0804872105

Wang, P., Chung, T.S., 2015. Recent advances in membrane distillation processes: Membrane development, configuration design and application exploring. J. Memb. Sci. 474, 39–56. doi:10.1016/j.memsci.2014.09.016

Wang, Q., Bai, Y., Xie, J., Jiang, Q., Qiu, Y., 2016. Synthesis and filtration properties of polyimide nanofiber membrane/carbon woven fabric sandwiched hot gas filters for removal of PM 2.5 particles Qiannan. Powder Technol. 292, 54–63. doi:10.1016/j.powtec.2016.01.008

Wang, X., Cheng, W., Wang, D., Ni, X., Han, G., 2019. Electrospun polyvinylidene fluoride-based fibrous nanocomposite membranes reinforced by cellulose nanocrystals for efficient separation of water-in-oil emulsions. J. Memb. Sci. 575, 71–79. doi:10.1016/j.memsci.2018.12.057

Wang, X., Jin, Z., Zhou, D., 2011. Study on color field of overall display design for home textile. Adv. Text. Mater. Pts. 1–3, 332–334, 510–513. doi:10.4028/www.scientific.net/AMR.332-334.510

Wang, Z., Wu, Z., Zhang, Y., Meng, J., 2017. Hyperbranched-polyol-tethered poly (amic acid) electrospun nanofiber membrane with ultrahigh adsorption capacity for boron removal. Appl. Surf. Sci. 402, 21–30. doi:10.1016/j.apsusc.2017.01.070

Washburn, E.W., 1921. The dynamics of capillary flow. Phys. Rev. 17, 273–283. doi:10.1103/PhysRev.17.273

Woo, Y.C., Chen, Y., Tijing, L.D., Phuntsho, S., He, T., Choi, J.S., Kim, S.H., Shon, H. K., 2017. CF4 plasma-modified omniphobic electrospun nanofiber membrane for produced water brine treatment by membrane distillation. J. Memb. Sci. 529, 234–242. doi:10.1016/j.memsci.2017.01.063

Woods, J., Pellegrino, J., Burch, J., 2011. Generalized guidance for considering pore-size distribution in membrane distillation. J. Memb. Sci. 368, 124–133. doi:10.1016/j.memsci.2010.11.041

Wu, C., Yan, H., Li, Z., Lu, X., 2015. Ammonia recovery from high concentration wastewater of soda ash industry with membrane distillation process. Desalin. Water Treat. doi:10.1080/19443994.2015.1010233

Wu, J., An, A.K., Guo, J., Lee, E.-J., Farid, M.U., Jeong, S., 2017. CNTs reinforced super-hydrophobic-oleophilic electrospun polystyrene oil sorbent for enhanced sorption capacity and reusability. Chem. Eng. J. 314, 526–536. doi:10.1016/j.cej.2016.12.010

Xie, Z., Duong, T., Hoang, M., Nguyen, C., Bolto, B., 2009. Ammonia removal by sweep gas membrane distillation. Water Res. 43, 1693–1699. doi:10.1016/j.watres.2008.12.052

Yan, K.K., Jiao, L., Lin, S., Ji, X., Lu, Y., Zhang, L., 2018. Superhydrophobic electrospun nanofiber membrane coated by carbon nanotubes network for membrane distillation. Desalination. 437, 26–33. doi:10.1016/j.desal.2018.02.020

Yao, X., Hu, Y., Grinthal, A., Wong, T.-S., Mahadevan, L., Aizenberg, J., 2013. Adaptive fluid-infused porous films with tunable transparency and wettability. Nat. Mater. 12, 529–534. doi:10.1038/nmat3598

Yarin, A.L., Koombhongse, S., Reneker, D.H., 2001. Taylor cone and jetting from liquid droplets in electrospinning of nanofibers. J. Appl. Phys. 90, 4836–4846. doi:10.1063/1.1408260

Zarebska, A., Nieto, D.R., Christensen, K. V., Norddahl, B., 2014. Ammonia recovery from agricultural wastes by membrane distillation: Fouling characterization and mechanism. Water Res. 56, 1–10. doi:10.1016/j.watres.2014.02.037

6 Dual-Layer and Multibore Membranes for Membrane Distillation

Jian Zuo, Kang-Jia Lu, and Peng Wang

6.1 DUAL-LAYER MEMBRANES

Dual-layer membranes are one type of membranes that consist of two layers from different materials. Figure 6.1 shows the schematic drawing of typical dual-layer membranes. One of the layers is the support layer, while the other layer is the functional layer (He et al., 2002; Xia et al., 2018). The dual-layer structure is advantageous over conventional single-layer or composite membranes because: (1) only a small amount of expensive materials with superior performance is used as the functional layer, while the support layer can be made from cost-effective polymers to minimize the material cost; (2) it is possible to combine the material advantages of each layer to develop a membrane with synergistic properties; (3) it is fabricated in one step, which is more convenient than composite membranes that require complex post-treatment steps to make. Since the development of dual-layer membrane fabrication technology in late 1970s (Henne et al., 1979), dual-layer membranes have found many applications in gas separation, biofuel separation, water reuse, membrane distillation (MD), pervaporation, and others (Fu et al., 2013; Li et al., 2002, 2004; Ong et al., 2016; Peng et al., 2012; Sun et al., 2010; Wang et al., 2009; Yanagimoto et al., 1987).

The fabrication of dual-layer membranes is via one-step simultaneous phase inversion method, normally known as coextrusion or cocasting. To fabricate dual-layer hollow fiber membranes, a coextrusion method is used, where two polymer dope solutions are extruded from a tri-orifice dual-layer spinneret simultaneously. Figure 6.2 shows a dual-layer spinneret, where the inner most channel is for bore fluid, the middle channel is for inner dope, and the outermost channel is for outer dope. During the spinning process, the bore fluid, inner dope, and outer dope are fed into the spinneret separately by three different pumps. Then, the solutions extruded out of the spinneret will pass through an air gap region before entering into the coagulation tank for phase inversion to take place. Next, a drum will take up the nascent fiber with a designated rotation speed. Finally, the fiber is stored in a water tank to complete the phase inversion process. Figure 6.3 illustrates the spinning process.

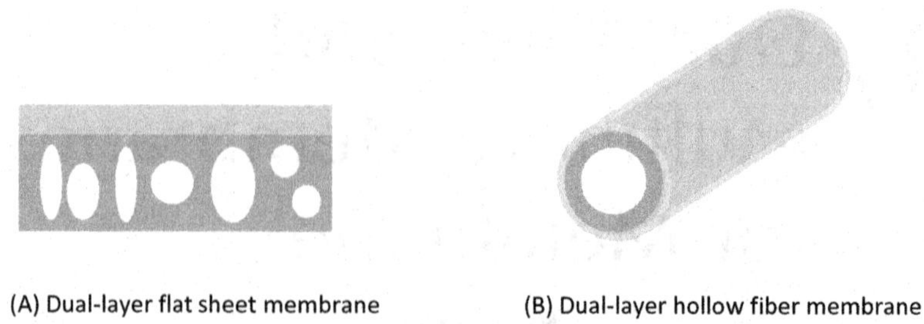

FIGURE 6.1 Schematic drawing of typical (A) flat sheet and (B) hollow fiber dual-layer membranes.

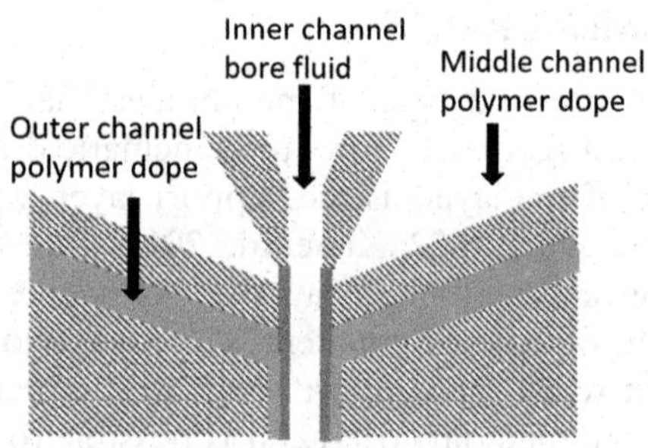

FIGURE 6.2 Schematic drawing of dual-layer spinneret.

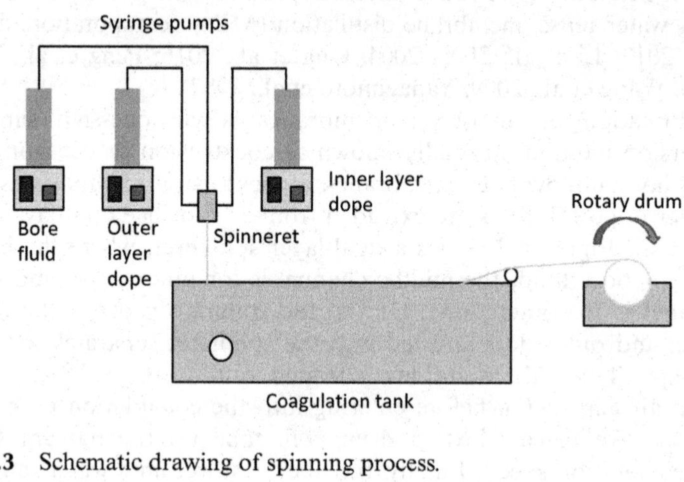

FIGURE 6.3 Schematic drawing of spinning process.

To fabricate dual-layer flat sheet membranes, a cocasting method is used. Generally, two casting knives with different thicknesses are used, where the thinner knife is used to cast the support layer and the thicker knife to cast the top layer (Fu et al., 2014; He et al., 2002; Kao et al., 2008; Xia et al., 2018). With the adjustment of the casting knife thickness, the membrane thickness can be controlled. Four steps are usually involved during the cocasting process: (1) pour the substrate dope solution on the casting plate, and use the thinner knife to cast about 4 cm in length; (2) pour the top layer dope solution on the substrate layer; (3) use the thicker knife to cast the top layer, and hold the two casting knives together to cast the dual-layer membrane; (4) immerse the as-cast membrane in a coagulation bath for phase inversion to take place. The nascent dual-layer membranes can then be freeze-dried for further usage.

6.1.1 Dual-Layer Membranes for MD Applications

For any MD application, the membrane plays a key role in determining the separation performance. The structure and chemistry of MD membrane are crucial parameters when designing a membrane. An ideal MD membrane not only needs to be hydrophobic to prevent wetting but also possesses a high permeation flux and good mechanical properties. To achieve such properties, many researchers have investigated different membrane structures and geometries. Among them, dual-layer membranes have emerged as promising designs to fulfil these requirements for MD applications.

Generally, in the field of MD applications, the dual-layer hollow fiber technology has been used to produce (1) morphologically controlled membranes, (2) membranes with enhanced mechanical integrity, and (3) hydrophobic/hydrophilic membranes. All these dual-layer hollow fiber membranes will contribute to the desirable MD membrane properties in one way or another. The rationale, design, formation mechanism, challenges, and advantages of these three types of dual-layer membranes will be discussed in detail.

6.1.1.1 Dual-Layer Membranes for Morphology Control

The permeation flux and wetting resistance of a MD membrane are highly dependent on the membrane morphology. Because of the coupled mass and heat transfer mechanism in a MD process, the effective driving force for permeation flux is the vapor pressure difference across the membrane. Thus, to improve the MD membrane permeation flux, one can either improve the mass transfer coefficient or reduce the heat transfer loss. The former can be achieved by tailoring the MD membrane structure to reduce mass transfer resistance. Usually, this can be realized by designing a membrane with a large pore size and porosity, open-cell pore structure, small membrane thickness, and tortuosity. The latter can be attained by reducing the membrane thermal conductivity. One of the methods is to increase the membrane porosity since air-filled pores have a much lower conductivity than the polymer matrix. Therefore, both methods require a membrane with a high porosity to increase permeation flux. Wang et al. have fabricated a single-layer hollow fiber membrane with finger-like macrovoids to increase membrane porosity (Wang et al., 2009). The resultant membrane shows a remarkable permeation flux of 79 kg/m^2 h at a feed temperature of 80 °C under a direct contact membrane distillation (DCMD) mode. This is attributed to the high membrane porosity and low tortuosity of the finger-like macrovoid structure. However, this design is not optimal, because there is always a concern of macrovoid structure on the membrane wetting and long-term stability.

Membrane wetting is the major enemy for a continuous MD operation. Not only can it deteriorate the membrane performance but also lose the membrane selectivity when the membrane pores are entirely wetted (Tomaszewska, 1999). When this occurs, the feed solution will be in direct contact with the permeate solution. Thus, membrane wetting should always be avoided. To improve the membrane wetting resistance, a membrane should have a high hydrophobicity, a small pore size, and a high tortuosity (Phattaranawik et al., 2003). In contrast to macrovoid structures, sponge-like structures could enhance membrane wetting resistance because of the narrow pore size distribution and higher tortuosity (Khayet and Matsuura, 2011). Teoh and Chung have fabricated hydrophobic sponge-like polyvinylidene fluoride-polytetrafluoroethylene (PVDF-PTFE) hollow fiber membranes for MD application (Teoh and Chung ,2009). The membranes exhibited a very narrow pore size distribution with a mean pore size of 0.116 μm and a maximum pore diameter of 0.212 μm with high wetting resistance and a stable long-term performance.

Therefore, there is a trade-off relationship on membrane structure requirements for permeation flux and wetting resistance. On the one hand, large pore sizes and porosity, open-cell pore structures, macrovoid structures, and small tortuosity are beneficial for a better permeation flux. On the other hand, small pore sizes and porosity, sponge-like and macrovoid-free structures, as well as large tortuosity are desirable for a higher wetting resistance. This trade-off relationship makes it difficult to achieve both high permeation flux and wetting resistance from a single configuration. Thus, dual-layer hollow fiber membranes may be employed for their dual morphological characteristics in one fiber. The structure of each layer can be separately controlled to achieve an optimized result.

Wang et al. have developed a dual-layer hollow fiber membrane consisting of a fully finger-like macrovoids inner layer and a sponge-like outer layer (Wang et al., 2011). Figure 6.4 demonstrates the schematic drawing of the dual-layer structure. The sponge-like outer layer helps to maintain a good wetting resistance, while the macrovoid inner layer aids to minimize the membrane tortuosity and mass transfer resistance. A dry-jet wet-spinning process was used to fabricate the fiber. The attainment of this structure was through the manipulation of inner and outer dope formulations. PVDF with a lower concentration was used in the inner dope to form macrovoid structures. On the other hand, PVDF with a higher concentration was used in the outer dope, which favored the formation of sponge-like structures. Since PVDF was the main polymer phase in both layers, there was no delamination

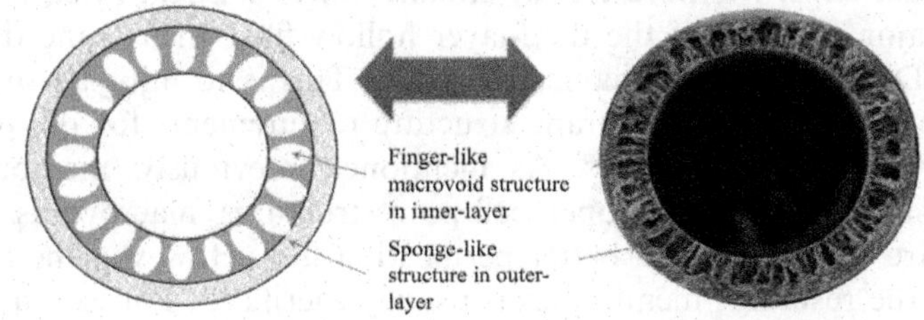

FIGURE 6.4 Schematic drawing of the dual-layer structure that consists of a fully finger-like macrovoids inner layer and a sponge-like outer layer (Wang et al., 2011). Copyright (2011) Elsevier.

between the two layers. To further enhance the outer layer wetting resistance, PTFE particles were added in the outer layer to increase its hydrophobicity. The resultant dual-layer membranes exhibited a superior permeation flux of 98.6 kg/m^2 h at 80 °C feed under DCMD mode. Moreover, the liquid entry pressure (LEP) and 200-h continuous performance showed a high membrane wetting resistance. Therefore, dual-layer membranes offer the opportunity to control the morphology of each layer and to optimize membrane performance.

In another work, Teoh et al. demonstrated the control of dual-layer hollow fiber morphology for better MD performance via incorporation of PTFE particles in the dope solution (Teoh et al., 2011). From their point of view, macrovoids are undesirable structure that have weak mechanical strength and are regarded as a major impediment for reliable membrane applications. Thus, they fabricated macrovoid free dual-layer hollow fiber membranes for seawater desalination. They utilized PTFE particles to suppress the formation of macrovoids and to enhance the outer surface hydrophobicity. The incorporation of 30 wt% PTFE in the PVDF matrix increased the dope viscosity and suppressed the macrovoid formation effectively. This was because that the diffusion rates of the bore fluid and external coagulation were retarded due to the high dope viscosity (Ren et al., 2002; Tsai et al., 2000) because a delayed demixing process tended to form macrovoid-free structures. As a result, dual-layer hollow fibers with a macrovoid-free morphology and a thin outer layer were obtained. The resultant membranes showed a relatively high water contact angle of 114.5° and a porosity of 81.5%. They also exhibited a 24% flux enhancement compared to the single-layer membrane fabricated from the same dope formulation of the outer layer dope. In addition, there was no delamination observed from the dual-layer hollow fibers, which was critical for the long-term membrane performance. As PVDF was used in both the inner and outer layers, both layers were compatible with no delamination and interfacial resistance. Thus, by employing the dual-layer spinning technology, hollow fiber membranes with a desirable morphology and better performance can be tailored.

6.1.1.2 Dual-Layer Membranes for Enhanced Mechanical Integrity

Another major purpose of the dual-layer hollow fiber membrane design is to enhance MD membrane mechanical integrity. This is to mitigate the trade-off relationship between the membrane structure requirements for the permeation flux and mechanical properties. As mentioned previously, membranes with large pore sizes and porosity, open-cell pore structures, macrovoids, and small tortuosity are beneficial for a better permeation flux. However, the mechanical strength of the resultant membranes is usually sacrificed. For example, Tomaszewska has developed PVDF membranes with large cavities, pore size, and porosity by incorporating lithium chloride (LiCl) into the dope solution (Tomaszewska, 1996). The permeation fluxes of resultant membranes were significantly increased. Nonetheless, there was a drastic decrease in membrane mechanical properties with an increase of LiCl concentration from 0 to 1 wt%, despite the enhanced permeation flux. The deteriorated mechanical property was unfavorable for practical MD applications. Similarly, Bottino et al. have also studied the effect of LiCl in the formation of PVDF membrane (Bottino et al., 1988). The addition of LiCl in the dope solution changed the morphology of the resultant PVDF membrane remarkably. Porous structures with larger pores were formed with higher LiCl concentrations. This was explained by the higher precipitation rate of the dope solution during the phase inversion process, because of the high affinity between LiCl and water. Subsequently, the membrane permeation flux was increased considerably. However, a reduction in membrane mechanical strength was also observed, which was in line with Tomaszewska's work. The reduced mechanical strength might instigate membrane deformation or pore collapse during long-term operations.

To improve MD membrane mechanical properties, many efforts and research have been carried out. For example, Wang et al. and Teoh et al. fabricated mixed matrix membranes by incorporating cloisite clay or PTFE particles to reinforce the membrane mechanical strength (Teoh et al., 2011; Wang et al., 2009). Wang et al. controlled the PVDF hollow fiber morphology by changing it from sponge-like structures to circular nodule structures to enhance membrane mechanical strength (Wang et al., 2000). Yang et al. produced a PVDF membrane which was chemically cross-linked with a perfluoro compound of perfluoropolyether containing ethoxysilane terminal groups. The resultant membrane showed better mechanical strength (Yang et al., 2011). These methods have been proven to be effective to improve MD membrane mechanical properties. However, PVDF is the major matrix material for the structural support in these investigations. This leads to a limitation in the overall membrane mechanical strength by the intrinsic properties of PVDF. One of the solutions to overcome the limitation is to fabricate dual-layer membranes with different materials. In that case, PVDF will function as the hydrophobic separation layer, while another mechanically strong material will provide support. Thus, a dual-layer membrane could potentially synergize the material properties to achieve good permeation flux and mechanical properties simultaneously.

Zuo et al. have pioneered the development of dual-layer hollow fiber membranes cospun from PVDF and polyetherimide for MD applications (Zuo et al., 2017). The resultant hollow fibers showed an increment of about 350% in tensile strength than conventional single-layer MD hollow fibers. A good permeation flux of 45.8 kg/m^2 h at 70 °C was also achieved during vacuum membrane distillation (VMD) of seawater. The challenge to develop such dual-layer membranes includes: (1) the delamination between the two polymer materials and (2) the dense interface morphology between the two layers. The former will result in defective membranes, where the two layers are easily separated. The latter will increase membrane mass transfer resistance significantly, which would lead to a low permeation flux. These issues have to be addressed before a useful hollow fiber membrane is constructed for MD applications. According to studies on dual-layer hollow fiber membranes, the delamination and dense interface issues are probably resulted from (1) incompatibility between the two polymer materials, (2) different shrinkage rates between the two polymer dopes during phase inversion, and (3) dope formulations and their compatibility (Khayet et al., 2005; Li et al., 2002). Thus, an appropriate inner layer polymer material has to be selected in the first place.

Zuo et al. chose Ultem as the inner layer polymer because of its good intrinsic mechanical strength and compatibility with PVDF (Ong and Chung, 2012; Zuo and Chung, 2013). Although PVDF and Ultem are immiscible, their solubility parameters are close. This infers that a strong molecule interaction may exist between the two polymer layers. Figure 6.5 presents the PVDF/Ultem dual-layer hollow fiber morphology, where no delamination is observed. In addition to the polymer interaction, the spinning conditions also help to produce a delamination free dual-layer structure. Firstly, the outer layer dope was designed to have a lower concentration than the inner layer dope. As a result, the outer layer had a greater shrinkage rate during the phase inversion process. Thus, it could tightly wrap the inner layer and form a seamless interface. At the same time, since the inner layer material Ultem

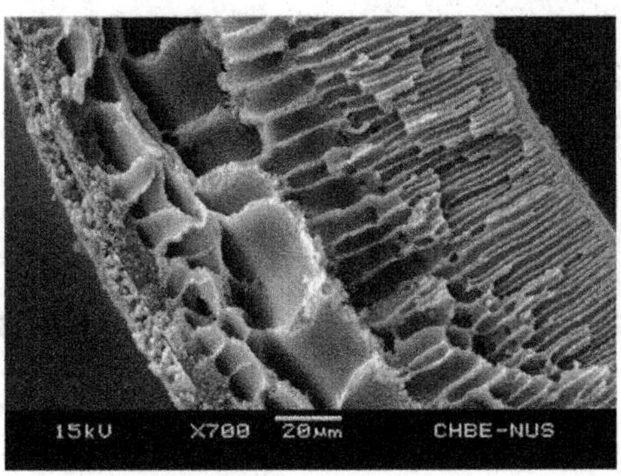

FIGURE 6.5 The cross-section image of dual-layer hollow fiber.

was mechanically strong and the outer layer material PVDF was stretchable, the outer layer adhered to the inner layer nicely. Moreover, the spinning parameters could also be manipulated to achieve a delamination-free membrane. For example, a ratio of the outer layer dope flow rate to the inner layer dope flow rate of around 0.125 was found to be helpful to form a delamination-free structure (Li et al., 2004). However, the resultant dual-layer hollow fiber had a rather low permeation flux due to the dense inner layer outer surface structure, which possessed a tremendously high mass transfer resistance.

Thus, the second challenge to be solved is the dense interface morphology. Zuo et al. introduced aluminum oxide (Al_2O_3) nanoparticles into the inner layer dope to create microporosity on the surface (Zuo et al., 2017). Figure 6.6 compares the difference in the outer surface of the inner layer before and after the incorporation of Al_2O_3 nanoparticles. It is clear that many Al_2O_3 nanoparticles appear on the surface, which creates microporosity for water vapor transport. As a result, the modified dual-layer membrane showed a much enhanced permeation flux of 45.8 kg/m^2 h at 70 °C. With the successful development of dual-layer PVDF/Ultem hollow fiber membranes, their mechanical properties have also been characterized. They show a maximum tensile strength of 4.34 MPa, which is considerably higher than most PVDF hollow fiber membranes. This is primarily because of the dual-layer structure that offers the opportunity to include another mechanically strong material as the support layer. It reinforces the dual-layer hollow fiber with significantly enhanced mechanical properties.

6.1.1.3 Hydrophobic/Hydrophilic Dual-Layer Membranes

The third type of dual-layer membrane design is to fabricate a hydrophobic/hydrophilic dual-layer structure to enhance the membrane permeation flux.

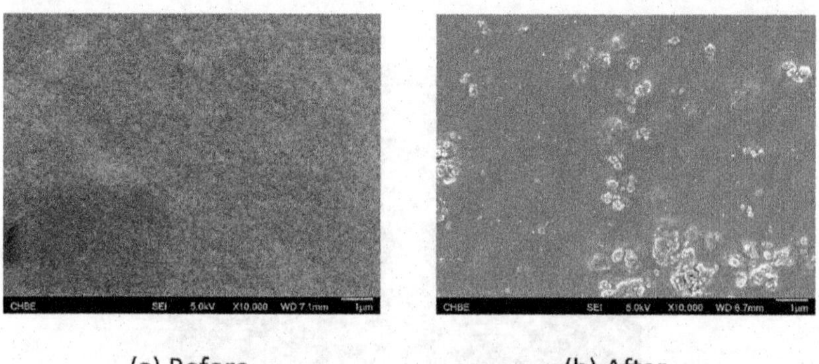

(a) Before (b) After

FIGURE 6.6 The inner layer outer surface image (a) before and (b) after the incorporation of 5 wt% Al_2O_3 (Zuo et al., 2017). Copyright (2017) Elsevier.

The rationale to include the hydrophilic layer is because it can be wetted by the permeate water easily. Consequently, the membrane functional layer thickness is reduced and the distance for vapor transport is minimized. Figure 6.7 illustrates the hydrophobic/hydrophilic dual-layer membranes, where the thin hydrophobic layer facing the feed is not wetted, while the hydrophilic layer facing the permeate side is wetted. During the MD process, the hydrophobic layer retains the liquid water feed, while water vapor will transport through the pores in the hydrophobic layer. Then, the water vapor will condense in the hydrophilic layer. Thus, the hydrophilic layer does not contribute to the vapor transport resistance much. As a result, the permeation flux would be enhanced because of the reduced mass transfer resistance. At the same time, the hydrophilic layer also supports the hydrophobic layer and provides the mechanical stability. Since Cheng and Wiersma (Cheng and Wiersma, 1982, 1983) introduced the first hydrophobic/hydrophilic configuration for MD application, many studies have demonstrated the potential and advantages of such hydrophobic/hydrophilic dual-layer design for MD applications.

Khayet, Matsuura, and their co-workers have fabricated hydrophobic/hydrophilic dual-layer membranes based on surface modifying macromolecules (SMMs) (Essalhi and Khayet, 2012; Khayet et al.2005; Khayet and Matsuura, 2003). The SMM is a hydrophobic fluoropolymer synthesized by polyurethane chemistry. It can be blended into the casting solution of hydrophilic host polymer (e.g., polyetherimide). Then, the solution is casted into flat sheet membranes via the phase inversion technique. When the polymer solution is equilibrated with air, the SMMs having a lower surface tension would concentrate at the air/solution interface. Thus, a composite membrane can be prepared in the single casting step. Following this method, Essalhi and Khayet have prepared a hydrophobic/hydrophilic dual-layer membrane for air gap membrane distillation (AGMD) and DCMD (Essalhi and Khayet.2012). The water permeation flux for DCMD was found to be 2.7–3.3 times higher than that for AGMD. This was mainly because that the hydrophilic layer was wetted during the DCMD operation. The water vapor transport path length was reduced to only the top hydrophobic layer of the membrane. Thus, the lower resistance and shorter transport pathway led to a higher permeation

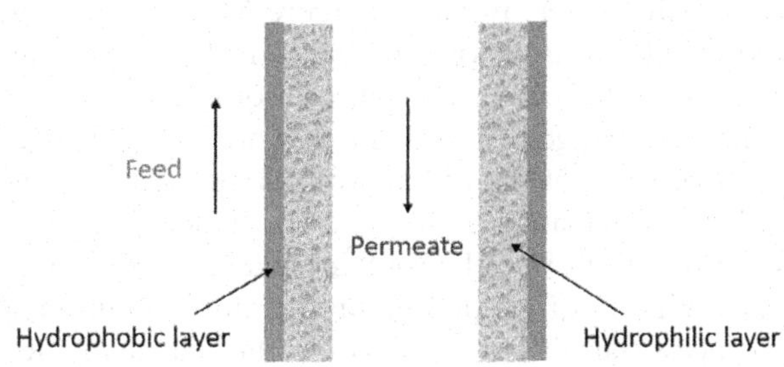

FIGURE 6.7 Schematic drawing of hydrophobic/hydrophilic membrane.

flux in DCMD. This demonstrates the effect of hydrophobic/hydrophilic dual-layer structure in enhancing the membrane permeation flux.

Chung and co-workers have fabricated hydrophobic/hydrophilic dual-layer hollow fiber membranes for MD applications (Bonyadi and Chung, 2007; Su et al., 2010). Bonyadi and Chung fabricated cospun hydrophobic/hydrophilic dual-layer hollow fiber membranes by incorporating hydrophobic and hydrophilic clay particles in each layer (Bonyadi and Chung, 2007). To eliminate the possible delamination issue, they used PVDF as the main polymer phase in both the outer and inner layer. Hydrophobic clay particles were added into the outer layer dope to increase its hydrophobicity. On the other hand, hydrophilic clay particles were added into the inner layer dope to tune its hydrophilicity. Furthermore, a hydrophilic polymer polyacrylonitrile (PAN) was added to the inner layer dope to enhance the wettability of the inner layer. Thus, the hydrophilic PAN polymer chain and hydrophilic clay particles would facilitate the water transfer from the permeate side to the interface of two membrane layers. Then, the inner layer could be totally wetted by permeate water. To validate the hydrophobicity and hydrophilicity of the two layers, water contact angle measurements were carried out. The outer layer showed a water contact angle of 140°. The high contact angle was attributed to the hydrophobic clay particles and the high membrane surface roughness caused by the porous fiber-like network. The inner layer exhibited a very low water contact angle of 50°, which confirmed the hydrophilic characteristic of the inner layer. A good flux of 55 kg/m^2 h was obtained at 90 °C in a DCMD test. The high flux attained was contributed from the dual-layer hydrophobic/hydrophilic structure that reduced vapor transport resistance.

Although a hydrophobic/hydrophilic dual-layer membrane has advantages of reducing vapor transport resistance, the thin hydrophobic functional layer faces a risk of membrane wetting. As discussed previously, membrane wetting could lead to the membrane performance decline or the loss of membrane rejection. Thus, membranes with better wetting resistances are always desired. Edwie et al. have developed a hydrophobic/hydrophilic dual-layer PVDF hollow fiber membrane with enhanced wetting resistance for MD applications (Edwie et al., 2012). To enhance the wetting resistance of the hydrophobic layer, two approaches were investigated, namely, (1) adding methanol to change the hydrophobic layer morphology and (2) incorporating fluorinated silica particles as a hydrophobic modifier. Figure 6.8 shows the membrane outer surface morphology without and with methanol in the dope. The average globule size was significantly decreased when methanol was added in the outer layer dope solution, where it changed from diameters of 0.714 ± 0.102 μm to 0.357 ± 0.051 μm. This was because of the different phase inversion duration. Generally, the size of globule has a positive correlation to the solid–liquid demixing time. For instance, a shorter time would result in smaller globule formation. In the current case, the addition of methanol in the dope solution accelerated the phase inversion process, because it reduced the thermal dynamic stability of the solution. Moreover, the addition of methanol reduced the viscosity of dope solutions. Thus, it kinetically accelerated the diffusion of the external coagulant

(a) Without methanol (b) With methanol

FIGURE 6.8 Surface morphology of the outer layer fabricated (a) without and (b) with methanol addition in the dopes (Edwie et al., 2012). Copyright (2012) Elsevier.

into the nascent fiber. Consequently, both effects contributed to the faster phase inversion process, which led to a smaller globule size. The smaller size globule had a denser structure that would help the wetting resistance of the resultant membrane. The stability of fabricated hydrophobic/hydrophilic dual-layer hollow fiber membranes was tested under a continuous DCMD operation for 5 days. The membrane maintained a separation factor of higher than 99.99% through the experiment. On the other hand, the separation factor of original membrane started to decrease after about 100 hours. This verified the stability and wetting resistance of the membrane fabricated by adding methanol. Regarding the second approach, self-synthesized fluorinated silica particles were added in the outer layer to enhance its hydrophobicity. It increased the water contact angle of the resultant membranes. However, the membranes' stability was not improved during the continuous operation. This was probably due to the hydrophilic hydroxyl groups on the silica particles that favored the occurrence of membrane wetting. Thus, the first approach is more effective than the second approach to enhance the wetting resistance of hydrophobic/hydrophilic dual-layer hollow fiber membranes.

6.2 MULTIBORE HOLLOW FIBER MEMBRANES FOR MD

Over the past years, various types of multibore hollow fibers (MBF) with different geometries and materials for numerous applications have been developed. For example, inge GmbH (part of BASF) commercialized a seven-bore hollow fiber polymeric membrane with a brand name Multibore for ultrafiltration (UF) (BASF, 2013). Later, Bu-Rashid and Czolkoss carried out a series of pilot plant experiments to evaluate the feasibility of this membrane as the pretreatment for seawater reverse osmosis (SWRO) desalination and very promising results were obtained (Bu-Rashid and Czolkoss, 2007). A 75% reduction in energy consumption was reported as compared with that of single-bore UF membranes. Zhu et al. successfully developed a ceramic hollow fiber membrane with four channels for oxygen separation. The breaking load of the resultant membrane was 3–6 times higher than that of a conventional single-channel membrane. The multichannel membrane also exhibited a higher oxygen flux as well as a good long-term stability (Zhu et al., 2014). Despite these geometries, membranes in round, hexagonal, triangular, and rectangular shapes with different numbers of bores as shown in Figure 6.9 have also been developed. Applications of multibore membranes in various processes including UF, MD, nanofiltration (NF), membrane bioreactor (MBR), and gas separation have been explored (Chang et al., 2019; Chi et al., 2017; Deng et al., 2012; Heijnen et al., 2012; Lu et al., 2016; Luo et al., 2014; Ma et al., 2017; Peng et al., 2011; Teoh et al., 2011; Wang et al., 2014; Wang and Chung, 2013; Zhu et al., 2014).

In this book chapter, we will focus on polymeric multibore hollow fibers that have been developed for membrane distillation applications. Due to the large pore size and high porosity in both bulk and surfaces, the MD

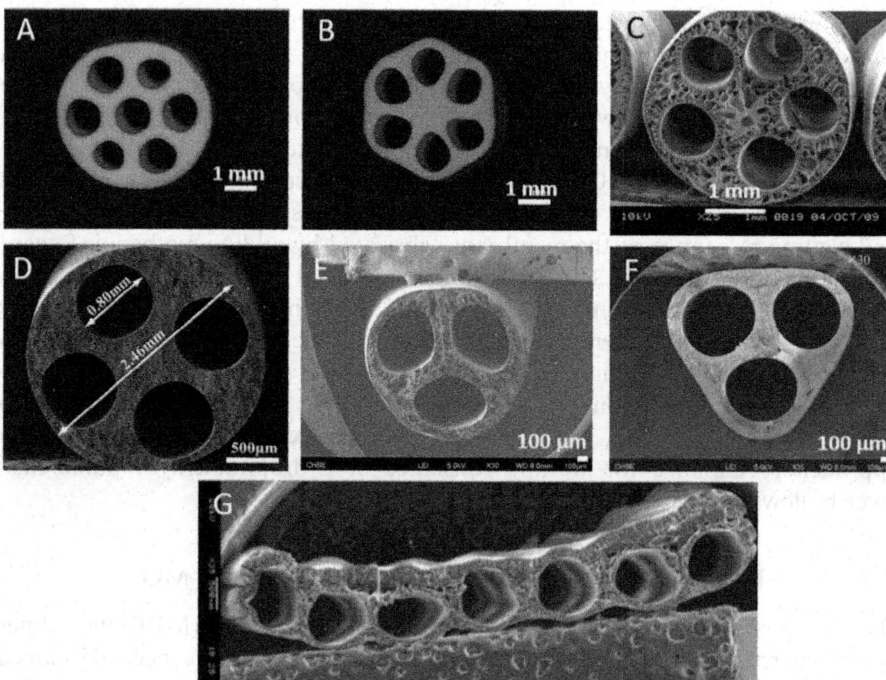

FIGURE 6.9 Hollow fiber membranes with (A) seven-bore round shape (Wang and Chung, 2013). Copyright (2013) American Chemical Society; (B) six-bore hexagonal shape (Wang and Chung, 2013). Copyright (2013) American Chemical Society; (C) five-bore round shape (Ma et al., 2017); (D) four-bore round shape (Zhu et al., 2014). Copyright (2014) John Wiley and Sons; (E) tri-bore round shape (Lu et al., 2016). Copyright (2016) Elsevier; (F) tri-bore triangular shape (Wang et al., 2014). Copyright (2014) Elsevier; (G) seven-bore rectangular shape (Peng et al., 2011). Copyright (2011) Elsevier.

membranes often suffer from weak mechanical properties in terms of tensile rigidity at both axial and radial directions. During an MD process, mechanical properties diminish further because of the elevated operation temperature (Wang and Chung, 2013). Multibore hollow fibers which not only provide good mechanical strength but also potentially possess high porosity have attracted increasing attention for MD applications and offer a promising solution to the poor mechanical stability problem of MD membranes.

6.2.1 Fabrication of MBF Membranes

Similar to the polymeric single-bore hollow fiber (SBF) MD membranes, multibore membranes are also fabricated via nonsolvent-induced phase separation (NIPS) method using specially designed spinnerets. Figure 6.10 shows the typical spinneret designs for seven-bore round, seven-bore rectangular, and tri-bore hollow fibers. In order to avoid the uneven distribution of polymer dope solutions, the dope and bore fluid solutions are usually injected from the top and side of the spinneret, respectively, as shown in Figure 6.10A. Figure 6.10B and 6.10C shows the bottom views of a round shape and a rectangular shape

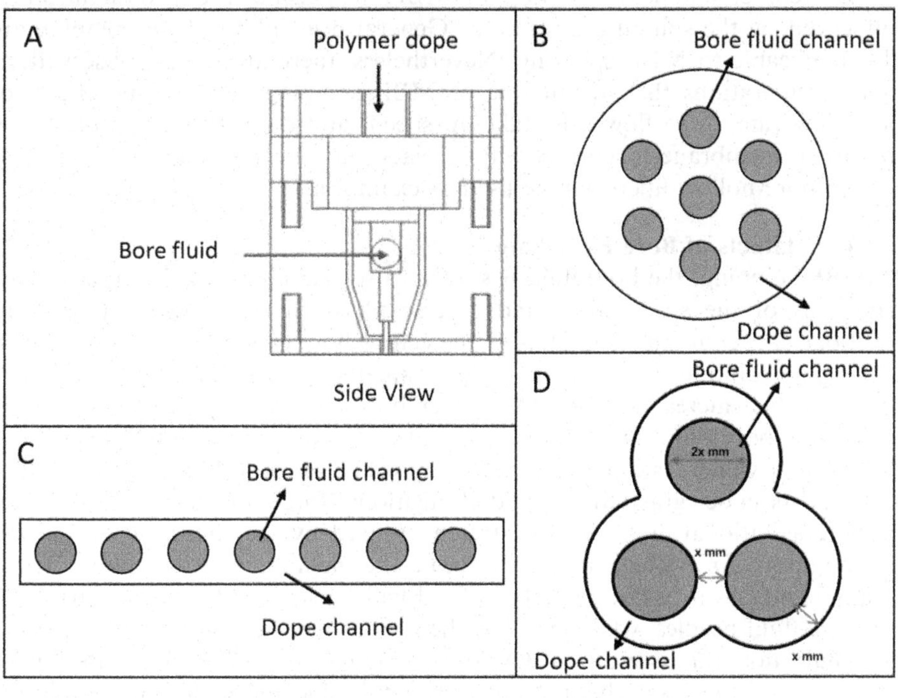

FIGURE 6.10 (A) Side view of a typical multibore spinneret; (Peng et al., 2011). Copyright (2011) Elsevier; bottom views of (B) seven-bore round, (C) seven-bore rectangular, and (D) tri-bore spinnerets (Wang et al., 2014). Copyright (2014) Elsevier.

seven-bore spinneret, respectively. In both designs, seven bore fluid channels are equally distributed in the dope flow channels. The dimensions of the dope and bore fluid channels, the distances between dope and bore fluid channels (i.e., wall thickness), and the gaps between bore fluid channels are important design parameters that directly affect the resultant membrane properties. For example, a certain distance between bore fluid channels must be employed to avoid the intrabore crossing of nascent fibers due to die swell phenomena. A large wall thickness and distance between bore fluid channels are also beneficial to the mechanical integrity of membranes. On the other hand, a large distance will result in a great mass transfer resistance which may cause a reduction in the vapor flux. Figure 6.10D shows a typical tri-bore spinneret with a blossom geometry. The tri-bore spinneret has three needles that distribute uniformly within the spinneret. It should be noted that this design is not to fabricate a membrane with a blossom geometry. The main purpose is to ensure a uniform extrusion gap for the polymer dope and also to generate an enough shear stress during the dope extrusion. Owing to the polymer chain relaxation and die swell effect, the nascent fiber is expected to exhibit a more regular contour. A similar phenomenon has been reported by Cufaz et al., where the circular fiber was obtained even a microstructured irregular spinneret was used (Çulfaz et al., 2010).

The geometry and morphology of MBF membranes are also significantly influenced by the spinning conditions. General principles in SBF spinning are also applicable to MBF spinning. Nevertheless, there are several observations and considerations that are unique to MBF spinning. The unique effects of bore flow rate, dope flow rate, take-up speed, and coagulant composition on multibore membrane formation will be elaborated in this chapter. Seven-bore and tri-bore hollow fibers will be used as examples.

6.2.1.1 Effects of Bore Flow Rate

In MBF spinning, the bore fluid is split into several channels. The uneven distributions of the shear stress and diverse phase inversion rates at different positions cause unique properties of the center bore channel. Figure 6.11 illustrates the evolution of the center bore channel as the bore flow rate decreases first and then increases back. The center bore channel apparently disappears at the low bore flow rate but reappears after increasing the flow rate back to the original value. As illustrated in Figure 6.11E and 6.11F, a radial direction stress exists under gravitational and elongational forces owing to (1) the membrane gravitational mass, (2) the stretch induced by the collection drum, and (3) the nonuniform phase inversion across the membrane cross section (Wang et al., 2004). As a result, more bore fluid will be squeezed into the surrounding bore-fluid needles, while less into the center bore-fluid needle if the overall bore fluid flow rate is reduced. Of course, the effects of bore flow rate on the formation of the bore channel are only applicable for multibore membrane geometries with a center bore such as five-bore and seven-bore round structures. This concern is eliminated in multibores membranes with even numbers

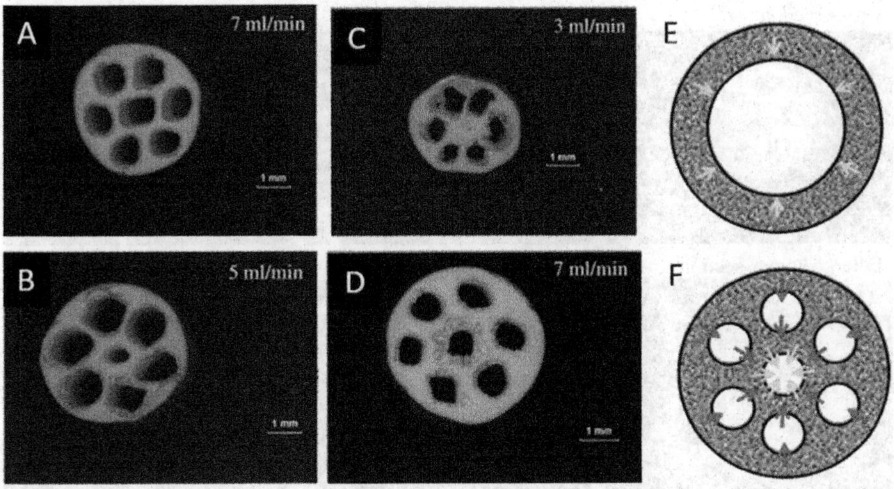

FIGURE 6.11 Evolution of cross-section morphology with different bore flow rates: (A) 7 ml/min, (B) 5 ml/min, (C) 3 ml/min, (D) 7 ml/min, (E) and (F): Stress direction of the polymer solution for one-needle and seven-needle spinnerets (Wang and Chung, 2012a). Copyright (2012) Elsevier.

of bores as well as three bores structures where all the bore channels are completely identical.

6.2.1.2 Effects of Dope Flow Rate

Unlike the single bore membrane that often shows a typical porous inner surface, the inner surface of the multibore membrane made by PVDF sometimes exhibits a loosely packed fiber-like network structure as shown in Figure 6.12. From the perspective of heat and mass transfer, this structure may promote a turbulence flow near the boundary layer, which suppresses temperature and concentration polarizations and hence, enhancing the effective driving force. This observation is ascribed to the big difference in hydraulic pressure between the spinning dope and the bore fluid. Owing to the special seven-needle spinneret design and viscous characteristics of the polymer dope, the pressure difference between the dope and bore fluid can be significantly higher than that of typical SBF spinning processes (Wang and Chung, 2012). Since the phase inversion of the PVDF polymer takes place at a much slower rate than other glassy polymer solutions, a liquid–liquid demixing starts at the surface and gradually proceeds to the cross section of the nascent fiber. Under this circumstance, the dope with a high pressure will facilitate the diffusion of the unsolidified polymer solution into the bore fluid in order to release pressure. Thereafter, the diffused polymer solution forms a loosely fiber-like network on the inner surface. As shown in Figure 6.12, the gradual disappearance of the loosely fiber-like network on the inner surface is observed as the dope flow

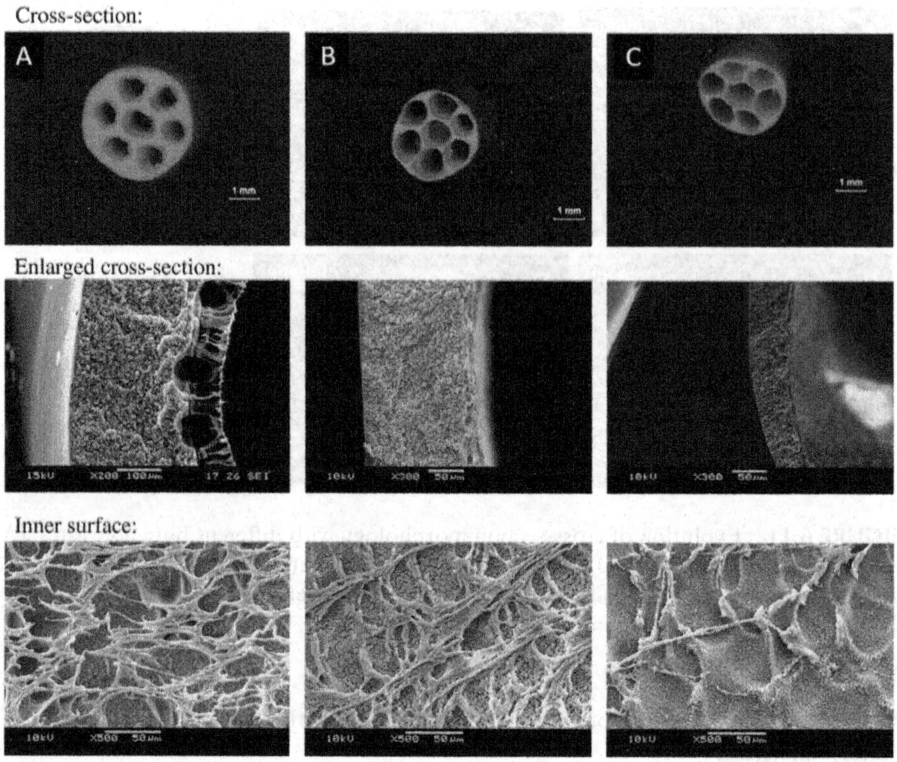

FIGURE 6.12 The cross-section and inner-surface morphologies of MBF membranes spun with different dope flow rates: (A) 14 ml/min, (B) 10 ml/min, and (C) 6 ml/min (Wang and Chung, 2012). Copyright (2012) Elsevier.

rate decreases because of the reduced dope pressure. It can also be seen that the bore size and seven-channel structure are well maintained as the dope flow rate decreases, while both the fiber wall thickness and the outer diameter of the fiber reduce.

Another interesting effect of dope flow rate has been reported for tri-bore hollow fiber spinning. It has been demonstrated that a lower dope flow rate is in favor of the formation of triangular shape hollow fiber (Chang et al., 2019; Wang et al., 2014). Figure 6.13 shows the geometrical evolution of tri-bore fibers from a round shape to a triangle shape by reducing the dope flow rate. This revolution is probably due to the stress balance among die swell, surface tension, and surface shrinkage. The surface tension which drives the formation of the circular shape is a function of solution surface energy and fiber perimeter, while the viscous force which favors the formation of triangle shape is influenced by the shear rate and solution rheological properties as well as the take-up stress. At a higher dope flow rate, the effect of die swell dominates the

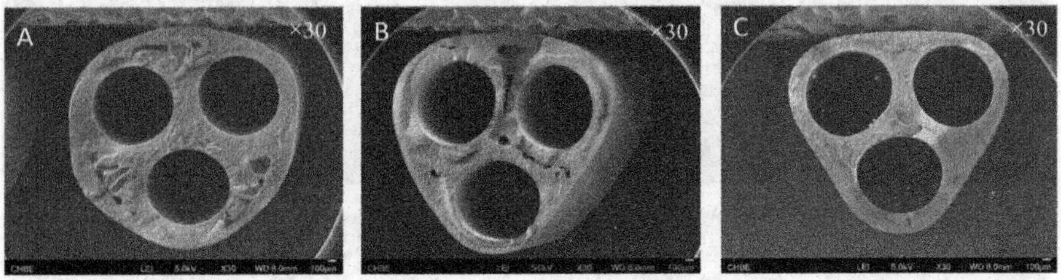

FIGURE 6.13 The SEM images of Matrimids tri-bore hollow fiber membranes using different dope compositions (Wang et al., 2014). Copyright (2014) Elsevier.

balance and thus forms a round shape. At a reduced dope flow rate, the viscous force and the stretching force due to the take-up drum become dominant and the triangle shape is therefore formed.

6.2.1.3 Effects of Take-Up Speed

High speed spinning is always preferred in industries because it not only increases throughput but also lower the production costs. However, various membrane properties including inner surface morphology, cross-sectional structure, and mechanical strength will be altered with high collection speeds. Figure 6.14 shows the cross-section and inner surface morphologies of MBF membranes spun with different take-up speeds. Both the outer diameter of membranes and the average diameter of bore channels decrease as the take-up speed increases. Meanwhile, the lotus-root structure gradually becomes a wheel-like structure as the take-up speed increases, as illustrated by Figure 6.14E and 6.14F.

Comparing with the sample spun from free fall, the wheel-like structure consists of a round-shape small center with a reduced diameter and shell-shape surrounding channels with a uniform wall thickness. With a more uniform wall thickness, the wheel-like structure could provide a better balance of the vapor transport efficiency, liquid wetting resistance, and mechanical properties. The relationship between the morphology of the central bore channel and take-up speed has also been reported by Bonyadi and Mackley. In their study on multibore microcapillary film (MCF) membranes, the size of the center bore channel decreased but the round shape could be well maintained at higher air gaps or take-up speeds, while those at two sides of rectangular membranes gradually changed from a round shape to an oval shape (Bonyadi and Mackley, 2012). Wang et al. believed that the transition of the membrane geometry was closely related to the stresses resulting from the high take-up speed and different speeds of phase inversion processes across the membrane. In their study, a relatively rapid phase inversion occurred at the outer skin because 50/50 IPA/water was

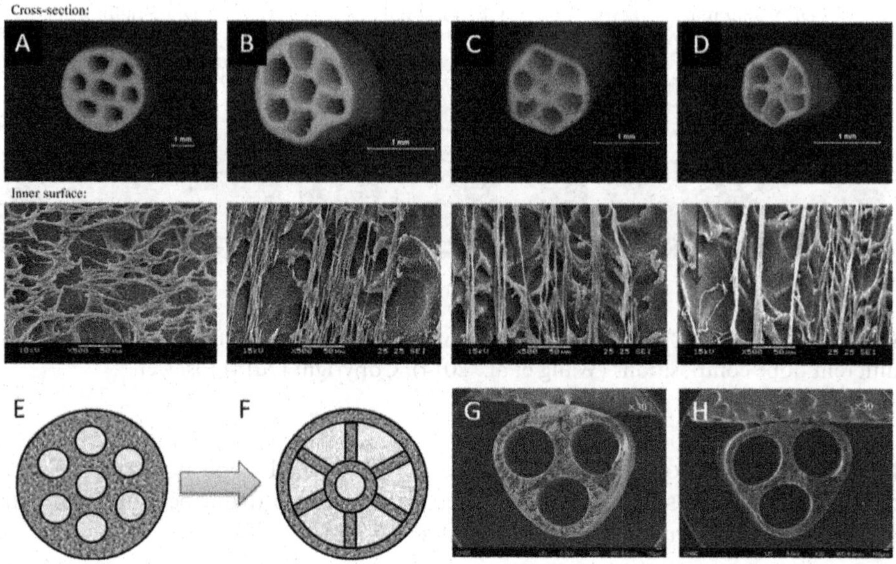

FIGURE 6.14 Morphologies of MBF membranes spun with different take-up speeds: (A) free fall, (B) 5 m/min, (C) 7.5 m/min, (D) 10 m/min, (E) and (F): Schematic illustration of geometry change with increasing take-up speed. The arrow represents the spinning direction (Wang and Chung, 2012). Copyright (2012) Elsevier. Tri-bore Matrimid hollow fiber membranes at take-up speeds of (G) free fall and (H) 150% free fall (Wang et al., 2014). Copyright (2014) Elsevier.

used as the outer coagulant and a relatively slow phase inversion took place at the bore center because 70/30 NMP/water was employed as the inner coagulant. A high take-up speed would induce different degrees of orientation and inward stress between the center and surrounding areas. As a consequence, a slow phase inversion took place at the fiber center that resulted in a loose morphology and small center bore channel (Wang and Chung, 2012).

In addition, there was a morphological evolution of the nanofiber structure on the inner surface of these MBF membranes. As shown in Figure 6.14, the quantity of nanofibers decreases while the length of a single nanofiber increases with an increase in the take-up speed. The decrease of nanofiber quantity is mainly caused by the faster phase inversion speed. At a high take-up speed, the nascent fiber elongates and the fiber dimension reduces rapidly during the air gap region, which will enhance the speed of phase inversion. This faster phase inversion could impede the convection of the polymer solution which forms nanofibers. In addition to the change in the phase inversion speed, the slightly higher bore fluid pressure observed during the experiment may also contribute to this morphological evolution. Due to the rapid elongation of the nascent hollow fiber, several shorter nanofibers may merge into one, which results in the increased length of a single nanofiber.

In tri-bore hollow fiber spinning, it has been reported that a higher take-up speed can facilitate the formation of a triangular geometry as shown in Figure 6.14G and 6.14H. This is because a large take-up speed can suppress the die-swell effect and enhance the elongational stretch.

6.2.2 Mechanical Properties and Performances of MBF and SBF Membranes

Table 6.1 compares the mechanical properties of MBF and SBF MD membranes from the literature. In all the cases, the maximum load of MBF membranes is significantly higher than that of SBF membranes due to the enlarged cross-sectional area. A high working load is desirable for membrane handling, module fabrication, and backwashing. A work from Wang and Chung shows that the burst pressure of MBF is also improved, indicating that the multibore structure can enhance fiber's mechanical properties in the hoop direction (Wang and Chung, 2013). On the other hand, the maximum tensile stress, strain, and Young's modulus of MBF are comparable to those of SBF. This is because the change of geometry has little impact on the membrane microstructure which is determined by the intrinsic properties of polymers, dope systems, and phase inversion conditions.

Table 6.1 also shows the MD performances of MBF and SBF membranes. It can be seen that in most cases SBF exhibits a slightly higher flux than MBF. This is because MBF membranes have interconnected sponge-like structures that reduce the effective membrane areas for vapor transport. On the other hand, since MBF membranes have superior mechanical properties, they can be fabricated with higher porosities which will reduce the mass transfer resistance and improve the vapor flux. Lu et al. have applied a mixture of isopropanol (IPA) and water as the external coagulant to increase the porosity and pore size of the tri-bore hollow fibers. With an IPA/water weight ratio of 6/4, a tri-bore membrane with an overall porosity of 86.1% and the maximum surface pore size of 327 nm was obtained. The resultant membrane not only exhibited excellent mechanical properties but also achieved a remarkable flux of 28 kg m^{-2} h^{-1} at 60 °C (Lu et al., 2016).

6.2.3 Comparing Multibore Membranes with Different Geometries

Multibore membranes with different geometries can be categorized into two groups by the outer contours, namely, rectangular and round shapes. The former contains bore channels aligning in a row and the latter has bore channels aligning around a circle center. The rectangular multibore membranes have the combined advantages offered by both hollow fiber and flat sheet membranes. They have a high packing density and improved mechanical durability and compressibility (Teoh et al., 2011). On the other hand, it was found that forming rectangular multibore membranes was very challenging because of the rapid die swell and chain relaxation. In addition, the cluster

TABLE 6.1

A comparison of mechanical properties and MD performances of MBF with SBF

Reference	Geometry	Max load (N)	Max tensile stress (MPa)	Max tensile strain (%)	Young's modulus (MPa)	Burst pressure (bar)	MD configuration	Feed temperature (°C)	Flux (kg m^{-2} h^{-1})
Wang and Chung, 2012	Seven-bore round	7.28 ± 0.47	0.92 ± 0.06	56.3 ± 3.9	28.9 ± 5.7	–	DCMD	80	28
	Single-bore	0.82 ± 0.06	0.91 ± 0.07	41.3 ± 6.6	27.9 ± 6.7	–	DCMD	80	36
Teoh et al., 2011	Seven-bore rectangular	2.87 ± 0.23	0.52 ± 0.04	121 ± 12	12.6 ± 2.9	–	DCMD	80	53.57
	Single-bore	0.52 ± 0.01	0.93 ± 0.03	143 ± 4	19.7 ± 2.5	–	DCMD	80	51.12
Lu et al., 2016	Tri-bore round	3.0 ± 0.2	1.8 ± 0.1	122.4 ± 8.5	30.5 ± 4.9	–	DCMD	60	17.2
	Single-bore	0.9 ± 0.1	1.6 ± 0.1	137.6 ± 12.3	27.5 ± 9.1	–	DCMD	60	19.9
Wang and Chung, 2013	Seven-bore round	13.33 ± 0.57	1.37 ± 0.03	48.4 ± 7.2	22.31 ± 3.8	2.83 ± 0.17	VMD	60	18
	Six-bore	15.7 ± 0.55	1.64 ± 0.05	41.9 ± 1.7	33.7 ± 7.29	3.07 ± 0.13	VMD	60	16.5
	Single-bore	1.31 ± 0.12	1.15 ± 0.01	41.5 ± 1.8	21.39 ± 3.3	2.57 ± 0.17	VMD	60	19
Chang et al., 2019	Tri-bore triangle	–	2.03 ± 0.02	102.0 ± 5.0	52.1 ± 1.5	–	VMD	60	37.7

agglomerations of rectangular membranes were more severe than that of round hollow fiber membranes as the latter could spread out more efficiently in a membrane module (Teoh et al., 2011; Wang and Chung, 2012). In contrast, lotus-root-like round shape MBFs are easier to fabricate, and they are more widely investigated and applied compared to rectangular membranes. Moreover, variations such as wheel shape or triangular membranes can also be fabricated using the round shape spinnerets by manipulating the spinning conditions. Such membranes may have higher packing densities, more uniform wall thicknesses, and smaller mass transfer resistances compared to round shape membranes.

Over the years, MBFs with different numbers of bore channels have been developed. Among them, seven-bore and tri-bore MBFs are more commonly studied than other geometries. Seven-bore MBFs exhibit excellent mechanical stabilities and good MD performances. However, the center bore channel is less efficiently used than the surrounding ones because the diffusion pathway in the center bore channel is longer (Wang and Chung, 2012). The design of tri-bore MBF eliminates the center bore. Thus, all the bore channels of tri-bore MBF have the same diffusion paths for vapor transfer. In addition, fewer channels of tri-bore hollow fibers also simplify the flow balance, spinneret design, and module fabrication (Lu et al., 2016).

Even though some conclusions can be drawn by comparing different works, it is still difficult to come out with a selection criterion due to the limited number of papers on MBFs. Direct comparisons of MBFs from different studies may be unfair because the geometry is not the only factor affecting membrane properties. Discrepancies on membrane morphology, diameter, porosity, and wall thickness should also be considered for a more convictive comparison. An unbiased comparison could possibly be achieved by a mathematical simulation in the future studies.

6.3 CONCLUSIONS

In this chapter, the development of dual-layer and multibore hollow fiber membranes for MD applications has been discussed. The dual-layer technique is used to fabricate (1) morphologically controlled membranes, (2) membranes with enhanced mechanical integrity, and (3) hydrophobic/hydrophilic membranes. All these dual-layer hollow fiber membranes will contribute to at least one of the desirable MD membrane properties. For instance, the hydrophobic/hydrophilic membranes could result in a high membrane flux owing to the reduced vapor transport resistance. Moreover, the hydrophobic/hydrophilic membranes could also enhance the membrane mechanical strength because the hydrophilic inner layer can provide an additional mechanical support. Thus, a dual-layer membrane provides the opportunity for a better control of membrane structure and chemistry, and the synergistic effects of each layer could lead to the optimal membrane performance.

In the second half of the chapter, MBFs with various geometries have been introduced and the unique effects of three spinning parameters, namely, the bore flow rate, dope flow rate, and take-up speed on the formation of MBF have been summarized. It is evident that multibore structures can significantly improve the mechanical strength of membranes, which brings many benefits to the membrane handling and module fabrication. In addition, MBFs can be made very porous in order to achieve high vapor fluxes. Finally, a comparison of different geometries is given in order to provide some guidance for the selection and fabrication of MBF MD membranes.

REFERENCES

BASF, 2013. inge GmbH, Multibore® membranes [WWW Document]. URL www.inge.basf.com/ev/internet/inge/en/content/inge/Produkte/Multibore_Membran (accessed 12.20.2018).

Bonyadi, S., Chung, T.S., 2007. Flux enhancement in membrane distillation by fabrication of dual layer hydrophilic-hydrophobic hollow fiber membranes. J. Memb. Sci. 306, 134–146. doi:10.1016/j.memsci.2007.08.034

Bonyadi, S., Mackley, M., 2012. The development of novel micro-capillary film membranes. J. Memb. Sci. 389, 137–147. doi:10.1016/j.memsci.2011.10.023

Bottino, A., Capannelli, G., Munari, S., Turturro, A., 1988. High performance ultrafiltration membranes cast from LiCl doped solutions. Desalination 68, 167–177. doi:10.1016/0011–9164(88)80052–3

Bu-Rashid, K.A., Czolkoss, W., 2007. Pilot tests of multibore UF membrane at addur SWRO Desalination Plant, Bahrain. Desalination 203, 229–242. doi:10.1016/j.desal.2006.04.010

Chang, J., Zuo, J., Lu, K.J., Chung, T.S., 2019. Membrane development and energy analysis of freeze desalination-vacuum membrane distillation hybrid systems powered by LNG regasification and solar energy. Desalination 449, 16–25. doi:10.1016/j.desal.2018.10.008

Cheng, D.Y., Wiersma, S.J., 1982. Composite membrane for a membrane distillation system. 4,316,772.

Cheng, D.Y., Wiersma, S.J., 1983. Composite membrane for a membrane distillation system. 4,419,242.

Chi, Y., Li, T., Wang, B., Wu, Z., Li, K., 2017. Morphology, performance and stability of multi-bore capillary La0.6Sr0.4Co0.2Fe0.8O3-δoxygen transport membranes. J. Memb. Sci. 529, 224–233. doi:10.1016/j.memsci.2017.02.010

Çulfaz, P.Z., Rolevink, E., van Rijn, C., Lammertink, R.G.H., Wessling, M., 2010. Microstructured hollow fibers for ultrafiltration. J. Memb. Sci. 347, 32–41. doi:10.1016/j.memsci.2009.10.003

Deng, J., Zhang, Y., Liu, J., Zhang, H., 2012. Preparation of three-bore hollow fiber charged nanofiltration membrane for separation of organics and salts. Water Sci. Technol. 65, 171–176. doi:10.2166/wst.2011.851

Edwie, F., Teoh, M.M., Chung, T.S., 2012. Effects of additives on dual-layer hydrophobic-hydrophilic PVDF hollow fiber membranes for membrane distillation and continuous performance. Chem. Eng. Sci. 68, 567–578. doi:10.1016/j.ces.2011.10.024

Essalhi, M., Khayet, M., 2012. Surface segregation of fluorinated modifying macromolecule for hydrophobic/hydrophilic membrane preparation and application in air gap and direct contact membrane distillation. J. Memb. Sci. 417–418, 163–173. doi:10.1016/j.memsci.2012.06.028

Fu, F.J., Sun, S.P., Zhang, S., Chung, T.S., 2014. Pressure retarded osmosis dual-layer hollow fiber membranes developed by co-casting method and ammonium presulfate (PAS) treatment. J. Memb. Sci. 469, 488–498. doi:10.1016/j.memsci.2014.05.063

Fu, F.J., Zhang, S., Sun, S.P., Wang, K.Y., Chung, T.S., 2013. POSS-containing delamination-free dual-layer hollow fiber membranes for forward osmosis and osmotic power generation. J. Memb. Sci. 443, 144–155. doi:10.1016/j.memsci.2013.04.050

He, T., Mulder, M.H.V., Strathmann, H., Wessling, M., 2002. Preparation of composite hollow fiber membranes: Co-extrusion of hydrophilic coatings onto porous hydrophobic support structures. J. Memb. Sci. 207, 143–156. doi:10.1016/S0376-7388(02)00118-7

Heijnen, M., Winkler, R., Berg, P., 2012. Optimisation of the geometry of a polymeric multibore ® ultrafiltration membrane and its operational advantages over standard single bore fibres. Desalin. Water Treat. 42, 24–29. doi:10.1080/19443994.2012.682968

Henne, W., Dunweg, G., Schmitz, W., Pohle, R., Lawitzki, F., 1979. Method of producing dialyzing membrane. US Patent 4164437

Kao, S.T., Teng, M.Y., Li, C.L., Hsieh, C.Y., Tsai, H.A., Wang, D.M., Lee, K.R., Lai, J. Y., 2008. Fabrication PC/PAN composite membranes by vapor-induced phase separation. Desalination 233, 96–103. doi:10.1016/j.desal.2007.09.031

Khayet, M., Matsuura, T., 2003. Application of surface modifying macromolecules for the preparation of membranes for membrane distillation. Desalination 158, 51–56.

Khayet, M., Matsuura, T., 2011. Membrane distillation principles and applications, 1st ed. Elsevier B.V., Amsterdam.

Khayet, M., Mengual, J.I., Matsuura, T., 2005. Porous hydrophobic/hydrophilic composite membranes: Application in desalination using direct contact membrane distillation. J. Memb. Sci. 252, 101–113. doi:10.1016/j.memsci.2004.11.022

Li, D.F., Chung, T.S., Wang, R., 2004. Morphological aspects and structure control of dual-layer asymmetric hollow fiber membranes formed by a simultaneous co-extrusion approach. J. Memb. Sci. 243, 155–175. doi:10.1016/j.memsci.2004.06.014

Li, D.F., Chung, T.S., Wang, R., Liu, Y., 2002. Fabrication of fluoropolyimide/polyethersulfone (PES) dual-layer asymmetric hollow fiber membranes for gas separation. J. Memb. Sci. 198, 211–223. doi:10.1016/S0376-7388(01)00658-5

Lu, K.J., Zuo, J., Chung, T.S., 2016. Tri-bore PVDF hollow fibers with a super-hydrophobic coating for membrane distillation. J. Memb. Sci. 514, 165–175. doi:10.1016/j.memsci.2016.04.058

Luo, L., Wang, P., Zhang, S., Han, G., Chung, T.S., 2014. Novel thin-film composite tri-bore hollow fiber membrane fabrication for forward osmosis. J. Memb. Sci. 461, 28–38. doi:10.1016/j.memsci.2014.03.007

Ma, C., Wu, X., Liu, Z., 2017. Performance and fouling characterization of a five-bore hollow fiber membrane in a membrane bioreactor for the treatment of printing and dyeing wastewater. Text. Res. J. 87, 102–109. doi:10.1177/0040517515624876

Ong, Y.K., Chung, T.S., 2012. High performance dual-layer hollow fiber fabricated via novel immiscibility induced phase separation (I2PS) process for dehydration of ethanol. J. Memb. Sci. 421–422, 271–282. doi:10.1016/j.memsci.2012.07.025

Ong, Y.K., Shi, G.M., Le, N.L., Tang, Y.P., Zuo, J., Nunes, S., Chung, T.S., 2016. Recent membrane development for pervaporation processes. Prog. Polym. Sci. 57, 1–31. doi:10.1016/j.progpolymsci.2016.02.003

Peng, N., Teoh, M.M., Chung, T.S., Koo, L.L., 2011. Novel rectangular membranes with multiple hollow holes for ultrafiltration. J. Memb. Sci. 372, 20–28. doi:10.1016/j.memsci.2011.01.022

Peng, N., Widjojo, N., Sukitpaneenit, P., Teoh, M.M., Lipscomb, G.G., Chung, T.S., Lai, J.Y., 2012. Evolution of polymeric hollow fibers as sustainable technologies:

past, present, and future. Prog. Polym. Sci. 37, 1401–1424. doi:10.1016/j.progpolymsci.2012.01.001

Phattaranawik, J., Jiraratananon, R., Fane, A.G., 2003. Heat transport and membrane distillation coefficients in direct contact membrane distillation. J. Membr. Sci. 212, 177–193.

Ren, J., Chung, T.S., Li, D., Wang, R., Liu, Y., 2002. Development of asymmetric 6FDA-2,6 DAT hollow fiber membranes for CO2/CH4 separation. 1. The influence of dope composition and rheology on membrane morphology and separation performance. J. Memb. Sci. 207, 227–240. doi:10.1016/S0376-7388(02)00251–X

Su, M., Teoh, M.M., Wang, K.Y., Su, J., Chung, T.S., 2010. Effect of inner-layer thermal conductivity on flux enhancement of dual-layer hollow fiber membranes in direct contact membrane distillation. J. Memb. Sci. 364, 278–289. doi:10.1016/j.memsci.2010.08.028

Sun S.P., Wang, K.Y., Peng, N., Hatton, T.A., Chung, T.S., 2010. Novel polyamide-imide/cellulose acetate dual-layer hollow fiber membranes for nanofiltration. J. Memb. Sci. 363, 232–242. doi:10.1016/j.memsci.2010.07.038

Teoh, M.M., Chung, T.S., 2009. Membrane distillation with hydrophobic macrovoid-free PVDF-PTFE hollow fiber membranes. Sep. Purif. Technol. 66, 229–236. doi:10.1016/j.seppur.2009.01.005

Teoh, M.M., Peng, N., Chung, T.S., Koo, L.L., 2011. Development of novel multichannel rectangular membranes with grooved outer selective surface for membrane distillation. Ind. Eng. Chem. Res. 50, 14046–14054. doi:10.1021/ie201292j

Tomaszewska, M., 1996. Preparation and properties of flat-sheet membranes from poly (vinylidene fluoride) for membrane distillation. Desalination 104, 1–11.

Tomaszewska, M., 1999. Membrane distillation. Environ. Prot. Eng. 25, 37–47.

Tsai, H.A., Li, L.D., Lee, K.R., Wang, Y.C., Li, C.L., Huang, J., Lai, J.Y., 2000. Effect of surfactant addition on the morphology and pervaporation performance of asymmetric polysulfone membranes. J. Memb. Sci. 176, 97–103. doi:10.1016/S0376-7388(00)00435–X

Wang, D., Li, K., Teo, W.K., 2000. Porous PVDF asymmetric hollow fiber membranes prepared with the use of small molecular additives. J. Memb. Sci. 178, 13–23. doi:10.1016/S0376–7388(00)00460–9

Wang, K.Y., Foo, S.W., Chung, T., 2009. Mixed matrix PVDF hollow fiber membranes with nanoscale pores for desalination through direct contact membrane distillation. Ind. Eng. Chem. Res. 4474–4483. doi:10.1021/ie8009704

Wang, K.Y., Li, D.F., Chung, T.S., Chen, S.B., 2004. The observation of elongation dependent macrovoid evolution in single-and dual-layer asymmetric hollow fiber membranes. Chem. Eng. Sci. 59, 4657–4660. doi:10.1016/j.ces.2004.06.035

Wang, P., Chung, T.S., 2012. Design and fabrication of lotus-root-like multi-bore hollow fiber membrane for direct contact membrane distillation. J. Memb. Sci. 421–422, 361–374. doi:10.1016/j.memsci.2012.08.003

Wang, P., Chung, T.S., 2013. A new-generation asymmetric multi-bore hollow fiber membrane for sustainable water production via vacuum membrane distillation. Environ. Sci. Technol. 47, 6272–6278. doi:10.1021/es400356z

Wang, P., Luo, L., Chung, T.S., 2014. Tri-bore ultra-filtration hollow fiber membranes with a novel triangle-shape outer geometry. J. Memb. Sci. 452, 212–218. doi:10.1016/j.memsci.2013.10.033

Wang, P., Teoh, M.M., Chung, T.S., 2011. Morphological architecture of dual-layer hollow fiber for membrane distillation with higher desalination performance. Water Res. 45, 5489–5500. doi:10.1016/j.watres.2011.08.012

Xia, Q.C., Liu, M.L., Cao, X.L., Wang, Y., Xing, W., Sun, S.P., 2018. Structure design and applications of dual-layer polymeric membranes. J. Memb. Sci. 562, 85–111. doi:10.1016/j.memsci.2018.05.033

Yanagimoto, T., 1987. Manufacuring of ultrafiltration membranes. Japanese Patent 62019205

Yang, X., Wang, R., Shi, L., Fane, A.G., Debowski, M., 2011. Performance improvement of PVDF hollow fiber-based membrane distillation process. J. Memb. Sci. 369, 437–447. doi:10.1016/j.memsci.2010.12.020

Zhu, J., Dong, Z., Liu, Z., Zhang, K., Zhang, G., Jin, W., 2014. Multichannel mixed-conducting hollow fiber membranes for oxygen separation. AIChE J. 60, 1969–1976. doi:10.1002/aic

Zuo, J., Chung, T.S., 2013. Design and synthesis of a fluoro-silane amine monomer for novel thin film composite membranes to dehydrate ethanol via pervaporation. J. Mater. Chem A. 1, 9814–9826. doi:10.1039/c3ta11728f

Zuo, J., Chung, T.S., O'Brien, G.S., Kosar, W., 2017. Hydrophobic/hydrophilic PVDF/ Ultem® dual-layer hollow fiber membranes with enhanced mechanical properties for vacuum membrane distillation. J. Memb. Sci. 523, 103–110. doi:10.1016/j.memsci.2016.09.030

7 Hydrophobic Modifications of Membranes with Improved Anti-Wetting Resistance

Dieling Zhao

7.1 INTRODUCTION

Hydrophobicity is an essential requirement of a membrane distillation (MD) membrane, with which the membrane is able to repel the feed solution and keep the membrane pores dry. However, progressive membrane wetting occurs in a continuous MD operation by vapor condensation and liquid penetration. This leads to inefficient MD performance with a low-quality permeate and is becoming one of the major challenges for MD commercialization (Alkhudhiri et al., 2012; Curcio and Drioli, 2005). Different approaches have been proposed to control wetting by means of process design and system optimization, such as pretreatment of feed solutions, regulations of operation conditions like temperature and flow rate (Rezaei et al., 2017, 2018; Wang and Chung, 2015). Nevertheless, extensive researches have also focused on molecular designs of membrane materials and structures in order to inherently prevent membranes from wetting (Eykens et al., 2017).

First of all, to lessen the possibility of pore wetting, hydrophobic (nonwetting) microporous polymeric or inorganic membranes have been employed in MD processes. Polymeric membranes are usually made from intrinsic or modified hydrophobic polymers with low surface energy, such as polytetrafluoroethylene (PTFE), polypropylene (PP), or polyvinylidene fluoride (PVDF), while inorganic ones are made from ceramics (Rezaei et al., 2018). Furthermore, enhanced surface hydrophobicity can be achieved by surface roughening, grafting of low surface energy materials, or combination of both methods (Enrico et al., 2015; Khayet and Matsuura, 2011).

The conventional method to assess the surface hydrophobicity of a MD membrane is by means of contact angle (θ_{CA}) measurements. Typically, a liquid droplet is dropped on a membrane surface and then its θ_{CA} is measured by a goniometer which is the angle between the wetted surface of the membrane and a tangent line drawn from the three-phase contact point to the curved drop face (Onsekizoglu et al., 2012). If $\theta_{CA} > 90°$, the membrane surface is considered as hydrophobic; otherwise, it is hydrophilic. If θ_{CA} reaches above 150°, the membrane is categorized as superhydrophobic (Figure 7.1). The advantage of this approach is its simplicity to perform. However, contact angle measurements are influenced by surface structure (e.g., roughness, pore size) and show hysteresis behavior (Rezaei et al., 2018). The static contact angle of a flat sheet membrane can be determined with the sessile drop method, by measuring the angle between the membrane surface and a water droplet using a contact angle goniometer. The hydrophobicity of a hollow fiber membrane can be determined by a force tensiometer which measures the force affecting the balance when a hollow fiber is brought in contact with a test liquid (Zhao and Jiang, 2018). The contact angle can then be calculated by using Equation (7.1) when the perimeter of the hollow fiber (P) and surface tension of the liquid (γ_1) are known.

$$\text{Wetting force} = \gamma_1 \mathrm{P} \cos\theta \tag{7.1}$$

Besides the contact angle, liquid entry pressure (LEP) is another primary metric for measuring the wettability of a MD membrane. While the contact angle is a surface property only, LEP is also affected by wettability inside of the pores. It is the minimal hydraulic pressure at which the feed solution penetrates into the hydrophobic membrane (Rezaei et al., 2018). LEP mainly depends on the maximum pore size and membrane hydrophobicity. The number is also directly related to the feed solution (i.e., solution concentration and liquid surface tension). The LEP value can be calculated by the Laplace–Young (Equation (7.2))

$$\mathrm{LEP} = \Delta P_{entry} = -B\gamma_1 \cos\theta / r_{\max} \tag{7.2}$$

where ΔP_{entry} is the entry pressure difference, B is the geometric pore coefficient based on the geometry of membrane pores (e.g., B is 1 for cylindrical pores, 0.4–0.6 for stretched membranes with a small curvature radius, and less than 1 for elliptical or irregularly shaped pores) (Saffarini et al., 2013; Warsinger et al., 2016a, 2016b, 2016c), γ_1 is the surface tension of the feed liquid, θ is the contact angle between the solution and membrane surface, and $r_{\max}$ is the radius

FIGURE 7.1 Classification of hydrophilicity and hydrophobicity of membranes according to their contact angles.

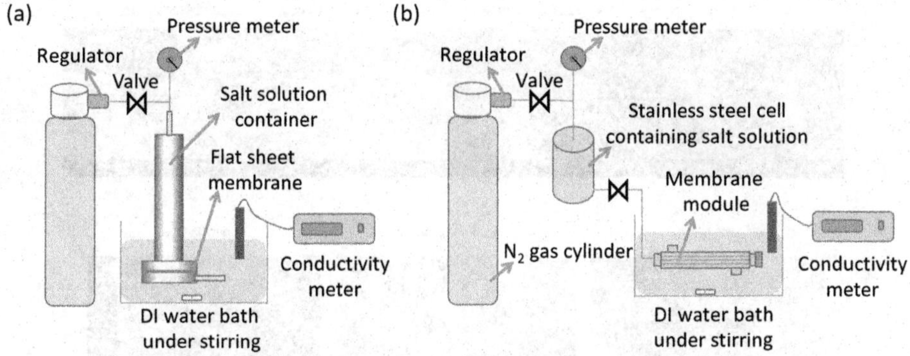

FIGURE 7.2 Scheme of the setups used to measure LEP of (a) flat sheet and (b) hollow fiber membranes. Reprinted with permission from Zhao et al. (2017). Copyright (2017) Elsevier.

of the maximum membrane pore size. According to this equation, membranes that possess high contact angles, small pore sizes, and low surface energy exhibit high LEP values for feed solutions with high surface tension.

Researchers usually report actual LEP values using setups as shown in Figure 7.2. Typically, the flat sheet or hollow fiber membrane is placed in a dead end cell or assembled in a module. The hydrophobic membrane surface is fed with a salt solution under a certain pressure provided by a compressed nitrogen tank. The LEP value is the minimum pressure when the conductivity of the permeation solution starts to increase significantly. In case of MD operation under fluctuating pressure and temperature, a membrane with LEP of minimum 2.5 bar is recommended (Eykens et al., 2017).

7.2 BULK MODIFICATIONS OF MEMBRANES

Among the materials investigated as MD membranes, hydrophobic copolymers such as PVDF-PTFE, poly(vinylidene fluoride-co-hexafluoropropylene) (PVDF-HEP), and poly(vinylidene fluoride-co-chlorotrifluoroethylene) (PVDF-CTFE) are preferred because of their advantages of easy fabrication, modification and scale-up, as well as relatively low costs. These fluoropolymers have characteristics of low surface tension, excellent chemical resistance, and high mechanical strength (García-Payo et al., 2009, 2010; Gugliuzza and Drioli, 2007; Lalia et al., 2014; Teoh et al., 2009a, 2009b, Teoh et al., 2011).

Teoh and coworkers have incorporated PTFE microparticles into PVDF dope solutions. The resultant membranes showed a high water contact angle of 130° (Teoh and Chung, 2009a, 2009b; Teoh et al., 2011). Garcia-Payo's group fabricated hollow fiber membranes using PVDF-HEP (García-Payo et al., 2009, 2010). In addition, the copolymer of tetrafluoroethylene (TFE) and 2,2,4-trifluoro-5-trifluoromethoxy-1,3-dioxole (TTD), namely, Hyflon AD (Solway

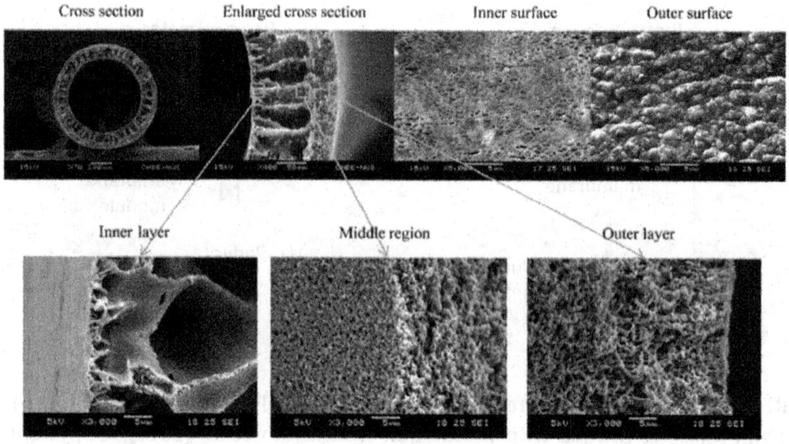

FIGURE 7.3 Structure of a dual-layer PVDF hollow fiber with a sponge-like outer layer and a fully finger-like inner layer. Reprinted with permission from Wang et al. (2011). Copyright (2011) Elsevier.

Plastics) has been used to prepare asymmetric membranes with contact angles higher than 120° (Gugliuzza and Drioli, 2007). Furthermore, Maab et al. have prepared MD membranes from aromatic fluorinated polyoxadiazoles and polytriazoles. The resultant membranes exhibited high porosity and superhydrophobicity with water contact angles up to 160° (Maab et al., 2012, 2013).

Membrane morphology also plays an important role in determining the wetting behavior of a MD membrane (Wang et al., 2009). For example, researchers have found that finger-like microvoids in the membrane matrix can reduce the LEP value. Since a faster solvent/nonsolvent exchange rate is responsible for the formation of microvoids, a fast demixing should be avoided in the wet/dry spinning process (Wang and Chung, 2015). Therefore, PTFE particles have been incorporated into the dope solutions not only to reduce the demixing rate during membrane formation but also to enhance the hydrophobicity. For example, Gryta and Barancewicz blended PVDF and PTFE together to form a sponge-like membrane structure with higher hydrophobicity (Gryta and Barancewicz, 2010). Wang et al. fabricated PVDF dual-layer hollow fibers with a sponge-like outer layer consisting of PTFE particles and a fully finger-like inner layer (Figure 7.3) to achieve high wetting resistance and vapor permeation simultaneously (Wang et al., 2011).

7.3 SURFACE MODIFICATIONS

7.3.1 Surface Modifications on Polymeric Membranes

The wetting property of a membrane is determined by both geometrical structure and surface chemistry. Therefore, apart from the aforementioned bulk modifications, many researches have focused on surface modification to manipulate the surface nonwettability of polymeric MD membranes. Generally, they can be classified into three categories, namely, hydrophobic polymeric coating, incorporation of surface modification macromolecules (SMMs), and plasma treatment (Ismail et al., 2018).

Various hydrophobic polymeric materials have been applied for surface coating on MD membranes, such as silicon rubber, styrene-butadiene, perfluoropolyether, and Teflon (Jin et al., 2008; Jung et al., 2016; Lu et al., 2016; Sanguineti et al., 2013; Shaulsky et al., 2017; Yang et al., 2011a). Jin et al. prepared a poly (phthalazinone ether sulfone ketone) (PPESK) hollow fiber membrane and coated its surface by silicone rubber and sol-gel polytrifluoropropylsiloxane. The contact angle and LEP of the resultant membrane increased to 100° and 0.12 MPa, respectively, while the permeation flux reduced remarkably if the silicone rubber solution had a high coating concentration (Jin et al., 2008). A hydrophilic polyamide (PA) membrane was also successfully coated with hydrophobic styrene-butadiene rubber and exhibited a LEP over 1.3 bar (Jung et al., 2016). Yang and coworkers modified the PVDF membrane surface with perfluoropolyether and obtained a membrane with a water contact angle of 115° and a LEP value of 3.9 bar (Yang et al., 2011b). Solvay has filed a patent to coat a perfluoropolyether polymer (Fluorolink) on the surface of polyethersulfone, regenerated cellulose, and PVDF membranes. The water contact angles and LEP of these three types of membranes exceeded 125° and 2.2 bar, respectively (Sanguineti et al., 2013). Lu et al. coated Teflon AF 2400 on a PVDF tri-bore hollow fiber by dip coating. The water contact angle of the resultant membrane increased from 105° to 150°, while the LEP value jumped from 0.58 bar to 1.4 bar (Lu et al., 2016).

The employment of SMMs helps to create composite membranes when being blended with hydrophilic base polymers. As the top skin facing the feed solution governs the membrane separation performance, the goals of forming composite membranes are to (1) make their top skin as thin as possible and (2) fasten the mass transfer and heat transfer of water vapor to the permeate side (Khayet and Matsuura, 2003a; Khayet et al., 2005; Qtaishat, 2009). SMMs are oligomeric fluoropolymers prepared via polyurethane chemistry and tailored with two polarity fluorinated groups. The membranes are prepared by the one-step phase inversion in which SMMs can migrate to the membrane surface during membrane formation and change the surface properties while maintaining the bulk properties. Additionally, only a small amount of SMMs are required. Khayet and Matsuura's group has extensively worked on SMMs modified MD membranes. During the phase inversion processes, hydrophobic and hydrophilic segments of SMMs moved to the top and bottom parts of the polyethylenimine (PEI) membranes, respectively. SMMs helped increase the LEP value from 0.1×10^5 Pa to above 2.94×10^5 Pa (Khayet and Matsuura, 2003b). Essalhi et al. also blended SMMs in PEI and obtained an MD membrane with the contact angle of the surface higher than the bottom (94.4° and 78.6°, respectively) (Essalhi and Khayet, 2012).

During the plasma treatment, the monomers on membrane surface are ionized to form free and active radicals, electrons, and ions which would subsequently adsorb, condense, and polymerize on the membrane surface. Meanwhile, the surface morphology may change since the surface is etched during the plasma treatment. Since the treatment changes the surface hydrophilicity in a nanometer scale, it does not affect the bulk membrane properties. As a consequence, the resultant composite membranes are similar to those by the SMM treatment (Wang and Chung, 2015; Yasuda, 1984). Besides, the resultant hydrophobic layer exhibits robust adhesion with the membrane surface. However, this method requires expensive equipment and extra care on the handling of toxic monomer gases. The key factors affecting the final membrane morphology and performance during the plasma treatment are (1) vapor pressure of monomer reactants, (2) voltage and power in the discharge reaction, and (3) temperature and (4) type of membranes (Eykens et al., 2017). Early in 1992, Kong et al. modified the surface of a cellulose nitrite membrane by plasma treatment of octafluorocyclobutane and achieved a membrane surface with a water contact angle as high as 120° (discharge power of 200 W for 5 min) (Kong et al., 1992). However, the flux decreased as a result of pore blocking when the discharge time was longer. In addition, the CF_4 plasma treatment introduced a moderate surface etching and abundant fluorine functional groups. Therefore, this approach could reduce the surface energy and enhance the surface roughness, thus making the membrane surface more hydrophobic (Chul Woo et al., 2017; Gancarz et al., 2015; Tian et al., 2015; Wei et al., 2012; Yang et al., 2014). Similarly, Wei et al. converted the surface of a hydrophilic polyethersulfone (PES) membrane to a hydrophobic one with a water contact angle higher than 120° (Wei et al., 2012).

7.3.2 Surface Modifications on Ceramic Membranes

Ceramic membranes are usually made from metal oxides such as alumina, zirconia, silica, titania, or a combination. They usually exhibit high chemical and thermal stability and thus are more appropriate for modification. In addition, their mechanical properties are relatively higher than polymeric membranes and do not easily be deformed during the long-term operation (da Silva Biron et al., 2018). However, ceramic membranes are hydrophilic due to the existence of abundant hydroxyl groups (–OH) which help absorb water into the pores. One must modify these hydroxyl groups via substitution with hydrophobic moieties so that the resultant ceramic membranes become hydrophobic for MD applications (Larbot et al., 2004). The modification methods for ceramic membranes are different from those for polymeric membranes because (1) the former can withstand higher temperatures than the latter and (2) their terminal functional groups are different. Hydrophobic ceramic membranes are usually prepared through surface coating using fluoroalkylsilanes (FAS). Organosilanes are silanes that contain at least one carbon–silicon bond. FAS is one type of organosilanes which also contain CF_2–CF_2 and CF_2–CF_3 groups to provide lower surface tension and hydrophobic characteristics (Krajewski et al., 2006). Generally, FAS is grafted on ceramic membrane surfaces via immersion, chemical vapor deposition (CVD), and sol-gel methods.

Among them, immobilization of organosilanes on membrane surfaces by immersion is most popular because of its relatively simpler procedure and shorter treatment time (Ahmad et al., 2015; Gazagnes et al., 2007; Koonaphapdeelert and Li, 2007; Krajewski et al., 2006; Kujawa et al., 2013; Larbot, 2004; Lu et al., 2009; Sah et al., 2004). In this process, organosilanes are first hydrolyzed in solvents so that active silanols are formed, which then react with hydroxyl groups on the surface of ceramic membranes. Since the coating layer is usually very thin (<2 μm), it is difficult to prepare a defect-free hydrophobic membrane. Moreover, the coating solution may penetrate into membrane pores and block pores. The effectiveness of this approach in terms of hydrophobicity enhancement is directly related to the type of functional groups in the organosilane molecules, hydrophobic tail length, immersion duration, hydroxyl group number on the membrane surface, surface roughness, and grafting multiplicity (Koonaphapdeelert and Li, 2007; Kujawski et al., 2016). Organosilanes with two or three silanols can further react with the adjacent molecules through Si–O–Si bonds which leads to form hydrophobic complexes (Fadeev and McCarthy, 2000). For example, Hendren et al. treated alumina ceramic membranes separately by perfluorodecyltriethoxysilane, trimethylchlrosilane, and trichloromethylsilane (Hendren et al., 2009). Their LEP increased from 246 kPa to ~550 kPa and contact angles jumped to as high as 140° after modifications. However, only the perfluorodecyltriethoxysilane coating did not result in pore blockage, thus its modified membrane had the highest water flux among those three membranes (Hendren et al., 2009).

Similar to the immersion method, covalent bonds can be formed between the hydroxyl groups on ceramic membrane surfaces and organosilanes by CVD because ceramic membranes can withstand modifications at high temperatures due to their inherently high thermal stability. Ahmad et al. heated the FAS solution to its boiling point during CVD so that its reactive vapor reacted with the hydroxyl groups on the membrane surface (Ahmad et al., 2015).

By means of the sol-gel method, a ceramic membrane is first dip-coated in a sol-gel containing FAS, followed by the pyrolysis of the ceramic membrane at high temperatures to remove the sol. Wang et al. coated a polydimethylsiloxane (PDMS) layer on β-sialon ceramic membranes (Wang et al., 2017). An increasing water contact angle up to 140° was obtained with growing pyrolysis temperature from 200 to 600 °C, but a further increase in pyrolysis temperature led to a decreased surface hydrophobicity since the polymer was burnt off at such high temperatures.

7.4 NANOMATERIALS MODIFIED HYDROPHOBIC MD MEMBRANES

Recently, nanomaterials have attracted significant attention in various fields including membrane modification. As mentioned above, surface chemistry changes the surface tension at the microscopic level, but the geometrical structure determines how the forces act upon the feed liquid. In the field of membrane distillation, researchers have devoted to mimicking the natural superhydrophobic surfaces through the formation of hierarchy surface structures with deployment of emerging nanomaterials (An et al., 2017; Baghbanzadeh et al., 2015; Boo et al., 2016; Dong et al., 2015; Efome et al., 2016; Fan et al., 2017; Hamzah and Leo, 2017; Hou et al., 2012; Lee et al., 2016, 2017; Ma et al., 2009; Meng et al., 2014a, 2014b; Rezaei and Samhaber, 2016; Sukitpaneenit and Chung, 2012; Woo et al., 2016).

Wang and his colleagues have fabricated nanoporous mixed matrix membranes by the addition of hydrophobic nanoclay particles (Wang et al., 2009). These PVDF/cloisite clay composite membranes prepared by dry-jet wet phase inversion had longer sustainability in continuous MD operations than the neat PVDF ones because the former had small surface pores and higher hydrophobicity than the latter. Similarly, Prince et al. prepared PVDF-hydrophobic clay nanocomposite membranes via electrospinning process (Prince et al., 2012). The addition of clay nanoparticles enhanced the water contact angle and LEP, respectively, from 128.0° to 154.2° and from 90 to 200 kPa.

Razmjou and his coworkers utilized sol-gel prepared fluorosilanized TiO_2 nanoparticles to modify MD membrane surface and obtained hierarchical structures with a significantly enhanced water contact angle up to 166° (Figure 7.4) (Razmjou et al., 2012). Lu et al. modified the surface of porous PVDF microfiltration membranes using 1H,1H,2H,2H-perfluorooctyl trichlorosilane (PFTS) and fluorinated SiO_2 nanoparticle suspensions via different methods. The resultant membranes displayed superhydrophobicity with a water contact angle up to 167.3° (Lu et al., 2016).

Lu et al. also added n-butylamine modified graphene oxide (GO-NBA) in PVDF membranes for direct contact membrane distillation. With a loading of 0.5 wt% GO-NBA, the LEP value of membranes increased by 67% together with enhancements in mechanical strength and burst pressure (Lu et al., 2017). Zhao and her colleagues modified the surface of PVDF membranes with fluorographite nanoparticles by filtration (Zhao et al., 2017). Both the superhydrophobicity of fluorographite and multilevel surface roughness gave rise to the high pore-wetting resistance in the continuous MD process for seawater desalination. Jafari's group prepared carbon quantum dots (CQDs)/PVDF electrospun membranes for air gap membrane distillation (AGMD)

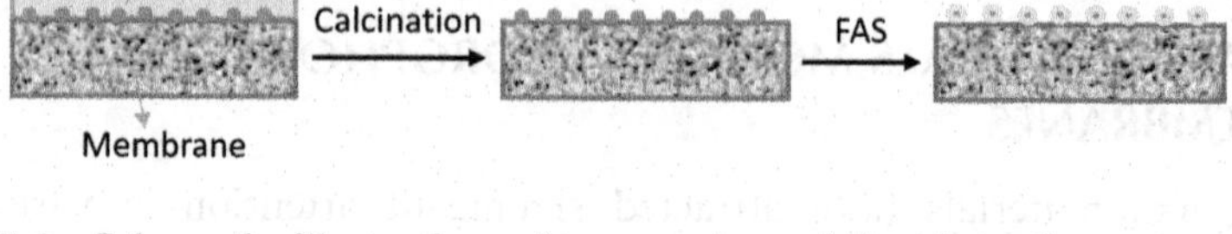

FIGURE 7.4 Schematic illustration of preparation of fluorinated nanoparticle-based membranes by sol-gel method.

(Jafari et al., 2018). CQDs were loaded in/on the nanofibrous. It resulted in a slightly decreased contact angle but a higher LEP. The resultant membranes exhibited more compact structures and rougher surfaces after adding CQDs compared to the neat PVDF membranes. Thus, the former had higher wetting resistance than the latter during the 60 h MD test (Jafari et al., 2018). Bhadra et al. immobilized PVDF membranes by means of detonation nanodiamonds (DNDs) which were carbon nanoparticles with a truncated octahedral architecture and a diameter of 2–8 nm. PVDF MD membranes embedded with only 2% DNDs had a flux enhancement by 118% in the sweep gas membrane distillation (Bhadra et al., 2014).

In recent years, CNTs have been extensively investigated to modify MD membranes due to their outstanding characteristics (Dumée et al., 2010, 2011; Fan et al., 2016; Gethard et al., 2011; Liao et al., 2013; Roy et al., 2014; Silva et al., 2015). Firstly, CNTs are highly hydrophobic and thus can potentially reduce the wetting tendency. In addition, the specific hollow structure of CNTs can improve the transport of vapor molecules (Figure 7.5). Dumée et al. investigated the assembly of CNTs via Van der Waals forces into bucky papers as self-supporting membranes (Dumée et al., 2011). Tijing et al. incorporated CNTs in nanofiber membranes, the resultant membranes showed bead formation and had more micro/nano surface roughness with higher hydrophobicity but their LEP values dropped slightly due to larger surface pores (Tijing et al., 2016). In another study, the incorporation of CNT increased water flux significantly (1.85 times) (Gethard et al., 2011). However, the authors found that the membranes had a high salt

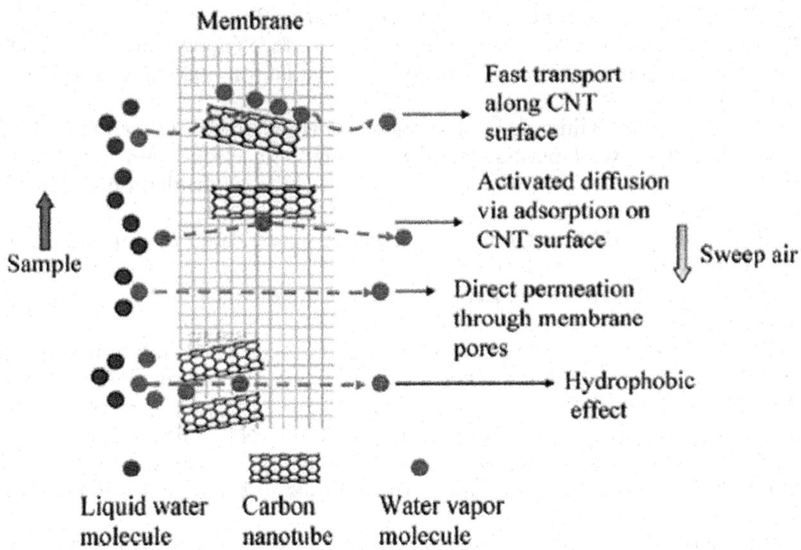

FIGURE 7.5 Carbon nanotube (CNT) incorporated MD membranes. Reprinted with permission from Gethard et al. (2011). Copyright (2011) American Chemical Society.

rejection (99%) to a low feed salinity (10 mg/L) but a low rejection (15%) to a higher feed salinity (34,000 mg/L). This might be due to the combined effects of high hydrophobicity and large flow channels of CNTs. Therefore, the advantages of embedding CNTs in MD membranes are questionable. In addition, CNTs have a relatively higher thermal conductivity, it may facilitate heat conduction to the permeate side, thus lowering the energy efficiency and permeate flux (Dumée et al., 2010). More fundamental studies are needed.

As the real feed compositions are more complex than those stimulated ones in MD studies, more works are now focusing on the development of amphiphobic or omniphobic MD membranes with the aid of nanomaterials. Generally, omniphobic MD membranes not only repel water but also other solvents and organics. This part will be introduced in the other chapters.

REFERENCES

Ahmad, N.A., Leo, C.P., Ahmad, A.L., Ramli, W.K.W., 2015. Membranes with great hydrophobicity: a review on preparation and characterization. Sep. Purif. Rev. 44, 109–134.

Alkhudhiri, A., Darwish, N., Hilal, N., 2012. Membrane distillation: a comprehensive review. Desalination 287, 2–18.

An, A.K., Lee, E.J., Guo, J., Jeong, S., Lee, J.G., Ghaffour, N., 2017. Enhanced vapor transport in membrane distillation via functionalized carbon nanotubes anchored into electrospun nanofibres. Sci. Rep. 7, 41562.

Baghbanzadeh, M., Rana, D., Lan, C.Q., Matsuura, T., 2015. Effects of hydrophilic silica nanoparticles and backing material in improving the structure and performance of VMD PVDF membranes. Sep. Purif. Technol. 157, 60–71.

Bhadra, M., Roy, S., Mitra, S., 2014. Nanodiamond immobilized membranes for enhanced desalination via membrane distillation. Desalination 341, 115–119.

Boo, C., Lee, J., Elimelech, M., 2016. Engineering surface energy and nanostructure of microporous films for expanded membrane distillation applications. Environ. Sci. Technol. 50, 8112–8119.

Chul Woo, Y., Chen, Y., Tijing, L.D., Phuntsho, S., He, T., Choi, J.S., Kim, S.H., Kyong Shon, H., 2017. CF4 plasma-modified omniphobic electrospun nanofiber membrane for produced water brine treatment by membrane distillation. J. Membr. Sci. 529, 234–242.

Curcio, E., Drioli, E., 2005. Membrane distillation and related operations – a review. Sep. Purif. Rev. 34, 35–86.

da Silva Biron, D., dos Santos, V., Zeni, M., 2018. Ceramic membranes applied in separation processes. Springer, Switzerland.

Dong, Z.Q., Ma, X.H., Xu, Z.L., Gu, Z.Y., 2015. Superhydrophobic modification of PVDF/SiO2 electrospun nanofiber membranes for vacuum membrane distillation. RSC Adv. 5, 67962–67970.

Dumée, L., Germain, V., Sears, K., Schütz, J., Finn, N., Duke, M., Cerneaux, S., Cornu, D., Gray, S., 2011. Enhanced durability and hydrophobicity of carbon nanotube bucky paper membranes in membrane distillation. J. Membr. Sci. 376, 241–246.

Dumée, L., Sears, K., Schütz, J., Finn, N., Huynh, C., Hawkins, S., Duke, M., Gray, S., 2010. Characterization and evaluation of carbon nanotube Bucky-Paper membranes for direct contact membrane distillation. J. Membr. Sci. 351, 36–43.

Efome, J.E., Rana, D., Matsuura, T., Lan, C.Q., 2016. Enhanced performance of PVDF nanocomposite membrane by nanofiber coating: a membrane for sustainable desalination through MD. Water Res. 89, 39–49.

Enrico, D., Aamer, A., Francesca, M., 2015. Membrane distillation: Recent developments and perspectives. Desalination 356, 56–84.

Essalhi, M., Khayet, M., 2012. Surface segregation of fluorinated modifying macromolecule for hydrophobic/hydrophilic membrane preparation and application in air gap and direct contact membrane distillation. J. Membr. Sci. 417–418, 163–173.

Eykens, L., De Sitter, K., Dotremont, C., Pinoy, L., Van der Bruggen, B., 2017. Membrane synthesis for membrane distillation: a review. Sep. Purif. Technol. 182, 36–51.

Fadeev, A.Y., McCarthy, T.J., 2000. Self-assembly is not the only reaction possible between alkyltrichlorosilanes and surfaces: monomolecular and oligomeric covalently attached layers of dichloro- and trichloroalkylsilanes on silicon. Langmuir. 16, 7268–7274.

Fan, X., Liu, Y., Quan, X., Zhao, H., Chen, S., Yi, G., Du, L., 2016. High desalination permeability, wetting and fouling resistance on superhydrophobic carbon nanotube hollow fiber membrane under self-powered electrochemical assistance. J. Membr. Sci. 514, 501–509.

Fan, Y., Chen, S., Zhao, H., Liu, Y., 2017. Distillation membrane constructed by TiO2 nanofiber followed by fluorination for excellent water desalination performance. Desalination 405, 51–58.

Gancarz, I., Bryjak, M., Kujawski, J., Wolska, J., Kujawa, J., Kujawski, W., 2015. Plasma deposited fluorinated films on porous membranes. Mater. Chem. Phys. 151, 233–242.

García-Payo, M.C., Essalhi, M., Khayet, M., 2009. Preparation and characterization of PVDF-HFP copolymer hollow fiber membranes for membrane distillation. Desalination 245, 469–473.

García-Payo, M.C., Essalhi, M., Khayet, M., 2010. Effects of PVDF-HFP concentration on membrane distillation performance and structural morphology of hollow fiber membranes. J. Membr. Sci. 347, 209–219.

Gazagnes, L., Cerneaux, S., Persin, M., Prouzet, E., Larbot, A., 2007. Desalination of sodium chloride solutions and seawater with hydrophobic ceramic membranes. Desalination 217, 260–266.

Gethard, K., Sae-Khow, O., Mitra, S., 2011. Water desalination using carbon-nanotube-enhanced membrane distillation. ACS Appl. Mater. Interface 3, 110–114.

Gryta, M., Barancewicz, M., 2010. Influence of morphology of PVDF capillary membranes on the performance of direct contact membrane distillation. J. Membr. Sci. 358, 158–167.

Gugliuzza, A., Drioli, E., 2007. PVDF and HYFLON AD membranes: Ideal interfaces for contactor applications. J. Membr. Sci. 300, 51–62.

Hamzah, N., Leo, C.P., 2017. Membrane distillation of saline with phenolic compound using superhydrophobic PVDF membrane incorporated with TiO2 nanoparticles: separation, fouling and self-cleaning evaluation. Desalination. 418, 79–88.

Hendren, Z.D., Brant, J., Wiesner, M.R., 2009. Surface modification of nanostructured ceramic membranes for direct contact membrane distillation. J. Membr. Sci. 331, 1–10.

Hou, D., Wang, J., Sun, X., Ji, Z., Luan, Z., 2012. Preparation and properties of PVDF composite hollow fiber membranes for desalination through direct contact membrane distillation. J. Membr. Sci. 405–406, 185–200.

Ismail, A.F., Mahman, M.A., Othman, M.H.D., Matsuura, T., 2018. Membrane separation principles and applications. Elsevier, Netherlands.

Jafari, A., Kebria, M.R.S., Rahimpour A., Bakeri, G., 2018. Graphene quantum dots modified polyvinylidenefluride (PVDF) nanofibrous membranes with enhanced performance for air gap membrane distillation. Chem. Eng. Proc.: Proc. Inten. 126, 222–231.

Jin, Z., Yang, D.L., Zhang, S.H., Jian, X.G., 2008. Hydrophobic modification of poly (phthalazinone ether sulfone ketone) hollow fiber membrane for vacuum membrane distillation. J. Membr. Sci. 310, 20–27.

Jung, J., Shin, Y., Choi, Y.J., Sohn, J., Lee, S., An, K., 2016. Hydrophobic surface modification of membrane distillation (MD) membranes using water-repelling polymer based on urethane rubber. Desalin. Water Treat. 57, 10031–10041.

Khayet, M., Matsuura, T., 2003a. Progress in membrane surface modification by surface modifying macromolecules using polyethersulfone, polyetherimide and polyvinylidene fluoride basepolymers: applications in the separation processes ultrafiltration and pervaporation. Fluid/Particle Sep. J. 15, 9–21.

Khayet, M., Matsuura, T., 2003b. Application of surface modifying macromolecules for the preparation of membranes for membrane distillation. Desalination 158, 51–56.

Khayet, M., Matsuura, T., 2011. Membrane distillation: principles and applications. Great Britain: Elsevier.

Khayet, M., Mengual, J. I., Matsuura, T., 2005. Porous hydrophobic/hydrophilic composite membranes. J. Membr. Sci. 252, 101–113.

Kong, Y., Lin, X., Wu, Y., Chen, J., Xu, J., 1992. Plasma polymerization of octafluorocyclobutane and hydrophobic microporous composite membranes for membrane distillation. J. Appl. Polym. Sci. 46, 191–199.

Koonaphapdeelert, S., Li, K., 2007. Preparation and characterization of hydrophobic ceramic hollow fibre membrane. J. Membr. Sci. 291, 70–76.

Krajewski, S., Kujawski, W., Bukowska, M., Picard, C., Larbot, A., 2006. Application of fluoroalkylsilanes (FAS) grafted ceramic membranes in membrane distillation process of NaCl solutions. J. Membr. Sci. 281, 253–259.

Kujawa, J., Kujawski, W., Koter, S., Jarzynka, K., Rozicka, A., Bajda, K., Cerneaux, S., Persin, M., Larbot, A., 2013. Membrane distillation properties of TiO2 ceramic membranes modified by perfluoroalkylsilanes. Desalin. Water Treat. 51, 1352–1361.

Kujawski, W., Kujawa, J., Wierzbowska, E., Cerneaux, S., Bryjak, M., Kujawski, J., 2016. Influence of hydrophobization conditions and ceramic membranes pore size on their properties in vacuum membrane distillatiren of water–organic solvent mixtures. J. Membr. Sci. 499, 442–451.

Lalia, B.S., Guillen, E., Arafat, H.A., Hashaikeh, R., 2014. Nanocrystalline cellulose reinforced PVDF-HFP membranes for membrane distillation application. Desalination. 332, 134–141.

Larbot, A., Gazagnes, L., Krajewski, S., Bukowska, M., Kujawski, W., 2004. Water desalination using ceramic membrane distillation. Desalination. 168, 367–372.

Lee, E.J.J., An, A.K., He, T., Woo, Y.C., Shon, H.K., 2016. Electrospun nanofiber membranes incorporating fluorosilane-coated TiO2 nanocomposite for direct contact membrane distillation. J. Membr. Sci. 520, 145–154.

Lee, J.G., Lee, E.J., Jeong, S., Guo, J., An, A.K., Guo, H., Kim, J., Leiknes, T., Ghaffour, N., 2017. Theoretical modeling and experimental validation of transport and separation properties of carbon nanotube electrospun membrane distillation. J. Membr. Sci. 526, 395–408.

Liao, Y., Wang, R., Fane, A.G., 2013. Engineering superhydrophobic surface on poly(vinylidene fluoride) nanofiber membranes for direct contact membrane distillation. J. Membr. Sci. 44, 74–84.

Lu, J., Yu, Y., Zhou, J., Song, L., Hu, X., Larbot, A., 2009. FAS grafted superhydrophobic ceramic membrane. Appl. Surf. Sci. 255, 9092–9099.

Lu, K.J., Zuo, J., Chung, T.S., 2017. Novel PVDF membranes compromising n-butylamine functionalized graphene oxide for direct contact membrane distillation. J. Membr. Sci. 539, 34–42.

Lu, X., Peng, Y., Ge, L., Lin, R., Zhu, Z., Liu, S., 2016. Amphiphobic PVDF composite membranes for anti-fouling direct contact membrane distillation. J. Membr. Sci. 505, 61–69.

Ma, Z., Hong, Y., Ma, L., Su, M., 2009. Superhydrophobic membranes with ordered arrays of nanospiked microchannels for water desalination. Langmuir. 25, 5446–5450.

Maab, H., AlSaadi, A., Francis, L., Livazovic, S., Ghaffour, N., Amy, G.L., Nunes, S.P., 2013. Polyazole hollow fiber membranes for direct contact membrane distillation. Ind. Eng. Chem. Res. 52, 10425–10429.

Maab, H., Francis, L., Al-saadi, A., Aubry, C., Ghaffour, N., Amy, G., Nunes, S.P., 2012. Synthesis and fabrication of nanostructured hydrophobic polyazole membranes for low-energy water recovery. J. Membr. Sci. 423–424, 11–19.

Meng, S., Mansouri, J., Ye, Y., Chen, V., 2014a. Effect of templating agents on the properties and membrane distillation performance of TiO2-coated PVDF membranes. J. Membr. Sci. 450, 48–59.

Meng, S., Ye, Y., Mansouri, J., Chen, V., 2014b. Fouling and crystallisation behaviour of superhydrophobic nano-composite PVDF membranes in direct contact membrane distillation. J. Membr. Sci. 463, 102–112.

Onsekizoglu, P., 2012. Membrane distillation: principle, advances, limitations and future prospects in food industry. In: Zereshki S, editor. Distillation-advances from modeling to applications. Croatia: InTech.

Prince, J.A., Singh, G., Rana, D., Matsuura, T., Anbharasi, V., Shanmugasundaram, T.S., 2012. Preparation and characterization of highly hydrophobic poly(vinylidene fluoride)–clay nanocomposite nanofiber membranes (PVDF-clay NNMs) for desalination using direct contact membrane distillation. J. Membr. Sci. 397–398, 80–86.

Qtaishat, M., Khayet, M., Matsuura, T., 2009. Novel porous composite hydrophobic/hydrophilic polysulfone membranes for desalination by direct contact membrane distillation. J. Membr. Sci. 341, 139–148.

Razmjou, A., Arifin, E., Dong, G., Mansouri, J., Chen, V., 2012. Superhydrophobic modification of TiO2 nanocomposite PVDF membranes for applications in membrane distillation. J. Membr. Sci. 415–416, 850–863.

Rezaei, M., Samhaber, W.M., 2016. Wetting behaviour of superhydrophobic membranes coated with nanoparticles in membrane distillation. Chem. Eng. Trans. 47, 373–378.

Rezaei, M., Warsinger, D.M., Lienhard, J.H., Samhaber, W.M., 2017. Wetting prevention in membrane distillation through superhydrophobicity and recharging an air layer on the membrane surface. J. Membr. Sci. 530, 42–52.

Rezaei, M., Warsinger, D.M., Lienhard V, J.H.,Duke, M.C., Matsuura, T., Samhaber, W. M., 2018. Wetting phenomena in membrane distillation: mechanisms, reversal, and prevention. Water Res. 139, 329–352.

Roy, S., Bhadra, M., Mitra, S., 2014. Enhanced desalination via functionalized carbon nanotube immobilized membrane in direct contact membrane distillation. Sep. Purif. Technol. 136, 58–65.

Saffarini, R.B., Mansoor, B., Thomas, R., Arafat, H.A., 2013. Effect of temperature dependent microstructure evolution on pore wetting in PTFE membranes under membrane distillation conditions. J. Membr. Sci. 429, 282–294.

Sah, A., Castricum, H.L., Bliek, A., Blank, D.H.A., ten Elshof, J.E., 2004. Hydrophobic modification of γ-alumina membranes with organochlorosilanes. J. Membr. Sci. 243, 125–132.

Sanguineti, A., Di Nicolo, E., Campanelli, P., 2013. Composite membrane comprising layer of perfluoropolyether on hydrophilic substrate. US patent. WO 2013/164287 A1.

Shaulsky, E., Nejati, S., Boo, C., Perreault, F., Osuji, C.O., Elimelech, M., 2017. Postfabrication modification of electrospun nanofiber mats with polymer coating for membrane distillation applications. J. Membr. Sci. 530, 158–165.

Silva, T.L.S., Morales-Torres, S., Figueiredo, J.L., Silva, A.M.T., 2015. Multi-walled carbon nanotube/PVDF blended membranes with sponge-and finger-like pores for direct contact membrane distillation. Desalination 357, 233–245.

Sukitpaneenit, P., Chung, T.S., 2012. PVDF/nanosilica dual-layer hollow fibers with enhanced selectivity and flux as novel membranes for ethanol recovery. Ind. Eng. Chem. Res. 51, 978–993.

Teoh, M.M., Chung, T.S., 2009a. Membrane distillation with hydrophobic macrovoid-free PVDF-PTFE hollow fiber membranes. Sep. Purif. Technol. 66, 229–236.

Teoh, M.M., Chung, T.S., 2009b. Micelle-like macrovoids in mixed matrix PVDF-PTFE hollow fiber membranes. J. Membr. Sci. 338, 5–10.

Teoh, M.M., Chung, T.S., Yeo, Y.S., 2011. Dual-layer PVDF/PTFE composite hollow fibers with a thin macrovoid-free selective layer for water production via membrane distillation. Chem. Eng. J. 171, 684–691.

Tian, M., Yin, Y., Yang, C., Zhao, B., Song, J., Liu, J., Li, X.M., He, T., 2015. CF4 plasma modified highly interconnective porous polysulfone membranes for direct contact membrane distillation (DCMD). Desalination 369, 105–114.

Tijing, L.D., Woo, Y.C., Shim, W.G., He, T., Choi, J.S., Kim, S.H., Shon, H.K., 2016. Superhydrophobic nanofiber membrane containing carbon nanotubes for high-performance direct contact membrane distillation. J. Membr. Sci. 502, 158–170.

Wang, J.W., Li, X.Z., Fan, M., Gu, J.Q., Hao, L.Y., Xu, X., Chen, C.S., Wang, C.M., Hao, Y.Z., Agathopoulos, S., 2017. Porous β-sialon planar membrane with a robust polymer-derived hydrophobic ceramic surface. J. Membr. Sci. 535, 63–69.

Wang, K.Y., Foo, S.W., Chung, T.S., 2009. Mixed matrix PVDF hollow fiber membranes with nanoscale pores for desalination through direct contact membrane distillation. Ind. Eng. Chem. Res. 48, 4474–4483.

Wang, P., Chung, T.S., 2015. Recent advances in membrane distillation processes: membrane development, configuration design and application exploration. J. Membr. Sci. 474, 39–56.

Wang, P., Teoh, M. M., Chung, T. S., 2011. Morphological architecture of dual-layer hollow fiber for membrane distillation with higher desalination performance. Water Res. 45, 5489–5500.

Warsinger, D.M., Servi, A., Connors, G., Gonzalez, J., Swaminathan, J., Chung, H.W., Arafat, H.A., Gleason, K.K., Lienhard, J.H., 2016b. A novel air-cleaning method for membrane distillation. In: Proceedings of the AMTA Membrane Technology Conference & Exposition (MTC16). San Antonio, TX.

Warsinger, D.M., Servi, A., Van Belleghem, S., Gonzalez, J., Swaminathan, J., Kharraz, J., Chung, H.W., Arafat, H.A., Gleason, K.K., Lienhard, J.H., 2016c. Combining air recharging and membrane superhydrophobicity for fouling prevention in membrane distillation. J. Membr. Sci. 505, 241–252.

Warsinger, D.M., Swaminathan, J., Servi, A., Lienhard, J.H., 2016a. Maintenance of gas layers for fouling prevention on submerged surfaces. US Patent, US15/157, p. 663.

Wei, X., Zhao, B., Li, X.M., Wang, Z., He, B.Q., He, T., Jiang, B., 2012. CF4 plasma surface modification of asymmetric hydrophilic polyethersulfone membranes for direct contact membrane distillation. J. Membr. Sci. 407–408, 164–175.

Woo, Y.C., Tijing, L.D., Shim, W.G., Choi, J.S., Kim, S.H., He, T., Drioli, E., Shon, H. K., 2016. Water desalination using graphene-enhanced electrospun nanofiber membrane via air gap membrane distillation. J. Membr. Sci. 520, 99–110.

Yang, C., Li, X., Gilron, J., Kong, D., Yin, Y., Oren, Y., Linder, C., He, T., 2014. CF4 plasma modified superhydrophobic PVDF membranes for direct contact membrane distillation. J. Membr. Sci. 456, 155–161.

Yang, W.F., Zhang, Z.Q., Gu, Z.Y., Shen, X.X., Zhang, Q.F., 2011a. Research on the superhydrophobic modification of polyvinylidene fluoride membrane. Adv. Mater. Res. 197–198, 514–517.

Yang, X., Wang, R., Shi, L., Fane, A.G., Debowski, M., 2011b. Performance improvement of PVDF hollow fiber-based membrane distillation process. J. Membr. Sci. 369, 437–447.

Yasuda, H., 1984. Plasma polymerization for protective coatings and composite membranes. J. Membr. Sci. 18, 273–284.

Zhao, D., Zuo, J., Lu, K.J., Chung, T.S., 2017. Fluorographite modified PVDF membranes for seawater desalination via direct contact membrane distillation. Desalination 413, 119–126.

Zhao, T., Jiang, L., 2018. Contact angle measurement of natural materials. Colloids Surf. B: Biointerfaces. 161, 324–330.

8 Design of Omniphobic Interfaces for Membrane Distillation

Kang-Jia Lu, Yuanmiaoliang Chen, and Tai-Shung Chung

8.1 INTRODUCTION

Membrane distillation (MD) is a promising technology to reclaim various wastewater from a wide range of industries (An et al., 2016; Calabro et al., 1991; El-Abbassi et al., 2013; Hussain et al., 2015; Khaing et al., 2010; Mokhtar et al., 2015; Sivakumar et al., 2013). Comparing to the traditional distillation processes, MD is modular and more space-saving (Lawson and Lloyd, 1997). Thanks to its special separation mechanism, MD has theoretically 100% rejections to nonvolatile compounds and produces extremely pure water that can meet the high quality standards of industries such as semiconductors and pharmaceuticals (Drioli et al., 2015; Hilal et al., 2011; Lu et al., 2016). It is also less affected by feed concentration and has a lower tendency to be fouled than conventional pressure-driven membrane processes, making it especially suitable for reclamation of high salinity wastewater such as concentrated reverse osmosis (RO) brine (Camacho et al., 2013; Deshmukh et al., 2018; Lu et al., 2018a). In addition, it typically operates at mild temperatures of less than 90 °C and pressures of atmospheric pressure or below, which makes it possible to be powered by low-grade energy sources such as solar energy or industrial waste heat. This will significantly lower its energy cost and carbon footprint (Henderyckx, 1967).

In spite of various advantages, the implementation of MD for industrial wastewater reclamation must be conducted with precaution because contaminates in feed solutions may cause membrane wetting (Lu et al., 2018b; Rezaei et al., 2018). Wetting should be avoided in MD because it causes a direct mixing of feed and distillate and results in detrimental effects on MD performance. The wetting phenomenon is closely related to the feed composition, which varies significantly in different industries (Gryta, 2005; Rezaei et al., 2018; Warsinger et al., 2015). For example, when reclaiming high salinity brines, inorganic salt crystals precipitated from concentrated brines may grow on membrane surface, reduce its surface hydrophobicity, and eventually cause membrane wetting. Surfactant adsorption is a major origin of wetting in treating shale gas wastewater. Trace amounts of oil and grease in wastewater from petrochemical plants may attach to the hydrophobic membrane and cause pore wetting and blockage. Wastewater from pharmaceutical and semiconductor industries contains solvents and acids that lower the surface tensions of feed solutions and lead to membrane wetting (Gryta, 2005; Israelachvili, 2011; Rezaei et al., 2018; Warsinger et al., 2015). On the other hand, in many cases various types of foulants may coexist and act synergistically to accelerate the membrane wetting.

Designing an omniphobic membrane that remains unwetted when dealing with complex feed solutions is a potential solution to improve the antiwetting properties of MD membranes. "Omni" means "all," thus the omniphobic MD membrane should repel both high and low-surface-tension liquids. The omniphobic surface has been found in nature species. Springtail is able to breathe through its cuticle surface even when it is immersed in water, oil, and other liquids (Boban et al., 2018; Hensel et al., 2016, 2014, 2013; Kota et al., 2014; Zhu et al., 2017). When they are immersed under liquids, a protective air layer called plastron will be formed and serve as a gas reservoir to enable regular gas exchange. Studies show that the omniphobic nature of the springtail's cuticle surface is attributed to its hierarchical reentrant structure and the layer of low surface energy lipids (Hensel et al., 2016, 2014). Inspired by springtails, several successful designs of omniphobic surfaces have recently been reported by Hensel et al. (2014), Tsujii et al. (1997), Tuteja et al. (2007, 2008a, 2008b), and Zhu et al. (2017). The concept of omniphobicity has also attracted increasing attention from membrane researchers and several pioneering attempts have been done to fabricate omniphobic membranes for membrane distillation (Boo et al., 2016a, 2016b; Chen et al., 2018a, 2018b; Deng et al., 2018; Huang et al., 2017; Kadi et al., 2019; Lee et al., 2016b; Li et al., 2019; Lin et al., 2014; Lu et al., 2018b; Woo et al., 2017, 2018; Zheng et al., 2018).

In this chapter, we would like to present a comprehensive review of omniphobic membranes for membrane distillation. To begin with, fundamental knowledge of wetting on a textured surface and important design parameters for omniphobic surfaces will be discussed. Subsequently, cutting-edge and inspiring technologies to fabricate omniphobic surfaces will be elaborated. The chapter will then focus on innovative designs of omniphobic membranes and their successful applications in membrane distillation. We hope this chapter can provide guidance and inspirations to those interested readers who are keen to develop robust MD membranes with exceptional wetting resistance.

8.2 DESIGN AN OMNIPHOBIC SURFACE

8.2.1 Fundamentals of Wetting on a Textured Surface

The wettability of a textured surface is often quantified in terms of the apparent or measured contact angle θ^* (Figure 8.1) of a liquid droplet on the surface. Different from the equilibrium contact angle θ (Figure 8.2) which is an intrinsic property of a solid–vapor–liquid interface, θ^* is significantly influenced by the texture of the solid surface. Two distinct models developed by Wenzel (Wenzel, 1936) and Cassie and Baxter (Cassie and Baxter, 1944) are commonly applied

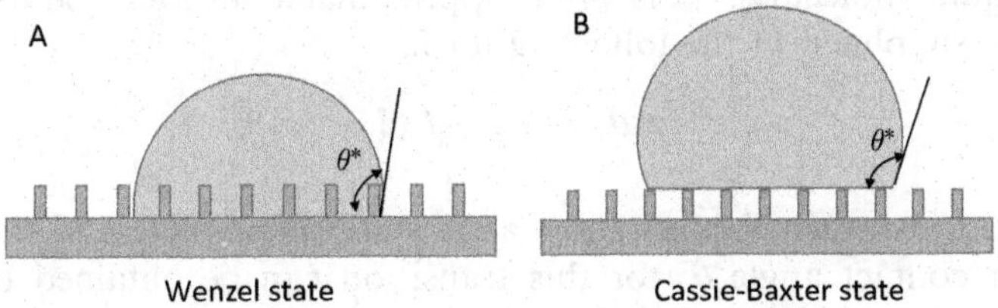

FIGURE 8.1 Liquid droplets on rough surfaces in a (A) Wenzel state and (B) Cassie–Baxter state.

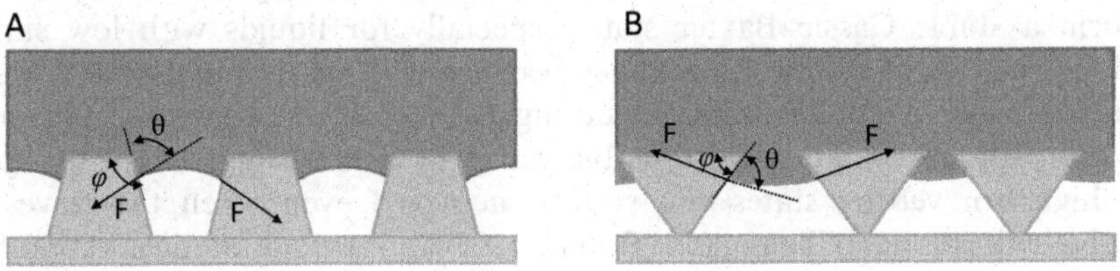

FIGURE 8.2 Possible liquid–vapor–solid interfaces on (A) pointed and (B) reentrant structures.

to predict the effects of roughness r (i.e., the ratio of the actual solid surface area to the projected area) on the measured contact angle. The Wenzel state shown in Figure 8.1A corresponds to a complete wetting state. It is widely adopted to describe the liquid–membrane interface in applications such as RO where a liquid transports through membrane pores. In a Wenzel model, roughness increases the area of the composite interface. Thus, it amplifies the effects of the intrinsic hydrophilicity or hydrophobicity.

$$\cos\theta^* = r\cos\theta \tag{8.1}$$

In a Cassie–Baxter state as shown in Figure 8.1B, a liquid droplet is retained on top of a solid–vapor–liquid interface. The nonwetting nature of the rough surface is caused by microscopic air pockets trapped below the liquid phase. The apparent contact angle in a Cassie–Baxter mode is calculated by the following equations:

$$\cos\theta^* = f_s\cos\theta + f_g\cos\theta_g \tag{8.2}$$

where f_s and f_g are the areas of solid–liquid and gas–liquid interfaces on a projected area of unity, respectively. For a surface with a morphology as shown Figure 8.1B, $f_s + f_g = 1$. θ and θ_g are equilibrium contact angles of the liquid droplet on solid and gas, respectively. Due to the low adhesion between gas and liquid molecules, θ_g is often approximated as 180°. Thus, Equation (8.2) can be simplified to the following form:

$$\cos\theta^* = -1 + f_s(1 + \cos\theta) \tag{8.3}$$

The transition between Wenzel and Cassie states is possible and the threshold equilibrium contact angle θ_c for this transition can be obtained by equating Equations (8.1) and (8.3):

$$\cos\theta_c = \frac{f_s - 1}{r - f_s} \tag{8.4}$$

Since $f_s < 1 < r$, θ_c should be greater than 90° for the transition from the Wenzel state to the Cassie–Baxter state to occur. This highlights the difficulty to form a stable Cassie–Baxter state, especially for liquids with low surface tensions such as alcohols and alkanes because it is nearly impossible for such liquids to achieve equilibrium contact angles over 90° on almost all the solids. However, studies on superhydrophobic leaves in nature suggest that unexpectedly high nonwetting states can still be achieved even when the leaves are slightly hydrophilic (Cheng and Rodak, 2010). Previous studies reveal that such phenomena origin from the reentrant surface texture which facilitates the formation of a composite interface.

A reentrant structure refers to a concave topography in which the liquid–solid contact area decreases as the liquid approaches the bottom surface. Figure 8.2A and 8.2B illustrates possible liquid–vapor–solid interfaces on pointed and reentrant structures, respectively. Despite the same equilibrium contact angles θ in both cases, the net tractions on the interfaces point in different directions. In general, if the equilibrium contact angle is smaller than the local texture angle φ as shown in Figure 8.2A, the net traction on the composite interface is pointing downward, exerting a pressure on the trapped air pocket and promoting imbibition of the liquid. By contrast, the net traction points away from the solid for a reentrant textured surface with $\varphi < \theta$, which can effectively prevent wetting (Chen et al., 2018a; Chhatre et al., 2010; Tuteja et al., 2008a).

Besides creating reentrant textures, one can also improve the antiwetting property by lowering the solid surface energy. A lower solid surface energy can reduce the attraction between liquid and solid surfaces and increase the equilibrium contact angle (Lu et al., 2018b; Tuteja et al., 2008a). It is also reported that a low surface energy coating can impede the formation of scales on membrane surfaces (Lu et al., 2016; Zhao et al., 2017).

8.2.2 Important Design Parameters

Two design parameters are important in engineering omniphobic surfaces, namely, the apparent contact angle θ^* and the robustness factor A^*. The former is the most straightforward and widely applied parameter to measure the surface wettability. A small θ^* (<90°) corresponds to a hydrophilic surface with high wettability, while a large $\theta^*(>90°)$ indicates a hydrophobic surface with a good antiwetting property. A surface with an exceptionally high $\theta^*(>150°)$ is called a superhydrophobic surface, which is desirable in membrane distillation. A^* is a dimensionless factor to assess the stability of a solid–liquid–vapor interface.

As mentioned in the preceding discussions, the value of θ^* depends highly on the surface geometrical factors and surface chemistry as shown in Equation (8.2). For surfaces with different textures, Equation (8.2) can be rewritten as different forms. For example, Equation (8.3) is a simplified Cassie–Baxter model for flat-topped pillars as shown in Figure 8.1B. However, it is not applicable to reentrant structures where the sum of f_s and f_g does not equal to one. Figure 8.3A shows a diagram of a composite interface on cylindrical fibers. Such texture is a typical reentrant structure found on woven, nonwoven, mesh, and electrospun membranes. For an area (A) with a unit axial length, f_s and f_g can be calculated as follows:

$$A = 2R + 2D \tag{8.5}$$

$$f_s = \frac{R \times (2\pi - 2\theta)}{A} \tag{8.6}$$

$$f_g = \frac{A - 2R\sin\theta}{A} \tag{8.7}$$

where R is the radius of the fiber and $2D$ is the distance between the two fibers. Let us define the dimensionless space ratio $D^* = \frac{R+D}{R}$, then the Cassie–Baxter equation can be rewritten as (Chhatre et al., 2010):

$$\cos\theta^* = -1 + \frac{1}{D^*}[(\pi - \theta)\cos\theta + \sin\theta] \tag{8.8}$$

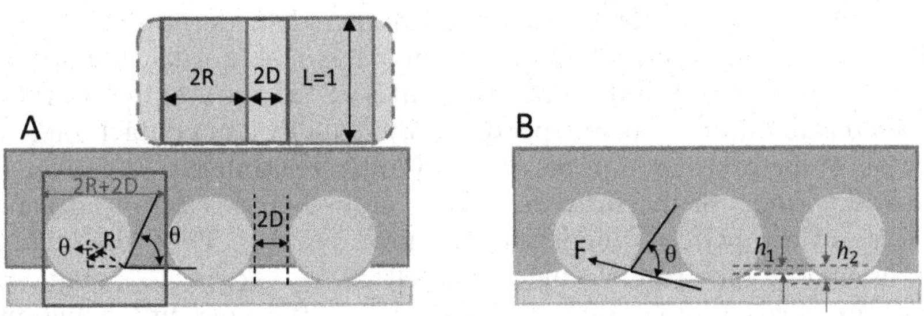

FIGURE 8.3 (A) An illustration of the composite interface on cylindrical fibers; the small graph on the top depicts the top view of cylindrical fibers with a unit length; (B) composite interface on cylindrical fibers considering the local sagging effect.

The effects of dimensionless space ratio D^*, radius of the surface texture R, distance between surface texture $2D$ and equilibrium contact angle θ on the value of apparent contact angle θ^* have been studied both mathematically and experimentally by Chhatre et al. (2010). In their study, contact angle measurements of a series of liquids were performed on a set of wire-mesh surfaces, which could provide systematically quantifiable cylindrical textures. θ was manipulated by dip coating the meshes using materials with different surface energies. The experimental results show that θ^* is indeed determined by θ and D^* as predicted by Equation (8.8). A larger θ and/or a greater D^* lead to a larger apparent contact angle, and vice versa. Individual structure parameters, namely, R and D, affect θ^* synergistically. In other words, the apparent contact angles are unaffected by the changes in R or D as long as the resultant D^* remains constant.

Though Equation (8.8) can accurately predict the apparent contact angle on a cylindrically textured surface, it can be only used when a Cassie–Baxter state is assured. Unfortunately, a Cassie–Baxter state is thermodynamically metastable (Tuteja et al., 2007, 2008a). As illustrated in Figure 8.3B, a transition to the Wenzel state is possible when the sagging height h_1 reaches the maximum height h_2 under the Laplace pressure or an externally applied pressure (Tuteja et al., 2008a). To quantify the tendency to resist the sagging of the composite interface, Chhatre et al. proposed a dimensionless robustness factor A^* for the cylindrical texture (Chhatre et al., 2010).

$$A^* = \frac{l_{cap}}{R(D^* - 1)}\left[\frac{(1 - \cos\theta)}{(D^* - 1 + 2\sin\theta)}\right] \tag{8.9}$$

where l_{cap} is the capillary length defined as $l_{cap} = \sqrt{\gamma_{lv}/\rho g}$, where γ_{lv} and ρg are the surface tension and specific weight of the liquid, respectively. $A^* \gg 1$ is preferred in order to obtain a robust solid–liquid–vapor interface. As reported by Chhatre et al., the robustness factor is inversely proportional to the radius of the surface texture R. Even at a constant space ratio D^*, a larger A^* is obtained on a fine mesh compared to that on a coarse mesh. Similarly, the composite interface becomes less stable when the space ratio factor D^* increases. In addition to the scales of the texture, the robustness of the composite interface also depends on the equilibrium contact angle θ. The value of A^* increases monotonically with an increase in θ.

In reality, most of the surfaces such as nonwoven or electrospun nanofiber surfaces are not perfect cylindrically textured surfaces. Thus, the quantitative analysis may not be suitable unless the effects of varying topographical and surface properties are considered. Nevertheless, the preceding equations and discussions still provide valuable insights to the design of omniphobic surfaces. The importance of the reentrant structure and the low surface energy has been echoed in novel fabrication technologies to be presented in the subsequent sections.

8.2.3 Novel Fabrication Methods of Omniphobic Surfaces

Ever since the mechanism of the omniphobic surface was revealed, scientists have tried various types of methods to engineer similar surfaces with remarkable repellences toward low-surface-tension liquids. The most typical strategy is the simultaneous incorporation of low surface energy chemistry and topographic features as illustrated in the preceding discussions. Instructed by this strategy, several omniphobic surfaces have been successfully developed.

Tuteja et al. synthesized a series of omniphobic electrospun surfaces by incorporating fluorodecyl polyhedral oligomeric silsesquioxane (fluoroPOSS) into polymethyl methacrylate (PMMA) solutions. The addition of fluoroPOSS not only reduced the surface energy but also facilitated the formation of a reentrant beads-on-string structure as shown in Figure 8.4A (Tuteja et al., 2008a, 2008b). In addition to this method, they also obtained omniphobic "micro-hoodoo" structure (Figure 8.4B) on flat silicon (Si) wafers by a silicon dioxide (SiO_2) deposition followed by a reactive ion etching of SiO_2 and a subsequent isotropic etching of Si via vapor-phase xenon difluoride (XeF_2) (Kota et al., 2014; Tuteja et al., 2007).

A surface with a special texture that shows great similarity to the "micro-hoodoo" structure was developed by Tan et al. via metal–organic framework (MOF) self-assembly (Tan et al., 2013). They referred the texture as "micro-mushrooms" (Figure 8.5). They first grew an array of NH_2-MIL-53 (Al) microneedles to an anodic aluminum oxide membrane. Next, the hydrophobicity of the microneedles was improved by reacting it with perfluorooctanoyl chloride. The modified membrane was then gently placed into an aqueous precursor solution. Due to the superhydrophobic nature of the membrane, it remained afloat on the solution surface, and the precursor solution would not contact with the microneedles except at their tips. As a result, mushroom-like

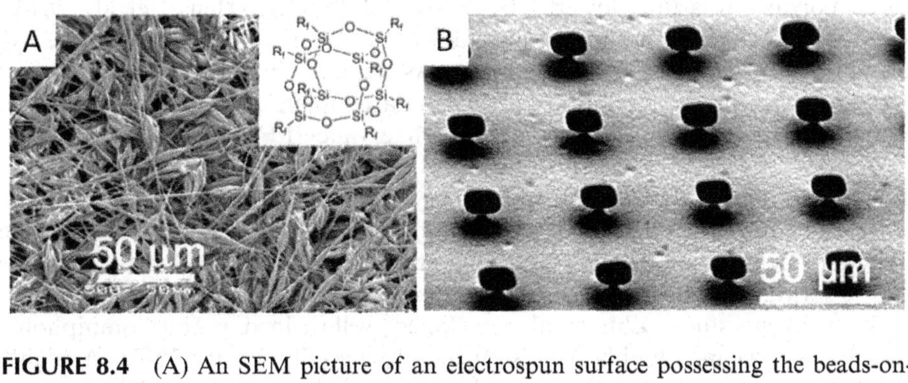

FIGURE 8.4 (A) An SEM picture of an electrospun surface possessing the beads-on-string morphology. The inset shows the molecular structure of the fluoroPOSS molecules; (B) an SEM picture of a micro-hoodoo surface (Tuteja et al., 2008a). Copyright (2008) National Academy of Sciences.

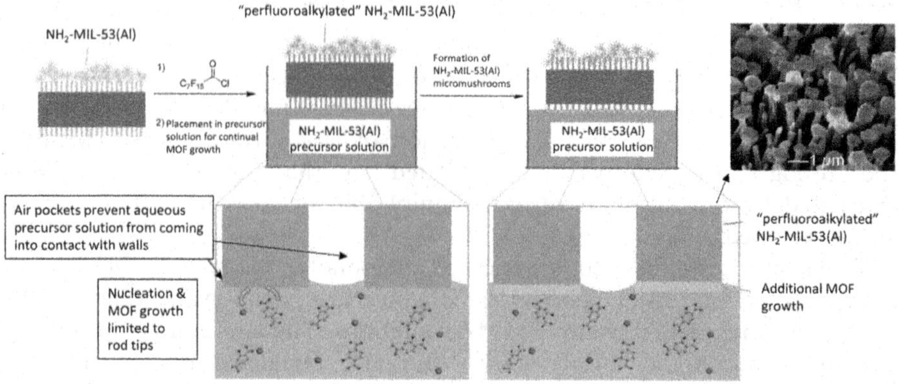

FIGURE 8.5 Graphical representation of NH2-MIL-53(Al) micromushroom synthesis through interfacial continual growth. The SEM graph shows the micromushroom structure of the omniphobic surface (Tan et al., 2013). Copyright (2013) American Chemical Society.

caps were formed on the needle stems through epitaxial MOF growth. After the surface energy of the mushroom caps was lowered with perfluorooctanoyl chloride, the resultant surface showed omniphobicity as reflected by a remarkable contact angle of 100° with n-hexadecane.

Hensel et al. adapted a nanoimprint lithography scheme to design delicate reentrant morphologies with a high resolution of a few tens of nanometers. They first applied precursor solutions to the springtails followed by UV-irradiation. The cured elastomeric mold was then gently peeled off and used as a prepatterned mold for a subsequent casting. With the aid of this technology, the skin structure of *Tetrodontophora bielanensis* (a European giant springtail) was translated into polymer replicas. The surface energy of the textured polymer was then lowered by Teflon AF coating (Hensel et al., 2013). Hensel et al. have also successfully fabricated an omniphobic membrane via a multistep reverse imprint lithography. As depicted in Figure 8.6, a two-tier silicon master mold was fabricated first as a template for feature replication. Subsequently, a perfluoropolyether dimethacrylate (PFPEdma) template was cast from the silicon master using a reverse imprint lithography approach. Finally, a poly(ethylene glycol) dimethacrylate (PEGdma) polymer membrane was cast from the PFPEdma template by the doctor-blade technique. The resultant membrane exhibited overhang structures that resembled the textures of the springtail cuticle surface.

In a recent study, Zhu et al. developed well-defined porous omniphobic membranes via microfluidic emulsion templating (Zhu et al., 2017). A highly uniform emulsion of silicone oil droplets in a polyvinyl alcohol (PVA) aqueous solution was prepared using a capillary microfluidic device and then deposited onto a glass substrate. Due to the different physicochemical natures of the oil

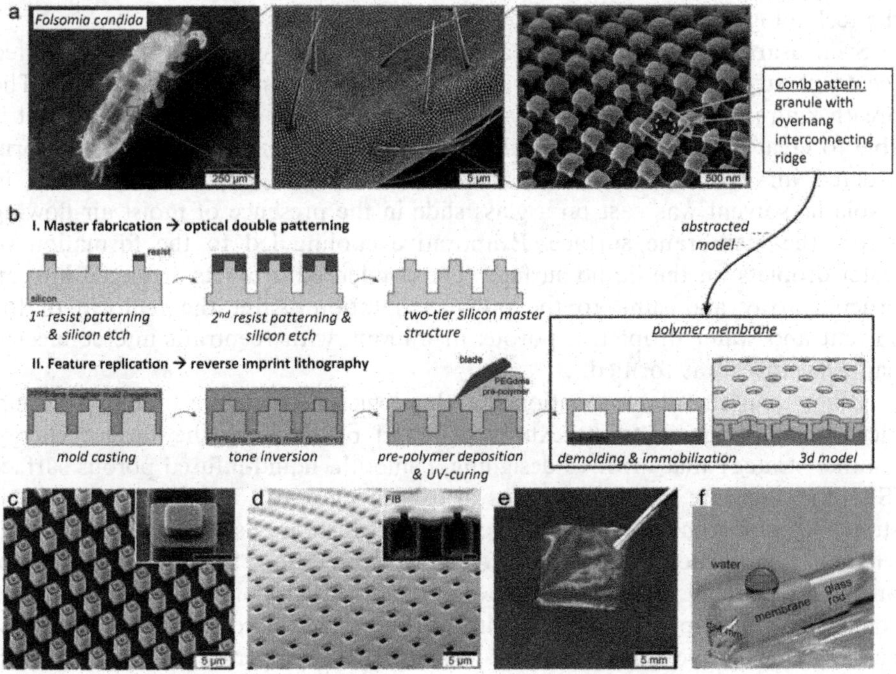

FIGURE 8.6 Springtail skin morphology and process scheme for manufacturing polymer membranes with similar structural features. a) Habitus image of Folsomia candida. Insets show scanning electron microscopy (SEMs) images of the characteristically contained bristles, granules, and ridges. The nanoscopic granules and interconnecting ridges form cavities, are arranged in a comb-like pattern, and provide a template for the developed polymer membranes. b) Process scheme for membrane fabrication: Firstly, a two-tier silicon master structure is fabricated by optical lithography. Secondly, the master structure serves as template for reverse imprint lithography. c) SEM image of the two-tier silicon master structure (inset: detailed view of a small pillar centered on a larger pillar, scale bar: 1 µm). d) SEM image (inset: cross section after focused ion beam preparation, scale bar: 1 µm) and e) photograph of the springtail-skin-inspired polymer membrane. f) Water droplet (colored with red dye) deposited on the membrane, which was transferred to a 4-mm diameter glass rod (Hensel et al., 2013). Copyright (2013) John Wiley & Sons, Inc.

and the aqueous solution, the oil droplets floated to the air–water interface and created a close-packed hexagonal monolayer array. As water evaporated, PVA solidified into densely packed honeycomb structures and the spherical oil droplets gradually deformed into reentrant monolayer structures. When the oil droplets were removed, a PVA membrane with a layer of well-defined reentrant microstructures was formed. Their study presents an innovative and scalable method that can precisely control the structural topographies. However, the resultant pore sizes of the membranes are of tens of micrometers. Further investigation on how to control the pore size is important in order to adopt the technology for MD membrane fabrication.

Srinivasarao et al. proposed a facile method to engineer interconnected reentrant structures based on breath figures (Srinivasarao et al., 2001). The breath figure refers to the fog created by exhaling onto a cold surface and it is able to create micro and nanoscale periodic structures (Bunz, 2006). To form ordered air bubbles at a membrane surface, a diluent of polystyrene in a volatile solvent was cast on a glass slide in the presence of moist air flowing across the membrane surface. Evaporative cooling led to the formation of water droplets on the liquid surface. The condensed droplets arranged into an ordered array and sunk to the polymer solution. After the removal of the solvent and water droplet, a porous membrane with reentrant interconnected nanostructures was formed.

Other than the above-mentioned technologies to fabricate omniphobic surfaces by creating reentrant textures followed by lowering the surface energy, another strategy that involves designing a smooth, liquid-infused porous surface (SLIPS) which are "liquid-like" has also been reported. The advantage of this strategy is that reentrant topographic features which are usually fragile and challenging to fabricate are no longer required. In addition, SLIPS do not involve pressure-dependent metastable states. Instead, a dynamic liquid–liquid–vapor contact line is formed by immobilizing perfluorinated fluid lubricants on micropatterned or nanotextured substrates, which leads to superior self-cleaning, anti-icing, and self-healing properties (Ma et al., 2013; Wong et al., 2011). However, the requirement of a mobile liquid which is immiscible to the contact liquid makes it unsuitable in the MD applications. Some researchers also proposed to coat a layer of "covalently attached liquids" by grafting flexible materials onto the substrates. Unfortunately, the feasibility of adopting this strategy in membrane distillation has not been explored yet due to the complexity and low permeability of the coating layer (Cheng et al., 2012; Wang and McCarthy, 2016).

8.3 OMNIPHOBIC MODIFICATIONS FOR MEMBRANE DISTILLATION

Omniphobic membranes which have been developed for robust membrane distillation will be summarized in this section. Depending on the types of supporting materials, they can be divided into three major categories, namely, inorganic membranes, phase inversed polymeric membranes, and electrospun nanofiber membranes.

8.3.1 Inorganic Membranes

Lin et al. developed the first omniphobic membrane for membrane distillation by modifying glass fiber membranes with silica nanoparticles (SiNPs) followed by surface fluorination and polymer coating (Lin et al., 2014). A five-step modification procedure was employed to grow SiNPs firmly on the glass fibers and to lower the surface energy using fluorinated alkyl silane (FAS). Figure 8.7 compares their morphology before and after the modification. The modified membrane exhibited larger contact angles toward not only water but also low-surface-tension organic solvents than a commercial polytetrafluoroethylene (PTFE) membrane. Both the omniphobic and PTFE membranes were tested in the direct contact membrane distillation (DCMD) experiments using feed solutions containing sodium dodecyl sulfate (SDS), and the former exhibited a much more stable performance than the latter.

Cheng et al. developed omniphobic membranes for MD by depositing zinc oxide (ZnO) on glass fiber membranes using a chemical deposition method, followed by surface fluorination and a polymer coating to lower the surface energy (Chen et al., 2018b). The resultant membrane showed remarkable contact angles of 152.8 ± 1.1° and 110.3 ± 1.9° toward water and ethanol, respectively. The DCMD tests using feeds containing 0.3 mM SDS also proved the enhanced stability after the omniphobic coating. Cheng et al. also fabricated omniphobic alumina hollow fiber membranes by depositing ZnO nanorods and nanoparticles on the membranes followed by surface fluorination using 1H,1H,2H,2H-perfluorodecyltrimethoxysilan (FAS 17) (Chen et al., 2018a). They found that both nanostructures enhanced the membrane omniphobicity due to the higher roughness, but the membrane coated with nanoparticles was more antiwetting than that coated with nanorods. Unlike nanoparticles, nanorods do not have reentrant structures. A concave composite interface is formed and the surface tension force is pointing downward as illustrated in Figure 8.8. As a result, the composite interface formed on

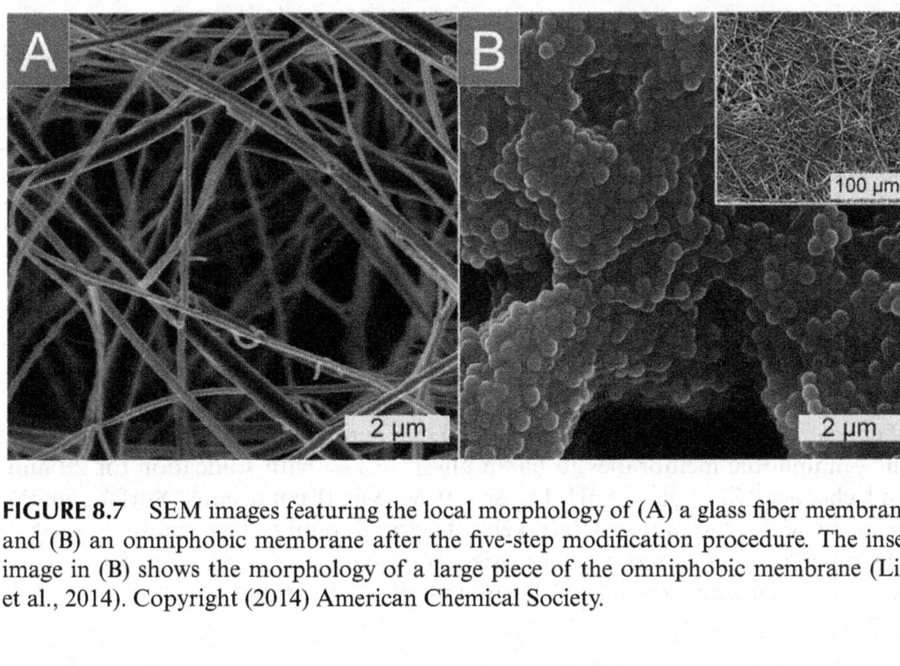

FIGURE 8.7 SEM images featuring the local morphology of (A) a glass fiber membrane and (B) an omniphobic membrane after the five-step modification procedure. The inset image in (B) shows the morphology of a large piece of the omniphobic membrane (Lin et al., 2014). Copyright (2014) American Chemical Society.

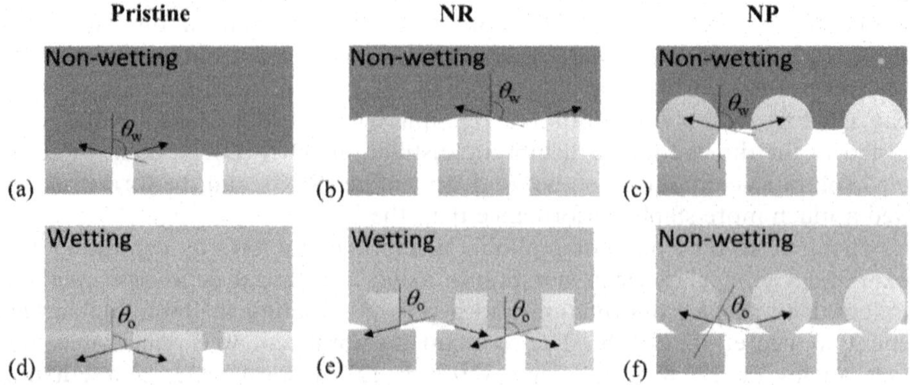

FIGURE 8.8 Schematic illustration of the liquid–air interfaces for (a–c) water and (d–f) a low surface tension liquid on (a, d) the pristine membrane (Pristine), (b, e) the membrane deposited by nanorods (NR), and (c, f) the membrane deposited by nanoparticles (NP) (Chen et al., 2018b). Copyright (2018) Elsevier Science Ltd.

nanorod structures is less stable compared to that on nanoparticles where the upturned net traction can effectively prevent the wetting to occur.

8.3.2 Phase Inversed Polymeric Membranes

Polymeric membranes that are fabricated via phase inversion method are commonly used in MD applications. Surface modifications of such membranes in order to improve their wetting resistance remain a hot topic. Polymeric membranes can be formed into different configurations such as flat sheet and hollow fiber. Generally speaking, decorating flat sheet membranes is relatively easier as compared to modifying hollow fibers because the former possess a smaller curvature than the latter. Thus, quite a few studies have reported successful omniphobic modifications on flat sheet MD membranes.

Boo et al. have performed an omniphobic coating on a commercial PVDF flat sheet membrane by SiNP deposition followed by perfluorodecyltrichlorosilane (FDTS) coating via vapor phase silanization (Boo et al., 2016a). For readers' information, the pristine PVDF membrane was first modified by electropositive (3-aminopropyl)-triethoxysilane (APTES) so that negatively charged SiNPs could attach onto the membrane surface via electrostatic interaction, as illustrated in the upper right of Figure 8.9. The mechanical and chemical durability of the SiNP and FDTS coating was assessed by subjecting the omniphobic membranes to harsh physical (i.e., bath sonication for 20 min) and chemical (i.e., pH 2, pH 12, or 1.0 M NaCl) conditions. Surface micrographs and contact angle data taken after the durability tests showed no obvious difference compared with the original omniphobic membrane. The wetting resistances of the pristine and omniphobic membranes were examined in

DCMD tests using (1) SDS, (2) mineral oil added feed solutions, and (3) pre-filtered real shale gas produced water. In all cases, the omniphobic membranes showed significantly more stable performance than the unmodified ones, indicating their great potential in treating low-surface-tension industrial wastewater. Recently, Li et al. proposed another approach where trimesoyl chloride (TMC) was employed as a bridge for binding SiNPs onto the PVDF surface via chemical interaction (Li et al., 2019). Figure 8.9 also elucidates this approach. Li et al. claimed that SiNPs deposition via chemical bounding was more robust than that via electrostatic interaction, and membranes prepared using the former method showed more stable performances in both impact-abrasive tests and long-term DCMD tests of 120 h.

Besides the preceding methods, Woo et al. recently reported a six-step layer-by-layer (LBL) assembly method to fabricate omniphobic membranes. Negatively charged silica aerogel and 1H,1H,2H,2H-perfluorodecyltriethoxysilane were attached on a PVDF flat sheet membrane successively using positively charged poly(diallyldimethylammonium chloride) as a interconnecting material (Woo et al., 2018). The modified membrane showed a remarkable water contact angle of 177.0 ± 0.4° and a sliding angle of 1.1 ± 0.3°, indicating its superhydrophobic and low hysteresis nature. The resultant membrane also exhibited extremely high contact angles toward low-surface-tension liquids including ethylene glycol, mineral oil, and methanol. Air gap membrane distillation (AGMD) performances of the pristine and modified membranes were tested using (1) real RO brine from the produced water of coal seam gas (CSG) and

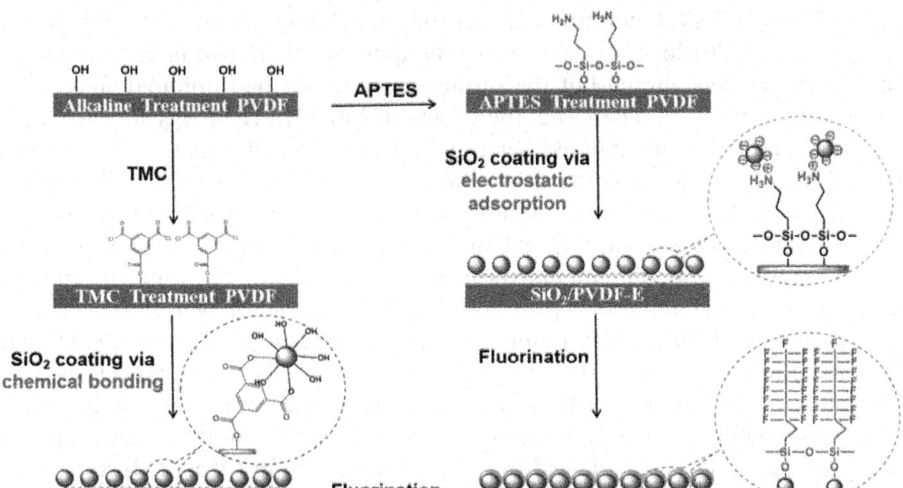

FIGURE 8.9 Schematic illustration of the preparation procedure of omniphobic PVDF membranes with SiO_2 nanoparticles coated on membrane surface via electrostatic adsorption and chemical bonding (Li et al., 2019). Copyright (2019) Elsevier Science Ltd.

(2) CSG RO brine with a gradual addition of SDS or humid acid as feed solutions. Even though the omniphobic membrane obtained a relatively lower flux, it gained a more stable performance and a better rejection than the unmodified one. The superior omniphobicity is attributed to the low-surface-tension coating and the formation of the hierarchical rough surface. In addition to the improvement in omniphobicity, the mechanical and thermal stability of the membrane has also been enhanced after the LBL assembly due to the strong adhesion between the assembled layers and the host polymer support.

Rather than employing monosized nanoparticles to create reentrant structures, Zheng et al. reported an omniphobic membrane with multilevel reentrant structures (Zheng et al., 2018). Multiscale microspheres were fabricated based on electrostatic interaction between electronegative SiNPs and electropositive polystyrene (PS) microspheres. Then, they were deposited onto a commercial PVDF porous substrate via spray coating. A chemical binding agent was utilized to enhance the adhesion between the microspheres and the PVDF support. Finally, the surface was fluorinated using FAS17 to reduce the surface energy. The resultant membrane exhibited outstanding water/hexadecane contact angles of 176°/138.4° and a water sliding angle of 7°. In the DCMD tests using a SDS stabilized hexadecane emulsion as the challenging feed solution, the omniphobic membrane showed an extremely stable flux and an outstanding rejection in contrast to other benchmark membranes.

Apart from the above-mentioned omniphobic MD membranes developed on flat sheet polymeric substrates, Lu et al. have successfully developed omniphobic hollow fiber membranes via silica nanoparticle deposition followed by Teflon AF 2400 coating (Lu et al., 2018b). Interestingly, the surface energy of the resultant fibers became much lower by dipping the original fibers in a very dilute Teflon AF 2400 coating solution (0.025 wt%) for a short coating time of 30 s (Lu et al., 2018b; Lu et al., 2016). By applying such coating conditions, the surface energy was altered but the surface morphology had minimal changes, as shown in Figure 8.10. Therefore, the effects of low surface energy and reentrant structure could be investigated separately. Contact angle results showed that both Teflon AF 2400 coating and omniphobic coating (i.e., SiNPs deposition followed by Teflon AF 2400 coating) can improve the repellence of membranes toward various types of liquids, but the effects of the latter are much more obvious. This observation highlights the significance of the reentrant structure to enhance surface omniphobicity. In addition, they also investigated the effects of nanoparticle size on surface omniphobicity by depositing SiNPs with different sizes as shown in Figure 8.10. Vacuum membrane distillation (VMD) tests demonstrated the exceptional wetting resistance of the membrane coated by 60 nm SiNPs as compared to membranes coated with larger SiNPs, indicating that small nanoparticles are preferred in order to form a robust Cassie–Baxter state.

8.3.3 Electrospun Nanofiber Membranes

Electrospinning has been demonstrated as an effective method in fabricating omniphobic surfaces (Lee et al., 2016a; Ma et al., 2005; Tuteja et al., 2008a,

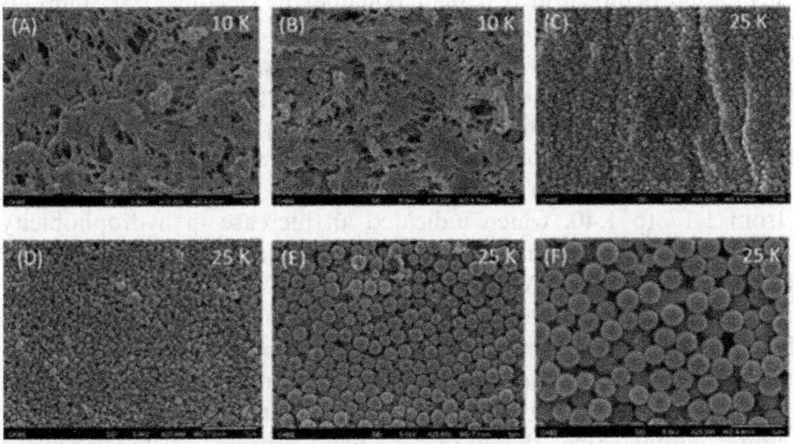

FIGURE 8.10 FESEM pictures of the outer membrane surfaces: (A) pristine membrane, (B) membrane coated with Teflon AF 2400, (C) membrane coated with 60 nm SiNPs, (D) membrane coated by both 60 SiNPs and Teflon AF 2400, (E) membrane coated by both 250 nm SiNPs and Teflon AF 2400, and (F) membrane coated by both 400 nm SiNPs and Teflon AF 2400 (Lu et al., 2018b). Copyright (2018) American Chemical Society.

2007). A well-designed electrospun fibrous structure provides an ideal substrate for developing omniphobic membranes, as the bottom-half of the electrospun fibrous network itself constitutes a reentrant structure (Tuteja et al., 2008a; Wang et al., 2016). However, the reentrant structure alone is a necessary but not sufficient condition for surface omniphobicity. In most cases, postmodification methods including lowering surface energy and increasing level of the reentrant structure are needed to impart omniphobic characteristics onto the electrospun fibrous substrates.

A CF_4 plasma treatment has been demonstrated as an effective method to improve surface hydrophobicity without sacrificing water flux of MD membranes (Yang et al., 2015, 2014). The increase in hydrophobicity arises from (1) suitable etching that enhances surface roughness, and (2) introduction of fluorine functional groups that reduce surface energy of membrane materials (Wei et al., 2012). Woo et al. have developed an omniphobic membrane via the combination of electrospinning and CF_4 plasma treatment (Woo et al., 2017). In this study, they treated the hot-pressed nanofiber membranes by CF_4 plasma and successfully reduced the surface energy by forming new CF_2–CF_2 and CF_3 bonds. At the optimal condition of 15 min CF_4 plasma treatment, the water contact angle increased from 133.6° to 160.6°, indicating a great improvement in hydrophobicity. The resultant membrane displayed omniphobic property with a wide range of low-surface-tension liquids, including methanol, mineral oil, and ethylene glycol. Due to such excellent repellency, the CF_4 plasma-treated

MD membrane demonstrated a high potential in treating feed solutions containing low-surface-tension substances.

Besides the hydrophobic plasma treatment, the addition of fluoroPOSS in the electrospinning dope solution has also been reported as a facile method to fabricate omniphobic nanofibrous membranes (Lu et al., 2018a). The addition of fluoroPOSS in electrospinning dopes did not change the morphology of nanofibers with regard to roughness or diameter, but only increased the F/C ratio from 1.17 to 1.40, which indicated an increase in hydrophobicity. The obtained fluoroPOSS-based omniphobic membrane successfully resisted surfactant-induced wetting in MD operations.

Constructing a hierarchical reentrant structure, typically refers a finer length scale reentrant texture on an underlying coarser one, is an alternative method to impart the omniphobic property onto membrane surfaces. By trapping air in multilevel textures, a hierarchical reentrant structure decreases the solid–liquid contact area, therefore enhancing wetting resistance toward low-surface-tension substances (Kota et al., 2013, 2012). Decorating the primary nanofibrous structure with spherical nanoparticles can effectively alter surface roughness and increase level of the reentrant structure. Silica and titanium dioxide nanoparticles are commonly used to create hierarchical structures (Lin et al., 2014; Razmjou et al., 2012). Lee et al. successfully attached negatively charged SiNPs on a positively charged nanofiber membrane via a simple dip-coating (Lee et al., 2016b). Cationic surfactant benzyltriethylammonium chloride (BTEAC) was added into the dope solution to render positive charged nanofibers, which facilitated the attachment of SiNPs. The first level of reentrant structure was endowed by the nanofibrous network, and the deposition of SiNPs introduced the secondary reentrant structure. Hydroxyl groups on SiNPs provided a platform for subsequent fluoroalkyl silanization to further reduce the surface energy of the membrane. The resultant membrane not only displayed high water contact angles with low-surface-tension liquids in air but also delivered stable MD performance with hypersaline feed solutions containing surfactants.

More recently, a simpler method that combines the construction of hierarchical structure and fluorination into one-step has been reported (Deng et al., 2018). After electrospinning, the electrospun PVDF nanofiber membranes were immediately immersed in fluorododecyltrichlorosilane (FTCS) solutions of different concentrations (0.5 wt%, 2 wt%, and 3.5 wt%) in order to obtain omniphobicity. Compared to other coating conditions, the membrane treated with a 2 wt% FTCS solution not only displayed high repellency to low-surface-tension liquids without an expense of membrane porosity but also stable DCMD performance when treating a surfactant contained feed solution. The omniphobicity was contributed by three factors: (1) the hierarchical reentrant structure provided by PVDF nanofibers, (2) the polymerized fluorododecyltrichlorosilane (PFTCS) buds on nanofiber surface (Figure 8.11), and (3) the extremely low surface energy of PFTCS.

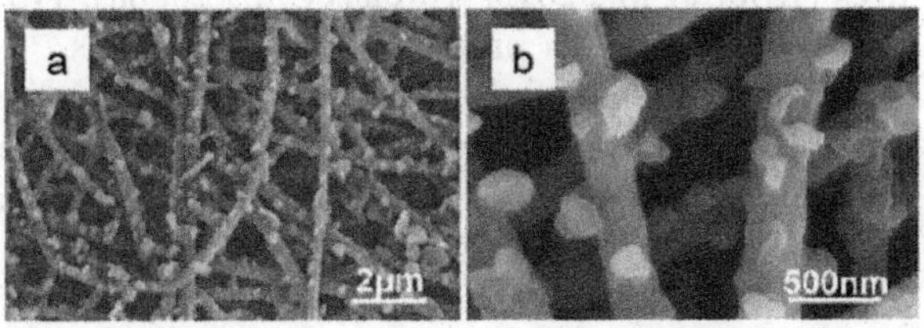

FIGURE 8.11 FESEM images of nanofibers of the PVDF membrane treated with a 2 wt % FTCS solution for 24 h (Deng et al., 2018). Copyright (2018) Elsevier Science Ltd.

8.4 ANTIWETTING AND ANTIFOULING PROPERTIES OF OMNIPHOBIC MEMBRANES

Improving antiwetting properties is one of the ultimate incentives of developing omniphobic membranes for membrane distillation. Previous discussions demonstrate that wetting resistance toward a wide range of liquids can be effectively enhanced by omniphobic modifications as evidenced by the increased liquid contact angles and stable MD performances using surfactant contained feed solutions (Boo et al., 2016a, 2016b; Chen et al., 2018a, 2018b; Deng et al., 2018; Huang et al., 2017; Kadi et al., 2019; Lee et al., 2016b; Li et al., 2019; Lin et al., 2014; Lu et al., 2018b; Woo et al., 2017, 2018; Zheng et al., 2018). In addition to membrane wetting, membrane fouling is another issue that largely impedes the industrialization of MD. Based on foulant types, membrane fouling in MD systems can be classified into three categories: (1) inorganic fouling, (2) organic fouling, and (3) biological fouling (Tijing et al., 2015). During MD operations, various foulants in feed solutions accumulate on membrane surfaces and inside membrane pores, resulting in blockage of membrane pores and degradation of vapor flux. Although omniphobic membranes have been proposed as a fouling mitigation strategy for MD, studies on antifouling performance of omniphobic membranes up today mainly focus on organic foulants such as humic acid (Woo et al., 2018) and oils (Boo et al., 2016a; Wang and Lin, 2017; Zheng et al., 2018). A less progress has been made on fouling resistance toward inorganic and biological fouling.

An omniphobic membrane with a hierarchical reentrant structure has been developed through a LBL assembly technique as described in Section 3.2 (Woo et al., 2018). To test the fouling resistance, this omniphobic composite membrane was utilized to treat a feed solution containing 20 mg/L humic acid. It remained uncontaminated even after a 72-h operation under AGMD, which could not be achieved by the neat PVDF membrane. Since the omniphobic membrane had a higher zeta potential compared to the neat PVDF, the strong electric repulsion was considered as the origin of its antifouling property. It was expected that omniphobic membranes could resist oil fouling in MD applications due to their in-air oleophobicity which has been reported in many literatures (Kota et al., 2013; Tuteja et al., 2008a). However, the situation is much more complicated in practical MD operations.

Zheng et al. found that the omniphobic membrane with a hierarchical reentrant structure displayed a better fouling resistance than pristine PVDF, FAS17-coated PVDF, and single-level reentrant structured PVDF membranes (Zheng et al., 2018). Their omniphobic membrane showed an extremely low sliding angle with a water/hexadecane emulsion, indicating high resistance toward low-surface-tension foulants. Owing to this super-repellency, neither apparent flux decline nor permeate conductivity increase was observed when it was challenged with a hexadecane emulsion feed solution stabilized with SDS. After the DCMD test, the omniphobic membrane still showed a high water contact angle of 168.3°, which means it remained uncontaminated during the MD operation. A similar result was reported by Elimelech and his group (Boo et al., 2016a), whose omniphobic membrane was able to treat shale gas waste water containing both mineral oil and surfactant, without the occurrence of wetting or fouling phenomenon.

In the above-mentioned two studies, omniphobic membranes have demonstrated superior anti-oil-fouling properties against water/oil emulsions that were stabilized with surfactants. A similar observation was reported by Wang et al. (Wang and Lin, 2017). Their omniphobic membrane also exhibited no fouling or wetting when a water/oil emulsion was stabilized by 100 mM Triton X-100. They found that the presence of surfactants played a critical role. Since amphiphilic surfactants consisted of both hydrophobic and hydrophilic groups, the hydrophobic groups of surfactants in the presence of oil would attach on the oil/water interface. Thus, the hydrophilic groups would be exposed and the oil droplets wrapped with surfactant molecules became hydrophilic. This reduces the hydrophobic–hydrophobic interaction between oil and membrane surface.

However, when treating a feed solution containing only crude oil without being stabilized by surfactants, omniphobic membranes were found to be fouled even though they displayed nonzero contact angles for oil both in air and underwater. This could be ascribed to the hydrophobic–hydrophobic interaction between oil droplets and the low-surface-energy coating of the omniphobic membrane (Figure 8.12). This hypothesis was further proven by the reduced adhesion force between oil and composite membranes consisting of hydrophobic substrate and hydrophilic top layer. In other words, omniphobic membranes made from low-surface-energy materials were in fact underwater oleophilic. The reentrant morphology on membrane surface could only prevent the penetration of oil droplets into membrane pores but not oil fouling on membrane surface.

To control oil fouling on membrane surface, hydrophilic surface modification is regarded as a widely used strategy. It is well known that, in contact with water, hydrophilic functional groups on membrane surface would form hydrogen bonds with water molecules, creating a hydration layer on membrane surface (Chen et al., 2010; Li et al., 2014; Wang et al., 2018). For oil droplets to attach on and spread over a hydrophilic surface, energy is required to

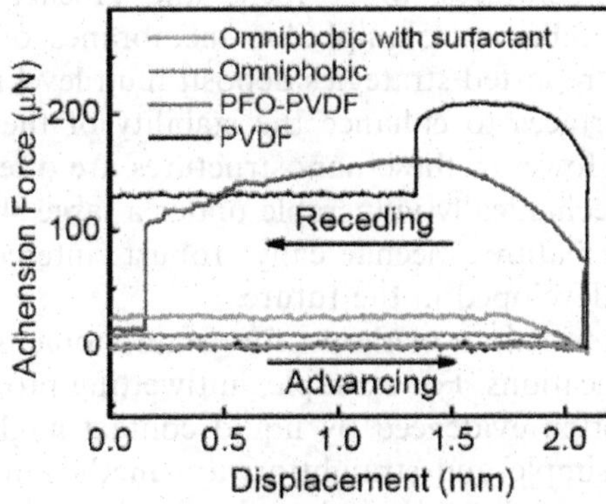

FIGURE 8.12 Force curves for the interactions between the PVDF membrane, the composite membrane, and the omniphobic membrane and a mineral oil droplet. The force curves display the forces measured by the tensiometer sensor as a function of probe displacement. All force curves were collected for the oil–membrane interaction under DI water, with the exception for the one that was collected in a Triton X-100 solution with the omniphobic membrane. The initial position of the oil droplet and the initial force applied on the oil droplet were both set as zero (Wang et al., 2017). Copyright (2017) Elsevier Science Ltd.

destroy the hydration layer on the surface. Therefore, the formed hydration layer serves as an energy barrier that is unfavorable for the oil attachment (Devi et al., 2005; Howarter and Youngblood, 2009; Yang et al., 2012; Zhang et al., 2014). Inspired by this principle, Huang et al. developed a Janus membrane comprising of an omniphobic substrate and a hydrophilic top layer, which exhibited both antiwetting and antifouling properties (Huang et al., 2017). The omniphobic substrate was fabricated by electrospinning a blend of poly(vinylidene fluoride-*co*-hexafluoropropylene) (PVDF-HFP) and cetyltrimethylammonium bromide (CTAB) followed by deposition of fluorinated SiNPs. Then a hydrophilic layer was developed by spraying a mixture of perfluorooctanoate, chitosan, and SiNPs on the omniphobic substrate. When treating an oil-in-water emulsion as the feed solution, the resultant Janus membrane exhibited underwater oleophobicity and provided a stable DCMD performance for 10 h without a loss of permeate flux or salt rejection.

8.5 CHALLENGES OF CURRENT APPROACHES

Omniphobic membranes have demonstrated great potential in reclaiming water from feed solutions containing low-surface-tension materials. Nevertheless, the existing synthesis technologies have several limitations that should be addressed in the future studies. First of all, the majority of current methods involve complex and time-consuming modification steps; some require harsh reaction conditions and hazardous reagents such as concentrated alkaline solutions at high temperatures. More facile and greener strategies must be invented in order to fabricate omniphobic membranes on a larger scale. In addition, most of the reported strategies deposit multilevel micro/nanoparticles on the membrane surfaces to enhance the stability of the solid–liquid–vapor composite interface. However, these nanostructures are often discretely distributed and might be mechanically vulnerable under a large shear stress. Alternative strategies that feature mechanically robust interconnected reentrant structures should be developed in the future.

Moreover, characterization methods for omniphobic surfaces should be tailored for MD applications. For example, antiwetting properties of omniphobic membranes are often evidenced by liquid contact angles measured in air. Even though it is a simple and straightforward method to assess the surface omniphobicity, it may not be able to describe the feed–membrane interface in real MD tests. Underwater contact angle measurements should be conducted when the feed solution contains water-immiscible liquids like oil. Furthermore, contact angles measured at elevated temperatures are highly recommended because the liquid surface tension is closely correlated to the temperature. In addition, in most of the current studies, omniphobic membranes exhibit lower fluxes compared to those of the pristine membranes due to the reduction in pore size. On the other hand, it is well known that a higher flux (i.e., a higher evaporation rate at the membrane surface) will result in a more severe concentration polarization and a higher possibility of wetting. Therefore, it is difficult to conclude whether the improved stability is due to the omniphobic coating or simply owing to the reduced flux. To demonstrate the superiority of omniphobic membranes, one should either adjust the operating parameters such as feed temperatures to achieve the same fluxes for both omniphobic and pristine membranes and then compare the wetting rates or conduct long-term performance tests to show the exceptional wetting resistance of omniphobic membranes.

8.6 CONCLUSIONS

A comprehensive review of the state-of-the-art omniphobic membranes for industrial wastewater treatment by membrane distillation has been provided in this chapter. By forming a stable solid–liquid–vapor interface, omniphobic membranes exhibit strong wetting resistances toward liquids with both high and low surface tensions. The most widely adopted strategy to design omniphobic surfaces is through the combination of a reentrant surface texture and a low-surface-energy coating. An outstanding omniphobic surface should have large apparent contact angles and great robustness factors toward various types of liquids. Novel technologies to fabricate robust omniphobic surfaces, and their advantages and disadvantages have been summarized in details.

Omniphobic membranes specifically developed for membrane distillation applications have been examined. These membranes could be divided into three categories based on their substrates, namely, inorganic membranes, phased inversed polymeric membranes, and electrospun nanofiber membranes. All the omniphobic MD membranes show superior wetting resistance in treating feed solutions containing salts and surfactants. Some also exhibit antifouling properties toward humid acid and surfactant-stabilized oil/water emulsions. These studies demonstrate the great potential of omniphobic membranes in reclaiming water from low-surface-tension industrial wastewaters. However, challenges in the current design approaches and imperfections in the present characterization methods are also discussed. It is believed that omniphobic MD membranes will be brought to the next level of R&D attention if the challenges can be successfully addressed.

ACKNOWLEDGEMENT

This chapter was reprinted from K. J. Lu, Y. Chen, T. S. Chung. Design of omniphobic interfaces for membrane distillation – A review. Water Res. 2019, 162, 64-77. Copyright (2019), with permission from Elsevier.

LIST OF SYMBOLS

A^*	Roughness factor
D	Distance between two cylindrical structures
D^*	Dimensionless space ratio
f_s, f_g	Area of solid–liquid and gas–liquid interface on a projected area of unity, respectively
r	Roughness ratio
R	Radius of the cylindrical texture
φ	Local texture angle
θ, θ_g	Equilibrium contact angle of liquid on solid and gas, respectively
θ_c	Threshold equilibrium contact angle
θ^*	Apparent contact angle

LIST OF ABBREVIATIONS

AGMD	Air gap membrane distillation
APTES	(3-Aminopropyl)-triethoxysilane
CSG	Coal seam gas
DCMD	Direct contact membrane distillation
FAS	Fluorinated alkyl silane
FAS17	1H,1H,2H,2H-perfluorodecyltrimethoxysilan
FTCS	Fluorododecyltrichlorosilane
FDTS	Perfluorodecyltrichlorosilane
fluoroPOSS	Fluorodecyl polyhedral oligomeric silsesquioxane
LBL	Layer-by-layer
MD	Membrane distillation
MOF	Metal–organic framework
PFPEdma	Perfluoropolyether dimethacrylate

PTFE	Polytetrafluoroethylene
PVA	Polyvinyl alcohol
PVDF	Polyvinylidene fluoride
RO	Reverse osmosis
SDS	Sodium dodecyl sulfate
Si	Silicon
SiO_2	Silicon dioxide
SiNP(s)	Silica nanoparticle(s)
SLIPS	Smooth, liquid-infused porous surfaces
TMC	Trimesoyl chloride
ZnO	Zinc oxide

REFERENCES

An, A.K., Guo, J., Jeong, S., Lee, E.J., Tabatabai, S.A.A., Leiknes, T.O., 2016. High flux and antifouling properties of negatively charged membrane for dyeing wastewater treatment by membrane distillation. Water Res. 103, 362–371.

Boban, M., Golovin, K., Tobelmann, B., Gupte, O., Mabry, J.M., Tuteja, A., 2018. Smooth, all-solid, low-hysteresis, omniphobic surfaces with enhanced mechanical durability. ACS Appl. Mater. Interfaces 10, 11406–11413. doi:10.1021/acsami.8b00521

Boo, C., Lee, J., Elimelech, M., 2016a. Omniphobic polyvinylidene fluoride (PVDF) membrane for desalination of shale gas produced water by membrane distillation. Environ. Sci. Technol. 50, 12275–12282. doi:10.1021/acs.est.6b03882

Boo, C., Lee, J., Elimelech, M., 2016b. Engineering surface energy and nanostructure of microporous films for expanded membrane distillation applications. Environ. Sci. Technol. 50, 8112–8119. doi:10.1021/acs.est.6b02316

Bunz, U.H.F., 2006. Breath figures as a dynamic templating method for polymers and nanomaterials. Adv. Mater. 18, 973–989. doi:10.1002/adma.200501131

Calabro, V., Drioli, E., Matera, F., 1991. Membrane distillation in the textile wastewater treatment. Desalination 83, 209–224. doi:10.1016/0011-9164(91)85096-D

Camacho, L.M., Dumée, L., Zhang, J., Li, J., Duke, M., Gomez, J., Gray, S., 2013. Advances in membrane distillation for water desalination and purification applications. Water. 5, 94–196. doi:10.3390/w5010094

Cassie, A.B.D., Baxter, S., 1944. Wettability of porous surfaces. Trans. Faraday Soc. 546–551. doi:10.1039/tf9444000546

Chen, L.H., Chen, Y.R., Huang, A., Chen, C.H., Su, D.Y., Hsu, C.C., Tsai, F.Y., Tung, K.L., 2018a. Nanostructure depositions on alumina hollow fiber membranes for enhanced wetting resistance during membrane distillation. J. Memb. Sci. 564, 227–236. doi:10.1016/j.memsci.2018.07.011

Chen, L.H., Huang, A., Chen, Y.R., Chen, C.H., Hsu, C.C., Tsai, F.Y., Tung, K.L., 2018b. Omniphobic membranes for direct contact membrane distillation: Effective deposition of zinc oxide nanoparticles. Desalination 428, 255–263. doi:10.1016/j.desal.2017.11.029

Chen, S., Li, L., Zhao, C., Zheng, J., 2010. Surface hydration: Principles and applications toward low-fouling/nonfouling biomaterials. Polymer. 51, 5283–5293. doi:10.1016/j.polymer.2010.08.022

Cheng, D.F., Urata, C., Yagihashi, M., Hozumi, A., 2012. A statically oleophilic but dynamically oleophobic smooth nonperfluorinated surface. Angew. Chemie Int. Ed. 51, 2956–2959. doi:10.1002/anie.201108800

Cheng, Y.T., Rodak, D.E., 2010. Is the lotus leaf superhydrophobic? Appl. Phys. Lett. 86, 1–3. doi:10.1063/1.1895487

Chhatre, S.S., Choi, W., Tuteja, A., Park, K.C., Mabry, J.M., McKinley, G.H., Cohen, R. E., 2010. Scale dependence of omniphobic mesh surfaces. Langmuir. 26, 4027–4035. doi:10.1021/la903489r

Deng, L., Ye, H., Li, X., Li, P., Zhang, J., Wang, X., Zhu, M., Hsiao, B.S., 2018. Self-roughened omniphobic coatings on nanofibrous membrane for membrane distillation. Sep. Purif. Technol. 206, 14–25. doi:10.1016/j.seppur.2018.05.035

Deshmukh, A., Boo, C., Karanikola, V., Lin, S., Straub, A.P., Tong, T., Warsinger, D.M., Elimelech, M., 2018. Membrane distillation at the water-energy nexus: Limits, opportunities, and challenges. Energy Environ. Sci. 11, 1177–1196. doi:10.1039/c8ee00291f

Devi, D.A., Smitha, B., Sridhar, S., Aminabhavi, T.M., 2005. Pervaporation separation of isopropanol/water mixtures through crosslinked chitosan membranes. J. Memb. Sci. 262, 91–99. doi:10.1016/j.memsci.2005.03.051

Drioli, E., Ali, A., Macedonio, F., 2015. Membrane distillation: recent developments and perspectives. Desalination 356, 56–84. doi:10.1016/j.desal.2014.10.028

El-Abbassi, A., Hafidi, A., Khayet, M., García-Payo, M.C., 2013. Integrated direct contact membrane distillation for olive mill wastewater treatment. Desalination 323, 31–38. doi:10.1016/j.desal.2012.06.014

Gryta, M., 2005. Long-term performance of membrane distillation process. J. Memb. Sci. 265, 153–159. doi:10.1016/j.memsci.2005.04.049

Henderyckx, Y., 1967. Diffusion doublet research. Desalination 3, 237–242. doi:10.1016/0011-9164(67)80014-6

Hensel, R., Finn, A., Helbig, R., Braun, H.G., Neinhuis, C., Fischer, W.J., Werner, C., 2013. Biologically inspired omniphobic surfaces by reverse imprint lithography. Adv. Mater. 26, 2029–2033. doi:10.1002/adma.201305408

Hensel, R., Helbig, R., Aland, S., Voigt, A., Neinhuis, C., Werner, C., 2013. Tunable nano-replication to explore the omniphobic characteristics of springtail skin. NPG Asia Mater. 5. doi:10.1038/am.2012.66

Hensel, R., Neinhuis, C., Werner, C., 2016. The springtail cuticle as a blueprint for omniphobic surfaces. Chem. Soc. Rev. 45, 323–341. doi:10.1039/c5cs00438a

Hilal, N., Kim, G.J., Somerfield, C., 2011. Boron removal from saline water: A comprehensive review. Desalination 273, 23–35. doi:10.1016/j.desal.2010.05.012

Howarter, J.A., Youngblood, J.P., 2009. Amphiphile grafted membranes for the separation of oil-in-water dispersions. J. Colloid Interface Sci. 329, 127–132. doi:10.1016/j.jcis.2008.09.068

Huang, Y.X., Wang, Z., Jin, J., Lin, S., 2017. Novel janus membrane for membrane distillation with simultaneous fouling and wetting resistance. Environ. Sci. Technol. 51, 13304–13310. doi:10.1021/acs.est.7b02848

Hussain, A., Minier-Matar, J., Janson, A., Adham, S., 2015. Treatment of produced water from oil & gas operations by membrane distillation, in: Proceedings of the 4th International Gas Processing Symposium. Elsevier B.V., pp. 285–292. doi:10.1016/B978-0-444-63461-0.50029-8

Israelachvili, J.N., 2011. Intermolecular and surface forces, 3rd ed. Elsevier, Amsterdam.

Kadi, K.E., Janajreh, I., Hashaikeh, R., Ahmed, R., 2019. Refinery processed water treatment via the low energy Direct Contact Membrane Distillation (DCMD). Oil Gas Sci. Technol. Rev. IFP Energies Nouv. 3, 1–9. doi:10.2516/ogst/2018077

Khaing, T.H., Li, J., Li, Y., Wai, N., Wong, F.S., 2010. Feasibility study on petrochemical wastewater treatment and reuse using a novel submerged membrane distillation bioreactor. Sep. Purif. Technol. 74, 138–143. doi:10.1016/j.seppur.2010.05.016

Kota, A.K., Choi, W., Tuteja, A., 2013. Superomniphobic surfaces: Design and durability. NPG Asia Mater. 6. doi:10.1557/mrs.2013.101

Kota, A.K., Kwon, G., Tuteja, A., 2014. The design and applications of superomniphobic surfaces. NPG Asia Mater 6, e109-16. doi:10.1038/am.2014.34

Kota, A.K., Li, Y., Mabry, J.M., Tuteja, A., 2012. Hierarchically structured superoleophobic surfaces with ultralow contact angle hysteresis. Adv. Mater. 24, 5838–5843. doi:10.1002/adma.201202554

Lawson, K.W., Lloyd, D.R., 1997. Membrane distillation. J. Memb. Sci. 124, 1–25. doi:10.1016/S0376-7388(96)00236-0

Lee, E.J., An, A.K., He, T., Woo, Y.C., Shon, H.K., 2016a. Electrospun nanofiber membranes incorporating fluorosilane-coated TiO 2 nanocomposite for direct contact membrane distillation. J. Memb. Sci. 520, 145–154. doi:10.1016/j.memsci.2016.07.019

Lee, J., Boo, C., Ryu, W.H., Taylor, A.D., Elimelech, M., 2016b. Development of omniphobic desalination membranes using a charged electrospun nanofiber scaffold. ACS Appl. Mater. Interfaces 8, 11154–11161. doi:10.1021/acsami.6b02419

Li, J., Guo, S., Xu, Z., Li, J., Pan, Z., Du, Z., Cheng, F., 2019. Preparation of omniphobic PVDF membranes with silica nanoparticles for treating coking wastewater using direct contact membrane distillation: Electrostatic adsorption vs. chemical bonding. J. Memb. Sci. 574, 349–357. doi:10.1016/j.memsci.2018.12.079

Li, X., Cai, T., Chung, T.S., 2014. Anti-Fouling behavior of hyperbranched polyglycerol-grafted poly(ether sulfone) hollow fiber membranes for osmotic power generation. Environ. Sci. Technol. 48, 9898–9907. doi:10.1021/es5017262

Lin, S., Nejati, S., Boo, C., Hu, Y., Osuji, C.O., Elimelech, M., 2014. Omniphobic membrane for robust membrane distillation. Environ. Sci. Technol. Lett. 1, 443–447. doi:10.1021/ez500267p

Lu, C., Su, C., Cao, H., Ma, X., Duan, F., Chang, J., Li, Y., 2018a. F-POSS based omniphobic membrane for robust membrane distillation. Mater. Lett. 228, 85–88. doi:10.1016/j.matlet.2018.05.126

Lu, K.J., Chen, Y., Chung, T.S., 2019. Design of omniphobic interfaces for membrane distillation – A review. Water Res. 162, 64–77. doi:10.1016/j.watres.2019.06.056.

Lu, K.J., Zuo, J., Chang, J., Kuan, H.N., Chung, T.S., 2018b. Omniphobic hollow-fiber membranes for vacuum membrane distillation. Environ. Sci. Technol. 52, 4472–4480. doi:10.1021/acs.est.8b00766

Lu, K.J., Zuo, J., Chung, T.S., 2016. Tri-bore PVDF hollow fibers with a super-hydrophobic coating for membrane distillation. J. Memb. Sci. 514, 165–175. doi:10.1016/j.memsci.2016.04.058

Ma, M., Mao, Y., Gupta, M., Gleason, K.K., Rutledge, G.C., 2005. Superhydrophobic fabrics produced by electrospinning and chemical vapor deposition. Macromolecules 38, 9742–9748. doi:10.1021/ma0511189

Ma, W., Higaki, Y., Otsuka, H., Takahara, A., 2013. Perfluoropolyether-infused nano-texture: A versatile approach to omniphobic coatings with low hysteresis and high transparency. Chem. Commun. 49, 597–599. doi:10.1039/c2cc37576a

Mokhtar, N.M., Lau, W.J., Ismail, A.F., Veerasamy, D., 2015. Membrane distillation technology for treatment of wastewater from rubber industry in Malaysia. Procedia CIRP. 26, 792–796. doi:10.1016/j.procir.2014.07.161

Razmjou, A., Arifin, E., Dong, G., Mansouri, J., Chen, V., 2012. Superhydrophobic modification of TiO2nanocomposite PVDF membranes for applications in membrane distillation. J. Memb. Sci. 415–416, 850–863. doi:10.1016/j.memsci.2012.06.004

Rezaei, M., Warsinger, D.M., Lienhard V, J.H., Duke, M.C., Matsuura, T., Samhaber, W. M., 2018. Wetting phenomena in membrane distillation: mechanisms, reversal, and prevention. Water Res. 139, 329–352. doi:10.1016/j.watres.2018.03.058

Sivakumar, M., Ramezanianpour, M., O'Halloran, G., 2013. Mine water treatment using a vacuum membrane distillation system. APCBEE Procedia. 5, 157–162. doi:10.1016/j.apcbee.2013.05.028

Srinivasarao, M., Collings, D., Philips, A., Patel, S., 2001. Three-dimensionally ordered array of air bubbles in a polymer film. Science 292, 79–83. doi:10.1126/science.1057887

Tan, T.T.Y., Reithofer, M.R., Chen, E.Y., Menon, A.G., Hor, T.S.A., Xu, J., Chin, J.M., 2013. Tuning omniphobicity via morphological control of metal-organic framework functionalized surfaces. J. Am. Chem. Soc. 135, 16272–16275. doi:10.1021/ja407896m

Tijing, L.D., Woo, Y.C., Choi, J.S., Lee, S., Kim, S.H., Shon, H.K., 2015. Fouling and its control in membrane distillation – A review. J. Memb. Sci. 475, 215–244. doi:10.1016/j.memsci.2014.09.042

Tsujii, K., Yamamoto, T., Onda, T., Shibuichi, S., 1997. Super oil-repellent surfaces. Angew. Chemie Int. Ed. English 36, 1011–1012. doi:10.1002/anie.199710111

Tuteja, A., Choi, W., Ma, M., Mabry, J.M., Mazzella, S.A., Rutledge, G.C., McKinley, G.H., Cohen, R.E., 2007. Designing superoleophobic surfaces. Science 318, 1618–1622. doi:10.1126/science.1148326

Tuteja, A., Choi, W., Mabry, J.M., McKinley, G.H., Cohen, R.E., 2008a. Robust omniphobic surfaces. Proc. Natl. Acad. Sci. 105, 18200–18205. doi:10.1073/pnas.0804872105

Tuteja, A., Choi, W., McKinley, G.H., Cohen, R.E., Rubner, M.F., 2008b. Design parameters for superhydrophobicity and superoleophobicity. MRS Bull. 33, 752–758. doi:10.1557/mrs2008.161

Wang, K., Hou, D., Wang, J., Wang, Z., Tian, B., Liang, P., 2018. Hydrophilic surface coating on hydrophobic PTFE membrane for robust anti-oil-fouling membrane distillation. Appl. Surf. Sci. 450, 57–65. doi:10.1016/j.apsusc.2018.04.180

Wang, L., McCarthy, T.J., 2016. Covalently attached liquids: instant omniphobic surfaces with unprecedented repellency. Angew. Chemie Int. Ed. 55, 244–248. doi:10.1002/anie.201509385

Wang, Z., Elimelech, M., Lin, S., 2016. Environmental applications of interfacial materials with special wettability. Environ. Sci. Technol. 50, 2132–2150. doi:10.1021/acs.est.5b04351

Wang, Z., Lin, S., 2017. Membrane fouling and wetting in membrane distillation and their mitigation by novel membranes with special wettability. Water Res. 112, 38–47. doi:10.1016/j.watres.2017.01.022

Warsinger, D.M., Swaminathan, J., Guillen-Burrieza, E., Arafat, H.A., Lienhard V., 2015. Scaling and fouling in membrane distillation for desalination applications: a review. Desalination 356, 294–313. doi:10.1016/j.desal.2014.06.031

Wei, X., Zhao, B., Li, X.M., Wang, Z., He, B.Q., He, T., Jiang, B., 2012. CF 4 plasma surface modification of asymmetric hydrophilic polyethersulfone membranes for direct contact membrane distillation. J. Memb. Sci. 407–408, 164–175. doi:10.1016/j.memsci.2012.03.031

Wenzel, R.N., 1936. Resistance of solid surfaces to wetting by water. Ind. Eng. Chem. 28, 988–994. doi:10.1021/ie50320a024

Wong, T.S., Kang, S.H., Tang, S.K.Y., Smythe, E.J., Hatton, B.D., Grinthal, A., Aizenberg, J., 2011. Bioinspired self-repairing slippery surfaces with pressure-stable omniphobicity. Nature 477, 443–447. doi:10.1038/nature10447

Woo, Y.C., Chen, Y., Tijing, L.D., Phuntsho, S., He, T., Choi, J.S., Kim, S.H., Shon, H. K., 2017. CF4 plasma-modified omniphobic electrospun nanofiber membrane for produced water brine treatment by membrane distillation. J. Memb. Sci. 529, 234–242. doi:10.1016/j.memsci.2017.01.063

Woo, Y.C., Kim, Y., Yao, M., Tijing, L.D., Choi, J.S., Lee, S., Kim, S.H., Shon, H.K., 2018. Hierarchical composite membranes with robust omniphobic surface using layer-by-layer assembly technique. Environ. Sci. Technol. 52, 2186–2196. doi:10.1021/acs.est.7b05450

Yang, C., Li, X.M., Gilron, J., Kong, D.-F., Yin, Y., Oren, Y., Linder, C., He, T., 2014. CF4 plasma-modified superhydrophobic PVDF membranes for direct contact membrane distillation. J. Memb. Sci. 456, 155–161. doi:10.1016/j.memsci.2014.01.013

Yang, C., Tian, M., Xie, Y., Li, X.M., Zhao, B., He, T., Liu, J., 2015. Effective evaporation of CF4 plasma modified PVDF membranes in direct contact membrane distillation. J. Memb. Sci. 482, 25–32. doi:10.1016/j.memsci.2015.01.059

Yang, J., Zhang, Z., Xu, X., Zhu, X., Men, X., Zhou, X., 2012. Superhydrophilic-superoleophobic coatings. J. Mater. Chem. 22, 2834–2837. doi:10.1039/c2jm15987b

Zhang, S., Wang, P., Fu, X., Chung, T.S., 2014. Sustainable water recovery from oily wastewater via forward osmosis-membrane distillation (FO-MD). Water Res. 52, 112–121. doi:10.1016/j.watres.2013.12.044

Zhao, D., Zuo, J., Lu, K.J., Chung, T.S., 2017. Fluorographite modified PVDF membranes for seawater desalination via direct contact membrane distillation. Desalination 413, 119–126. doi:10.1016/j.desal.2017.03.012

Zheng, R., Chen, Y., Wang, J., Song, J., Li, X.M., He, T., 2018. Preparation of omniphobic PVDF membrane with hierarchical structure for treating saline oily wastewater using direct contact membrane distillation. J. Memb. Sci. 555, 197–205. doi:10.1016/j.memsci.2018.03.041

Zhu, P., Kong, T., Tang, X., Wang, L., 2017. Well-defined porous membranes for robust omniphobic surfaces via microfluidic emulsion templating. Nat. Commun. 8. doi:10.1038/ncomms15823

9 Using Green Solvents to Fabricate Membrane Distillation Membranes

Jian Chang and Tai-Shung Chung

9.1 INTRODUCTION

Organic solvents are very commonly used in many steps of membrane fabrication. Many of these solvents are toxic and pose threat to human health and the environment, but it is hard to replace them because of their broadly satisfactory properties such as solubility, viscosity, polarity, and boiling points (Figoli et al., 2014; Xing et al., 2014). Common solvents to prepare polyvinylidene fluoride (PVDF) membranes for membrane distillation (MD) are dimethylformamide (DMF), dimethylacetamide (DMAc), and N-methyl-2-pyrrolidone (NMP). For example, Tomaszewska used DMF and DMAc to cast flat sheet membranes with a permeation flux of about 10 kg/m^2 h (Tomaszewska, 1996). Hou et al. employed DMAc to fabricate single-layer hollow fiber membranes with a permeation flux of about 20 kg/m^2 h (Hou et al., 2009). Wang et al. utilized NMP to produce dual-layer hollow fiber membranes with a permeation flux over 40 kg/m^2 h (Wang et al., 2011). However, all these three solvents are very toxic and can cause many diseases. Table 9.1 briefly summarizes their hazard statements and linked diseases (Casarett et al., 2001; Regulation (EC) No. 1272/2008, 2008). In contrast, triethyl phosphate (TEP), for example, is a much safer solvent (Byrne et al., 2016; Capello et al., 2007). It is only harmful when being swallowed. In this case, not only can workers be better protected, the treatment of waste effluent can also be simplified.

The major challenge when replacing a common toxic solvent with a green solvent is the difficulty to achieve competitive product performance (Figoli et al., 2014; Xing et al., 2014). So far, very few works have examined the effects of using green solvents in MD membrane fabrication. Typically, TEP is a solvent of PVDF and has been used in the following studies. Nejati et al. produced flat sheet PVDF membranes, but the porosity is all below 75% (Nejati et al., 2015). Their super thin membrane had a flux up to 40 L/m^2 h (LMH). However, no quantitative mechanical properties were reported. Fadhil et al. fabricated a flat sheet membrane made of copolymer PVDF-polyvinylidenefluoride–hexafluoropropylene (HFP). The membranes had relatively low fluxes even after the optimization of the copolymer and additive concentrations to improve the overall performance (Fadhil et al., 2016).

TABLE 9.1
The hazard statements and toxicology of common organic solvents

	(a) Regulation (EC) No 1272/2008, 2008		(b) Casarett and Doull's toxicology	
Solvent	Hazard codes	Hazard statements	Strength of evidence	Diseases linked to this toxicant
DMF	H360D	May damage the unborn child	Strong evidence	Acute hepatocellular injury (hepatitis); steatosis (fatty liver)
	H332	Harmful if inhaled	Good evidence	Pancreatitis
	H312	Harmful in contact with skin	Limited evidence	Testicular cancer
	H319	Causes serious eye irritation		
DMAc	H360D	May damage the unborn child	Limited evidence	Acute hepatocellular injury (hepatitis)
	H332	Harmful if inhaled		
	H312	Harmful in contact with skin		
NMP	H319	Causes serious eye irritation	Good evidence	Fetotoxicity (miscarriage/spontaneous abortion, stillbirth)
	H315	Causes skin irritation	Limited evidence	Low birth weight/small for gestational age/intrauterine growth retardation
TEP	H302	Harmful if swallowed	–	–

Chang et al. also prepared flat sheet PVDF membranes with the highest flux below 13 LMH (Chang et al., 2014). Lastly, there was only one work before 2018 that studied hollow fiber PVDF membranes using TEP (Chang et al., 2017). The authors conducted a fundamental study on membrane formation using TEP and uncovered the major difficulties when using TEP to replace common toxic solvents in hollow fiber fabrication for MD applications. The solutions to overcome the difficulties were also demonstrated. As a result, both the environmental and economic goals can be pursued in membrane manufacture.

9.2 PREPARATION OF PVDF HOLLOW FIBER MEMBRANES

A PVDF/NMP system was also chosen as a benchmark for comparison. NMP was chosen over DMF and DMAc because the former is less toxic than the latter. Three solvent systems were studied: (D1) NMP 88 wt%; (D2) NMP/TEP 44/44 wt%; and (D3) TEP 88 wt%. The specific hollow fiber spinning conditions are tabulated in Table 9.2, and the resultant fibers are named as M1–M9, respectively.

TABLE 9.2
Spinning conditions for PVDF hollow fiber membranes

	M1	M2	M3	M4	M5	M6	M7	M8	M9
Dope composition (PVDF/ NMP/TEP wt%)	12/88/0	12/44/44	12/0/88						
Bore fluid composition (wt%)	NMP/water 40/60			TEP/water 40/60				TEP/ water 30/70	TEP/ water 20/80
External coagulant (wt%)	IPA/water 30/70				Water 100	TEP/ water 30/70	TEP/ water 60/40	TEP/water 30/70	

9.3 PHASE SEPARATION BEHAVIOR OF PVDF/SOLVENT SYSTEMS

Figure 9.1 shows the ternary phase diagram of PVDF/solvent/water systems at 25 °C. The binodal curves denote the limit of the homogenous phase. When the solution is brought across the binodal curve, liquid–liquid demixing takes place (Cheng et al., 1999; Yilmaz and McHugh, 1986). Therefore, the binodal curve is essentially a liquid–liquid demixing boundary where the stable region is on its left and the unstable region is on its right. The binodal curve of TEP is on the left side of NMPs, which means the PVDF/TEP system requires a less nonsolvent to induce phase inversion. This could be because TEP is a slightly weaker solvent for PVDF compared to NMP (Liu et al., 2012; Piezotech Arkema,

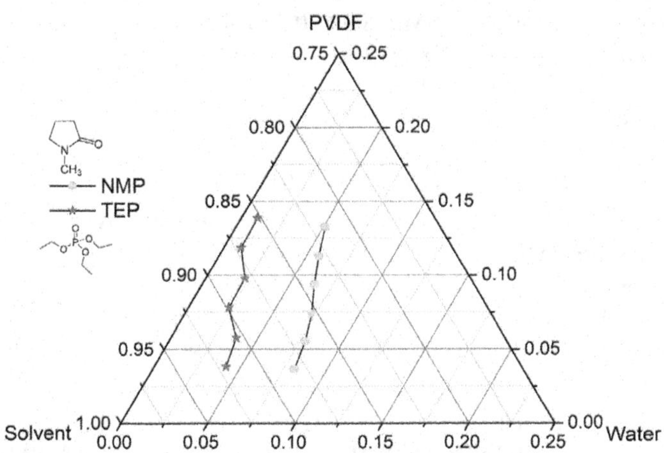

FIGURE 9.1 Phase diagram of PVDF/solvent/water systems (Chang et al., 2017). Copyright (2017) Elsevier.

1993). This is also supported by a comparison of their solubility parameters shown in Table 9.3(a) (Barton, 1983; Bottino et al., 1988). TEP has a slightly more different solubility parameter (22.2 $MPa^{1/2}$) from PVDF (23.2 $MPa^{1/2}$) than NMP (22.9 $MPa^{1/2}$). As a result, TEP is slightly less compatible with PVDF. This could potentially favor a more porous structure as the polymer chains may exist in a more coiled state and leave more space among them. Previous studies also suggested that PVDF membranes made from a weak solvent like TEP tended to have a sponge structure (Marino et al., 2017; Yeow et al., 2004) or a honeycomb structure (Bottino et al., 1991).

The weak solvent power of TEP is also supported by the molecular simulation results of the two PVDF/solvent systems. It is clear in Figure 9.2(a) that PVDF polymer chains are more curled in TEP, with a smaller radius of gyration (18.9 Å vs. 25.7 Å) as shown in Table 9.3(b), which may be attributed to the larger molecular size of TEP and its slightly weaker interactions with PVDF polymer chains (−21.72 kJ/mol vs. −23.71 kJ/mol) for each PVDF repeat unit. In addition, the polymer chains display a smaller mean squared displacement (MSD) in TEP than in NMP as shown in Figure 9.2(b), which also proves the weak solvent power of TEP.

The kinetics of phase inversion behavior is shown in Figure 9.3. The camera captures the exact position of the precipitation front when time elapses. It is clear that water intrudes much slower in the TEP solution. This may arise from

TABLE 9.3

a) Hansen solubility parameters at 25 °C (Barton, 1983; Bottino et al., 1988). b) Molecular simulation results of the PVDF/solvent systems. c) Dynamic (shear) viscosity of the dope solutions for hollow fiber spinning

(a)

Solvent/Polymer	Water	NMP	TEP	IPA	PVDF
δ_d (dispersion force)	15.6	18.0	16.8	15.8	17.0
δ_p (dipole force)	16.0	12.3	11.5	6.1	12.1
δ_h (hydrogen bonding force)	42.3	7.2	9.2	16.4	10.2
δ_{sp} (solubility parameter)	47.8	22.9	22.2	23.5	23.2

(b)

	12/88 PVDF/NMP	12/88 PVDF/TEP
Molecular size of solvent ($Å^3$)	104.7	183.2
Gyration of polymer (Å)	25.7 ± 1.9	18.9 ± 1.7
Self-diffusivity of solvent (cm^2/s)	8.02 ± 0.17 ($\times10^{-7}$)	1.43 ± 0.21 ($\times10^{-7}$)
Interactions between polymer and solvent (kJ/mol)	−23.71 ± 0.68	−21.72 ± 0.71

(c)

Dope solution	D1 (12/88 PVDF/NMP)	D2 (12/44/44 PVDF/NMP/TEP)	D3 (12/88 PVDF/TEP)
Dynamic viscosity (Pa s)	10.9 ± 0.04	12.7 ± 0.05	15.0 ± 0.02

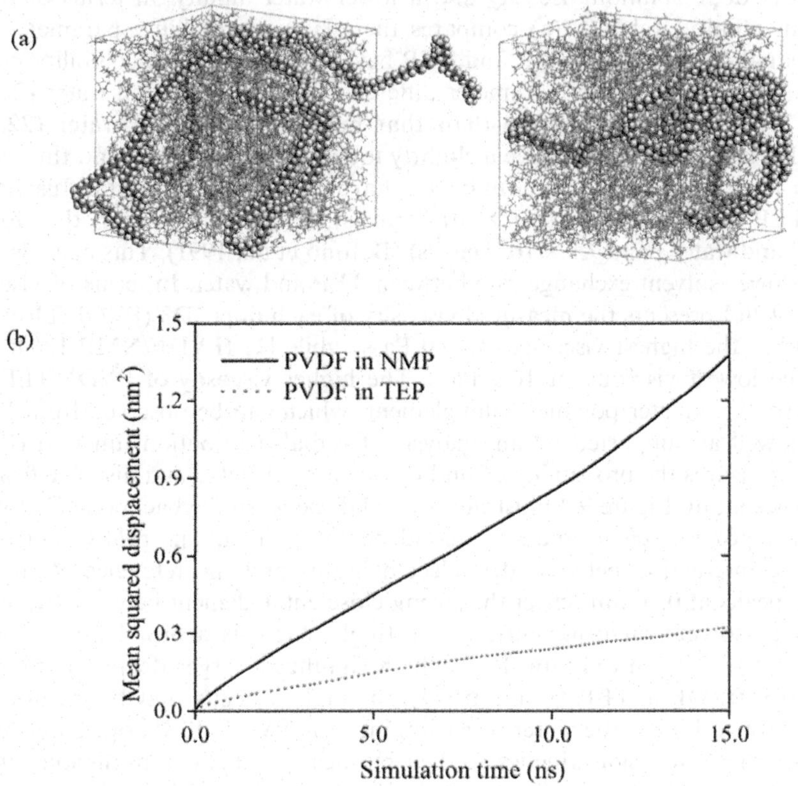

FIGURE 9.2 a) Snapshots from the molecular dynamics simulations showing the PVDF chain conformation in NMP (left) and TEP (right) solvents b) Mean squared displacement (MSD) for the PVDF polymer chains in NMP and TEP solvents (Chang et al., 2017). Copyright (2017) Elsevier.

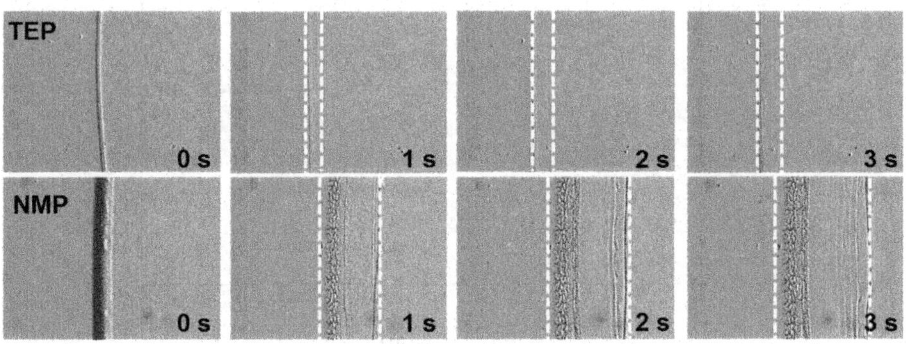

FIGURE 9.3 Phase inversion phenomena of PVDF/solvent systems in water (Chang et al., 2017). Copyright (2017) Elsevier.

a higher dope solution viscosity and a lower water affinity. In terms of water–solvent affinity, Table 9.3(a) compares their Hansen solubility parameters and individual components. NMP and TEP have very comparable solubility parameters. Since the solubility parameter difference between TEP and water (22.2 vs. 47.8 $MPa^{1/2}$) is slightly more than that between NMP and water (22.9 vs. 47.8 $MPa^{1/2}$), TEP might have a slightly lower affinity to water and thus retard water from intrusion. Additionally, the harmonic mean value of diffusivity for the TEP and water pair (8.7×10^{-6} cm^2/s) is slightly less than that for the NMP and water pair (9.1×10^{-6} cm^2/s) (Bottino et al., 1991). This may also lead to a slower solvent exchange rate between TEP and water. In terms of viscosity, Table 9.3(c) presents the measured viscosity of each dope. D3 (PVDF/TEP 12/88 wt%) has the highest viscosity of 15.0 Pa s, while D1 (PVDF/NMP 12/88 wt%) has the lowest viscosity of 10.9 Pa s. The higher viscosity of PVDF/TEP may arise from a greater polymer entanglement, which can be observed from Figure 9.2 as well as supported by an analysis of radial distribution function (RDF). RDF measures the probability of finding one atom located at a distance from the reference atom. Figure 9.4 illustrates the RDF between the backbone C atoms in the first polymer chain and the second polymer chain. The peaks indicate the binding interaction between the selected atoms and the reference atoms. The sharp peaks at 0.56 nm reflect the strong close entanglement between two chains for both solvent environments. Interestingly, there is a broad hump located from 1.5 to 2.5 nm only for TEP, which signifies a larger degree of entanglement for PVDF in TEP. Consequently, the high viscosity may be the dominating factor to hinder the water intrusion into the PVDF/TEP dope solution.

The phase inversion kinetics studied by means of light transmission through cast membranes also confirm the above observation. When phase inversion

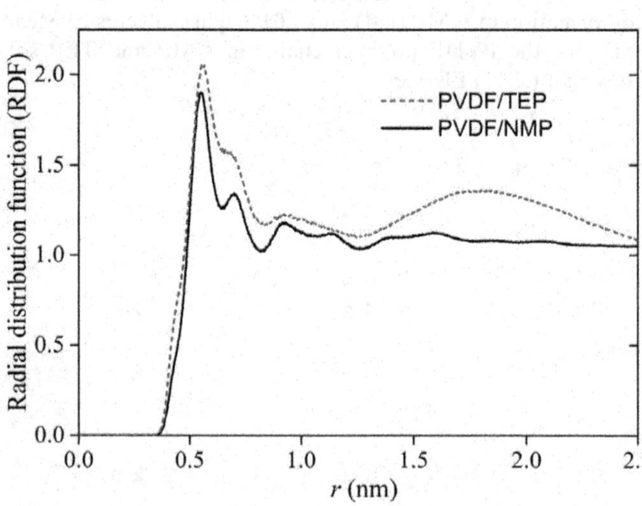

FIGURE 9.4 Radial distribution functions (RDFs) between the C atoms in different polymer chains (Chang et al., 2017). Copyright (2017) Elsevier.

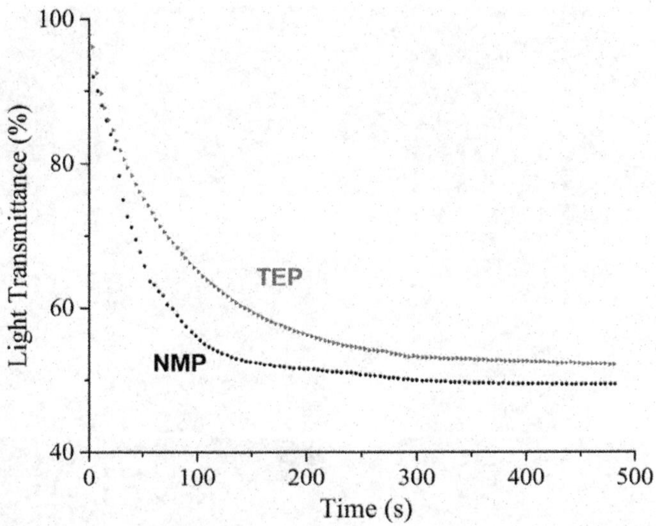

FIGURE 9.5 The decline of light transmittance through flat-sheet membrane films cast from NMP and TEP dope solutions and then immediately immersed in IPA/water 30/70 wt% (Chang et al., 2017). Copyright (2017) Elsevier.

progresses, the thin nascent membrane gradually loses its translucency measured by a spectrophotometer. The decline rate of light intensity is usually proportional to the speed of phase inversion (Feng et al., 2016; Yang et al., 2008). As shown in Figure 9.5, the TEP solution shows a tardier decay of light transmittance, indicating that the PVDF/TEP solution has a slower phase inversion kinetics than the PVDF/NMP solution. As a result, the former favors to form a porous sponge-like membrane structure than the latter.

9.4 EFFECTS OF SOLVENT COMPOSITION IN DOPE SOLUTIONS

Figure 9.6 shows the field-emission scanning electron microscopy morphology of hollow fibers spun from dopes with different solvent compositions. From M1 to M3, the spinning conditions are the same except the solvent composition in dopes changes from NMP 88 wt% to NMP/TEP 44/44 wt% and then to TEP 88 wt%, respectively. A clear transition of porosity can be observed. M1 has the most dense outer and inner surfaces and cross section, while M3 is the most porous. Figure 9.7 shows that the overall porosity follows the order of M3 (89.7%) > M2 (79.5%) > M1 (72.4%). In addition, as shown in Table 9.4, the surface porosity of M3 is 20–30 times higher than that of M1. The disparity of porosity may result from the effects of a weaker solvent and a higher viscosity of the PVDF/TEP solution, as discussed previously. Moreover, the diffusivity of TEP in water (6.3×10^{-6} cm^2/s) is much less than water in TEP (13.7×10^{-6} cm^2/s) (Yang et al., 2008), which favors a water-induced spinodal decomposition and

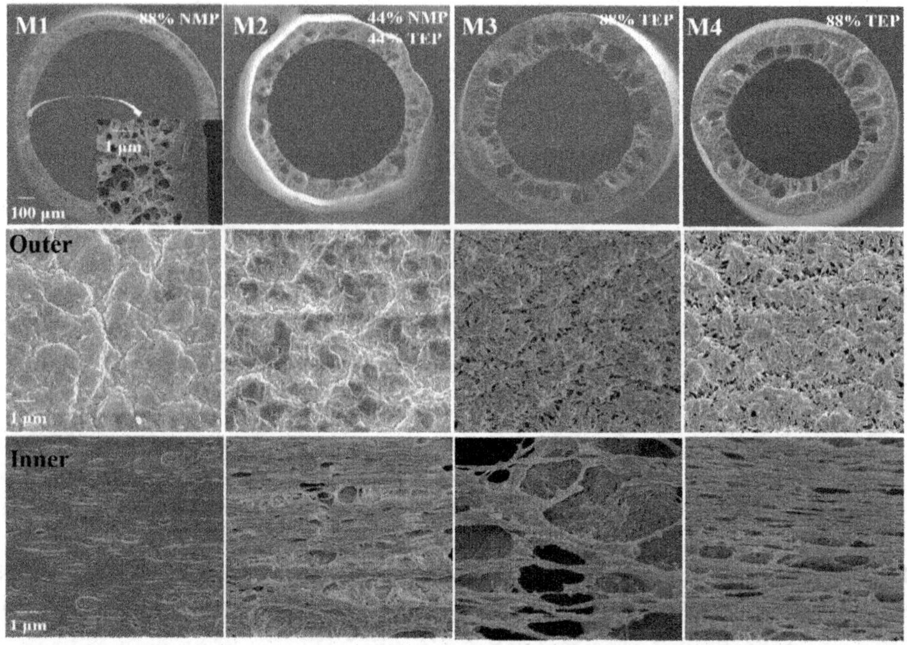

FIGURE 9.6 Cross-section and outer/inner surface morphologies of hollow fiber membranes spun with different solvent composition (Bore fluid: NMP/water 40/60 wt%; external coagulant: IPA/water 30/70 wt%) (Chang et al., 2017). Copyright (2017) Elsevier.

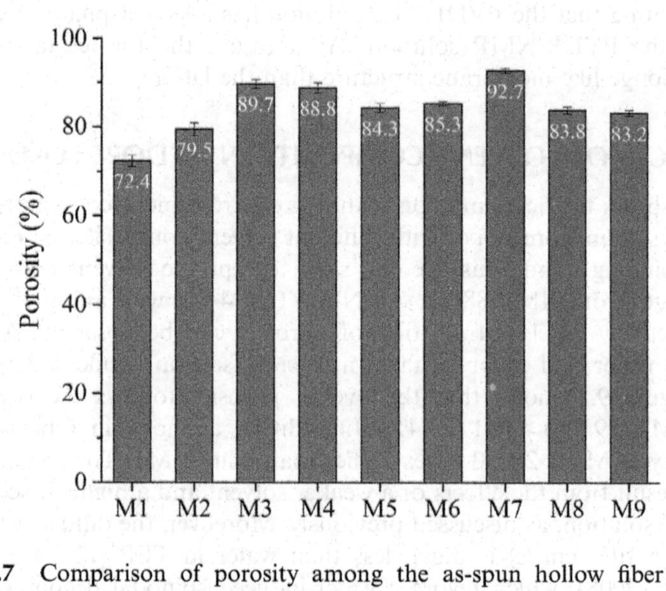

FIGURE 9.7 Comparison of porosity among the as-spun hollow fiber membranes (Chang et al., 2017). Copyright (2017) Elsevier.

TABLE 9.4

Comparison of hollow fibers spun using different solvent composition

	Flux (kg/m² h)	Rejection (%)	Overall porosity (%)	Outer surface porosity (%)	Inner surface porosity (%)	Contact angle (°)	Elongation at break (%)	Max tensile stress (MPa)	Young's modulus (MPa)	LEP (bar)
M1	<1	99.99	72.4 ± 1.2	0.49 ± 0.05	2.67 ± 0.3	95.42 ± 9.6	306 ± 6.3	6.29 ± 0.08	119 ± 4.8	3.1 ± 0.1
M2	15.1 ± 0.2	99.99	79.5 ± 1.4	0.50 ± 0.04	7.75 ± 0.3	102.8 ± 1.3	153 ± 9.4	2.72 ± 0.10	58.9 ± 3.4	2.8 ± 0.1
M3	21.8 ± 0.4	99.99	89.7 ± 0.9	16.9 ± 0.16	57.1 ± 0.9	113.6 ± 6.0	106 ± 7.6	0.76 ± 0.01	15.0 ± 0.9	0.2 ± 0.05
M4	23.5 ± 0.2	99.99	88.8 ± 1.1	13.4 ± 0.19	25.1 ± 0.5	113.6 ± 4.8	96.2 ± 10	0.87 ± 0.08	18.9 ± 1.3	0.2 ± 0.05
M5	10.7 ± 0.6	99.99	84.3 ± 0.9	0.88 ± 0.06	17.4 ± 0.2	89.89 ± 2.1	174 ± 5.4	1.78 ± 0.04	36.2 ± 1.6	2.5 ± 0.05
M6	19.9 ± 0.5	99.99	85.3 ± 0.4	3.18 ± 0.04	38.7 ± 0.4	102.4 ± 1.8	152 ± 6.5	1.35 ± 0.06	32.3 ± 1.0	1.6 ± 0.05
M7	18.4 ± 0.8	99.99	92.7 ± 0.5	3.05 ± 0.08	11.5 ± 0.2	143.0 ± 2.8	98.7 ± 5.2	0.42 ± 0.02	9.34 ± 1.0	0.8 ± 0.05
M8	19.3 ± 0.3	99.99	83.8 ± 0.8	2.14 ± 0.05	35.5 ± 0.8	103.8 ± 3.5	127 ± 3.3	1.60 ± 0.06	35.3 ± 2.3	1.8 ± 0.05
M9	18.0 ± 0.5	99.99	83.2 ± 0.6	1.63 ± 0.09	9.69 ± 0.3	99.12 ± 4.0	173 ± 4.3	2.08 ± 0.04	41.5 ± 3.2	2.0 ± 0.05

a more porous structure. Table 9.4 tabulates their direct contact membrane distillation (DCMD) performance at 60 °C, all three hollow fibers have good salt rejections of about 99.99%, but the water vapor flux rises considerably from M1 (<1 kg/m^2 h) to M3 (22 kg/m^2 h) because of the increased porosity. Clearly, TEP can be a very effective solvent to fabricate PVDF membranes with a high flux.

Table 9.4 also shows that the contact angle increases from M1 to M3 due to the effect of higher surface roughness (Razmjou et al., 2012; Zhang et al., 2005). However, a dramatic drop in LEP is also observed. The LEP value of M1 is 3.1 bar, but it decreases to 0.2 bar in M3. In addition, the mechanical properties in terms of the maximum tensile stress, Young's modulus, and elongation at break all drop substantially from M1 to M3, as shown in Table 9.4, due to the overly large pore size and macrovoids. Therefore, more efforts are needed to remedy the low LEP value and mechanical weakness in order to produce useful PVDF membranes using TEP for MD.

To further apply the principle of green chemistry, the bore fluid of M3 is changed from a NMP/water mixture to a TEP/water mixture in M4. Figure 9.6 and Table 9.4 compare their morphology and physicochemical properties. They have similar mechanical properties and overall morphologies. However, an obvious change lies at the inner surface morphology of M4, which becomes less porous. Because TEP is a weaker solvent than NMP, this leads to a faster phase inversion at the inner surface and results in a less porous surface in M4. In short, TEP favors to produce membranes with a more porous structure and a higher vapor flux. However, the mechanical integrity is greatly compromised. The next few sections attempt to solve this dilemma.

9.5 EFFECTS OF TEP CONCENTRATION IN THE EXTERNAL COAGULANT

The addition of isopropyl alcohol (IPA) into the external water coagulant is a common practice to induce the delay-demixing process (Fadhil et al., 2016; Nejati et al., 2015; Zuo et al., 2017). It often leads to a more porous membrane structure suitable for MD applications. While it can improve the permeation flux, the mechanical properties of the resultant membranes are often weakened. To overcome it, one may change the external coagulant to water or a TEP/water mixture. As a result, the phase inversion process is expedited so that the nascent porous membrane is rapidly solidified with reasonably good mechanical properties. According to Fadhil et al. (2016), the TEP/water mixture has a milder effect of delay demixing than the IPA/water mixture. An additional benefit of replacing IPA by TEP in the external coagulant is to simplify the downstream wastewater treatment. Since the wastewater only comprises TEP and water, it would be easier to treat it than to treat that containing three components (i.e., water, TEP, and IPA).

To search for an appropriate TEP concentration in the external coagulant, a range of 0 to 60 wt% TEP in water is studied. Figure 9.8 shows the morphology of the hollow fiber membranes when the coagulant changes from water

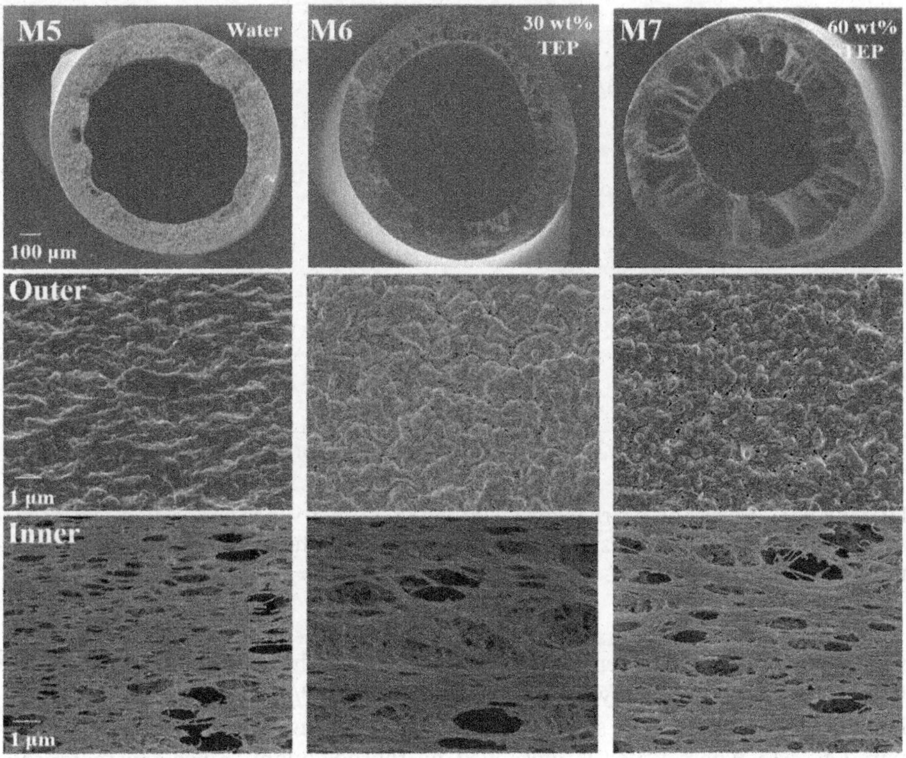

FIGURE 9.8 Cross-section and outer/inner surface morphologies of hollow fiber membranes spun with different external coagulants (Dope solution: PVDF/TEP 12/88 wt%; bore fluid: TEP/water 40/60 wt%) (Chang et al., 2017). Copyright (2017) Elsevier.

to 30/70 TEP/H_2O and 60/40 TEP/H_2O. Firstly, the outer surface pore size becomes larger from membranes M5 to M7 because a softer coagulant tends to induce more delay demixing and results in a more porous outer surface. Secondly, the cross sections turn out to be largely distinct. Large macrovoids gradually evolve with an increase in TEP concentration. Since almost all macrovoids are inward pointed (Widjojo and Chung, 2006), the macrovoids observed in M7 are formed because of the bore fluid intrusion. The cause of this phenomenon is probably due to the fact that the much stronger delay demixing at the outer surface allows the bore fluid to intrude deep into the cross section. In contrast, the external coagulants of M5 and M6 are relatively stronger so that the membranes solidify faster enough to counter the bore fluid intrusion. The porosity measurements shown in Figure 9.7 and Table 9.4 also reflect the increasing trend of porosity from M5 to M7.

Figure 9.9 displays the variations of mechanical properties and permeation flux under DCMD tests as a function of external coagulant composition, while Table 9.4 gives the detailed quantitative values. All three aspects of

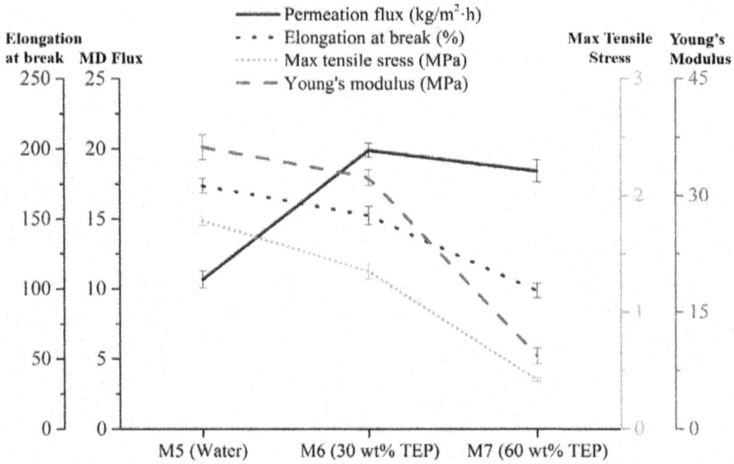

FIGURE 9.9 Effects of external coagulant on MD water vapor permeation flux, elongation at break, maximum tensile stress, and Young's modulus (Chang et al., 2017). Copyright (2017) Elsevier.

mechanical properties decrease from M5 to M7 when the external coagulant becomes weaker. Especially for M7 where 60/40 TEP/H_2O is used as the coagulant, the maximum tensile stress and Young's modulus drop by nearly 75% compared to M5 where water is employed as the coagulant. Since the membrane structure becomes overly porous, it actually impairs the mechanical integrity. On the other hand, the permeation flux shows a different trend. It increases from M5 to M6 but decreases from M6 to M7. With a higher porosity, the permeation flux is expected to increase from M5 to M6. However, the huge macrovoids in M7 actually widen the cross-section thickness (M7, 324 μm; M6, 166 μm) and thus prolong the vapor traveling distance. As a result, the permeate flux dips slightly.

Among these three membranes (i.e., M5 to M7), M6 has most balanced MD performance. It has the highest flux but slightly lower mechanical properties compared to M5. Therefore, the 30/70 TEP/H_2O is used as the external coagulant in the subsequent studies in order to search for the optimal bore fluid composition.

9.6 EFFECTS OF TEP CONCENTRATION IN THE BORE FLUID

The TEP concentration in the bore fluid is 40 wt% for M6. It is further varied from 20 to 60 wt% in the following investigation. However, hollow fibers spun from bore fluids containing 50 and 60 wt% TEP are not reported here because the spinning process is not stable. This is because when the bore fluid is too soft, it is hard to form the nascent membrane. The dope solution flows through the spinneret in a very fluidic state due to the extra-low degree of solidification. The spinning process is frequently discontinued due to the fiber breakage.

The range of TEP concentration in the bore fluid is therefore focused on 20, 30, and 40 wt%. Figure 9.10 illustrates their hollow fiber morphology. From M6 to M8 and M9, the TEP concentration decreases. As expected and consistent with the observation in Figure 9.8, the inner surfaces become denser and the outer surfaces remain nearly the same. However, the morphological change in the cross section is interesting. When the TEP concentration deceases to 20 wt%, a macrovoid-free membrane structure is formed. This is because when the coagulation strength becomes higher with a lower TEP concentration in the bore fluid, the membrane precipitates faster. The rapid solidification prevents a large amount of bore fluid from penetrating into the cross section during phase inversion. Thus, macrovoids are not readily formed. A macrovoid-free membrane should further improve the mechanical properties and make the membranes longer lasting. Figure 9.11 confirms the enhancement in mechanical properties from M6 to M8 and M9. Comparing M8 to M9, the maximum tensile stress is improved by over 50% to 2.08 MPa, while the Young's modulus is also improved to be greater than 40 MPa.

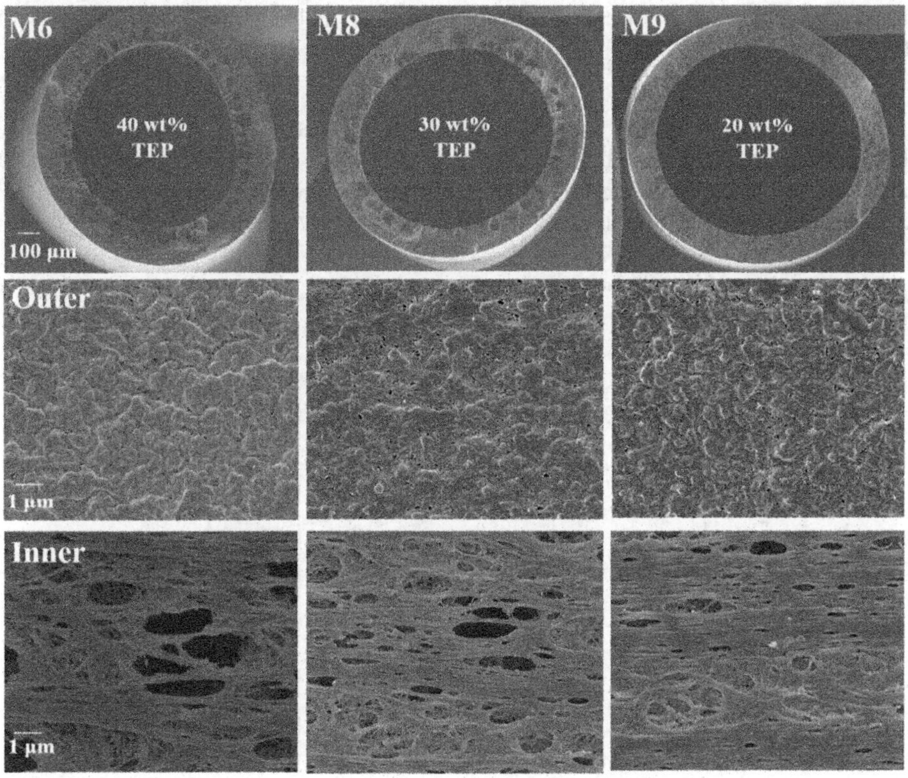

FIGURE 9.10 Cross-section and outer/inner surface morphologies of hollow fiber membranes spun with different bore fluids (Dope solution: PVDF/TEP 12/88 wt%; external coagulant: TEP/water 30/70 wt%) (Chang et al., 2017). Copyright (2017) Elsevier.

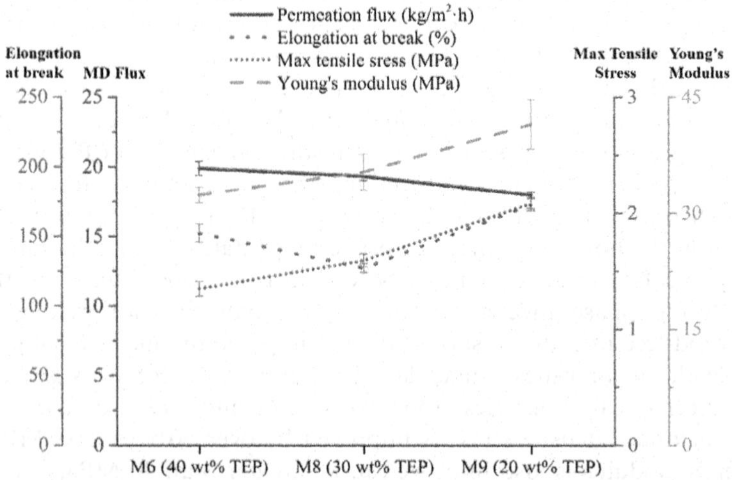

FIGURE 9.11 Effects of bore fluid on MD water vapor permeation flux, elongation at break, Maximum tensile stress, and Young's modulus (Chang et al., 2017). Copyright (2017) Elsevier.

As illustrated in Figure 9.11, the permeation fluxes of M6, M8, and M9 vary between 18 and 20 kg/m^2 h. The flux reduction from M6 to M8 and M9 can be attributed to the slightly denser inner surfaces and fewer macrovoids. Figure 9.7 also shows that the membrane porosity drops slightly from 85.3% to 83.8% and

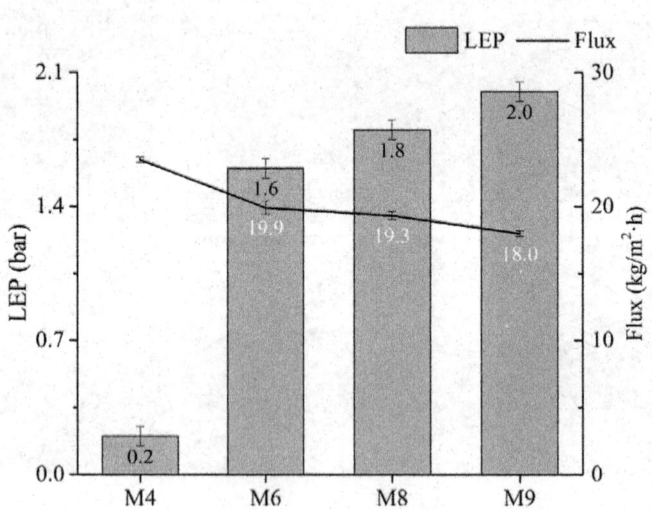

FIGURE 9.12 Experimental measurements of liquid entry pressure (LEP) and water vapor permeation flux (Chang et al., 2017). Copyright (2017) Elsevier.

83.2%, respectively, indicating the vapor permeation resistance across the membranes is mildly increased from M6 to M8 and M9. However, the wetting resistance of M8 and M9 is encouraging. As presented in Table 9.4 and Figure 9.12, the LEP values are much improved from M6 to M8 and M9 owing to the denser inner surfaces. Despite a slightly lower MD flux in the resultant membranes, a higher LEP value is greatly important for real MD operations (García-Payo et al., 2000).

Considering the overall effects of TEP concentration on permeation flux, mechanical strength, and LEP, a concentration of 30 wt% TEP might be an optimum choice for the bore fluid. Therefore, the hollow fiber membrane M8 fabricated from the optimal conditions, external coagulation of TEP/water 30/70 wt% and bore fluid of TEP/water 30/70 wt%, shows the most balanced flux of 19.3 kg/m^{2}h, tensile strength of 1.6 MPa, and LEP of 1.8 bar.

9.7 CONCLUSION

By comparing to one of the most common solvents, NMP, the phase inversion thermodynamics and kinetics of the PVDF/TEP system have been closely investigated. PVDF/TEP dope solutions naturally promote a more porous membrane structure. All membranes fabricated have porosity higher than 80%, with the highest value of 92.7%. This advantage is attributed to a higher viscosity and a lower water affinity. The dope viscosity tends to be higher as TEP is a weak solvent for PVDF. The lower affinity to water is confirmed by the comparison of solubility parameters.

To avoid unsatisfactory compromises on mechanical strength and LEP values caused by the use of TEP, the chemical compositions of external coagulant and bore fluid are further optimized. When both the external coagulant and bore fluid contain 30 wt% TEP in water, the resultant hollow fiber membranes have a promising permeation flux of about 20 kg/m^2 h and an LEP of 1.8 bar. Clearly, the newly developed PVDF hollow fiber membranes spun from the green solvent of TEP at optimal conditions have achieved high MD performance in terms of permeation flux, robust mechanical strength, and strong antiwetting property.

In addition, all dope solutions investigated are simply PVDF/solvent binary systems. No additive or post-treatment is involved. The simple fabrication method would smoothen out the complexity in scale-up and ensure favorable for commercialization.

This is the first attempt to fabricate MD hollow fiber membranes using TEP. It may provide useful insights to green membrane production for many years to come.

REFERENCES

Barton, A.F.M., 1983. CRC handbook of solubility parameters and other cohesion parameters, CRC Press, Boca Raton, FL.

Bottino, A., Camera-Roda, G., Capannelli, G., Munari, S., 1991. The formation of microporous polyvinylidene difluoride membranes by phase separation. J. Membr. Sci. 57 (1), 1–20.

Bottino, A., Capannelli, G., Munari, S., Turturro, A., 1988. Solubility parameters of poly(vinylidene fluoride). J. Poly. Sci. Part B: Poly. Phys., 26 (4), 785–794.

Byrne, F. P., Jin, S., Paggiola, G., Petchey, T. H. M., Clark, J. H., Farmer, T. J., Hunt, A. J., Robert McElroy, C., Sherwood, J., 2016. Tools and techniques for solvent selection: green solvent selection guides. Sustain. Chem. Process. 4 (1), 7.

Capello, C., Fischer, U., Hungerbühler, K., 2007. What is a green solvent? A comprehensive framework for the environmental assessment of solvents. Green Chem., 9 (9), 927–934.

Casarett, L.J., Doull, J., Klaassen, C.D., 2001. Casarett and Doull's toxicology: the basic science of poisons, McGraw-Hill Medical Pub. Division, New York.

Chang, H.H., Tsai, C.H., Wei, H.C., Cheng, L.P., 2014. Effect of structure of PVDF membranes on the performance of membrane distillation. Membr. Water Treat. 5, 41–56.

Chang, J., Zuo, J., Zhang, L., O'Brien, G. S., Chung, T.S., 2017. Using green solvent, triethyl phosphate (TEP), to fabricate highly porous PVDF hollow fiber membranes for membrane distillation. J. Membr. Sci. 539, 295–304.

Cheng, L.-P., Lin, D.-J., Shih, C.-H., Dwan, A.-H., Gryte, C. C., 1999. PVDF membrane formation by diffusion-induced phase separation-morphology prediction based on phase behavior and mass transfer modeling. J. Polym. Sci. B: Polym. Phys. 37 (16), 2079–2092.

Fadhil, S., Marino, T., Makki, H. F., Alsalhy, Q. F., Blefari, S., Macedonio, F., Nicolò, E.D., Giorno, L., Drioli, E., Figoli, A., 2016. Novel PVDF-HFP flat sheet membranes prepared by triethyl phosphate (TEP) solvent for direct contact membrane distillation. Chem. Eng. Process. Process Intensif. 102, 16–26.

Feng, Y., Han, G., Zhang, L., Chen, S.-B., Chung, T.S., Weber, M., Staudt, C., Maletzko, C., 2016. Rheology and phase inversion behavior of polyphenylenesulfone (PPSU) and sulfonated PPSU for membrane formation. Polymer, 99, 72–82.

Figoli, A., Marino, T., Simone, S., Di Nicolò, E., Li, X. M., He, T., Tornaghi, S., Drioli, E., 2014. Towards non-toxic solvents for membrane preparation: a review. Green Chem. 16 (9), 4034–4059.

García-Payo, M. C., Izquierdo-Gil, M. A., Fernández-Pineda, C., 2000. Wetting study of hydrophobic membranes via liquid entry pressure measurements with aqueous alcohol solutions. J. Colloid Interface Sci. 230 (2), 420–431.

Hou, D., Wang, J., Qu, D., Luan, Z., Ren, X., 2009. Fabrication and characterization of hydrophobic PVDF hollow fiber membranes for desalination through direct contact membrane distillation. Sep. Purif. Technol. 69 (1), 78–86.

Liu, F., Tao, M.-M., Xue, L.-X., 2012. PVDF membranes with inter-connected pores prepared via a Nat-ips process. Desalination 298, 99–105.

Marino, T., Blefari, S., Di Nicolò, E., Figoli, A., 2017. A more sustainable membrane preparation using triethyl phosphate as solvent. Green Process. Synth. 6 (3), 295–300.

Nejati, S., Boo, C., Osuji, C. O., Elimelech, M., 2015. Engineering flat sheet microporous PVDF films for membrane distillation. J. Membr. Sci. 492, 355–363.

Piezotech Arkema, 1993. PVDF Active solvents and PVDF latent solvents. Available from: www.piezotech.fr/fr/5-technique-piezoelectric-eap-electroactive-polymer/news/news-40-pvdf-active-solvents-and-pvdf-latent-solvents.html (accessed 07.01.2019).

Razmjou, A., Arifin, E., Dong, G., Mansouri, J., Chen, V., 2012. Superhydrophobic modification of TiO2 nanocomposite PVDF membranes for applications in membrane distillation. J. Membr. Sci.415–416, 850–863.

Regulation (EC) No 1272/2008. 2008. Official Journal of the European Union. Available from: http://eur-lex.europa.eu/LexUriServ/LexUriServ.do?uri=OJ:L:2008:353:0001:1355:en:PDF (accessed 07.01.2019).

Tomaszewska, M., 1996. Preparation and properties of flat-sheet membranes from poly (vinylidene fluoride) for membrane distillation. Desalination 104 (1), 1–11.

Wang, P., Teoh, M. M., Chung, T.S., 2011. Morphological architecture of dual-layer hollow fiber for membrane distillation with higher desalination performance. Water Res. 45 (17), 5489–5500.

Widjojo, N., Chung, T.S., 2006. Thickness and air gap dependence of macrovoid evolution in phase-inversion asymmetric hollow fiber membranes. Ind. Eng. Chem. Res. 45 (22), 7618–7626.

Xing, D. Y., Chan, S. Y., Chung, T.S., 2014. The ionic liquid [EMIM]OAc as a solvent to fabricate stable polybenzimidazole membranes for organic solvent nanofiltration. Green Chem. 16 (3), 1383–1392.

Yang, Q., Chung, T.S., Chen, S. B., Weber, M. 2008. Pioneering explorations of rooting causes for morphology and performance differences in hollow fiber kidney dialysis membranes spun from linear and hyperbranched polyethersulfone. J. Membr. Sci. 313 (1), 190–198.

Yeow, M. L., Liu, Y. T., Li, K., 2004. Morphological study of poly(vinylidene fluoride) asymmetric membranes: effects of the solvent, additive, and dope temperature. J. Appl. Polym. Sci. 92 (3), 1782–1789.

Yilmaz, L., McHugh, A.J., 1986. Analysis of nonsolvent–solvent–polymer phase diagrams and their relevance to membrane formation modeling. J. Appl. Polym. Sci. 31(4), 997–1018.

Zhang, H., Lamb, R., Lewis, J., 2005. Engineering nanoscale roughness on hydrophobic surface – preliminary assessment of fouling behaviour. Sci. Technol. Adv. Mater. 6 (3–4), 236.

Zuo, J., Chung, T.S., O'Brien, G. S., Kosar, W., 2017. Hydrophobic/hydrophilic PVDF/Ultem® dual-layer hollow fiber membranes with enhanced mechanical properties for vacuum membrane distillation. J. Membr. Sci. 523, 103–110.

Section B

Hybrid Systems

10 Membrane Distillation Crystallizer

Lin Luo and Tai-Shung Chung

10.1 INTRODUCTION

Membrane technologies such as reverse osmosis (RO) and membrane distillation (MD) have been employed for wastewater treatment and seawater desalination to overcome the global water scarcity (Amy et al., 2017). Currently, RO plants account for 60% of desalination plants to produce high-quality potable water. However, desalination technologies must further enhance the water recovery, lower the energy consumption, and overcome the issue of discharging concentrated brine (Kim et al., 2016; Sanmartino et al., 2017). New strategies are developed to harvest the values of waste concentrated brines such as RO-MD (Martinetti et al., 2009), forward osmosis-MD (Wang et al., 2011a; Zhang et al., 2014), MD – crystallization (MDC) (Amy et al., 2017; Eumine Suk and Matsuura, 2006; Ji et al., 2010; Khayet and Matsuura, 2011; Wang and Chung, 2015; Zhu et al., 2009), RO – pressure retarded osmosis (PRO) (Achilli et al., 2009; Han et al., 2015), etc. Another strategy to mitigate the negative impact of brine disposal is by realizing the objective of zero liquid discharge (El-Abbassi et al., 2013), which is to enable nearly complete water recovery and eliminate brine disposal by only producing two streams: water and salts. The concept of zero liquid discharge can be accomplished by inventing a new desalination technology or combining the well-established processes into a hybrid system. Generally, the implementation of the hybrid desalination technology may be more advantageous because additional processes can be directly integrated into existing facilities (El-Bourawi et al., 2006). Among many available techniques, the integration of membrane distillation with a crystallization unit (i.e., MDC) could serve as a potential solution to realize the zero liquid discharge (Drioli et al., 2012). Since the mining industry is also facing problems of mineral depletion, water shortage, and high energy consumption (Mudd, 2010), harvesting salts from MD brines may be an alternative approach to produce value-added minerals with relatively low energy requirements.

In a MDC system, pure water can be recovered by the MD operation, resulting in the concentration of feed solution, and the concentrated brine drives the coupled crystallization process in an external crystallizer for solute crystal production. As early as 1989, Wu and Drioli have reported the MDC system for concentrated salt solutions, where crystallization was realized outside the MD membrane module (Wu and Drioli, 1989). Afterward, there have been many attempts to develop MDC systems to treat inorganic salt solution, wastewater and pharmaceutical solution (Edwie and Chung, 2013, 2012; Gryta, 2002; Kim et al., 2018; Wu et al., 1991).

10.2 CRYSTALLIZERS IN MDC

10.2.1 Crystallization

Crystallization from a solution can be considered in a three-step mechanism (Figure 10.1): nucleation, nuclei growth, and growth cessation (Drioli et al., 2015). In the nucleation stage, solute molecules aggregate to form crystal nuclei. Then nuclei grow to become macroscopic crystals, then redissolving and aggregation may happen to reach a more thermodynamically stable state. Finally, macroscopic crystals grow up to a certain size and the growth terminates.

The driving force for crystallization is supersaturation, which is a nonequilibrium state with a higher amount of solute than its solubility dissolved in the solution. The rates of nucleation and nuclei growth increase with the supersaturation degree at a given temperature. For a supersaturated solution at temperature T, its relative supersaturation factor σ can be calculated as follows (Zarkadas and Sirkar, 2004b):

$$\sigma = \frac{C - C^*}{C^*} \tag{10.1}$$

where C is the actual solute concentration and C^* is the saturated concentration at temperature T. During the crystallization process, the precise control of supersaturation is critical in achieving the desired and reproducible features in terms of crystal size, size distribution, morphology, purity, and so on. However, the crystallization process can also be influenced by other conditions, such as the presence of impurities and fluid dynamics.

To generate salts from a brine solution, many crystallization methods have been developed, such as cooling crystallization (Edwie and Chung, 2013; Gryta, 2002; Zarkadas and Sirkar, 2004a), evaporative crystallization (Geertman, 2000), antisolvent crystallization, etc. Among them, cooling crystallization is the most common method in MDC, considering the high temperatures of MD eluents. For most solutes, for example, NaCl in water, the salt solubility increases as temperature elevates (Figure 10.2) (Luo et al., 2017). When cooling happens,

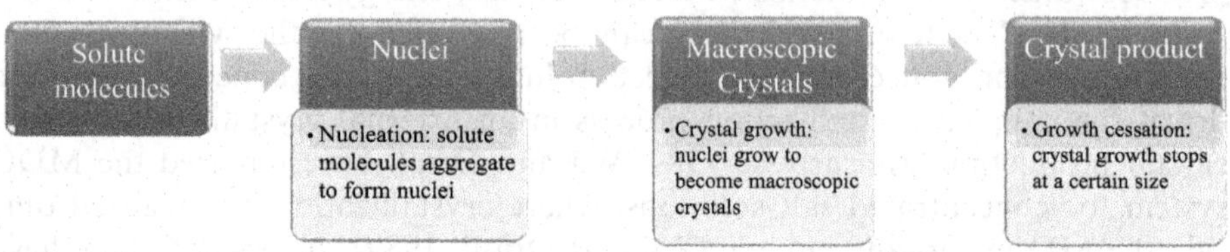

FIGURE 10.1 The steps of a crystallization process.

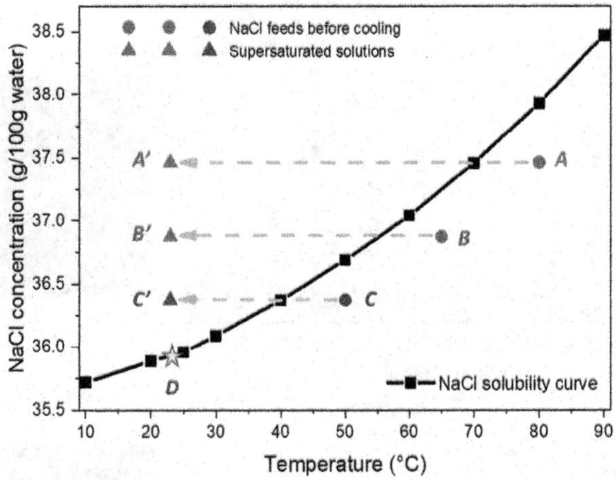

FIGURE 10.2 The solubility curve of NaCl in water and cooling of NaCl feed solutions. Revised from Luo et al. (2017). Copyright (2017) American Chemical Society.

the solution becomes supersaturated, which drives the excess dissolved salt to precipitate out in the form of crystals. As shown in Figure 10.2, through rapid cooling, theoretically unsaturated NaCl feed solutions A, B, and C with different NaCl concentrations in different high temperatures travel into the supersaturated zone and become states A′, B′, and C′, respectively, at room temperature. Supersaturation depletion happens with crystal generation and growth, and states A′, B′, and C′ will finally become stable at the saturated state D (a NaCl concentration of 26.4% at room temperature). Cooling crystallization could be carried out in stirred vessels or tubes in a batch or continuous mode. Both conventional and novel crystallizers were utilized and investigated in the MDC processes.

10.2.2 Conventional Crystallizers

The most commonly coupled crystallizer is the traditional stirred tank installed with a stirring bar, a cooling coil, or a cooling jacket. For instance, the concentrated feed solution from MD was transferred into a cooling tank in a batch mode for NaCl crystallization, by cooling and stirring the retentate solution. In a previous study (Edwie and Chung, 2012), NaCl crystal properties and crystal size distribution (CSD) were investigated on a batch-type crystallizer. With the initial solution concentration of around 27 wt% at 60 °C, the salt crystallization was induced by decreasing the solution temperature at different cooling rates (Figure 10.3E). As shown in Figure 10.3, the CSD of NaCl crystals formed under natural cooling shifts toward larger sizes as compared to that of rapid cooling. The average crystal sizes are 225 μm and 85 μm for samples under natural and rapid cooling, respectively. These results can be explained from

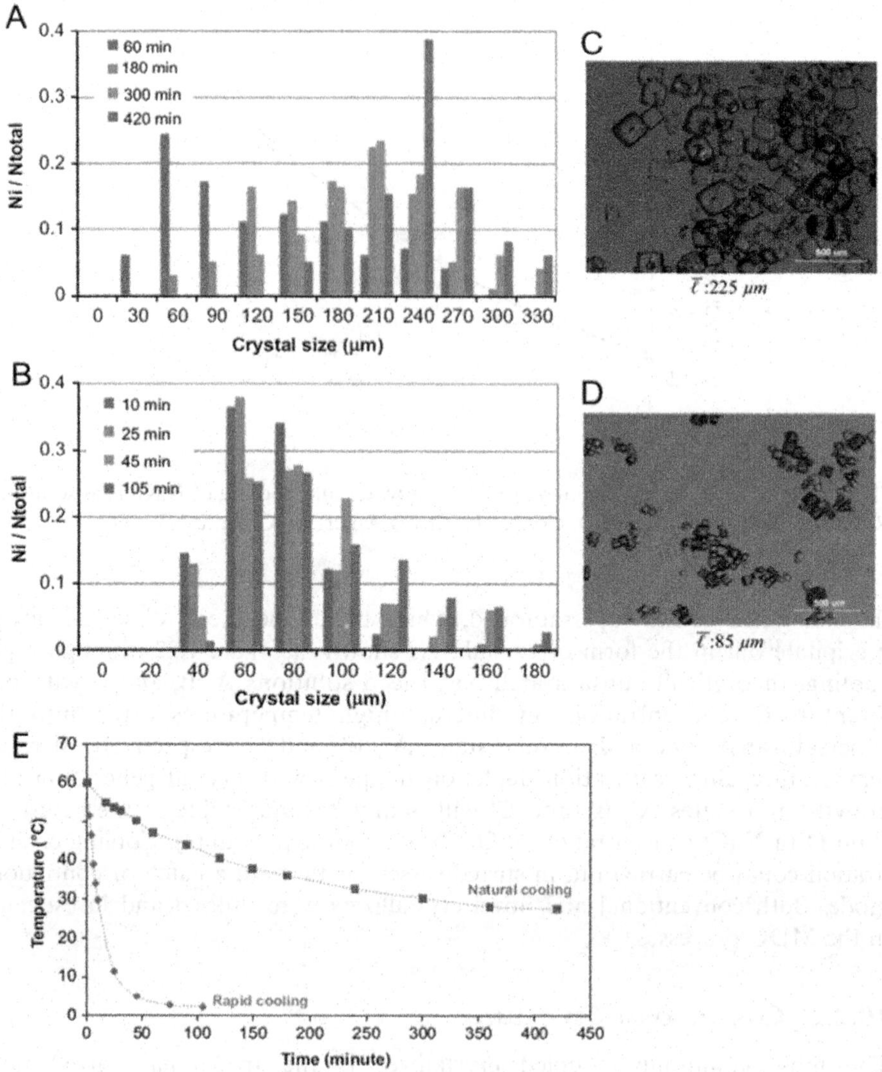

FIGURE 10.3 Effects of cooling rate on CSD: (A) Natural cooling and (B) rapid cooling. The microscopic images of salt crystals: (C) Natural cooling (420 min) and (D) rapid cooling (105 min). (E) Temperature profiles of MD retentate solutions during crystallization (Edwie and Chung, 2012). Copyright (2012) Elsevier.

the nucleation and growth aspects. Since nucleation is driven by the degree of supersaturation, while growth is controlled by diffusion. A faster cooling would result in formation of smaller size crystals as compared to a slower cooling which allows the growth of the crystal nuclei to form larger size crystals.

Similar designs could also be found elsewhere (Chan et al., 2005; Chen et al., 2014b; Ji et al., 2010). In the Gryta's work, three crystallization tanks were installed in the MDC setup for the purpose of continuous retentate cooling, crystal separation, and mother liquor reheating, separately (Gryta, 2002). Continuous operation could also be realized by directly returning the retentate solution from the outlet of a MD module back to the crystallizer for cooling (Curcio et al., 2001). For a MDC system using traditional cooling tanks, the scaling-up of the MD unit can be easily achieved by increasing the membrane area or module numbers, while the scaling-up of the stirred tank crystallizer could be difficult due to the impossibility of maintaining the same hydrodynamic conditions when the tank size is increased (Chabanon et al., 2016).

10.2.3 Solid Hollow Fiber Cooling Crystallizer

Driven by the transition from the batch mode to a continuous operation mode and the ease of scaling up, new crystallizer concepts are expected. Since the control of supersaturation is of the primary importance for a crystallization process, a novel crystallizer, namely, solid hollow fiber cooling crystallization (SHFCC) has been developed for salt crystals extraction. As proposed by Prof. Sirkar and his co-workers, the basic concept of SHFCC involves polymeric hollow fibers serving as heat-exchanging tubes, but the hollow fiber dimensions are orders of magnitude smaller than those conventional metal tubes for heat exchange (Chen et al., 2014a, 2016a, 2016b; Zarkadas and Sirkar, 2004b, 2007). As illustrated in Figure 10.4, the hot concentrated brine solution is pumped through the lumen side of hollow fibers, where rapid cooling takes place by the

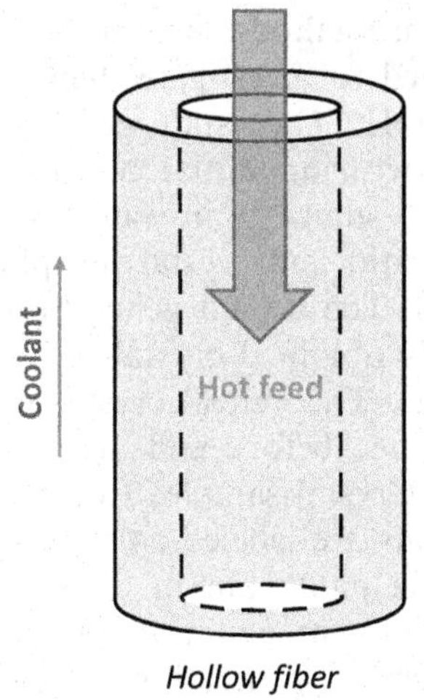

FIGURE 10.4 Basic concept of SHFCC (Luo et al., 2017). Copyright (2017) American Chemical Society.

coolant circulating on the shell side. The brine temperature exiting from the hollow fiber becomes much lower than its initial temperature so that supersaturation occurs. Owing to the solubility change, the excess dissolved salt crystallizes and precipitates out from the supersaturated solution at a high rate. This technology has been tested successfully for both low solubility systems such as paracetamol in water (Zarkadas and Sirkar, 2007) and high solubility systems such as KNO_3 in water and salicylic acid in ethanol (Zarkadas and Sirkar, 2004b). The combination of fast cooling via SHFCC and the downstream mixer also realized the successful operation of paracetamol crystallization at 30–40 °C below the metastable zone limit (Zarkadas and Sirkar, 2007). Also, the utilization of polymeric hollow fibers provides advantages of greater corrosion resistance, geometric flexibility, and ease of manufacturing and scaling-up (Zarkadas and Sirkar, 2004a). In addition, the extraordinarily large surface area to volume ratio of hollow fiber modules improves the overall crystallization performance in terms of heat transfer efficiency and precise temperature control of the crystallization solution (Zarkadas and Sirkar, 2004b).

Thus, comparing with the conventional crystallizer (e.g., mixed suspension mixed product removal (MSMPR)), the SHFCC crystallizer could produce crystals with a nucleation rate of 2–3 magnitude orders higher, crystal mean size of 3–4 times smaller, and a narrower crystal size distribution (CSD) (Chen et al., 2016b; Zarkadas and Sirkar, 2004b). Nonporous polymeric hollow fibers made from PVDF, PP, etc., could be applied as heat exchangers in SHFCC. In a previous study (Luo et al., 2017), during the SHFCC experiments of salt crystallization from NaCl brine and concentrated synthetic seawater, the influences of operating parameters and downstream agitation methods on the crystal properties were investigated.

10.2.3.1 Downstream Agitation Methods

Different downstream agitation methods, such as magnetic stirring and ultrasonication, have been employed to investigate their effectiveness on salt crystallization and properties in SHFCC crystallizer. When a pure NaCl solution is used as the feed solution with an initial NaCl concentration of 27.25%, which corresponds to the NaCl solubility in water at 70 °C, Figure 10.5 shows the NaCl crystal generation ratio and crystal morphologies as a function of downstream agitation method. The crystal generation ratio (Rc) is defined as the ratio of the experimentally obtained crystal weight (m_t) to the theoretical calculated crystal weight (m_{tot}). Theoretical crystal weight (m_{tot}) is calculated from the NaCl solubility change before and after the cooling for a certain amount of the feed solution. Crystal samples from magnetic stirring, ultrasonication, and control conditions are labeled as stirred, sonicated, and control, respectively. NaCl crystals precipitate out in different rates after cooling by hollow fibers and applying different agitation methods. Magnetic stirring is the fastest one in facilitating crystal generation with a Rc value approaching 1 within just 3 min, indicating all the excessive dissolved salt in the supersaturated solution precipitates out in the form of crystals. A further increase in stirring time beyond 3 min cannot improve the crystal harvest any more. In

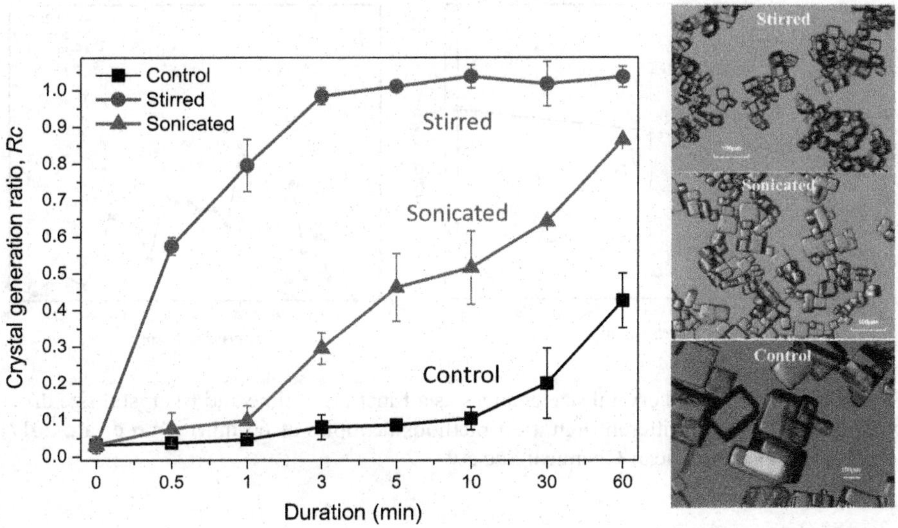

FIGURE 10.5 NaCl crystal generation ratios and crystal morphologies under different agitation methods (crystal morphologies samples at 60 min) (Luo et al., 2017). Copyright (2017) American Chemical Society.

contrast, the crystal generation ratios of the sonicated and control samples increase much slower. Their *Rc* values reach only 0.86 and 0.43 at 60 min, separately. The microscope observation on salt crystals reveals that all crystals are transparent and have a cubical shape which is the typical structure of NaCl crystals. Crystals also form clusters in small groups, especially for those produced by the stirring method. This phenomenon arises from the fact that mechanical agitation facilitates heterogeneous nucleation (Li and Kaner, 2006), and also crystals in a stirred suspension are susceptible to grow via an agglomeration mechanism (Linnikov, 2014).

In terms of crystal properties, the produced crystals are preferred to have a desired dimension with a narrow CSD (Edwie and Chung, 2013; Zarkadas and Sirkar, 2004b). Figure 10.6a displays the evolution of crystal size (i.e., the calculated CED based on the particle area from microscope images) with time. The crystal size of the control sample increases from the initial value of 64.3 ± 20.3 μm at 0 min to 128.8 ± 49.4 μm at 60 min due to the gradual crystal growth under quiescent conditions. While crystals produced by stirring and ultrasonication exhibit obviously smaller sizes of 36.9 ± 13.6 μm and 34.6 ± 16.6 μm at 60 min, respectively. In addition, their crystal sizes are insensitive to the agitation duration for crystallization. Compared to the crystal size of 350–400 μm by traditional forced circulation crystallizers (Choi, 2005; Geertman, 2000) or 60–80 μm by conventional stirred tank crystallizer (Edwie and Chung, 2013; Tun et al., 2005), SHFCC coupled with downstream magnetic stirring could produce salt crystals of a much smaller size and narrower CSD.

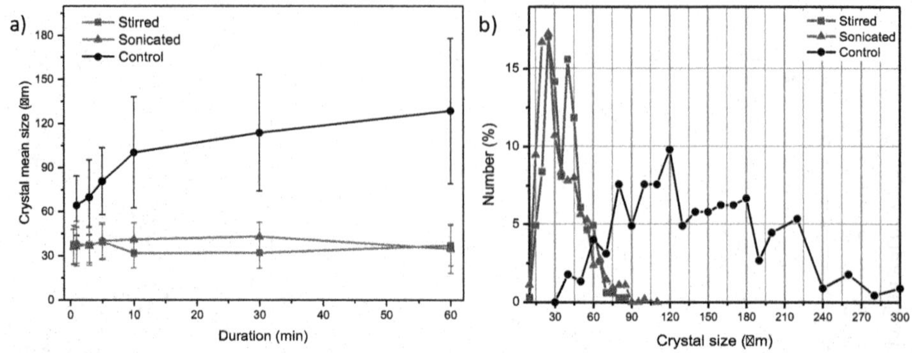

FIGURE 10.6 a) NaCl crystal size changes as a function of time and b) crystal size distributions (CSD) under different agitation methods (samples at 60 min) (Luo et al., 2017). Copyright (2017) American Chemical Society.

The produced CSD is a function of (1) the total number of nuclei formed during the nucleation stage, (2) the duration for agitation and crystallization, (3) the degree of supersaturation, and (4) the final system temperature and pressure (De Yoreo and Vekilov, 2003; Srinivasan et al., 1995). In SHFCC, salt nuclei may start to form when the chilled supersaturated feed solution flows out from the lumen side of the hollow fibers. For the control sample, the growth of nuclei and crystals is obvious since the crystal size increases gradually as a function of time. However, it has a relatively slow depletion rate of supersaturation due to the limited mass transfer rate under quiescent conditions. As a consequence, it has a low crystal generation ratio *Rc*. The situation changes for the magnetic stirred sample. The agitation-induced mechanical disturbances speed up the onset of nucleation in the supersaturated solution (Mullin and Raven, 1961). Agitation has been reported to enhance the mass transfer and aid the diffusional transport step during nucleation, but impede or disrupt the continual growth of subnuclei (Kamalipour et al., 2017; Li and Kaner, 2006; Mullin and Raven, 1961). Therefore, the agitation by magnetic stirring facilitates the nucleation step but disrupts the crystal growth, leading to a high rate of crystal generation and a small crystal size. Likewise, ultrasonication has similar effects on crystallization. Ultrasonic waves produce acoustic cavitation microbubbles in solutions, which repeatedly grow and collapse in synchrony. These phenomena induce large shearing forces and trigger the nucleation in supersaturated solutions (Yuichi et al., 2013). However, the large shearing forces can also break down the growing crystals and thus increase the number of nuclei, leading to the formation of monodispersed crystals of minimal sizes (Srinivasan et al., 1995), as shown in Figure 10.6b. The CSDs of NaCl crystals generated by different methods are presented in Figure 10.6b. Compared to the control sample, the CSDs of NaCl crystals formed at 60 min under magnetic stirring and ultrasonication conditions exhibit smaller sizes and narrower distributions. Their mean crystal sizes are 128.8 μm, 36.9 μm, and 34.6 μm for control, stirred, and sonicated samples, respectively.

Overall, both ultrasonication and magnetic stirring produce uniform crystals with a small size of around 35 μm. However, magnetic stirring induces the fastest crystal generation (Rc = 1 at 3 min) due to the enhanced mass transfer in the diffusion-controlled crystallization process.

10.2.3.2 Influences of Operation Conditions

The influence of feed NaCl concentration on crystal generation is studied by cooling down feed solutions with different saturation temperatures (T_s). The relative supersaturation factor can be calculated based on Equation (8.1). Also, the influence of stirring intensity on crystal generation is investigated by adjusting the stirring speed when mixing a chilled supersaturated feed solution with a T_s of 70 °C. As the stirring speed increases from 200 to 800 rpm, the Reynolds number of the stirring vessel increases from 515 to 2062. Figure 10.7 presents nucleation rates (r) of NaCl crystals and the salt crystal productivity (p) from different conditions, the estimated nucleation rates are within the range of 1.10×10^7 to 1.02×10^9 crystals per m^3 per second. The achieved highest nucleation rate of 1.02×10^9 crystals per m^3 per second is 3–4 times higher than that produced by a conventional cooling crystallizer under the batch mode (Edwie and Chung, 2013). As shown in Figure 10.7, the increase in the degree of feed supersaturation or the stirring speed accelerates the nucleation rate obviously, since the overall nucleation rate is driven by the degree of feed supersaturation and facilitated by the agitation-enhanced diffusion. Thus, stirring the most concentrated feed solution (T_s = 70 °C) at the highest speed (800 rpm) could produce the highest crystal nucleation rate of 1.02×10^9 crystals per m^3 per second.

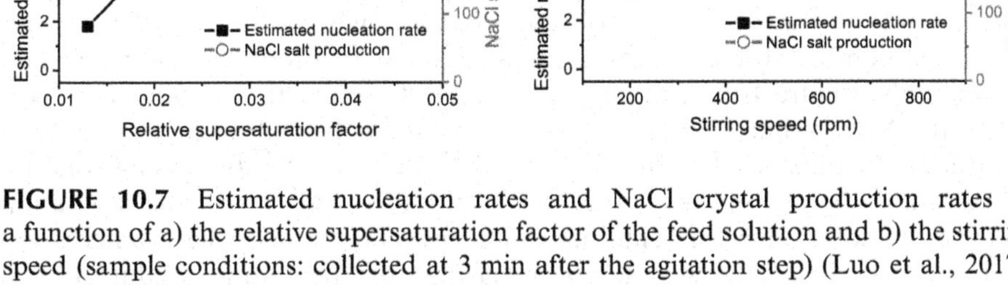

FIGURE 10.7 Estimated nucleation rates and NaCl crystal production rates as a function of a) the relative supersaturation factor of the feed solution and b) the stirring speed (sample conditions: collected at 3 min after the agitation step) (Luo et al., 2017). Copyright (2017) American Chemical Society.

On the other hand, a higher nucleation rate also contributes to a higher salt crystal production rate. More salts could be produced from the higher concentrated supersaturated solution due to the larger amount of salt dissolved inside (Figure 10.7a). However, the NaCl crystal production rate is less affected by the stirring speed, as illustrated in Figure 10.7b. Although the nucleation rate seems to be relatively low when stirring at 200 rpm, the overall crystal growth is still mainly determined by the degree of feed supersaturation. Consequently, stirring at 200, 500, and 800 rpm results in comparable salt productions. Overall, a higher NaCl concentration in the feed solution can bring a higher crystal generation rate in terms of nucleation and crystal generation ratios due to the increased degree of supersaturation. A magnetic stirring speed of 500 rpm is sufficient to overcome the diffusional resistance during crystallization and produce both high nucleation and salt precipitation rates.

10.3 MDC PROCESS

Membrane distillation is a thermally driven separation process which involves the transport of water vapor across a microporous hydrophobic membrane (Eumine Suk and Matsuura, 2006; Khayet, 2008; Souhaimi and Matsuura, 2011; Wang and Chung, 2015; Wang et al., 2011b). It can handle high salt concentrations at relatively low operating temperatures. The hydrophobic MD membrane only allows the vapor passage but retains all nonvolatile solutes, thus the process has a theoretical rejection of 100% to solid or nonvolatile substances (Khayet and Matsuura, 2011; Seo et al., 2018). Also, the process can be powered by solar energy, geothermal energy, or industrial waste energy to reduce the overall energy consumption (Shim et al., 2015; Wang and Chung, 2012). In a hybrid MDC system, pure water can be recovered by the MD operation, resulting in a supersaturated salt solution, which drives the coupled crystallization process. Additionally, MDC is attractive in crystal production owing to its optimal control of the supersaturation level that enables the formation of higher quality crystals than other conventional technologies such as cooling or evaporative crystallization. There are many attempts to develop MDC systems for the treatment of brine solutions (Edwie and Chung, 2012, 2013; Gryta, 2002; Kim et al., 2018).

10.3.1 Simultaneous Membrane Distillation – Crystallization (SMDC)

In the study of Edwie et al., MD and conventional cooling crystallizer unit have been combined and operated simultaneously to produce fresh water and salt crystals. Figure 10.8 demonstrates the MDC setup. In a continuous operation, the NaCl saline solution in the stirred tank is heated up to a preset temperature and transferred to the feed side of a MD module. Increasing solution temperature slightly increases the NaCl solubility and lowers the saturation state of the solution before entering the membrane module where water evaporation occurs and the salt solution leaving the membrane module is still undersaturated. The salt solution is circulated through the shell side of the

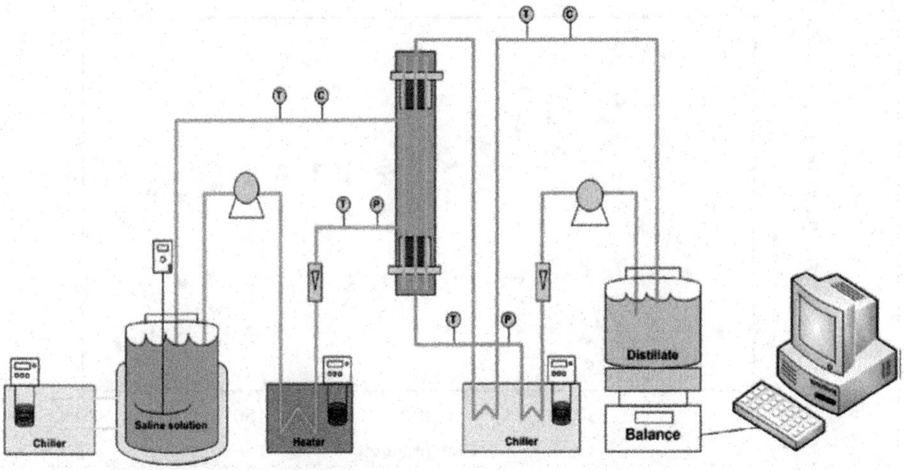

FIGURE 10.8 Schematic diagram of membrane distillation – crystallization (MDC) setup (Edwie and Chung, 2012). Copyright (2012) Elsevier.

module while the retentate solution with a higher salinity is returned back to the crystallizer in which it is cooled down and become oversaturated. As a result of cooling and stirring, the supersaturated NaCl ions are crystallized from the solution until their concentration is reduced to 26.4% at room temperature. With this unique design, the salt concentration at the inlet of the MD module could be maintained and allowed a stable operation in a long-term operation, for example, 5000 min (Edwie and Chung, 2013).

10.3.1.1 Effects of Feed Temperatures

Feed temperature has been found to influence the operation flux and crystal properties in SMDC. Figure 10.9a plots evolution of the vapor permeation flux during the SMDC operation at different feed temperatures. As mentioned previously, the feed concentration to the MD module is constant in the SMDC process as the water evaporation and diffusion across the membrane is offset by the NaCl crystallization. Thus, the membrane flux is expected to be stable over the entire SMDC operation. However, Figure 10.9a clearly shows that the vapor permeation flux are only stable when the feed temperature is lower than 50 °C while increasing the temperature to 60 and 70 °C causes a gradual reduction of membrane flux. This is because the feed temperature also governs the scaling formation in the SMDC process. NaCl has a positive solubility coefficient which means the salt solubility increases with increasing solution temperature. Moreover, the reduction of the feed solution viscosity with increasing temperature increases the heat transfer coefficient at the feed boundary layer that increases the convective heat transfer from the bulk to the membrane surface. These retard scaling formation on the membrane. Also, the increment of vapor permeation flux with increasing feed temperature leads

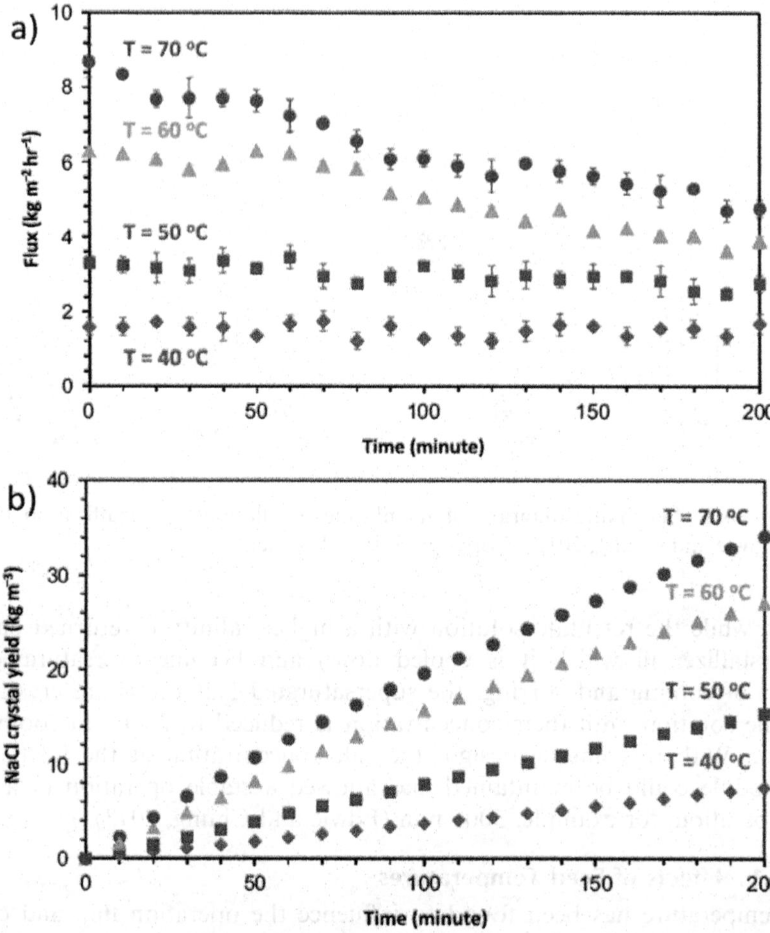

FIGURE 10.9 a) The evolution of permeation flux and b) yield of NaCl crystals during the SMDC operation at various feed temperatures (Edwie and Chung, 2013). Copyright (2013) Elsevier.

to a higher concentration of the nonvolatile salt near the evaporation surface. Meanwhile, the latent heat uptake from the surrounding increases with increasing evaporation rate and hence, lowering the temperature near the evaporation surface. The intensifications of both concentration and temperature polarizations may cause oversaturation of the concentrated feed solution and trigger salt crystallization in the membrane module.

The feed temperature has also been found to increase the NaCl crystal yield and decrease the average crystal size due to the high supersaturation degree. As shown in Figure 10.9b, the production of salt crystals conforms to the distillate production exhibiting an increasing trend with increasing feed

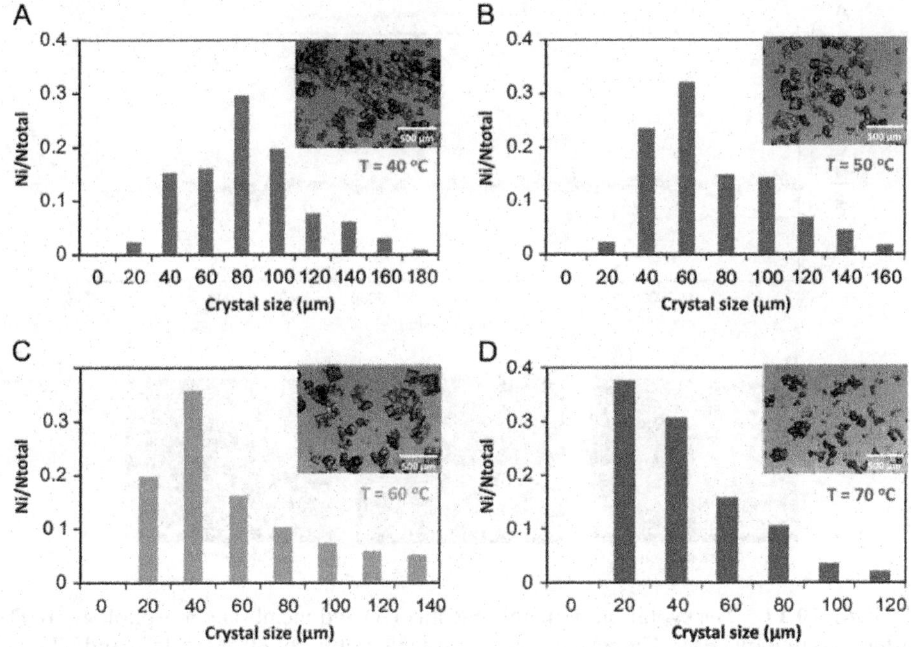

FIGURE 10.10 Crystal size distribution of NaCl crystals at different SMDC feed temperatures at residence time of 60 min. (A) 40 °C, (B) 50 °C, (C) 60 °C, (D) 70 °C (Edwie and Chung, 2013). Copyright (2013) Elsevier.

temperature and operation duration. The NaCl yield is controlled by the mass of water evaporated from the feed solution. The highest yield of 34 kg NaCl per m^3 of feed solution is achieved when the feed temperature is 70 °C after a 200-min operation.

Figure 10.10 presents the effects of feed temperature on CSD when the crystallizer residence time is fixed at 60 min. It can be seen that the CSD of NaCl crystals formed under a higher feed temperature shifts toward a smaller size as compared to that of a lower feed temperature. The average size reduces from 87.4 μm to 48.82 μm for crystals formed under feed temperatures of 40 °C and 70 °C, respectively. These results can be attributed to a higher rate of water evaporation at a higher feed temperature. As the crystallizer operating temperature is fixed at 25 °C, the higher water evaporation leads to a higher supersaturation of the MD retentate solution entering the crystallizer. The calculated nucleation rates in SMDC under 40 °C to 70 °C are in the range of 2.21×10^9 to 3.54×10^{10} crystals per m^3 of the crystallizing solution.

10.3.1.2 Continuous SMDC Operation

The stability of membrane flux during the continuous SMDC operation is also investigated. Figure 10.11A shows that the membrane flux is stable over

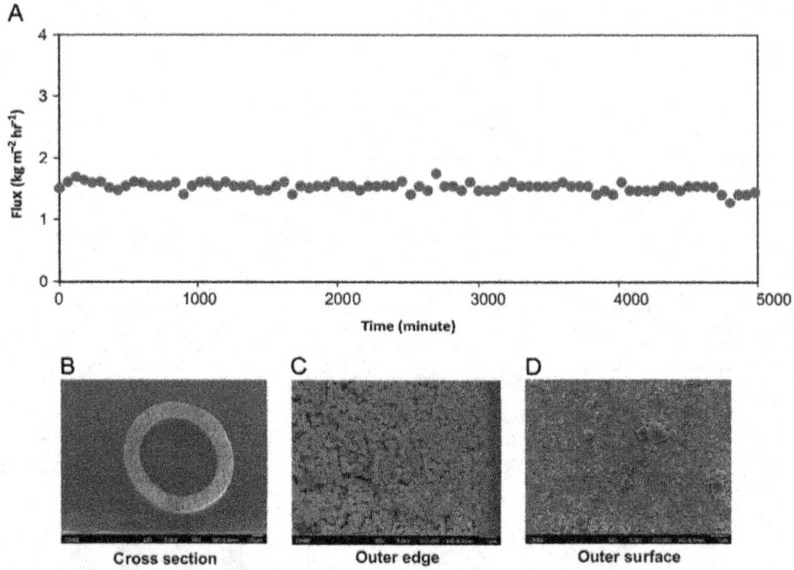

FIGURE 10.11 The evolution of membrane flux (A) and membrane morphology (B–D) after a continuous SMDC operation at a feed temperature of 40 °C (Edwie and Chung, 2013). Copyright (2013) Elsevier.

5000 min with the distillate conductivity ranging from 0.9 μS/cm to 2.6 μS/cm when the feed temperature is fixed at 40 °C. The membrane morphologies after the continuous SMDC operation are shown in Figure 10.11B–D. The cross section and outer edge morphologies are indifferent from the as-spun membrane but a few of salt crystals deposited on the membrane surface. Overall, the experimental results have verified the feasibility of the SMDC process for simultaneous productions of the fresh water and inorganic salt from the saturated brine.

10.3.2 Integrated Membrane Distillation – Solid Hollow Fiber Cooling Crystallization (MD-SHFCC) System

A hybrid system of MD and SHFCC has been successfully developed for simultaneous production of fresh water and inorganic salt crystals from NaCl brine solutions (Luo et al., 2018). In the study of Luo et al., porous PVDF hollow fiber membranes consisting of PTFE particles were spun for the MD subsystem (membrane distillation hollow fiber, MDHF), while nonporous PVDF hollow fibers were fabricated as heat exchangers for the SHFCC subsystem (membrane crystallization hollow fiber, MCHF).

The schematic of the MD-SHFCC experimental setup is shown in Figure 10.12, consisting of a direct contact MD on the left and a SHFCC unit on

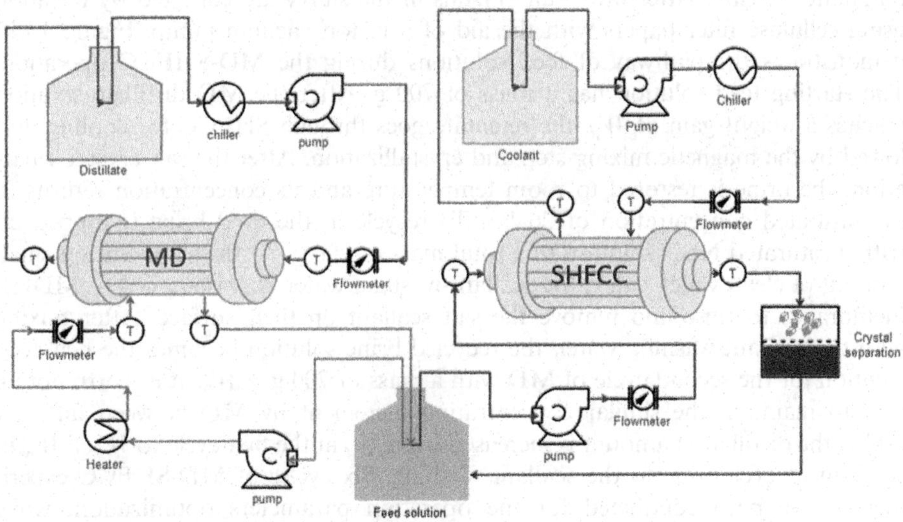

FIGURE 10.12 MD-SHFCC desalination system (Luo et al., 2018). Copyright (2018) Elsevier.

the right. In the MD unit, the NaCl feed solution is heated up to various temperatures and circulate through the shell side of MDHF membranes by a circulation pump. Simultaneously, the distillate stream (DI water) is circulated counter-currently in the lumen. The solution temperatures are maintained by temperature circulators and the weight change of the distillate is monitored by a balance and recorded by a computer. After being circulated in MD, the feed solution is concentrated due to water evaporation. Because the elevated feed temperature increases the NaCl solubility, the retentate has a higher salinity but remains undersaturated in most cases. Thereafter, the hot retentate is pumped into the SHFCC unit for cooling. In other words, it is pumped through the lumen side of MCHF hollow fibers where rapid cooling takes place by the coolant (DI water) circulating on the shell side. The brine temperature exiting from the hollow fiber becomes much lower than its initial temperature so that supersaturation occurs. Owing to the solubility change, the excess dissolved salt crystallizes and precipitates out from the supersaturated solution at a high rate.

The MD-SHFCC experiments have been carried out in a semicontinuous way. The NaCl concentration of the feed solution is fixed at 26.4 wt%. Firstly, the feed solution is heated up to a feed inlet temperature of and circulated in MD for fresh water production. Afterward, the hot concentrated feed solution or retentate is immediately transferred to the SHFCC section for cooling crystallization. The outlet temperature of the retentate in each SHFCC run is controlled to be around 23 °C before entering the mixing step. Then the chilled brine is collected in a plastic beaker, immediately followed by the downstream magnetic stirring. Afterward, salt crystals in the slurry are collected by filtration using cellulose filter papers with the aid of a piston vacuum pump. Figure 10.13 demonstrates the pathway of feed solutions during the MD-SHFCC operation. The starting feed solution has a mass of 700 g. After the MD distillate solution reaches a weight gain of W_d, the retentate goes through SHFCC for cooling, followed by the magnetic mixing step and crystallization. After the salt crystal separation, the brine is restored to room temperature and its concentration returns to the saturated concentration of 26.4%. To recycle it, the used brine is topped up with a saturated NaCl solution to a total mass of 700 g. At the same time, a small amount of clean water, called the scallant-washing water W_w, is applied to MDHF membranes to rinse and remove the salt scallant on their surface. After mixing with the scallant-washing water, the recycled brine solution becomes the new feed solution for the second cycle of MD with a mass of 700 g + W_w. It is worth noting that to maintain the similar concentration increment by MD between different cycles, the distillation amount is increased from W_d in the first cycle to $W_d + W_w$ in subsequent cycles due to the scallant washing. Six cycles of MD-SHFCC experiments have been conducted for the operation parameters optimization, while 15 cycles (or 10 hours) are carried out for a relatively long-term study.

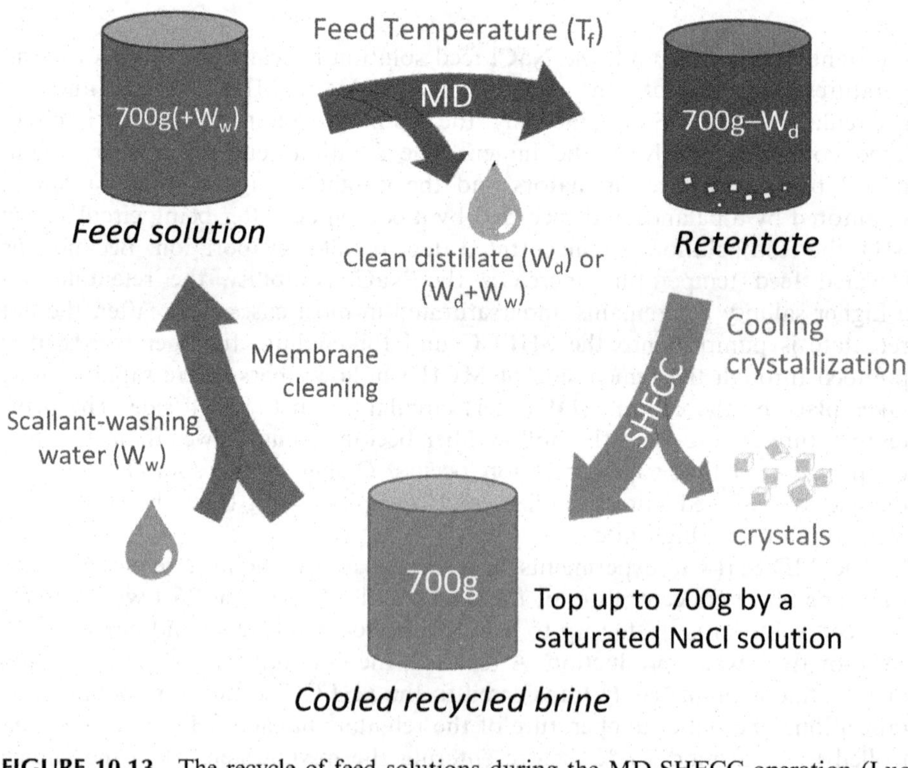

FIGURE 10.13 The recycle of feed solutions during the MD-SHFCC operation (Luo et al., 2018). Copyright (2018) Elsevier.

10.3.2.1 MD-SHFCC Operations at Various Feed Temperatures

Results of six-cycle MD-SHFCC operations at various feed temperatures have been presented in Figure 10.14. As the distillation amount (W_d) varies from 7.1 to 14 g, the corresponding NaCl concentration in the concentrated feed solution also increases from 26.67% to 26.94%. Figure 10.14a and b shows the MD flux and distillate conductivity at 60 °C as a function of time in six cycles, and the dotted arrow lines indicate the starting of each cycle. Figure 10.14a exhibits that the MD flux fluctuates within 15–17.2 kg/m²h with a very low distillate conductivity of <3 μS/cm when the hybrid

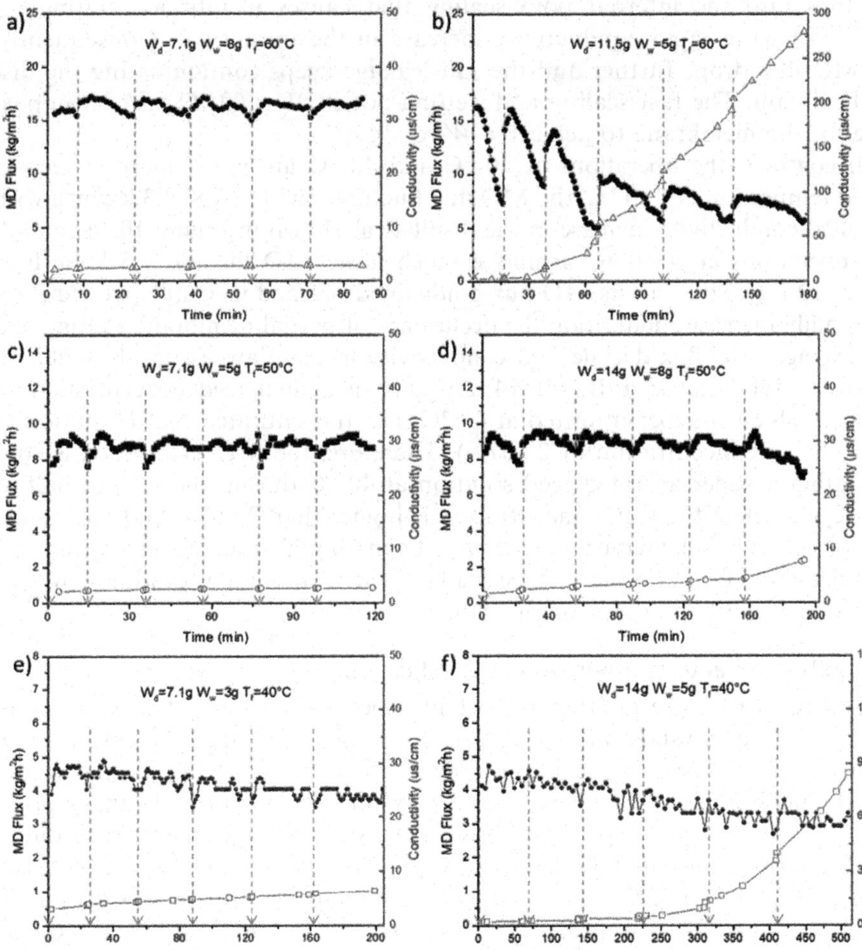

FIGURE 10.14 Six cycles of MD-SHFCC operations at a feed temperature of 60 °C (a, b), 50 °C (c,d), and 40 °C (e,f) (Luo et al. 2018). Copyright (2018) Elsevier.

system is operated at a distillation amount of 7.1 g (W_d = 7.1 g) and a washing water amount of 8 g (W_w = 8 g). However, the MD flux exhibits a sharp decline during the first 3 cycles but with a good recovery at the first few cycles when W_d = 11.5 g and W_w = 5 g, as illustrated in Figure 10.14b. The flux decline becomes progressively irreversible and the conductivity continuously increases because of partial membrane wetting and salt leakage. A comparison of Figure 10.14a and 10.14b indicates that a higher W_d requires a longer operation time for the MD membrane to concentrate the feed solution to a higher concentration, which enhances membrane surface scaling. Pore blockage by the salt scallant on the membrane surface should be the main reason of the flux decline in early cycles, and the washing step could help to recover the flux effectively. Gradually, the surface scaling develops into the internal pore scaling that causes membrane wetting, as reflected by the rapid conductivity increase in the third cycle. Consequently, the MD flux drops further and the salt leakage keeps contaminating the distillate stream. The fast scaling and wetting potential at 60 °C make it impossible for the membrane to achieve a W_d of 14 g.

In contrast, the operations at 50 °C and 40 °C are much more stable. At a feed temperature of 50 °C, the MD flux fluctuates within 7.8–9.3 kg/m^{2}h with a limited conductivity increase in the distillate as shown in Figure 10.14c and d. The operations at = 40 °C exhibit a much lower MD flux of 3–5 kg/m^{2}h in Figure 10.14e and f, but its MD flux gradually drops and its distillate conductivity steadily increases, indicating the occurrence of partial membrane wetting and salt leakage. The flux decline and conductivity increase are especially visible in Figure 10.14f because a W_d of 14 g results in a final feed concentration of 26.94%, which is supersaturated at 40 °C (i.e., the saturated NaCl solution at 40 °C has a concentration of 26.67%). Therefore, the MD membrane is concentrating a supersaturated feed solution at 40 °C during the second half of each cycle where the feed concentration is higher than 26.67% and has a high risk of scaling and membrane wetting. Even though the regular membrane washing could clean up the surface scallant, the membrane is gradually wetted over a long operation duration of ~8 h.

10.3.2.2 Characterizations of Salt Crystals

Crystal properties are another important aspect in hybrid system operations. The produced crystals are preferred to have a desired dimension with a narrow crystal size distribution (Edwie and Chung, 2013; Zarkadas and Sirkar, 2004b). Figure 10.15a shows the crystal size distributions and morphologies of NaCl crystals produced from MD-SHFCC experiments with different distillation amounts (W_d) at T_f = 40 °C. The average crystal size increases from 38.4 ± 11.5 to 42.9 ± 14.6 and 44.5 ± 14.5 μm from operations with a W_d of 7.1, 11.5, and 14 g, respectively. As W_d increases from 7.1 to 14 g, the final salt concentration in the retentate after MD rises from 26.67% to 26.94%. Accordingly, the amount of excess dissolved NaCl rises from 4.8 to 9.8 g per kg water in the chilled retentate after SHFCC at 23 °C, suggesting a higher degree of supersaturation. At higher supersaturation, the chemical potential

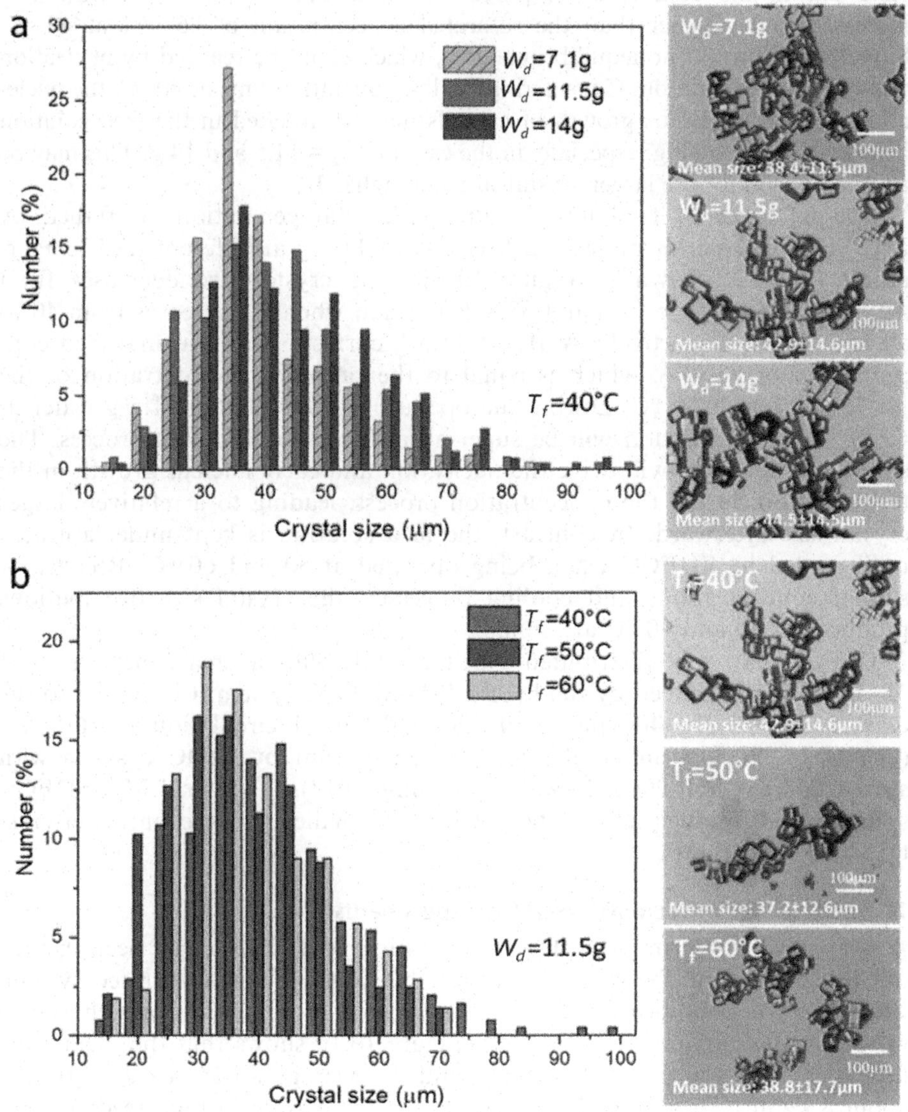

FIGURE 10.15 The size distribution and morphologies of salt crystals produced at (a) T_f = 40 °C after different distillation amounts, (b) various feed temperatures with W_d = 11.5 g (Luo et al. 2018). Copyright (2018) Elsevier.

difference between the NaCl liquid and solid states is larger. Therefore, there are higher driving forces for both nucleation and nuclei growth processes that lead to a significantly faster salt production and a relatively larger salt crystal size. Moreover, the whole feed solution is maintained at a supersaturated state during the concentration process of MD because its concentration is increased to be higher than the saturated concentration of 26.67% at 40 °C. Supersaturation is a nonequilibrium state, which could be relieved by nucleation and crystal precipitation. Considering the long operation time at 40 °C, the nucleation and early nucleus growth of crystals may be initiated in the feed solution before SHFCC cooling, especially in the case of $W_d = 11.5$ and 14 g. This may be another reason for the larger crystal size at a higher W_d.

The influence of feed temperature on crystal generation is studied by MD-SHFCC experiments with a fixed W_d of 11.5 g at different feed temperatures (T_f). As shown in Figure 10.15b, the crystal size decreases from 42.9 ± 14.6 to 37.2 ± 12.6 and 38.8 ± 17.7 μm when T_f increases from 40 to 50 and 60 °C, respectively. A W_d of 11.5 g corresponds to a final feed concentration of 26.84%, which is equal to the saturated concentration of the NaCl solution at 50 °C. During the operation of distillating 11.5 g water at 40 °C, the feed solution will be supersaturated during the MD process. The supersaturated state will drive the nucleation and early nucleus growth in the feed solution during the concentration process, leading to a relatively larger crystal size afterward. In contrast, the feed solution is kept undersaturated until cooled by SHFCC when being operated at 50 and 60 °C. Because of similar concentrations and cooling processes, the crystal size distributions produced at 50 and 60 °C are similar.

Overall, narrowly distributed NaCl crystals with a small mean size of 35–45 μm are produced by the hybrid MD-SHFCC system. Compared to the crystal size of 350–400 μm by traditional forced circulation crystallizers (Geertman, 2000; Tun et al., 2005) or 60–80 μm by MDC systems with stirred tanks (Choi, 2005; Edwie and Chung, 2013), the hybrid MD-SHFCC system could produce salt crystals of a much smaller size and narrower crystal size distribution.

10.3.2.3 Fifteen-Cycle MD-SHFCC Experiments

A relatively long-term MD-SHFCC operation of 15 cycles has been carried out to demonstrate the stability and feasibility of the new integrated system. The operation conditions of $W_d = 14$ g, $W_w = 8$ g, $T_f = 50$ °C are chosen to maximize the harvest crystal weight. Figure 10.16 shows that the membrane flux appears as wavelike fluctuations, and the average water flux over 10 h is about 8 kg/m^{2}h. It is impressive considering the extremely high feed concentration of more than 26.4%. The distillate conductivity increment is moderately controlled within an acceptable range of <22.5 μS/cm (or a NaCl concentration of 0.01 g/L in the distillate), indicating the high purity of the distilled water. Over the 15 cycles, the NaCl crystal yield is continuously increasing to about 64 g per kg feed (or 1.5 kg/m^{2}h). The NaCl crystal yield could be further improved if the MD membrane area is increased or the semicontinuous operation could be developed into a continuous process. Overall, the experimental results have verified the feasibility of the MD-SHFCC system for simultaneous productions of fresh water and inorganic salt from the high salinity brine.

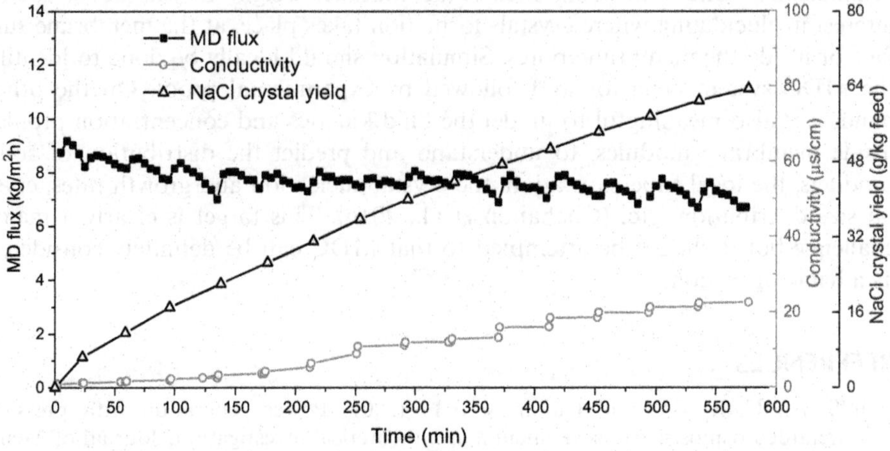

FIGURE 10.16 The evolution of membrane flux, distillate conductivity, and crystal yield during a 15-cycle MD-SHFCC operation with $T_f = 50$ °C, $W_d = 14$ g, $W_w = 8$ g (Luo et al. 2018). Copyright (2018) Elsevier.

10.4 CONCLUSION AND PROSPECTS

The combination of MD and crystallization technologies offers new opportunities in simultaneous productions of fresh water and inorganic salt crystals from brine streams. The energy and economic aspects of MDC also need to be considered. The specific energy consumption of MD crystallizer units is determined by the feed solution condition, operation temperatures, distillation amounts, flow rates, etc. In addition, the energy consumption of the integrated system could be further reduced when utilizing alternative energy sources such as waste heat, solar energy, cold energy from the regasification of liquefied natural gas (LNG), etc.

Significant innovations may happen in terms of new design, rationalization, and optimization of MDC productions. More systematic analyses of possible advantages or drawbacks are needed, regarding the introduction of an innovative membrane unit, clear protocols and comparison indexes of choosing the best materials and operative conditions, accurate modeling for scale-up, etc. In terms of MD unit, the problem of developing new microporous MD membranes with high hydrophobic characteristics and stability, with a narrow distribution of pore size and improved structural and morphological characteristics, is a crucial aspect. In addition, all feasible approaches devoted to increase the efficiency of MD membrane devices by reducing both concentration and temperature polarization phenomena, to enhance mass and heat transfer coefficients, to control the fouling problems and related drawbacks (clogging, loss of hydrophobicity, etc.) are expected to be matter of further investigations for researchers working on this area (Curcio and Drioli, 2005). Also, developing of modeling approaches for the MDC process could be significant in the study of understanding crystallization and improving membrane module design. It will be of major interest in elucidating where crystals formation takes place, at the membrane surface or inside the membrane pores. Simulation should ideally be done to identify the MDC system behavior and followed by experimental proof. On the other hand, it is also meaningful to model the fluid kinetics and concentration profiles inside membrane modules, to understand and predict the distribution of solid products, the local supersaturation, the crystal nucleation and growth rates, crystal size distribution, etc. (Chabanon et al., 2016). This target is clearly a major challenge but it should be attempted to that MDC can be definitely considered as a liable operation.

REFERENCES

Achilli, A., Cath, T.Y. and Childress, A.E. (2009) Power generation with pressure retarded osmosis: An experimental and theoretical investigation. Journal of Membrane Science 343(1), 42–52.

Amy, G., Ghaffour, N., Li, Z., Francis, L., Linares, R.V., Missimer, T. and Lattemann, S. (2017) Membrane-based seawater desalination: Present and future prospects. Desalination 401, 16–21.

Chabanon, E., Mangin, D. and Charcosset, C. (2016) Membranes and crystallization processes: State of the art and prospects. Journal of Membrane Science 509, 57–67.

Chan, M.T., Fane, A.G., Matheickal, J.T. and Sheikholeslami, R. (2005) Membrane distillation crystallization of concentrated salts – Flux and crystal formation. Journal of Membrane Science 257(1–2), 144–155.

Chen, D., Singh, D., Sirkar, K.K. and Pfeffer, R. (2016a) Continuous preparation of polymer coated drug crystals by solid hollow fiber membrane-based cooling crystallization. International Journal of Pharmaceutics 499(1–2), 395–402.

Chen, D., Singh, D., Sirkar, K.K., Zhu, J. and Pfeffer, R. (2014a) Continuous Polymer Nanocoating on Silica Nanoparticles. Langmuir 30(26), 7804–7810.

Chen, G., Lu, Y., Krantz, W.B., Wang, R. and Fane, A.G. (2014b) Optimization of operating conditions for a continuous membrane distillation crystallization process with zero salty water discharge. Journal of Membrane Science 450, 1–11.

Chen, X., Su, Y., Reay, D. and Riffat, S. (2016b) Recent research developments in polymer heat exchangers – A review. Renewable and Sustainable Energy Reviews 60, 1367–1386.

Choi, B.S. (2005) Crystallization of Sodium Chloride from a Concentrated Calcium Chloride-Potassium Chloride-Sodium Chloride Solution in a CMSMPR Crystallizer: Observation of Crystal Size Distribution and Model Validation, Ann Arbor. ProQuest Dissertations Publishing.

Curcio, E., Criscuoli, A. and Drioli, E. (2001) Membrane crystallizers. Industrial and Engineering Chemistry Research 40(12), 2679–2684.

Curcio, E. and Drioli, E. (2005) Membrane distillation and related operations – A review. Separation and Purification Reviews 34(1), 35–86.

De Yoreo, J.J. and Vekilov, P.G. (2003) Principles of Crystal Nucleation and Growth. Reviews in Mineralogy and Geochemistry 54(1), 57–93.

Drioli, E., Di Profio, G. and Curcio, E. (2012) Progress in membrane crystallization. Current Opinion in Chemical Engineering 1(2), 178–182.

Drioli, E., Profio, G.D. and Curcio, E. (2015) Membrane-Assisted Crystallization Technology, pp. 151–191, London Imperial College Press.

Edwie, F. and Chung, T.S. (2012) Development of hollow fiber membranes for water and salt recovery from highly concentrated brine via direct contact membrane distillation and crystallization. Journal of Membrane Science 421, 111–123.

Edwie, F. and Chung, T.-S. (2013) Development of simultaneous membrane distillation–crystallization (SMDC) technology for treatment of saturated brine. Chemical Engineering Science 98, 160–172.

El-Abbassi, A., Hafidi, A., Khayet, M. and García-Payo, M.C. (2013) Integrated direct contact membrane distillation for olive mill wastewater treatment. Desalination 323, 31–38.

El-Bourawi, M.S., Ding, Z., Ma, R. and Khayet, M. (2006) A framework for better understanding membrane distillation separation process. Journal of Membrane Science 285(1–2), 4–29.

Eumine Suk, D. and Matsuura, T. (2006) Membrane-Based Hybrid Processes: A Review. Separation Science and Technology 41(4), 595–626.

Geertman, R. (2000) Sodium chloride: Crystallization Reference Module in Chemistry, Molecular Sciences and Chemical Engineering."Encyclopedia of Separation Science", Ian D. Wilson 4127–4134. Academic Press/Cambridge, Massachusetts

Gryta, M. (2002) Concentration of NaCl solution by membrane distillation integrated with crystallization. Separation Science and Technology 37(15), 3535–3558.

Han, G., Zhang, S., Li, X. and Chung, T.-S. (2015) Progress in pressure retarded osmosis (PRO) membranes for osmotic power generation. Progress in Polymer Science 51, 1–27.

Ji, X., Curcio, E., Al Obaidani, S., Di Profio, G., Fontananova, E. and Drioli, E. (2010) Membrane distillation-crystallization of seawater reverse osmosis brines. Separation and Purification Technology 71(1), 76–82.

Kamalipour, M., Dehghani, S.A.M., Naseri, A. and Abbasi, S. (2017) Role of agitation and temperature on calcium sulfate crystallization in water injection process. Journal of Petroleum Science and Engineering 151, 362–372.

Khayet, M.Membrane Distillation. In: Li N.N., Fane A. G., Ho W.S. W., Matsuura T., editors. Advanced Membrane Technology Applications, Hoboken, USA: John Wiley & Sons, Inc.; 2008, p. 297–369.

Khayet, M.S. and Matsuura, T. (2011) Membrane Distillation: Principles and Applications, Elsevier, Amsterdam; Boston.

Kim, J., Kim, J. and Hong, S. (2018) Recovery of water and minerals from shale gas produced water by membrane distillation crystallization. Water Research 129, 447–459.

Kim, Y.D., Thu, K., Ng, K.C., Amy, G. and Ghaffour, N. (2016) A novel integrated thermal-/membrane-based solar energy-driven hybrid desalination system: Concept description and simulation results. Water Research 100, 7–19.

Li, D. and Kaner, R.B. (2006) Shape and Aggregation Control of Nanoparticles: Not Shaken, Not Stirred. Journal of the American Chemical Society 128(3), 968–975.

Linnikov, O.D. (2014) Mechanism of precipitate formation during spontaneous crystallization from supersaturated aqueous solutions. Russian Chemical Reviews 83(4), 343.

Luo, L., Chang, J. and Chung, T.-S. (2017) Cooling crystallization of sodium chloride via hollow fiber devices to convert waste concentrated brines to useful products. Industrial & Engineering Chemistry Research 56(36), 10183–10192.

Luo, L., Zhao, J. and Chung, T.-S. (2018) Integration of membrane distillation (MD) and solid hollow fiber cooling crystallization (SHFCC) systems for simultaneous production of water and salt crystals. Journal of Membrane Science 564, 905–915.

Martinetti, C.R., Childress, A.E. and Cath, T.Y. (2009) High recovery of concentrated RO brines using forward osmosis and membrane distillation. Journal of Membrane Science 331(1), 31–39.

Mudd, G.M. (2010) The environmental sustainability of mining in Australia: Key mega-trends and looming constraints. Resources Policy 35(2), 98–115.

Mullin, J.W. and Raven, K.D. (1961) Nucleation in agitated solutions. Nature 190 (4772), 251.

Sanmartino, J.A., Khayet, M., García-Payo, M.C., El-Bakouri, H. and Riaza, A. (2017) Treatment of reverse osmosis brine by direct contact membrane distillation: Chemical pretreatment approach. Desalination 420, 79–90.

Seo, D.H., Pineda, S., Woo, Y.C., Xie, M., Murdock, A.T., Ang, E.Y.M., Jiao, Y., Park, M.J., Lim, S.I., Lawn, M., Borghi, F.F., Han, Z.J., Gray, S., Millar, G., Du, A., Shon, H.K., Ng, T.Y. and Ostrikov, K. (2018) Anti-fouling graphene-based membranes for effective water desalination. Nature Communications 9(1), 1–12.

Shim, W.G., He, K., Gray, S. and Moon, I.S. (2015) Solar energy assisted direct contact membrane distillation (DCMD) process for seawater desalination. Separation and Purification Technology 143, 94–104.

Souhaimi, M.K. and Matsuura, T. (2011) Membrane Distillation.

Srinivasan, R., Shirgaonkar, I.Z. and Pandit, A.B. (1995) Effect of Sonication on Crystal Properties. Separation Science and Technology 30(10), 2239–2243.

Tun, C.M., Fane, A.G., Matheickal, J.T. and Sheikholeslami, R. (2005) Membrane distillation crystallization of concentrated salts – Flux and crystal formation. Journal of Membrane Science 257(1), 144–155.

Wang, K.Y., Teoh, M.M., Nugroho, A. and Chung, T.S. (2011a) Integrated forward osmosis-membrane distillation (FO-MD) hybrid system for the concentration of protein solutions. Chemical Engineering Science 66(11), 2421–2430.

Wang, P. and Chung, T.S. (2012) A conceptual demonstration of freeze desalination-membrane distillation (FD-MD) hybrid desalination process utilizing liquefied natural gas (LNG) cold energy. Water Research 46(13), 4037–4052.

Wang, P. and Chung, T.S. (2015) Recent advances in membrane distillation processes: Membrane development, configuration design and application exploring. Journal of Membrane Science 474, 39–56.

Wang, P., Teoh, M.M. and Chung, T.S. (2011b) Morphological architecture of dual-layer hollow fiber for membrane distillation with higher desalination performance. Water Research 45(17), 5489–5500.

Wu, Y. and Drioli, E. (1989) The behaviour of membrane distillation of concentrated aqueous solution. Water Treatment 4, 399–415.

Wu, Y., Kong, Y., Liu, J., Zhang, J. and Xu, J. (1991) An experimental study on membrane distillation-crystallization for treating waste water in taurine production. Desalination 80(2), 235–242.

Yuichi, Y., Masatomo, S., Hisashi, Y. and Yuji, G. (2013) Ultrasonication: An efficient agitation for accelerating the supersaturation-limited amyloid fibrillation of proteins. Japanese Journal of Applied Physics 52(7S), 07HA01.

Zarkadas, D.M. and Sirkar, K.K. (2004a) Polymeric hollow fiber heat exchangers: An alternative for lower temperature applications. Industrial & Engineering Chemistry Research 43(25), 8093–8106.

Zarkadas, D.M. and Sirkar, K.K. (2004b) Solid hollow fiber cooling crystallization. Industrial & Engineering Chemistry Research 43(22), 7163–7180.

Zarkadas, D.M. and Sirkar, K.K. (2007) Cooling crystallization of paracetamol in hollow fiber devices. Industrial & Engineering Chemistry Research 46(10), 2928–2935.

Zhang, S., Wang, P., Fu, X. and Chung, T.S. (2014) Sustainable water recovery from oily wastewater via Forward Osmosis – Membrane Distillation (FO-MD). Water Research 52, 111–121.

Zhu, A., Christofides, P.D. and Cohen, Y. (2009) Effect of thermodynamic restriction on energy cost optimization of RO membrane water desalination. Industrial & Engineering Chemistry Research 48(13), 6010–6021.

11 Freeze Desalination-Membrane Distillation Hybrid Systems

Jian Chang, Kang-Jia Lu, Peng Wang, and Tai-Shung Chung

11.1 INTRODUCTION

Freeze desalination (FD) refers to the process in which fresh water is extracted by harvesting and melting the ice crystals from the chilled feed (seawater or brackish water) (Manwell and McGowan, 1994). In this process, the saline water is cooled and frozen at controlled conditions. Almost all the salts are rejected by the ice crystals generated and as a result, fresh water can be obtained from the melted ice. It is an emerging desalination technology because of its insensitivities to fouling problems and the provision of cold energy from processes such as regasification of liquefied natural gas (LNG) in recent years (Nebbia and Menozzi, 1968; Johnson, 1976; Shirai et al., 1999; Wang and Chung, 2012). The cold energy is the heat absorbed from ambient surrounding when converting LNG (−162 °C) back to natural gas at LNG terminals (Miyazaki et al., 2000; Wang and Chung, 2012). It is one of high-quality energy sources to cool other media, but most cold energy during LNG regasification has been wasted until today. Because the latent heat of ice fusion is only one-seventh of the latent heat of water vaporization (Rahman et al., 2006), it is much easier and more cost-effective to induce and harvest ice crystals from seawater by the cold energy in FD than to produce potable water by means of evaporation in many distillation processes (Cao et al., 2015; Wang and Chung, 2012).

Considering the characteristics and mechanisms of FD and membrane distillation (MD) processes, both systems require large quantities of high-quality cold energy. For the FD process, cold energy is needed to cool and crystallize the pre-cooled saline feed solution; while in the MD process, cold energy is required to continuously cool and maintain the low temperature of the distillate stream. Therefore, the large amount of cold energy from regasification of LNG can be effectively used in the design of FD-MD hybrid systems to minimize the overall energy consumption.

There are two categories of freezing processes in FD, namely, direct and indirect contact freezing (Lu and Xu, 2010). The former mixes seawater with a liquid refrigerant directly. The refrigerant vaporizes and escapes from the top of the freezer. At the same time, seawater is cooled and then freezes. This process has a high production rate, but the requirements of the refrigerant are stringent. It should be nontoxic, chemically inactive, immiscible in water, and resistant to forming hydrates (Rahman et al., 2006). In contrast, indirect freezing uses walls to isolate the refrigerant from seawater (Zvinowanda et al., 2014). Heat exchange happens through the crystallizer walls with adjustable cooling conditions (Mandri et al., 2011; Wang and Chung, 2012). An intermediate coolant may be utilized to pass the cold energy from LNG to seawater at an appropriate temperature to facilitate ice production without incurring salts. Some previous studies used −10 °C or −15 °C for simplicity (Badawy, 2015; Efrat, 2011; Wang and Chung, 2012), there is a need to find the optimal coolant temperature to maximize the water production while minimizing the salt attachment and energy consumption.

Various crystallizer configurations such as static layer crystallizer (Rich et al., 2010), progressive crystallizer (Fujioka et al., 2013), and dynamic layer crystallizer (Rich et al., 2012) have been employed to investigate the ice formation in FD. In a static layer growth system, the seawater stays stagnant on a surface, the driving forces for heat and mass transfers are free convection and gravity. Thus, it would take a long time, usually more than 10 h per batch, to form a reasonable amount of ice crystals. In a progressive crystallizer, only one single ice crystal is present, which is at one end of the vessel. The ice–solution interface gradually moves toward the other end. However, the speed of the freezing front is highly constrained. A fast moving freezing front would inevitably cause concentration polarization. In a dynamic growth system, seawater flows down along a cold surface and ice crystals form underneath the flowing film. One of the major obstacles in this process is the salt entrainment (Reddy et al., 2009). The brine tends to be trapped in the interstices of ice crystals due to the capillary force. Different from aforementioned approaches, we aim to use a simple batch reactor and have a deeper understanding of each basic parameter that would benefit FD development and FD-MD integration.

There are two major challenges in the FD technology. One is the amount of salts adsorbed on ice crystals, while the other is the amount of washing water to remove these salts. Since ice crystals generated in FD have very small sizes, they have an enormous specific surface area. Their surfaces are usually covered by a slew of salts due to the strong adhesive force between ice and brine. Therefore, one must wash ice crystals to remove these salts. The amount of washing water has been overlooked in most studies, but we would revisit this crucial issue and optimize the amount of washing water. Therefore, this chapter may provide useful insights and operational data for the entire FD process. In addition to these challenges, FD also faces the difficulties of being accepted as a new technology from industries, especially when it is perceived as costly and mechanically complex (Randall and Nathoo, 2015).

11.2 BACKGROUND AND THEORY OF FD

11.2.1 Pivotal Role of Washing in FD

Freezing pure water and saline water are distinctly different. Without the disturbance of soluble ions, the ice crystals grow on top of one another. Their sizes can be very large. Visibly, the ice could form an entire chunk in the cylinder. However, water does not freeze in the same manner when many ions are present. Instead, dendritic ice crystals (Lorain et al., 2001; Tleimat, 1980) will occupy the space and form a slurry.

Brine adheres to ice crystals considerably due to the massive interfacial tension (Shwartz and Probstein, 1968). A thorough separation of ice and brine requires accessory procedures other than simply draining off or normal filtration. The design of washing columns has been actively investigated (Harriott et al., 1969; Shwartz and Probstein, 1968) and used in pilot plants (Efrat, 2011). The ice slurry is introduced to the bottom of the washing column that is filled with clean water. Ice crystals float up by the buoyancy force and the brine trapped in interstices would be displaced. Clean ice can then be harvested at the top of the washing column. However, there exists a serious dilemma regarding whether to use agitation and pulsing. Without them, the mass transfer would be extremely slow because the buoyant movement is severely inhibited by the adherence force between ice and water. If agitation and pulsing are adopted, their contributions to longitudinal mixing would disrupt the concentration gradient distribution (Sherwood and Brian, 1966). Once the upper portion of the column contains a high amount of salts, the column is no longer able to purify ice. In contrast, washing ice crystals with the aid of vacuum filtration can drain out salty water within seconds. Not only can it lower the salt impurity on ice crystals to be almost negligible but also significantly reduce the energy consumption for the washing procedures.

11.2.2 Supercooling and Nucleation

Supercooling is the process where a liquid is cooled below its freezing point without crystallizing. For instance, pure water has to be cooled below 0 °C to initiate freezing. Although 0 °C is widely understood as its freezing point, it is rather an equilibrium point during the freezing process. The real temperature for water to start freezing depends on when nucleation happens. According to crystallization kinetics, a trace amount of ice lattice larger than the critical size must be generated (Tiller, 1992). In other words, there is a minimum embryo-crystal size (i.e., the critical radius) for the embryo to grow spontaneously. In fact, for a given temperature there will be a critical radius. The relationship among them is reflected by the Kelvin equation (Tleimat, 1980):

$$r^* = \frac{2\sigma T_f}{L_f \Delta T} \tag{11.1}$$

where r^* is the critical radius, T_f is the freezing temperature, L_f is the latent heat of fusion, ΔT is the temperature difference departed from the freezing point, and σ is the interfacial tension between water and ice that is estimated to be 22 erg/cm^2. The critical radius is then plotted relative to water temperature in Figure 11.1(a).

Once ice has the critical radius, nucleation may take place by either homogeneous or heterogeneous nucleation. Theoretically, a cluster of about 25 water

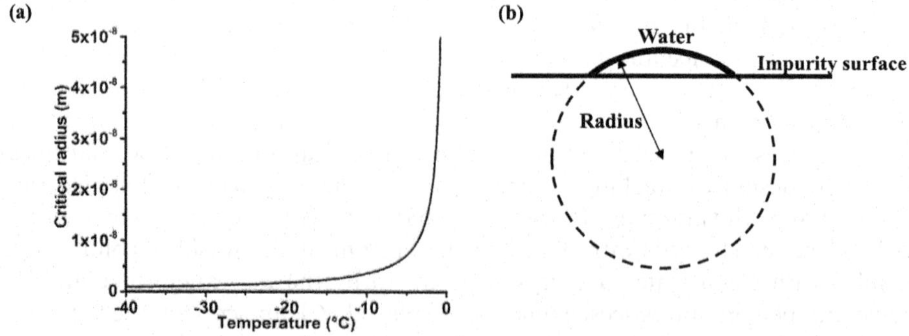

FIGURE 11.1 (a) The critical radius relative to temperature. (b) A water cluster's radius when it wets a surface (Chang et al., 2019). Copyright (2019) Elsevier.

molecules would form ice at −41 °C via homogeneous nucleation. In reality, it is difficult to cool the water down to near this temperature without forming ice because the nucleation barrier is largely reduced by the existence of impurities that lead to heterogeneous nucleation. As depicted in Figure 11.1(b), when water molecules wet an impurity surface, a small portion can actually correspond to a much larger radius (Muldrew and McGann, 1999). As a result, this cluster of water molecules is likely to stabilize themselves at a much higher temperature and spontaneously set off the heterogeneous nucleation. The ice crystal would then grow irreversibly as long as the cooling condition is maintained.

Additionally, the supercooling condition can be interrupted when nucleation is triggered by external nucleus (Randall and Lewis, 2009). If any seed ice is added into the supercooled solution, the status of supercooling would cease immediately and a large amount of ice crystals would form shortly. This process releases so much heat that the system temperature would rise shortly to the equilibrium point, that is, freezing point. In the absence of a seed, the temperature at which ice crystallization will occur is highly stochastic (random). Seeding helps to overcome this variability because ice crystallization will occur at the same temperature with seeding (Randall et al., 2012).

11.2.3 Freezing-Point Depression

Freezing-point depression is a phenomenon where the freezing point of a solvent is decreased by the presence of solutes (Gooch, 2011; Halde, 1980). It is noted that the lowering of the freezing point does not depend on the nature of the solutes. Instead, it is only proportional to the number of dissolved molecules and ions. The expression of freezing-point depression is generally given as:

$$\Delta T_f = -K_{FP} mo \tag{11.2}$$

where ΔT_f is the freezing point depression, K_{FP} is the cryoscopic constant which is 1.86 K kg/mol for water, and *mo* is the molality (mol solute/1000 g solvent).

Thus, for seawater with a salt concentration of 35 g/kg, the freezing point is depressed to −2.1 to −2.2 °C. However, once the solution starts freezing, it gets more and more concentrated. The freezing point is therefore further depressed. Figure 11.2 illustrates this process on a NaCl-water phase diagram. Route 1 stands for the cooling of saline water from room temperature to its freezing point. Route 2 is the process of supercooling explained in Section 11.2.2. The temperature quickly jumps back to the freezing point in Route 3 when nucleation takes place. Then the entire system proceeds along the equilibrium curve as shown by Route 4. If seeding is applied at the point when seawater is just cooled to the freezing point, supercooling would be forestalled and the process paths become 1 and 4 instead.

11.3.1 Determination of Major Parameters of FD

The major process parameters that greatly affect the FD unit performance are washing procedure, coolant temperature, process duration, supercooling, and agitation (Rahman et al., 2006). Each of them was investigated under *ceteris paribus* conditions. (i.e., with other conditions remaining the same).

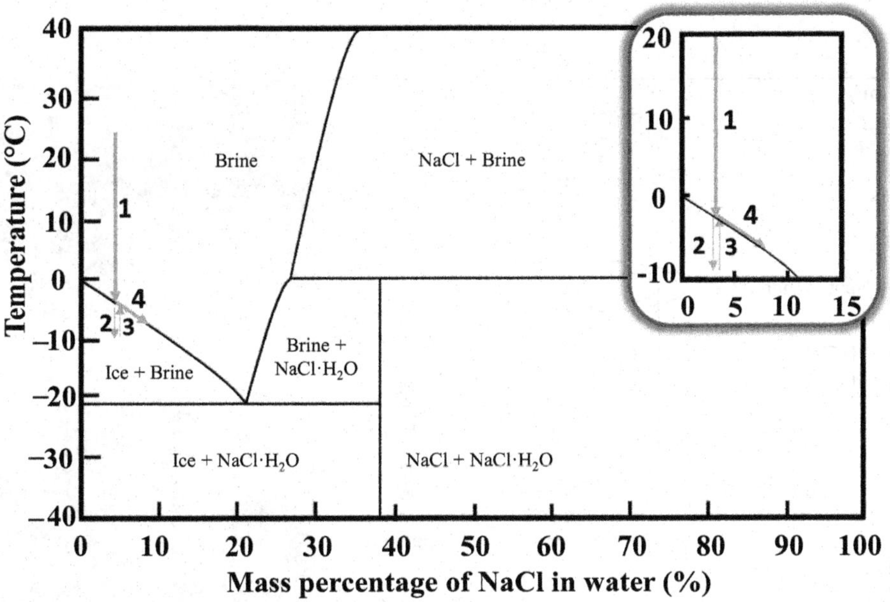

FIGURE 11.2 The operating process of freeze desalination (FD) illustrated on the phase diagram of sodium chloride solutions (Chang et al., 2019). Copyright (2019) Elsevier.

For comparison, a base case was established where a NaCl solution (35 g/kg) of 1 kg was cooled for 1 h without supercooling, agitated by the stirrer at 100 rpm and a coolant temperature of −8 °C. Among the parameters, washing procedure outweighs other parameters because the product water must contain minimal impurities and the salts adhered onto the ice crystals' surfaces actually account for most of the impurities (Kobayashi et al., 1996). Figure 11.3 shows the setup of the ice crystallization unit (Wang and Chung, 2012). Figure 11.4 displays the schematic diagram and the image of the vacuum filtration unit.

11.3.2 Washing Procedures

11.3.2.1 The Amount of Washing Water without Recycling

A NaCl solution (35 g/kg) or seawater of M_0 (1 kg) was precooled in a freezer to around 0 °C before being poured into the crystallizing cylinder. A certain amount of seed ice (M_s) was used to trigger the nucleation at its freezing point (−2.1 °C) or a certain supercooled temperature. The process was timed starting from the onset of nucleation. After a specified freezing duration, the entire ice slurry was transferred to a washing unit which is a vacuum filtration unit consisting of a metal mesh. The brine was drained off from the slurry by applying vacuum filtration for 10 s to completely remove the brine. The raw ice was weighed (M_i). Then, chilled tap water (0 °C) was used to rewash the raw ice. For each rewash, a designated amount of tap water was used with the

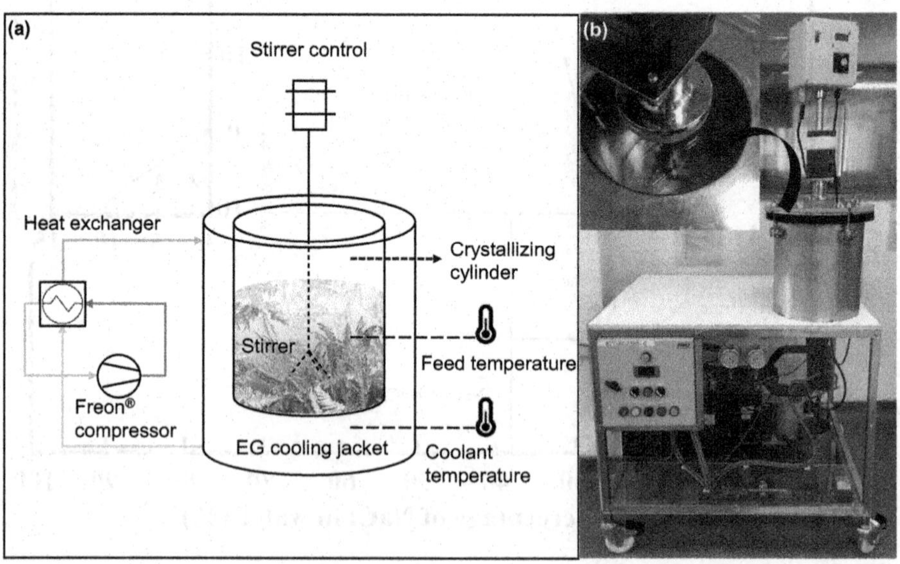

FIGURE 11.3 Schematic design and images of the indirect contact freeze desalination unit (Chang et al., 2019). Copyright (2019) Elsevier.

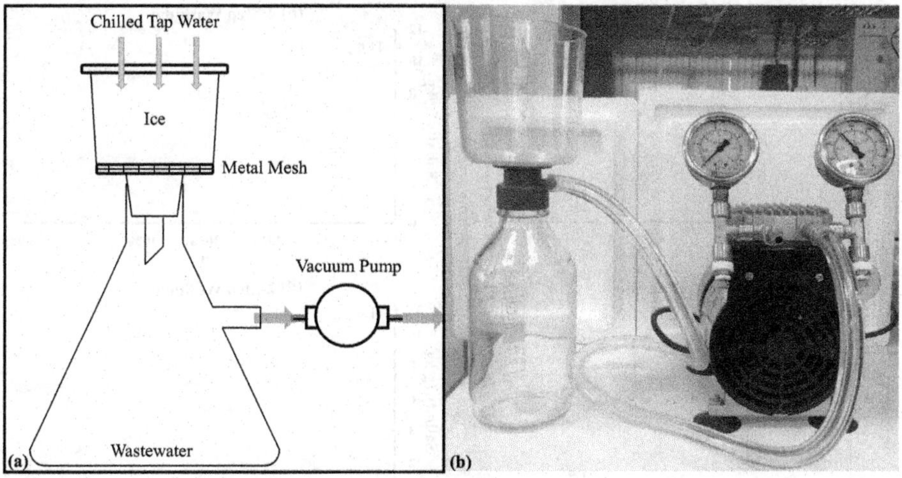

FIGURE 11.4 Schematic design and images of the vacuum filtration set and the washing unit (Chang et al., 2016). Copyright (2016) Elsevier.

aid of vacuum filtration for 10–20 s. Eventually, the product ice (M_{pi}) was transferred to a melting unit to melt naturally. The mass of washing water was represented as a percentage (η) of the raw ice (M_i), that is, $\eta \cdot M_i$.

Figure 11.5(a) shows the salinity of the product ice as a function of η (i.e., the mass ratio of chilled tap water to raw ice) with a stepwise increment of 10%. The salinity deceases after each step of washing. After six steps, the ice salinity meets the WHO potable water standard (<0.5 g/kg). In other words, to meet this standard, the amount of washing water should be about 60 wt% of the raw ice. This translates to that about 40% of the raw ice can be delivered as the product ice, considering a portion of the clean ice will be consumed as washing water. To simplify the washing procedure, the η value is increased in each step and the effects are depicted in Figure 11.5(b)–(d). Two conclusions may be drawn from these results. Firstly, the required washing steps can be reduced with an increase in stepwise η value and produce ice that meets the WHO standard. Secondly, regardless of the η value, the total amount of washing water is all the same, which is 60 wt% of the raw ice. This proves the proposition that the total η value has to be about 60 wt%. It may be noted that the simplest case of one-step washing with an η value of 60 wt% is not included in the figure. That is because the resultant ice has a salinity of 1.101 g/kg. Therefore, at least a two-step washing should be applied.

It is worth noting that during washing, all measures were taken to minimize agglomeration of ice crystals. For instance, any compression or other external forces should be avoided on the ice bed. This is also part of the reasons why centrifugal separation is not chosen to separate wastewater from the ice in the washing step.

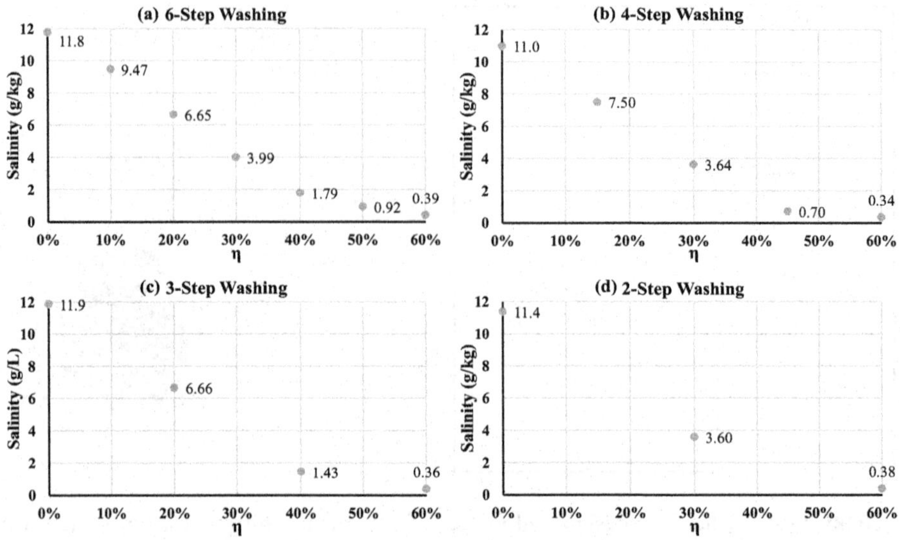

FIGURE 11.5 The effect of different washing procedures on ice salinity as a function of η (the mass ratio of chilled tap water to raw ice) (Chang et al., 2016). Copyright (2016) Elsevier.

11.3.2.2 The Amount of Washing Water with Recycling

To reduce the amount of tap water for ice washing, the used washing water is recycled to wash ice. A simple two-step washing is investigated. As illustrated in Figure 11.6, the wastewater from the first step of ice washing has a high salinity close to the concentrated feed left in the reactor, whereas the one from the second step of ice washing has a salinity of about 11.47 g/kg. The latter is about 80% lower in salinity than the former. It is because the porous ice bed exposes most of the crystal surfaces. After a full contact, the first stream of tap water is able to dissolve most of the surface impurities by the time when it flows away from the bed. Since there is a small amount of residues left, the wastewater from the second step is much less concentrated. Therefore, the wastewater from the second step of ice washing (i.e., referred to as the wastewater 2) could be reused to wash the raw ice. To demonstrate the feasibility, experiments were conducted as presented in Figure 11.7. The wastewater from the second step of ice washing (i.e., wastewater 2) in batch A is reused to wash the raw ice in batch B. As a result, the tap water needed in batch B is only 40 wt% of the raw ice, which is a reduction of 20%. When combining these two batches together, one can expect that the total amount of tap water for ice washing is reduced to about 50 wt% of the raw ice.

If the wastewater from the third step of ice washing in batch B (i.e., referred to as the wastewater 3 in Figure 11.7) is reused, then the amount of washing water may be further reduced to below 50 wt% of the raw ice. The

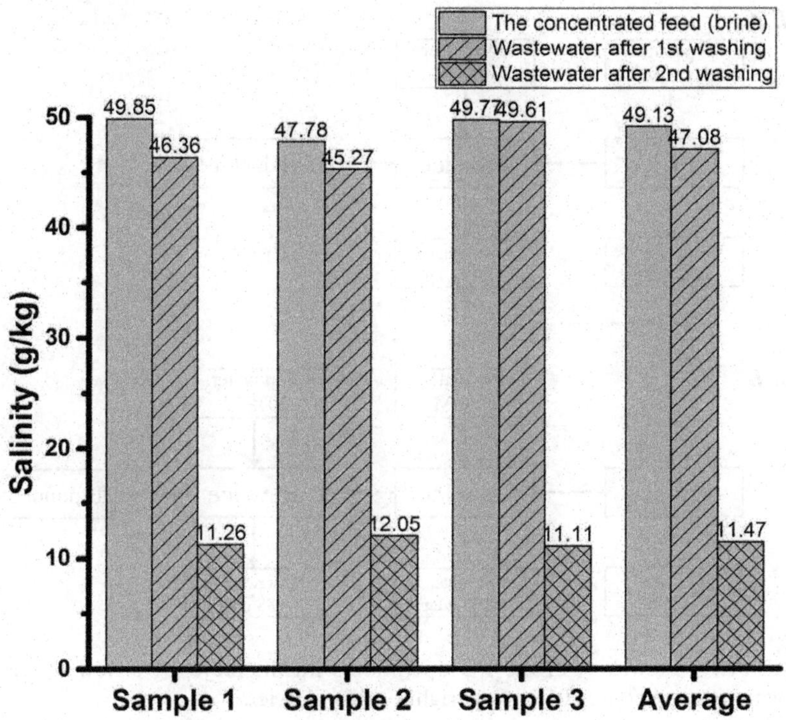

FIGURE 11.6 Salinities of brine and wastewater after the first and second step washing from FD (Chang et al., 2016). Copyright (2016) Elsevier.

reuse of washing water from more than two batches of ice washing will be studied in the future. In addition, the concentrated wastewater left from FD may be recyclable using MD because it is less sensitive to the feed concentration (Khayet and Matsuura, 2011; Wang and Chung, 2012, 2015).

11.3.3 Effects of Coolant Temperature

As illustrated in Sections 11.2.2 and 11.2.3, the feed solution undergoes a freezing process at the equilibrium point (i.e., freezing point). Therefore, the driving force for freezing is the temperature difference between the feed solution and the circulating coolant. A large temperature difference would facilitate a higher ice production rate. However, an excessive driving force would distort the freezing process. When the coolant temperature is lower than −10 °C, a thick ice layer would quickly form on the cylinder wall within minutes. The fast agglomeration of ice may trap salts inside, which are difficult to be washed away. In addition, the dense ice layer is difficult to be peeled off from the cylinder wall. Thus, a very large temperature difference is not desirable. On the contrary, a very small

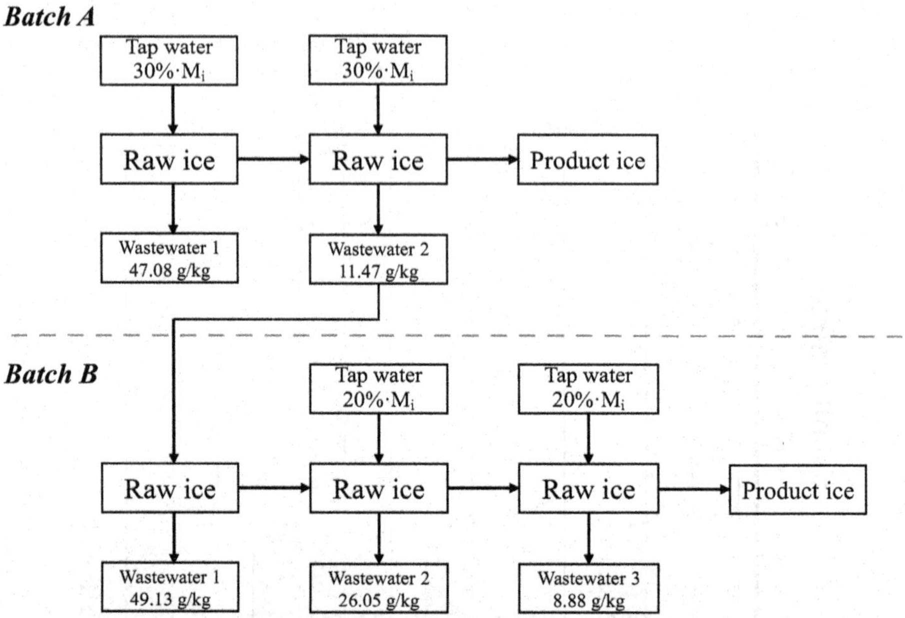

FIGURE 11.7 Flowchart of the FD operation with the recycling of low concentrated wastewater (Chang et al., 2016). Copyright (2016) Elsevier.

temperature difference would prolong the ice production process. Hence, an optimal coolant temperature should be identified.

Figure 11.8 shows the ice recovery rate and salinity as a function of coolant temperature. The experiments were conducted using seed ice of 1 g at −2.1 °C with an agitation speed of 100 rpm for 1 h and then washed optimally using chilled tap water equal to 50 wt% of the raw ice with recycling. A sufficiently high coolant flow rate was used throughout the experiments so that the coolant temperature increased a negligible amount after passing the cooling jacket. The resultant ice salinity varies in the range of 0.3–0.4 g/kg when the coolant temperature is higher than or equal to −8 °C. However, the ice salinity increases to 0.81 g/kg at −9 °C, which is higher than the WHO purity standard, because the ice layer is formed on the cylinder wall. The trapped concentrated salt solution between the ice crystals cannot be easily removed.

Figure 11.8 also shows that the ice recovery rate increases from 8.1% to 23% when the coolant temperature drops from −5 °C to −9 °C. However, the ice recovery rate only increases 1.9% from −8 °C to −9 °C, whereas it increases around 4.3% per degree of temperature drop in the range of −5 °C to −8 °C. The former is due to the fact that the ice layer on the cylinder wall impairs the heat exchange between the brine and the coolant, while the latter arises from the fact that ice crystals can hardly form a layer on the wall when the coolant temperature is at −8 °C or above. The incorporation of a scraper

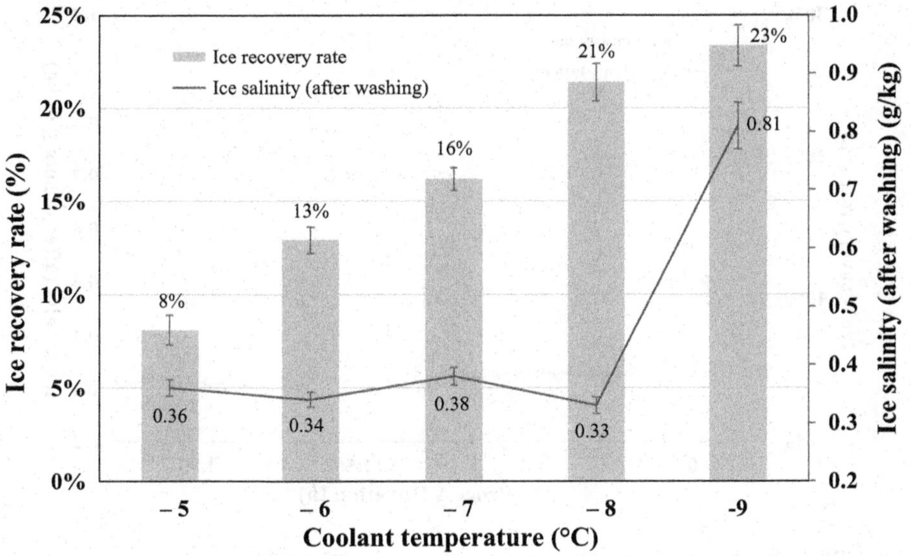

FIGURE 11.8 Ice recovery rate and salinity as a function of coolant temperature (freezing duration: 1 h; seed ice: 1 g at −2.1 °C; agitation: 100 rpm; optimized washing: chilled tap water = 50 wt% of the raw ice) (Chang et al., 2016). Copyright (2016) Elsevier.

to constantly remove the ice layer will be investigated in future studies. Thus, to maximize the ice production while maintaining high purity water, −8 °C should be the best coolant temperature.

11.3.4 Effects of Freezing Duration

Apart from coolant temperature, freezing duration also determines the amount of ice produced. Generally, a longer duration would yield a higher ice production, but the purity of product ice may be sacrificed at a very high yield. There is an optimal duration to achieve both high recovery and purity.

Figure 11.9 shows the effect of freezing duration on ice recovery and salinity under the same condition as Figure 11.8 except the coolant temperature was set at −8 °C for various durations. When the duration was shorter than 1.75 h, the ice recovery increased with time and the salinity remained in the acceptable range. Results of very long durations are not presented here because the resultant ice is very impure with a salinity higher than 1 g/kg.

When increasing the freezing duration, the ice appearance in the reactor changes sharply. When about 20–30% of the seawater freezes to ice, an ice bed can be easily seen floating on top of the brine. Unlike an iceberg that rises above the sea, the ice bed remains below the brine surface due to the strong capillary effect. When the ice amount increases to roughly 40 wt% of the feed, it fully occupies the brine volume and therefore only a uniform ice slurry can

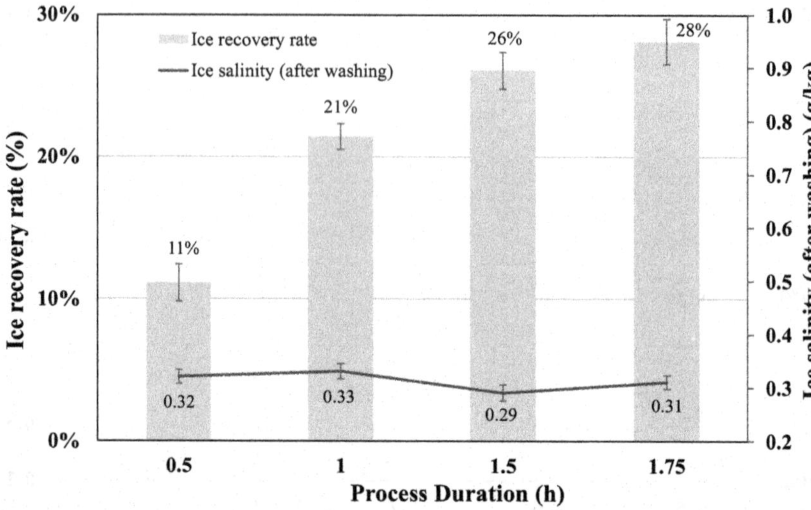

FIGURE 11.9 Ice recovery rate and salinity as a function of freezing duration (coolant temperature: −8 °C for 1 h; seed ice: 1 g at −2.1 °C; agitation: 100 rpm; optimized washing: chilled tap water = 50 wt% of the raw ice) (Chang et al., 2016). Copyright (2016) Elsevier.

be observed. Based on experimental results, converting 55% of the seawater to ice would be close to its limit and any further cooling would increase the recovery in a much slower manner. Furthermore, when it is higher than 55%, all the remaining brine would be absorbed into the voids of the porous ice bed due to the strong capillary effect. As a result, the entire system loses its fluidity and causes severe difficulties in subsequent transferring steps. In other words, it is suggested to convert only up to 50–55% of the seawater to ice in a single stage FD, which corresponds to an ice recovery rate of 25–27.5% after considering the amount of washing water. Therefore, the optimal freezing duration is around 1.5 h.

11.3.5 Effects of Seeding at Different Temperatures

Without seeding, the feed NaCl solution starts homogeneous nucleation at −9.8 °C. The metastable zone (Myerson and Ginde, 2002) is therefore between −9.8 °C and −2.1 °C. The seed could be introduced at any solution temperature in the metastable zone. Figure 11.10 shows the effect of seeding temperature on ice recovery and salinity. A low seeding temperature increases the ice production almost linearly while maintaining a reasonable purity. When homogeneous nucleation happens at −9.8 °C, ice crystals form rapidly in a short time. Thus, the exclusion of salts from ice is compromised. The ice formed at this temperature has a salinity above the acceptable range.

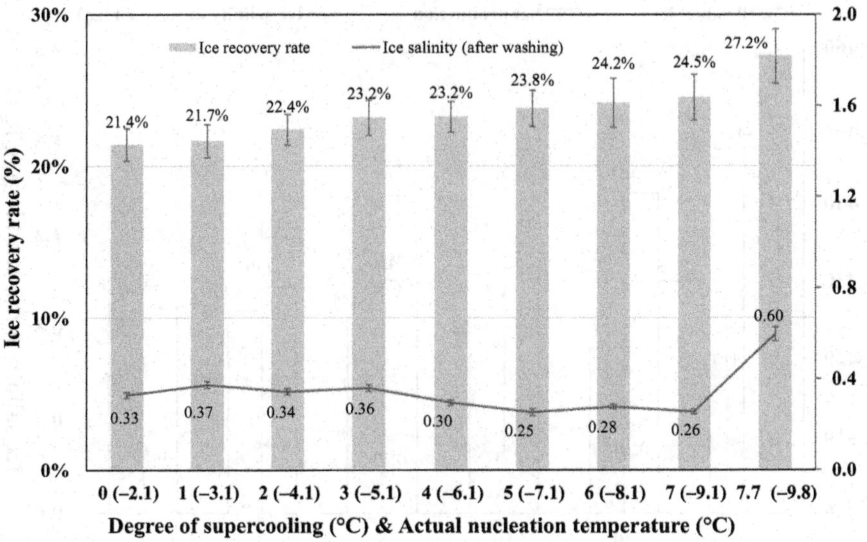

FIGURE 11.10 Ice recovery rate and salinity as a function of the degree of supercooling (freezing duration: 1 h; seed ice: 1 g at 2.1 °C; agitation: 100 rpm; optimized washing: chilled tap water = 50 wt% of the raw ice) (Chang et al., 2016). Copyright (2016) Elsevier.

Although a higher recovery rate can be achieved by seeding at a lower temperature, it is actually at the cost of more energy consumption. For instance, it took 42 min to cool the solution from −2.1 °C to −9.8 °C. Since the freezing duration lasted another 1 h, the entire duration was actually extended to nearly 1.75 h. However, its ice recovery rate of 27.2% was very comparable to that of 28.1% in Figure 11.9, which was conducted at −8 °C for 1.75 h. In addition, the former has an ice salinity much higher than the latter (0.6 vs. 0.31 g/kg) and the WHO standard (0.5 g/kg). Therefore, the degree of supercooling should be smaller than 7 °C. In addition, cooling the solution below its freezing point becomes harder and harder because its heat capacity also increases simultaneously (Holten et al., 2012). One should not keep lowering the solution temperature when it is already able to produce ice. With a small amount of seed ice, it can easily eliminate the waiting period toward lower temperatures and induce a smooth transit between phase changes.

11.3.6 Effects of Seeding Amount

Figure 11.11 shows the amount of ice production and its salinity as a function of the amount of seed ice (M_s) added into the feed at −2.1 °C. The more the seeds result the more the ice (M_{pi}). Surprisingly, after subtracting the mass of the seed ice, the net ice production decreases with an increase in the amount of seed ice because of the redundancy of excessive seeding.

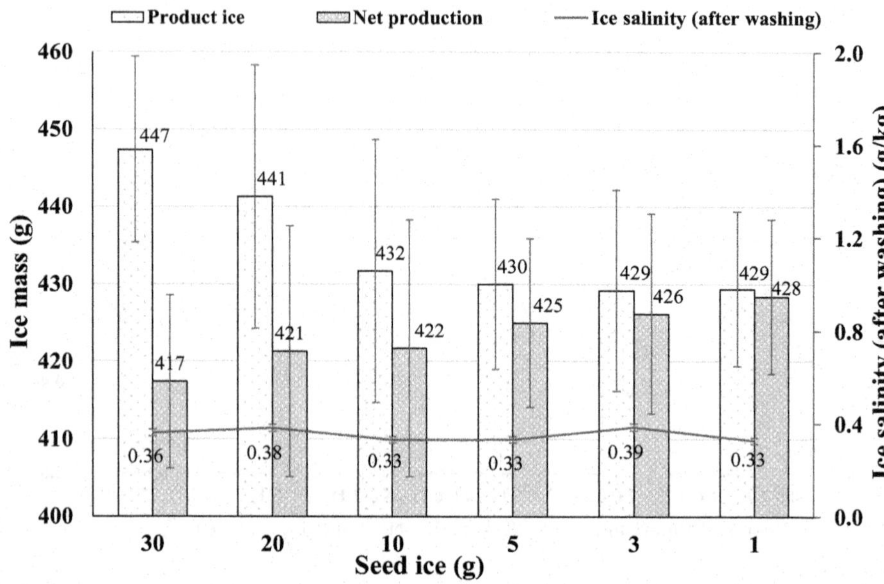

FIGURE 11.11 Ice mass and salinity as a function of seeding amount (coolant temperature: −8 °C; freezing duration: 1 h; seed ice: 1 g at −2.1 °C; agitation: 100 rpm; optimized washing: chilled tap water = 50 wt% of the raw ice) (Chang et al., 2016). Copyright (2016) Elsevier.

To understand why more seeding results in a lower net ice production, the role of seeding needs to be elaborated. Essentially, it is to trigger the onset of nucleation, so a fraction of seed ice is enough to serve the purpose. Assuming the minimum amount is $M_{s\text{-}min}$, then the extra portion of the seed ice (i.e., $M_s - M_{s\text{-}min}$) is redundant to activate the nucleation. In addition, it occupies the space for the ice formation from the feed. Therefore, as shown in Figure 11.11, 1 g of seed ice is enough in the current case. This amount can be further reduced. Nonetheless, it is needless because 1 g is already reasonably minuscule. The ice-seeding amount is thus recommended to be 0.1 wt% of the feed solution, which should be practical and economical.

11.3.7 Effects of Agitation

Freezing is a highly heat-sensitive process where a large amount of heat has to be exchanged between the cylinder and the seawater crystallizer instantaneously. Apart from the thermal conductivity of the cylinder wall, mixing in the system also plays a crucial role.

As shown in Figure 11.12, the ice recovery rate improves from 18.7% to 21.4% when speeding up the stirrer from 50 rpm to 100 rpm. This is because a better agitation increases the heat transfer rate and reduces the concentration polarization. When a freezing process starts, water molecules are integrated

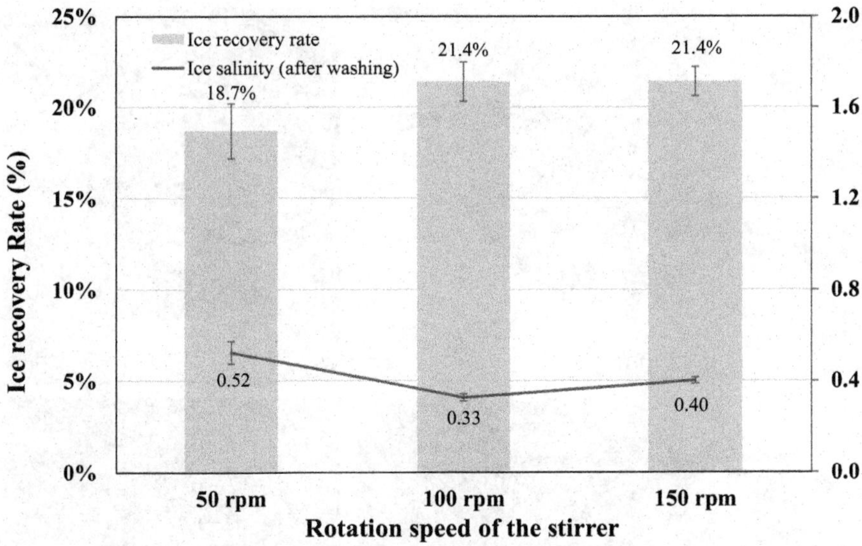

FIGURE 11.12 Ice recovery rate and salinity as a function of agitation (coolant temperature: −8 °C; freezing duration: 1 h; seed ice: 1 g at −2.1 °C; optimized washing: chilled tap water = 50 wt% of the raw ice) (Chang et al., 2016). Copyright (2016) Elsevier.

into ice crystal lattices and the adjacent salt concentration rises (Gay et al., 2003). Without enough mixing, concentration polarization takes place and makes the crystal growth difficult. Therefore, increasing agitation improves the ice recovery. Meanwhile, the ice salinity drops from 0.52 g/kg to 0.33 g/kg because the crystal surfaces are no longer contiguous with locally concentrated brine. However, the ice recovery rate remains nearly the same (21.4%) if a further increase in agitation to 150 rpm. This indicates that 100 rpm is sufficient to mitigate the localized concentration polarization. Hence, a moderate amount of agitation, such as 100 rpm, would be feasible and advisable.

11.3.8 Morphology of Ice Crystals

Figure 11.13 shows that ice crystals obtained from various conditions appear in clusters of amorphous shapes. Their sizes vary widely, but most of them are in the range of 0.3–1 mm. The formation of amorphous ice crystals can be comprehended according to the Kossel model of crystal growth (Mullin, 2001), as illustrated in Figure 11.14. Crystals grow the fastest at kinks and gradually complete the surface. A fresh crystal could commence on the surface or at the corners, but its growth rate is much slower than those at kinks. Therefore, the maximum growth rate takes place theoretically when the surface is full of kinks. Since collisions among crystals in turbulent freezing conditions keep generating numerous kinks, it produces the ice crystals with irregular surfaces and amorphous morphologies very fast.

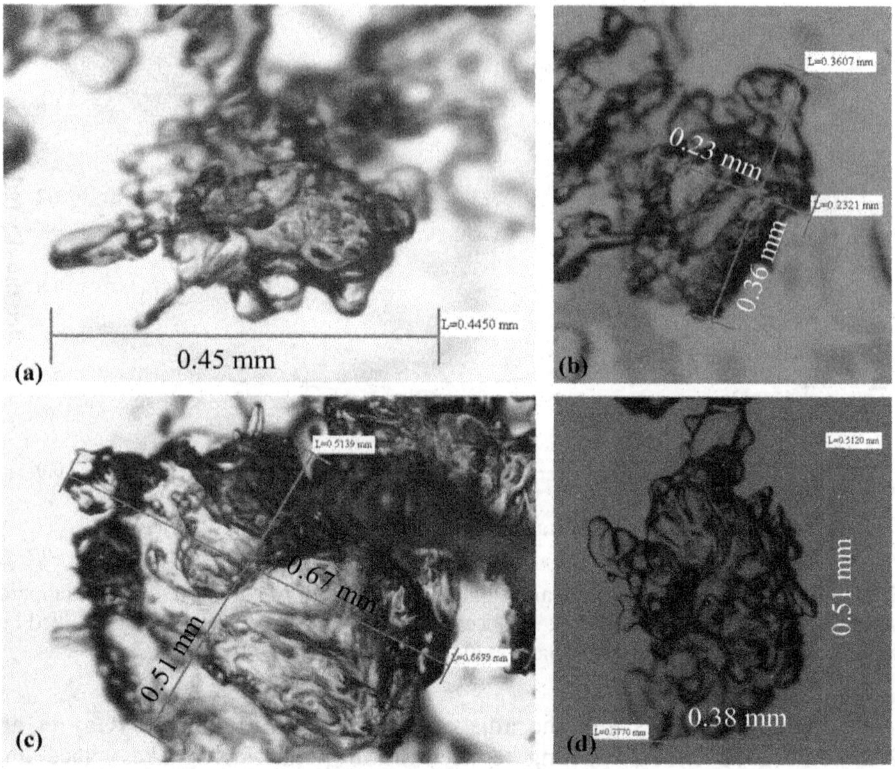

FIGURE 11.13 The morphologies of ice crystals (feed: 35 g/kg NaCl solution; coolant temperature: −8 °C; freezing duration: 1 h; agitation: 100 rpm; seed ice: 1 g at −2.1 °C; optimized washing: chilled tap water = 50 wt% of the raw ice) (Chang et al., 2016). Copyright (2016) Elsevier.

Many studies in literatures preferred a much slower ice production because they aimed to form larger ice crystals with an easier washing step. As a result, their ice crystals must grow very slowly in static environments (Rahman et al., 2006; Williams et al., 2015) and took hours or even days for freezing duration (Badawy, 2015; Mahdavi et al., 2011; Williams et al., 2015). The speedy ice growth comes partially from the smaller crystal sizes. It might appear harder to be washed, but the dilemma is alleviated since the proposed vacuum-filtration washing is highly efficacious. Besides, small ice crystals add a bonus to the melting step as they melt faster.

11.3.9 Iron Crystallizer

The aforementioned studies were conducted using a crystallizer consisting of a glass cylinder wall. A metal one with the same dimension was also

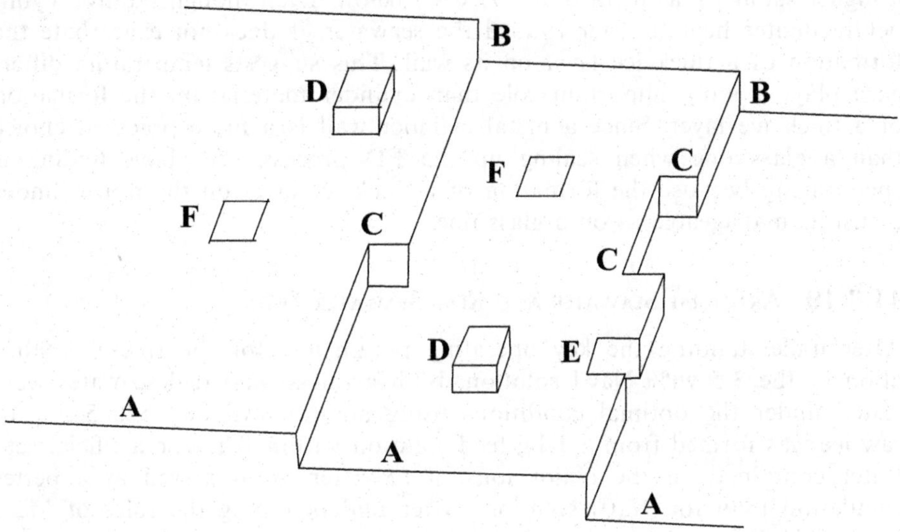

FIGURE 11.14 Kossel's model of a growing crystal surface: (A) flat surfaces, (B) steps, (C) kinks, (D) surface-adsorbed growth units, (E) edge vacancies, and (F) surface vacancies (Mullin, 2001).

studied for comparison. Since the thermal conductivity of iron is around 50–80 times of glass (Tritt, 2004), it was expected to expedite the freezing process. As shown in Table 11.1, the iron cylinder generally increases the ice recovery rate by 1.2–1.8 times. Surprisingly, the lowest possible coolant temperature ensuring an acceptable purity is found to be around −8 °C, which is the same as that for the glass crystallizer. Similarly, a thick ice layer with

TABLE 11.1

A comparison of the effects of cylinder materials on ice recovery and salinity*

Run	Cylinder material	Coolant temperature (°C)	Process duration (h)	Ice recovery rate (%)	Ice salinity (after washing) (g/kg)
1	Glass	−8	0.5	11	0.32
2	Iron	−8	0.5	20	0.39
3	Glass	−8	1	21	0.33
4	Iron	−8	1	27	0.32
5	Glass	−9	1	23	0.81
6	Iron	−9	1	30	0.65

* Feeds: 1 kg NaCl solution (35 g/kg); seed ice: 1 g at −2.1 °C; agitation: 100 rpm; optimized washing: chilled tap water = 50 wt% of the raw ice.

a higher salinity was formed at −9 °C or below. Even though the iron cylinder facilitates heat transfer toward the seawater, it does not exacerbate the formation of a thick ice layer on its wall. This suggests temperature difference plays a more important role than cylinder material on the formation of a thick ice layer. Since a metal cylinder wall is a more practical choice than a glass one when scaling up the FD process, the above finding is encouraging because the formation of a thick ice layer on the iron cylinder is just as manageable as on a glass one.

11.3.10 Artificial Seawater and Real Seawater Tests

After understanding the key operating parameters for the freeze desalination of the 3.5 wt% NaCl solution, both artificial and real seawater were tested under the optimal conditions using an iron cylinder. 520–525 g of raw ice was formed from a 1 kg feed solution within 1 h. The artificial seawater contained all the major ions of seawater, so it served as a better simulation than the NaCl solution. After understanding the roles of Mg^+, Ca^{2+}, and K^+, real seawater that has a much more complicated composition was finally tested. Table 11.2 tabulates the composition of artificial seawater and real seawater.

After deducting the mass of seed and washing water, it translates to a recovery rate of about 26%. This value agrees with the result of 26.7% obtained from the NaCl solution. Table 11.3 lists the compositions of the melted ice. They have the total dissolved solid (TDS) of 310 ppm to 320 ppm, these values are very close to the ice made from the NaCl solution (i.e., 0.33 g/kg or 330 ppm). Thus, it affirms that the FD technology is promising to desalinate real seawater.

Figure 11.15 compares the relative percentages of major ions Mg^+, Ca^{2+}, and K^+ in feeds and product ice in details. Interestingly, the relative percentage of each ion is similar in both the feed and product ice. This suggests that none of these ions preferentially remains in the product ice because salt

TABLE 11.2
The compositions of artificial seawater and real seawater

(a)

Artificial seawater composition (Subow, 1931)

Salt	NaCl	$MgCl_2$	$MgSO_4$	$CaCl_2$	KCl	$NaHCO_3$	NaBr	Total
g/kg	26.518	2.447	3.305	1.141	0.725	0.202	0.083	34.421

(b)

Real seawater composition

Component	Ca	Mg	Na	K	Ba	Sr	CO_3	HCO_3
mg/kg	396.15	1485.45	9895.1	791.35	0.099	7.9225	24.1	107.4
Component	SO_4	Cl	F	NO_3	B	SiO_2	CO_2	TDS
mg/kg	2872.25	19793.5	2	1	4.79	4.95	0	35386.05

TABLE 11.3
Ice compositions, after washing ice made from artificial seawater and real seawater*

(a)
Ice salinity of artificial seawater tests

Component	Na^+	Mg^{2+}	Ca^{2+}	K^+	TDS
ppm	92.96 ± 7.92	11.8 ± 0.99	5.56 ± 0.45	5.6 ± 1.1	315.52 ± 23.86

(b)
Ice salinity of real seawater tests

Component	Na^+	Mg^{2+}	Ca^{2+}	K^+	TDS
ppm	94.57 ± 15.74	12.2 ± 1.94	6.99 ± 0.80	3.58 ± 2.11	321.45 ± 52.74

* Feeds: 1 kg artificial seawater or real seawater; freeze duration: 1 h using an iron cylinder; coolant temperature: −8 °C; seed ice: 1 g at −2.1 °C; optimized washing: chilled tap water = 50 wt% of the raw ice.

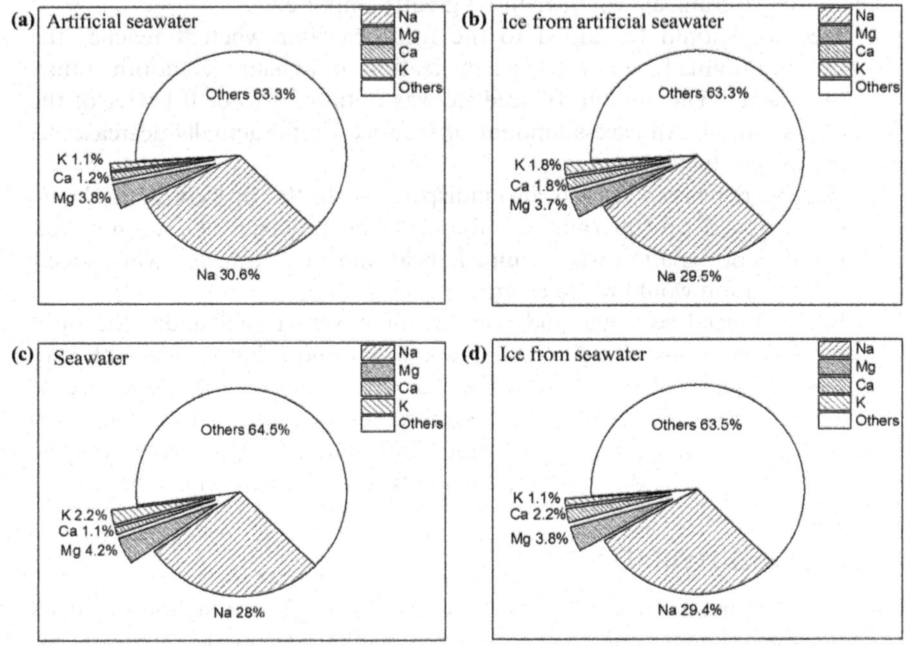

FIGURE 11.15 Compositions of (a) artificial seawater, (b) ice produced from artificial seawater, (c) real seawater, and (d) ice produced from real seawater (Chang et al., 2016). Copyright (2016) Elsevier.

ions in the solution mainly adhere onto the ice surfaces and they are not trapped in the ice lattices. The nontrapping behaviors are important and desirable for freeze desalination of real seawater, especially in the washing process. Therefore, the process parameters optimized provide good guidelines for the further development of the FD process.

11.3.11 Summary of FD

1. The importance of an effective washing procedure outweighs many other operating parameters. Vacuum filtration was found to effectively separate ice from the washing water. The amount of washing water needed was optimized to be 50 wt% of the raw ice.
2. While LNG is the source of cold energy, an intermediate coolant would be used to cool the seawater. The optimal coolant temperature should be around −8 °C. When using a colder coolant, a thick ice layer would likely form on the cylinder wall and thus reduce the purity of the product ice.
3. The optimal freezing duration was found to be 1 h for an iron crystallizer and 1.5 h for a glass crystallizer. With a longer duration, the amount of ice would significantly reduce the solution's fluidity. Heat and mass transfer would then be greatly impaired.
4. Seed ice should be added to the feed seawater when it reaches the freezing point. It would trigger nucleation and ensure a smooth transit to freezing. The amount of seed ice was optimized to be 0.1 wt% of the feed solution. An excess amount of seed ice would actually decrease the net ice production.
5. Homogeneous mixing was also indispensable in this process. A moderate stirrer speed of 100 rpm was found to be practical and economical. Insufficient agitation would impede heat and mass transfer, while excessive agitation would waste energy.
6. Both artificial seawater and real seawater were tested under the optimized conditions. The product ice's amount and salinity were both very close to those obtained from the 3.5 wt% NaCl solution. Raw ice of 520–525 g was formed from a 1 kg feed solution within 1 h, which corresponds to a recovery rate of about 26%. Thus, the process parameters optimized can be directly used for the freeze desalination of seawater.

11.4 FD-MD Hybrid Systems

While FD has many advantages over the conventional desalination processes, it is vulnerable to high feed concentrations (Wankat, 1973). By contrast, membrane distillation is able to treat feed concentrations that are a few times higher than the seawater concentration without much compromise on separation performance (Wang and Chung, 2015). According to the Raoult's law, the vapor pressure only drops 3.6% when the seawater concentration is doubled. Given this benefit, MD is highly suitable for treating the concentrated brine from the

FD step. A drawback of MD is its high energy demand. Nevertheless, it can be operated with the aid of renewable heat resources such as solar energy. It is a clean energy resource, and more importantly, it is able to generate electricity, a high-exergy energy carrier (Müller et al., 2011). Not only can it be used to heat up the feed solution but also power pumps and many other equipment. Additionally, using electricity will make it easier for the MD operators to control the heating duty. Although the direct solar water heating has a higher efficiency than photovoltaic efficiency, it has a drawback by fluctuating the feed with random temperatures (Qtaishat and Banat, 2013; Wang et al., 2017). Another means to generate electricity is to harvest energy from the LNG regasification process. In the study conducted by Dorosz et al., it was found that a direct expansion-organic Rankine cycle hybrid system (DE-ORC) reached the highest exergetic efficiency of 36%, that is, 380 kJ/kg (Dorosz et al., 2018).

Among the four types of MD configurations, vacuum membrane distillation (VMD) has a much lower heat loss as vacuum is applied to the permeate side of the membrane, so it is chosen to be an example of the FD-MD analysis. The membrane used is in the geometry of tri-bore hollow fibers fabricated with the green solvent TEP (Chang et al., 2019). The membrane has an optimum permeation flux of 51.7 kg/m^2·h for a feed temperature of 70 °C. The flowchart for the system is shown in Figure 11.16.

11.4.1 The Source and the Consumption Requirements of Cold Energy

The FD-MD hybrid system requires cold energy and electricity. The cold energy can be provided by the regasification process of LNG. While the minimum work to liquefy the natural gas is in the range of 1500–3500 kJ/kg, the actual physical exergy is about 1040 kJ/kg due to the irreversibility in the liquefaction process (Dorosz et al., 2018). It should be noted that exergy is

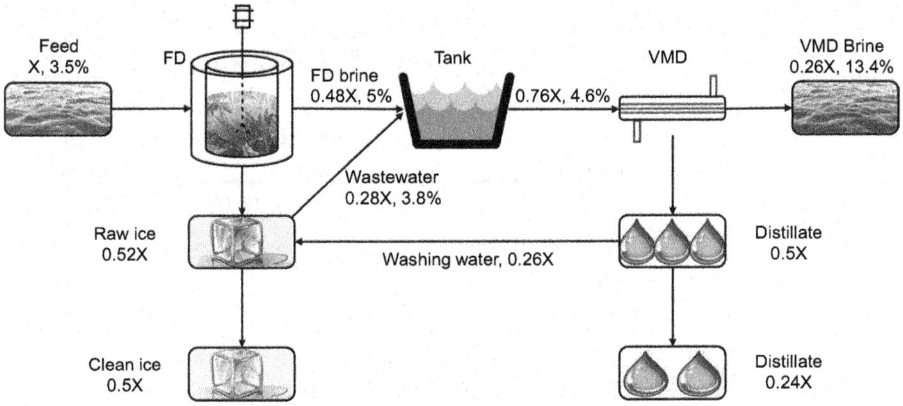

FIGURE 11.16 Flowchart of the FD-VMD hybrid system (Chang et al., 2019). Copyright (2019) Elsevier.

used here instead of energy in order for a fair calculation. Exergy is only the available part of energy, and the unavailable part is called anergy (Bolland, 1991; Schaschke, 2014). In other words, a conservative and realistic evaluation of the readily available amount of cold energy from the regasification process of LNG is 1040 kJ/kg.

Then we calculate the cold energy required in the whole process. According to the flowchart in Figs. 11.16 and 11.17, a 3.5 wt% NaCl solution of X kg is fed to the FD setup. The total amount of cold energy required here includes (1) cooling the feed solution from room temperature to its freezing point and (2) removing the heat of ice formation. Because of freezing-point depression, the feed solution will only start to freeze when the temperature reaches −2.1 °C. In other words, the feed solution needs to be cooled from 25 °C to −2.1 °C before it freezes. Taking the water heat capacity of 4.18 kJ/kg K, the energy consumption for this cooling step is equal to 4.18 × (25 + 2.1)X = 113.31X kJ. The heat release during the ice formation is 333.55 kJ/kg so the energy required to remove the heat of 0.52X kg ice formation is 0.52X × 333.55 = 173.45X kJ. Thus, the total required cold energy in this step equates to (113.31 + 173.45)X = 286.76X kJ.

Besides the FD step, the VMD step also consumes cold energy. The distillate steam, 0.5X kg, needs to be condensed and cooled. Specially, part of the distillate, 0.26X kg, must be cooled from 70 °C to 0 °C so that it can be used to wash the raw ice without melting it rapidly. Since the heat of condensation is 2257 kJ/kg, the total condensation energy for the distillate steam is equal to 0.5X × 2257 = 1128.5X kJ. In addition, the energy to cool part of it, 0.26X kg, from 70 °C to 0 °C is 0.26X × 4.18 × 70 = 76.1X kJ. Combining the energy consumption for both FD and VMD steps, the total cold energy requirement is equal to (286.76 + 1128.5 + 76.1)X = 1491X kJ.

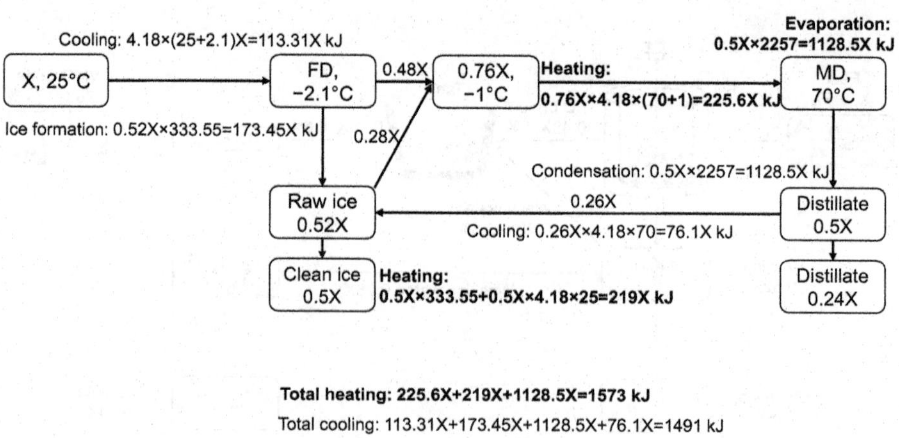

FIGURE 11.17 Flowchart of energy calculations for the FD-VMD hybrid system (Chang et al., 2019). Copyright (2019) Elsevier.

To sum up, when X kg of 3.5 wt% salt solution is passed to the proposed hybrid system, the cold energy needed is 1491X kJ and it can be supplied by the regasification of 1.43X kg LNG (i.e., 1.43 = 1491/1040).

11.4.2 The Source and the Consumption Requirements of Electrical Energy

Similar to a normal distillation process, VMD requires a large amount of thermal energy. Although the thermal energy can be harvested from low-exergy energy carriers like fossil fuels, it is also viable and actually advantageous to use one of the high-exergy energy carriers, that is, electricity. It can be used for versatile purposes including heating, circulation, and generating vacuum. The sources of electricity considered in this FD-VMD system are solar panels and the regasification process of LNG. Figure 11.18 presents the solar panels on a regular day in Singapore.

Figure 11.19 presents the monthly electricity generation results of the solar panels (18.9 m^2) in the year 2017. The average value was 327 kW h. Therefore, the hourly generation was 85.4 kJ/m^2. This value was calculated based on the energy literally generated in the daytime but normalized across 24 hours. In addition, since the data were collected throughout an entire year, all the possible

FIGURE 11.18 (a) Solar panels on rooftop. (b) SUNNY BOY solar inverter. (c) Preventive maintenance (Chang et al., 2019). Copyright (2019) Elsevier.

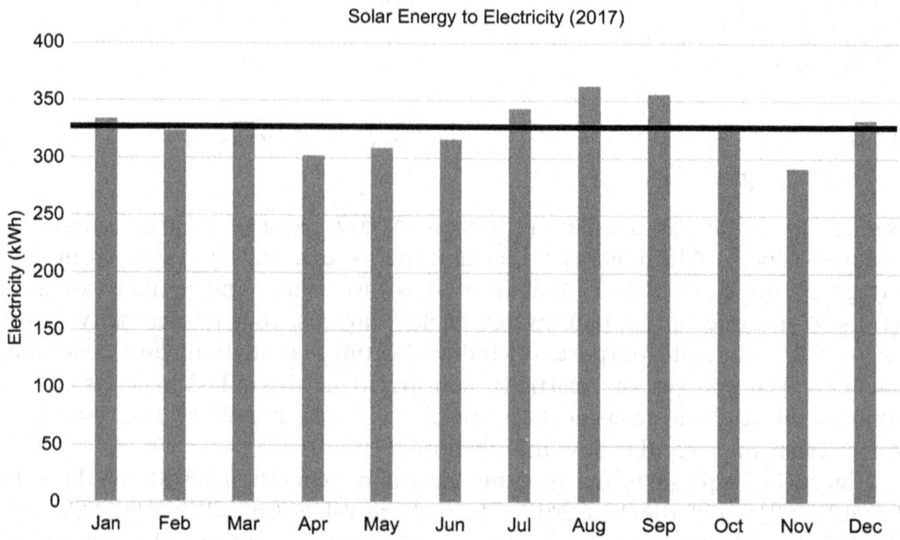

FIGURE 11.19 Monthly power generation of the solar panels in 2017 (National University of Singapore, Singapore) (Chang et al., 2019). Copyright (2019) Elsevier.

inclement weather conditions had been included such as rain and overcast. Since the current solar system has a solar inverter that fills the generated electricity into the existing power grid instantaneously. Thus, there is no need for a battery to store the energy. It also means the average value 85.4 kJ/m^2 can be considered being generated every single hour throughout the year regardless of weather conditions. Therefore, this value is used.

Alternatively, electricity can also be harvested from the LNG regasification process. Dorosz et al. calculated in detail the power generation rate of four common systems and found out the DE-ORC hybrid system was the most efficient. It is able to generate 380 kJ/kg, yielding an efficiency value of 36.2% (Dorosz et al., 2018). However, using the regasification of LNG for electricity generation should not be confused with using it for cold energy exploitation in Section 11.3.5. Physical exergy has two components as thermal and pressure components (Kanbur et al., 2017). For example, when the cold energy is drawn on, the thermal component is exploited; when LNG is used in direct expansion through a turbine, its pressure component is made use of instead. Therefore, one can harvest either cold energy of 1040 kJ/kg or electricity of 380 kJ/kg from the LNG regasification process.

Finally, we calculate the electrical energy required in the whole process, namely, heating, circulation, and generating vacuum. Following the flowchart in Figure 11.16, 0.76X kg of mixed streams of saline water needs to be heated to 70 °C and vaporized in the VMD step. The temperature in the tank is estimated to be −1 °C. Therefore, the heating energy is equal to 0.76X × 4.18 ×

(70 + 1) = 225.6X kJ. In addition, the energy required to melt the clean ice and heat it up to 25 °C is 0.5X × 333.55 + 0.5X × 4.18 × 25 = 219X kJ. Since out of the 0.76X kg, 0.5X kg vaporizes and it consumes latent heat of 0.5X × 2257 = 1128.5X kJ, the total heating energy required in the FD-VMD system is 1573X kJ (i.e., 1573 = 225.6 + 219 + 1128.5).

Besides the heating energy, the system also consumes electricity for circulation pumps and generating vacuum. However, various literatures have suggested these energy requirements are negligible compared to the heating energy. The estimated values are below 1.2 kWh/m^3 (i.e., 4.32X kJ) for circulation pumps and below 1.3 kWh/m^3 (i.e., 4.68X kJ) for vacuum pumps (Cabassud and Wirth, 2003; Lee and Kim, 2013, 2014). Compared to the total heating energy of 1573X kJ, the sum of these two is less than 1% (i.e., (4.32 + 4.68)/1573 = 0.57%)). Therefore, it is safe to use the values of the total heating energy to represent the total electricity consumption.

Dividing the total energy consumption by either solar energy supply rate (85.4 kJ/m^2) or LNG regasification supply rate (380 kJ/kg), one can yield the following results. If solar panels are chosen to be the source of electrical energy, the area required is 1573X/85.4 = 18.4X m^2; if the LNG regasification process is used instead, the mass of the required LNG is 1573X/380 = 4.14X kg. Particularly, in the current lab-scale capacity (18.9 m^2), 1.03 kg saline water (3.5 wt%) can be treated by the hybrid system hourly, it translates to hourly and daily produce 0.76 kg and 18 kg potable water, respectively, with a water recovery of 74%.

Although solely relying on solar panels to power up large-scale operations might require a large footprint, it may be more feasible to employ the LNG method if it is available. Alternatively, solar panels can be accompanied with the LNG method or another means of energy sources.

11.4.3 Heat Integration

Some of the heating and cooling duty in the system can be integrated so that the total energy consumption is further reduced, as shown in Figure 11.20. In particular, the vaporized steam of VMD is of high temperature, so it can be used to heat up the VMD feed tank and melt the clean ice. To make sure the heat exchanging is sufficiently efficient, the minimum temperature difference for heating VMD feed can be assumed to be 15 °C (Towler and Sinnott, 2008; Varbanov et al., 2012). Thus, the VMD feed stream will be heated to 55 °C at no cost. Under this condition, the partial heating duty is decreased from 225.6X kJ to 47.7X kJ (= 0.76X × 4.18 × 15) because we only need to heat it up from 55 °C to 70 °C. Furthermore, the clean ice can also be melted and heated up for free. Meanwhile, the partial cooling duty is decreased from 1128.5X kJ to 731.5X kJ because of a saving on heating energy of 396.9 X kJ (= 225.6X − 47.7X + 219X) due to the heat exchange integration.

As a result, the total cooling duty becomes 1491X − 396.9X = 1094X kJ that can be supplied by the regasification of 1.05X kg LNG, while the total

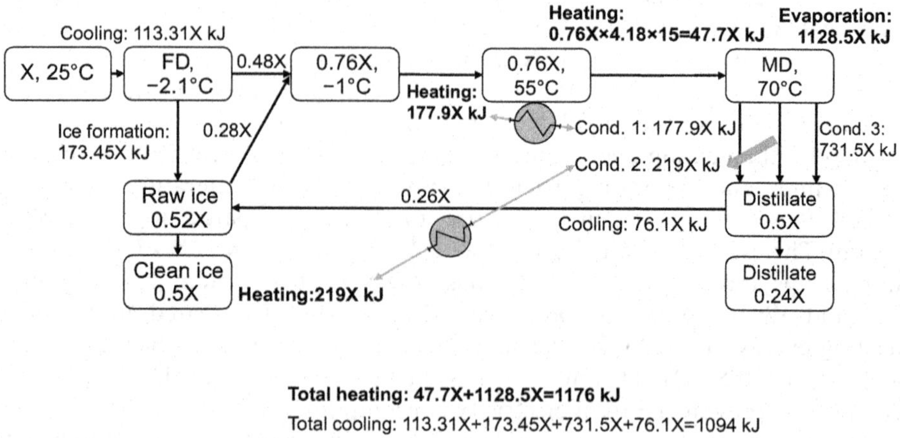

FIGURE 11.20 Flowchart of energy calculations for the FD-VMD hybrid system with heat integration (Chang et al., 2019). Copyright (2019) Elsevier.

heating duty drops to 1573X − 396.9X = 1176X kJ which can be supplied by solar panels of 13.8X m^2 or the regasification of 3.10X kg LNG.

11.5 CONCLUSION

By combining FD and MD processes and adopting the optimized operation parameters, the hybrid FD-MD system has been successfully demonstrated. FD, being the first step, is used to harvest clean ice by freezing the simulated seawater. To ensure the purity of the product ice, the raw ice is washed by about one-half of the VMD distillate. To treat 1 kg seawater by FD, the cold energy could be provided by the regasification process of 1.26 kg LNG. The concentrated FD brine is further treated by VMD to increase the total water recovery. For each 1 kg of seawater fed to the system, the total water recovery is 0.74 kg per hour. The heating duty could be covered by either solar panels of 13.8 m^2 or the regasification of 3.10 kg LNG if the system is fully integrated.

NOMENCLATURE

K_{FP} cryoscopic constant
L_f latent heat of fusion
mo molality (mol solute/1000 g solvent)
M_0 mass of feed solution
M_i mass of raw ice
M_{pi} mass of product ice
M_s mass of seed ice
r^* critical radius
T_f freezing temperature

ΔT temperature difference departed from freezing point
ΔT_f freezing point depression

GREEK SYMBOLS

σ interfacial tension between water and ice
η percentage of washing water's mass relative to raw ice
φ ice recovery rate

LIST OF ABBREVIATION

WHO World Health Organization

REFERENCES

Badawy, S.M., 2015. Laboratory freezing desalination of seawater. Desalination Water Treat. 57, 1–8.

Bolland, O., 1991. A comparative evaluation of advanced combined cycle alternatives. J. Eng. Gas Turbines Power 113 (2), 190–197.

Cabassud, C. and Wirth, D., 2003. Membrane distillation for water desalination: How to chose an appropriate membrane? Desalination 157, 307–314.

Cao, W., Beggs, C. and Mujtaba, I.M., 2015. Theoretical approach of freeze seawater desalination on flake ice maker utilizing LNG cold energy. Desalination 355 (0), 22–32.

Chang, J., Zuo, J., Lu, K.J. and Chung, T.S., 2016. Freeze desalination of seawater using LNG cold energy. Water Research. 102, 282–293.

Chang, J., Zuo, J., Lu, K.-J. and Chung, T.S., 2019. Membrane development and energy analysis of freeze desalination-vacuum membrane distillation hybrid systems powered by LNG regasification and solar energy. Desalination 449, 16–25.

Dorosz, P., Wojcieszak, P. and Malecha, Z., 2018. Exergetic analysis, optimization and comparison of lng cold exergy recovery systems for transportation. Entropy 20, 59.

Efrat, T., 2011. Utilizing available "coldness" from Liquefied Natural Gas (LNG) regasification process for seawater desalination, IDA World Congress.

Fujioka, R., Wang, L.P., Dodbiba, G. and Fujita, T., 2013. Application of progressive freeze-concentration for desalination. Desalination 319 (0), 33–37.

Gay, G., Lorain, O., Azouni, A. and Aurelle, Y., 2003. Wastewater treatment by radial freezing with stirring effects. Water Res. 37 (10), 2520–2524.

Gooch, J.W., 2011. Encyclopedic Dictionary of Polymers. Gooch, J.W. (ed), p. 326, Springer New York, New York, NY.

Halde, R., 1980. Concentration of impurities by progressive freezing. Water Res. 14 (6), 575–580.

Harriott, P., Leinroth, Jr., J.P., Von Berg, R.L. and Wiegandt, H.F., 1969. Comments on "an analysis of counterwashers for freeze-distillation desalination" by Shwartz and Probstein. Desalination 6 (1), 117–119.

Holten, V., Bertrand, C.E., Anisimov, M.A. and Sengers, J.V., 2012. Thermodynamics of supercooled water. J. Chem. Phys. 136 (9), 094507.

Johnson, W.E., 1976. State-of-the-art of freezing processes, their potential and future. Desalination. 19 (1–3), 349–358.

Khayet, M., Matsuura, T., 2011. Membrane Distillation: Principles and Applications.

Elsevier, Amsterdam.

Kanbur, B.B., Xiang, L., Dubey, S., Choo, F.H., Duan, F., 2017. Cold utilization systems of LNG: A review. Renew. Sustainable Energy Rev. 79, 1171–1188.

Kobayashi, A., Shirai, Y., Nakanishi, K. and Matsuno, R., 1996. A method for making large agglomerated ice crystals for freeze concentration. J. Food Eng. 27 (1), 1–15.

Lee, J.-G. and Kim, W.-S., 2013. Numerical modeling of the vacuum membrane distillation process. Desalination 331, 46–55.

Lee, J.-G. and Kim, W.-S., 2014. Numerical study on multi-stage vacuum membrane distillation with economic evaluation. Desalination 339, 54–67.

Lorain, O., Thiebaud, P., Badorc, E. and Aurelle, Y., 2001. Potential of freezing in wastewater treatment: soluble pollutant applications. Water Res. 35 (2), 541–547.

Lu, Z. and Xu, L., 2010. Freezing desalination process. Encyclopedia of Desalination and Water Resources Thermal Desalination Processes.http://www.desware.net/sample-chapters/d04/d08-063.pdf

Mahdavi, M., Mahvi, A.H., Nasseri, S. and Yunesian, M., 2011. Application of freezing to the desalination of saline water. Arab. J. Sci. Eng. 36 (7), 1171–1177.

Mandri, Y., Rich, A., Mangin, D., Abderafi, S., Bebon, C., Semlali, N., Klein, J.P., Bounahmidi, T. and Bouhaouss, A., 2011. Parametric study of the sweating step in the seawater desalination process by indirect freezing. Desalination 269 (1–3), 142–147.

Manwell, J.F. and McGowan, J.G., 1994. Recent renewable energy driven desalination system research and development in North America. Desalination 94 (3), 229–241.

Miyazaki, T., Kang, Y.T., Akisawa, A. and Kashiwagi, T., 2000. A combined power cycle using refuse incineration and LNG cold energy. Energy 25 (7), 639–655.

Muldrew, K. and McGann, L.E., 1999. Physical aspects of freezing, University of Alberta, cryobiology – A short course. Available from: http://people.ucalgary.ca/~kmuldrew/cryo_course/course_outline.html, (accessed 16.04.23).

Müller, A., Kranzl, L., Tuominen, P., Boelman, E., Molinari, M. and Entrop, A.G., 2011. Estimating exergy prices for energy carriers in heating systems: Country analyses of exergy substitution with capital expenditures. Energy Build. 43, 3609–3617.

Mullin, J.W., 2001. Crystallization (Fourth Edition). Mullin, J.W. (ed), pp. 216–288, Butterworth-Heinemann, Oxford.

Myerson, A.S. and Ginde, R., 2002. Handbook of Industrial Crystallization (Second Edition), Butterworth-Heinemann, Woburn.

Nebbia, G. and Menozzi, G.N., 1968. Early experiments on water desalination by freezing. Desalination 5 (1), 49–54.

Qtaishat, M.R. and Banat, F., 2013. Desalination by solar powered membrane distillation systems. Desalination 308, 186–197.

Rahman, M.S., Ahmed, M. and Chen, X.D., 2006. Freezing-melting process and desalination: I. Review of the state-of-the-art. Sep. Purif. Rev. 35 (2), 59–96.

Randall, D.G. and Lewis, A.E., 2009. Seeding for selective salt recovery during Eutectic Freeze Crystallization. International Mine Water Conference, 19–22 October, Pretoria, South Africa, 639–646.

Randall, D.G. and Nathoo, J., 2015. A succinct review of the treatment of Reverse Osmosis brines using Freeze Crystallization. J. Water Process Eng. 8, 186–194.

Randall, D.G., Nathoo, J., Genceli-Güner, F.E., Kramer, H.J.M., Witkamp, G.J. and Lewis, A.E., 2012. Determination of the metastable ice zone for a sodium sulphate system. Chem. Eng. Sci. 77, 184–188.

Reddy, S.T., Kramer, H.J.M., Lewis, A.E. and Nathoo, J., 2009. Investigating factors that affect separation in a eutectic freeze crystallisation process. Abstracts of the International Mine Water Conference, Document Transformation Technologies, 649–655.

Rich, A., Mandri, Y., Bendaoud, N., Mangin, D., Abderafi, S., Bebon, C., Semlali, N., Klein, J.P., Bounahmidi, T., Bouhaouss, A. and Vessler, S., 2010. Freezing desalination of sea water in a static layer crystallizer. Desalination Water Treat. 13 (1–3), 120–127.

Rich, A., Mandri, Y., Mangin, D., Rivoire, A., Abderafi, S., Bebon, C., Semlali, N., Klein, J.-P., Bounahmidi, T., Bouhaouss, A. and Veesler, S., 2012. Sea water desalination by dynamic layer melt crystallization: Parametric study of the freezing and sweating steps. J. Crystal Growth 342 (1), 110–116.

Schaschke, C., 2014. A Dictionary of Chemical Engineering, Oxford University. Oxford, United Kingdom.

Sherwood, T.K. and Brian, P.L.T., 1966. Research on Saline Water Conversion by Freezing, pp. 29–43, U.S. Govt. Print. Off., Washington.

Shirai, Y., Wakisaka, M., Miyawaki, O. and Sakashita, S., 1999. Effect of seed ice on formation of tube ice with high purity for a freeze wastewater treatment system with a bubble-flow circulator. Water Res. 33 (5), 1325–1329.

Shwartz, J. and Probstein, R.F., 1968. An analysis of counterwashers for freeze-distillation desalination. Desalination 4 (1), 5–29.

Subow, N. N., 1931. Oceanographical Tables. U.S.S.R., p. 208, Oceanogr. Institute, Hydro-meteorol. Com., Moscow.

Tiller, W.A., 1992. The Science of Crystallization: Macroscopic Phenomena and Defect Generation, Cambridge University Press.

Tleimat, B.W., 1980. Principles of Desalination (Second Edition). Laird, K.S.S.D.K. (ed), pp. 359–400, Academic Press.

Towler, G. and Sinnott, R., 2008. Chemical Engineering Design, Principles, Practice and Economics of Plant and Process Design, Elsevier, Amsterdam, The Netherlands.

Tritt, T.M., 2004. Thermal Conductivity: Theory, Properties, and Applications, Springer, New York.

Varbanov, P. S., Fodor, Z. and Klemeš, J. J., 2012. Total Site targeting with process specific minimum temperature difference (ΔT_{min}). Energy 44 (1), 20–28.

Wang, F., Wang, S., Li, J., Xia, D. and Liu, J., 2017. Seawater desalination with solar-energy-integrated vacuum membrane distillation system, J. Water Reuse Desalin. 7, 16–24.

Wang, P. and Chung, T.S., 2012. A conceptual demonstration of freeze desalination-membrane distillation (FD-MD) hybrid desalination process utilizing liquefied natural gas (LNG) cold energy. Water Res. 46 (13), 4037–4052.

Wang, P. and Chung, T.S., 2015. Recent advances in membrane distillation processes: Membrane development, configuration design and application exploring. J. Membr. Sci. 474, 39–56.

Wankat, P.C., 1973. Desalination by natural freezing. Desalination. 13, 147–157.

Williams, P.M., Ahmad, M., Connolly, B.S. and Oatley-Radcliffe, D.L., 2015. Technology for freeze concentration in the desalination industry. Desalination. 356 (0), 314–327.

Zhao, S., Zou, L., Tang, C.Y. and Mulcahy, D., 2012. Recent developments in forward osmosis: Opportunities and challenges. J. Membr. Sci. 396, 1–21.

Zvinowanda, C.M., Zikalala, N.N., Mtombeni, T., Maree, J.P., Louw, W.J. and Oosthuizen, F., 2014. Process optimization of freeze desalination of brine using HybridICE(TM) pilot plant. Desalination Water Treat. 52 (34–36), 6315–6324.

12 Pressure Retarded Osmosis-Membrane Distillation

Zhen Lei Cheng and Gang Han

12.1 INTRODUCTION

Fueled by exponential growth in world population, water supply and energy consumption, global water and energy demands have risen rapidly in tandem with exacerbated environmental issues. These problems necessitate the exploration of alternative freshwater and renewable energy resources. The integration of membrane distillation (MD) and pressure retarded osmosis (PRO) processes may provide compelling answers to the global challenges of water and energy scarcity (Ghaffour et al., 2014; Loeb and Norman, 1975; Logan and Elimelech, 2012).

Leveraging on the principle of naturally induced osmosis, PRO facilitates the conversion and capture of the free energy of mixing between two solutions with different salinities into useful work (Gerstandt et al., 2008; Han et al., 2015a; Lee et al., 1981; Loeb and Norman, 1975). In this process, water transports from a low-salinity feed solution across a semipermeable membrane to a high-salinity draw solution against a hydraulic pressure. The increased volume of the diluted draw solution is a form of mechanical energy, which can be converted to electricity through hydro turbines or recovered via other energy recovery devices (Gerstandt et al., 2008; Lee et al., 1981; Loeb and Norman, 1975; Ramon et al., 2011). The estimated global osmotic energy that can be generated from the mixing of ocean and fresh river water is approximately 1750–2000 TWh per year (Thorsen and Holt, 2009). Additional osmotic energy is projected when using high-salinity water sources that provide greater salinity gradients (Achilli et al., 2014; Chung et al., 2012; Kim et al., 2012). In addition, PRO is a theoretically carbon-free technology as there is no CO_2 emission during the energy production. However, the relatively high costs of pretreatment and pumping, membrane fouling and scaling, low extractable osmotic energy due to the small salinity gradient, and potential environmental impacts of discharging the brackish water may render PRO difficult to generate a sizable net power via conventional mixing of seawater and river water (Efraty, 2013; Lin et al., 2014a; Sikdar, 2014; Wan and Chung, 2015).

On the other hand, MD is a thermally driven membrane process where the heated feed stream evaporates and transports across the membrane, and eventually gets condensed on the cold permeate side (Curcio and Drioli, 2005; Edwie and Chung, 2013; Francis et al., 2014; Gryta et al., 2001). Compared to reverse osmosis (RO) for desalination, MD is less sensitive to the feed salt concentration or the osmotic pressure of the feed. Thus, it can treat feed streams with higher total dissolved solids (TDS). In addition, the modest operating temperature used allows MD to utilize sustainable or alternative energy resources such as solar energy, geothermal energy, or waste heat. The amount of worldwide rejected waste heat is about 9400 TWh per year with most of the waste heat in streams below 150 °C or dissipated as radiation, which could be considered as a potential energy supply to MD (Lin et al., 2014b). Nonetheless, the use of MD for direct seawater desalination is still challenging because of its relatively high energy intensity, membrane fouling and wetting (Curcio and Drioli, 2005; Edwie and Chung, 2013; Francis et al., 2014; Gryta et al., 2001).

By integrating the PRO and MD processes, the proposed advanced PRO-MD system enables the sustainable generation of osmotic power and clean water (Han et al., 2015b; Lee et al., 2015; Lin et al., 2014a). Such a hybrid system may not only avoid the shortcomings of each individual process but also strength their advantages. Compared to a typical MD process, the net energy consumption for desalination is lowered by the power generated from PRO, while the circulated draw solution is almost foulant-free due to the design of integrated-loop process configuration or the effort of pretreatment already taken in upstream processes which reduces the membrane replenishment costs. From the perspective of PRO, the draw solution is an abundantly available and low-cost residual from the MD unit but with high purity and salinity. Therefore, ultrahigh osmotic energy is achievable and the PRO membrane fouling on the draw solution side could be well controlled. Since the PRO unit is able to achieve a high water recovery rate through forward osmosis, the feed solution stream could be effectively concentrated and its discharge volume would be much reduced. In addition, the system integration could mitigate the disposal problems of the MD concentrated brine and the resultant brackish water from the normal PRO. Furthermore, the elimination of additional chemical usage in pretreatment and possible reuse of impaired wastewater make this integrated process more environmentally friendly. Nevertheless, one possible drawback of the PRO-MD hybrid process might be the relatively high energy consumption in the MD unit unless waste heat and other renewable energy such as solar power are used. Draw solutes from the circulated draw solution may get lost in the PRO unit due to the reverse salt diffusion, which would reduce the osmotic driving force and the corresponding water flux. However, this problem can be minimized by the strategic process operation (e.g., occasional purging/replenishment to adjust the stream concentration at a designated level) and the employment of effective PRO membranes with excellent salt rejections.

In summary, the PRO-MD hybrid process possesses the following merits: (1) lower operation costs for desalination, (2) higher water recovery, (3) less membrane fouling, and (4) minimal negative environmental impacts. Based on different process configurations, there are two types of hybrids generally classified in the literature, namely, close-loop and open-loop, and their details will be elaborated in the following sections.

12.2 PROCESS CONFIGURATIONS

12.2.1 Close-Loop Process

The close-loop PRO-MD hybrid process, also known as osmotic heat engine (OHE), is a promising technology that can produce electricity from waste heat sources (Logan and Elimelech, 2012). This process comprises a power generation stage (i.e., PRO) that converts the energy released from the controlled mixing of two solutions with different concentrations into useful work and a thermal separation stage (i.e., MD) where waste heat is used to regenerate the low and high concentration solutions, which are then recirculated back to the energy production stage (i.e., PRO) (McGinnis et al., 2007). A representative study was carried out by Lin et al. (Lin et al., 2014b) and a simplified schematic of their proposed PRO-MD hybrid system is shown in Figure 12.1, where HX and PX represent heat exchanger and pressure exchanger that are used to prompt the thermal/mechanical efficiency of respective operating stage, respectively. A hydro-turbine (TB) in parallel with PX is used to convert the energy of mixing to useful work. It is noted that the additional heat source (i.e., the block "H") and the heat sink (i.e., the block "C") are associated with MD to reflect the working principle of

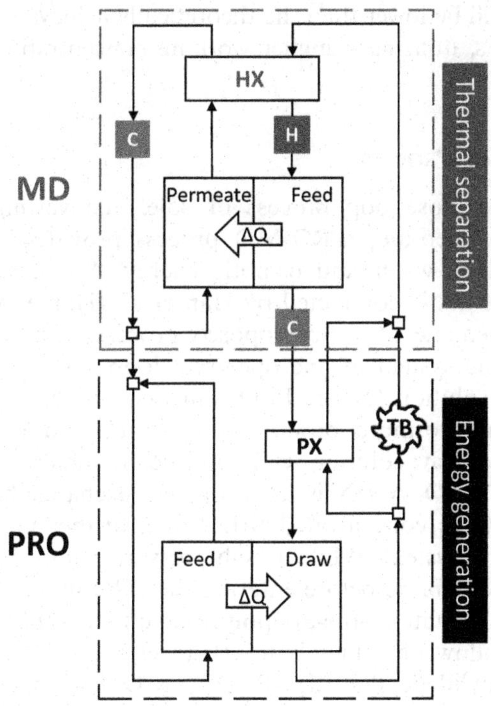

FIGURE 12.1 Schematic diagram of a PRO-MD close-loop hybrid system (Lin et al., 2014b). Copyright (2014) American Chemical Society.

a generic heat engine, which absorbs heat from the heat source, converts part of the absorbed heat to useful work through the turbine and releases the remaining heat to the heat sinks.

To achieve a more holistic assessment of the viability of this hybrid process, Lin et al. conducted a systematic analysis by performing a module-scale modeling to couple the heat and mass transfer between the MD and PRO stages for waste heat source ranging from 40 to 80 °C and working concentrations of 1, 2, and 4 M NaCl (Lin et al., 2014b). Subsequently, the potential of the process is assessed to identify the major factors controlling the system performance and to examine the thermodynamic limit of the energy conversion efficiency, under both limited and unlimited mass and heat transfer kinetics. In both cases, the relative flow rate between the MD permeate and feed streams is identified as an important operation parameter and there is an optimal relative flow rate that maximizes the overall energy efficiency of the PRO-MD process for given operating temperatures and concentrations. In the case of unlimited mass and heat transfer kinetics, the energy efficiency of the system can be analytically determined based on thermodynamics. Their study indicates that the hybrid PRO-MD system can theoretically achieve an energy efficiency of 9.8% (81.6% of the Carnot efficiency) with hot and cold operating temperatures of 60 and 20 °C, respectively, and a working solution of 1.0 M NaCl. When mass and heat transfer kinetics are limited, conditions that more closely represent actual operations, the practical energy efficiency will be lower than the theoretically achievable efficiency. In such practical operations, utilizing a higher working concentration will yield greater energy efficiency.

12.2.2 Open-Loop Process

Different from the close-loop process of solely converting waste heat into useful energy, the open-loop PRO-MD process provides a dual function of producing both clean water and osmotic energy. A conceptual experimental demonstration was firstly conducted by Han et al. (Han et al., 2015b) and the corresponding schematic of their proposed process is shown in Figure 12.2. They configured the system of one open-end loop using freshwater or wastewater as the feed solution for the PRO unit and one closed loop including a PRO unit to generate osmotic power through a TB and a MD unit to reconcentrate the diluted draw solution while produce potable water and then circulate it back to the PRO. A HX is used for heat exchange between the diluted draw solution and the concentrated MD brine, further increasing the energy saving of the hybrid system. While a high-pressure pump (HP) is to pressurize the concentrated MD brine before entering the PRO unit.

By employing the state-of-the-art polyethersulfone (PES) thin-film composite (TFC) PRO hollow fiber membrane (Han et al., 2016; Zhang and Chung, 2013) and polyvinylidene fluoride (PVDF) MD flat sheet membrane (Han et al., 2015b) and using synthetic freshwater and real wastewater as feeds, the technical feasibility of the PRO-MD hybrid system and the key parameters governing the system operation have been systematically investigated. More

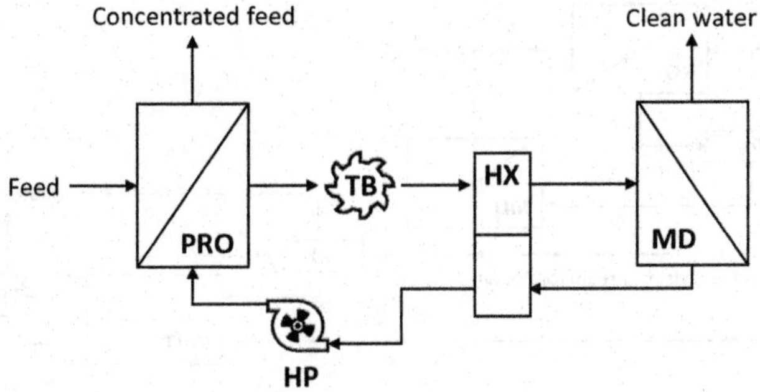

FIGURE 12.2 Schematic diagram of a PRO-MD open-loop hybrid system (Han et al., 2015b). Copyright (2015) RSC Publications.

specifically, the main operation variables are (1) the salinity and quality of the feed water and draw solution, (2) the power density of the PRO membrane and its salt rejection, (3) water recovery rate of the feed and the dilution effects of the draw solution (or the degree of mixing) in the PRO unit, (4) water flux and salt rejection in the MD unit, and (5) the fouling and scaling behaviors of each unit. These parameters are correlated with one another and could be optimized over a wide operational range to enhance the effectiveness of the hybrid system. As a result, it is found that when employing a 2 M NaCl MD concentrate as the draw solution, ultrahigh power densities of 31 W/m^2 and 9.3 W/m^2 can be achieved by the PRO unit using deionized water and real wastewater as feeds, respectively. Simultaneously, high-purity (i.e., ~100% rejection) potable water with a flux of 32.5–63.1 $L/(m^2\ h)$ can be produced by the MD unit at 40–60 °C almost without any detrimental effects of fouling. In case of using wastewater as the feed, the water flux decreases fast with increasing the percentage of water recovery because of membrane fouling. As a result, power density drops from 7.8 W/m^2 to 4.5 W/m^2 and further to 2.7 W/m^2 when the water recovery is increased to 10% and 70%, respectively. These results suggest that the water recovery cannot be too high in order to achieve the power density of 5 W/m^2, set by Statkraft as the commercially viable benchmark (Thorsen and Holt, 2009). Thus, regular cleaning is required to maintain the productivity of the PRO membrane.

Another process configuration of the open-loop hybrid system is to link the feed inlet of MD with the discharged brine from RO. In this case, MD produces more clean water by achieving a greater water recovery rate than the stand-alone RO, while its highly concentrated brine could be used as a draw solution in the subsequent PRO unit. The resultant RO-MD-PRO hybrid system has been adopted in a Korean R&D project entitled "Global MVP" (M for MD, V for valuable resource recovery, and P for PRO), which was launched in 2013 (Kim et al., 2016; Lee et al., 2019). The corresponding process diagram is schematized in Figure 12.3, where the corresponding components of HX and HP are not included for simplicity. The division stream from the RO brine exiting

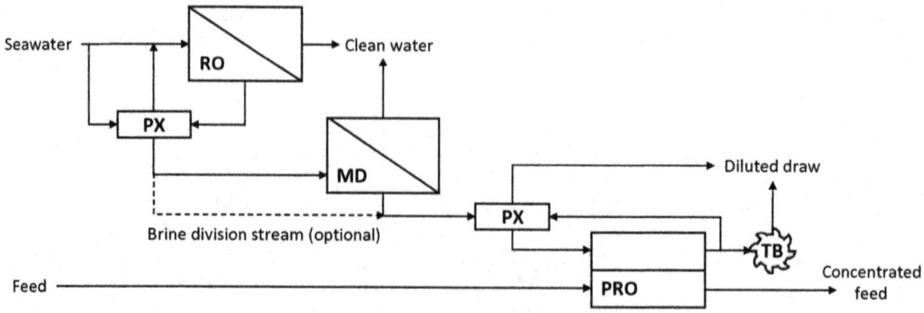

FIGURE 12.3 Schematic diagram of a RO-MD-PRO open-loop hybrid system (Kim et al., 2016). Copyright (2016) Elsevier.

from its PX, shown as the dashed line, is optional based on the operating conditions.

Lee et al. pioneered a simulation study of the hybrid system only integrating a multistage vacuum membrane distillation (MVMD) and PRO without considering the upstream RO unit (Lee et al., 2015). The MVMD unit employs a recycling flow scheme for the continuous production of both distillate water and highly concentrated brine. By theoretically assessing the distillate and power production of the MVMD-PRO system with respect to the inlet feed flow rate and recycling flow ratio in MVMD, it is found that the production of distillate water slightly increases with a decrease in the recycling flow when the inlet feed flow rate is constant. The maximum possible brine concentration from the MVMD unit is 1.9 M NaCl at an inlet feed flow rate of 3 kg/min and a 90% recycling flow. Under this condition, a peak power density of 9.7 W/m^2 is achieved in the PRO unit at 13 bar using river water as the feed solution with flow rates of 0.5 kg/min for both feed and draw streams.

A more comprehensive study including both the upstream RO unit and the downstream MD-PRO integration was later introduced by Kim's group (Kim et al., 2016). They analyzed the commercial feasibility of such a RO-MD-PRO hybrid configuration by evaluating the performance indicators, such as power generation and specific energy consumption (SEC), and the design and operating factors, such as the RO discharged brine division ratio (BDR), the plant dimension ratio of (MD + PRO) to RO as well as the cost of the MD heating source. Overall, the RO-MD-PRO hybrid system outperforms a stand-alone RO in terms of reducing SEC and alleviating the disposal and environmental impacts of the concentrated RO brine. An increase in BDR has positive effects on the efficiency of the hybrid system by raising the water production of MD and the energy generation of PRO. At lower BDR values, the SEC is greatly influenced by the PRO plant dimension. Nonetheless, both the MD and PRO plant dimensions almost equally contribute to the efficiency of the hybrid system with the increase of BDR. Last but not least, the cost of the MD heating source acts a dominant role in determining the overall process efficiency based on their analysis. If the required heat supply cannot be reimbursed, the hybrid system will be unfavorable at any BDR value. Thus, it is crucial to set the RO-MD-PRO hybrid system in locations where free or low-cost thermal energy is easily accessible. Similar findings have also been acquired in another study (Choi et al., 2017).

Few years later, the same group revised the previously proposed RO-MD-PRO hybrid configuration (i.e., Figure 12.3) by setting BDR equal to 1 and replacing the TB with a PX in order to transfer the pressure to the seawater intake stream in favor of greater mechanical efficiency (Chae et al., 2018). To provide a holistic understanding of the hybrid process, they proposed a new dimensionless performance index ($N_{r/c}$) to assess the energy efficiency of RO-PRO (i.e., RO-MD-PRO omitting the MD unit) and RO-MD-PRO under different water recovery rates of RO/MD and dilutive factors of PRO. This performance index ($N_{r/c}$) physically quantifies the ratio of the total energy generated by PRO to the total energy consumed by RO/MD. It is found that the typical water recovery rate of 0.4–0.6 for RO is not effective in terms of the energy efficiency of hybrid processes ($N_{r/c} < 0.3$). Both higher water recovery rate of RO/MD and larger dilutive factor of PRO, raised by a corresponding more concentrated inlet brine, favor $N_{r/c}$. However, this may impose a challenge in membrane design and process operation as a high water recovery easily causes fouling/scaling of RO/MD membranes, and subsequently more robust PRO membrane modules with a less pressure drop are needed to handle the brine. Regular cleaning/maintenance and special precautions have to be paid when operating the hybrid system at such harsh conditions. It also should be noted that a higher water flux across the PRO membrane does not always guarantee a higher energy efficiency when it comes to PRO-hybridized processes, as opposed to the stand-alone PRO. Lastly, the simulation results once more confirm that RO-MD-PRO generally has a higher energy efficiency than RO-PRO if an inexpensive heating source, such as waste heat, is used. The heating cost of MD is found to have a great negative impact on $N_{r/c}$ under the low water recovery rate range (<0.6), while it is somewhat insensitive to $N_{r/c}$ under the high water recovery rate range (>0.7).

12.3 MEMBRANE DESIGN AND OPERATION

For the hybrid process, the membrane design and operation of PRO and MD units generally follows the guideline in their respective applications. In terms of the MD operation, the aforementioned close-loop process (i.e., OHE) adopts the configuration of direct contact MD (DCMD). While other operation configurations such as vacuum MD (VMD), air gap MD (AGMD), and sweeping gas MD (SGMD) are also applicable to the open-loop process. For readers' interest, Chapter 2 provides a comprehensive overview as well as a detailed comparison among these four MD processes.

The representative membrane morphologies (Figure 12.4) and properties (Table 12.1) can be found in the study by Han et al. (Han et al., 2015b), where they employed the PES-TFC hollow fiber membrane for PRO and the PVDF flat sheet membrane for MD. Figure 12.4(a)–(c) shows the PRO

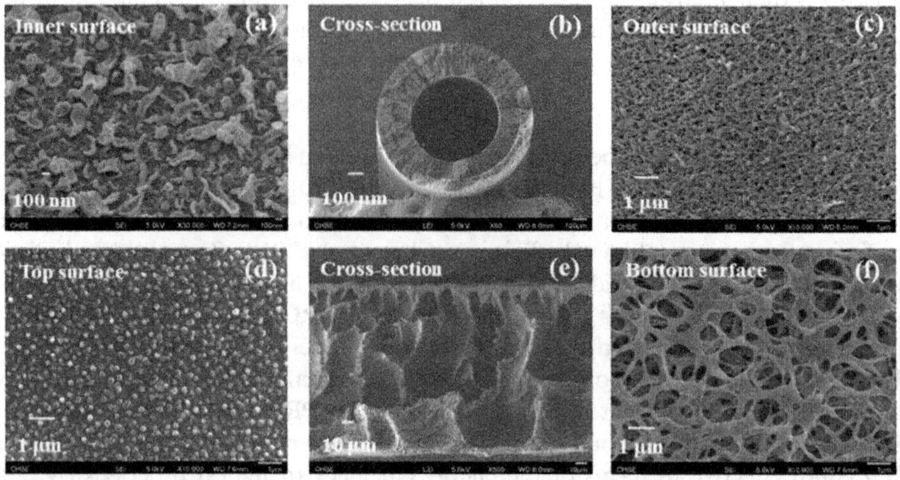

FIGURE 12.4 The surface and cross-section morphology of the PES-TFC PRO hollow fiber membrane (a–c), and the PVDF MD flat sheet membrane (d–f) (Han et al., 2015b). Copyright (2015) RSC Publications.

TABLE 12.1

The specifications of the PES-TFC PRO hollow fiber membrane and PVDF MD flat sheet membrane (Han et al., 2015b). Copyright (2015) RSC Publications

	Water permeability, *A* (L m^{-2} h^{-1} bar^{-1})	Salt rejection, *R* (%) (2000 ppm @ 5 bar)	Salt permeability *B*(L m^{-2} h^{-1})	Structural parameter, *S* (μm)	OD/ID (μm)	Burst pressure (bar)
PES-TFC PRO hollow fiber membrane	3.3 ± 0.3	97.5 ± 0.2	0.31 ± 0.3	450	973/ 562	>20
	Mean pore size (μm)	**Water contact angle (°)**	**Porosity (%)**	**Selective layer**		**Effective area per module (cm^2)**
PVDF MD flat sheet membrane	168.4 ± 1.6	86.3 ± 1.3	78.6 ± 2.3	Top surface		15.2

membrane morphology. It consists of a thin polyamide selective layer on the inner surface of a porous PES fiber substrate. Macroscopically, the PES-TFC membrane possesses good concentricity and a medium dimension in diameter (i.e., OD is 973 μm) with a relatively thick cross section of ~200 μm. Microscopically, this membrane has a straight finger-like macrovoid structure near the inner cross section and a highly porous open-cell microstructure in the outer surface which is capable of reducing water transport resistance. As a result, this PES-TFC membrane not only possesses outstanding mechanical strength but also exhibits promising transport characteristics. Figure 12.4(d)–(f) shows the images of the PVDF flat sheet membrane for the MD unit. It possesses a relatively thin and dense top surface, an open-pore bottom surface, and a porous cross section full of macrovoids. The thin top layer provides the separation function, while the porous cross section could offer both a low transport resistance and reduced thermal conductivity across the membrane due to a large porosity of up to ~78%. The porous bottom surface provides uninterrupted transport channels for water vapor transport. In order to maintain the vacuum of the membrane module at the permeate side and provide better wetting resistance, the PVDF membrane has a relatively small mean pore size of ~168 nm. Besides, the MD membrane is relatively hydrophobic with a water contact angle of 86.3° due to the hydrophobic nature of the PVDF material and the membrane surface morphology.

As the PRO-MD hybrid process often handles concentrated brines at elevated temperatures, special measures should be taken during the membrane design and operation. Thanks to the nature of the close-loop system (i.e., OHE) and the pretreatment in the upstream of the open-loop process (e.g., RO-MD-PRO), the brine stream entering MD/PRO is theoretically foulant-free. This saves a significant amount of efforts in tackling various types of fouling, such as biological, particulate, and colloidal fouling, and is considered to be an advantage of the hybrid process as mentioned earlier in the introduction section. However, inorganic scaling is particularly concerned when operating MD at high water recovery rate and feed salinity. It is noted that the solubility and crystal formation of salts vary widely over the temperature range relevant to MD. Moreover, the solubility of individual salts may be either positively or negatively correlated with temperature. For instance, the solubility of NaCl increases while those of $CaCO_3$, $Mg(OH)_2$, and $Ca_3(PO_4)_2$ decrease with increasing temperature (Gryta, 2002; Morel and Hering, 1993). This negative correlation might bring the concentrated RO discharged brine even closer to the saturation point during MD, increasing the risk of scaling. Other study shows that a higher temperature also reduces the induction periods of some salts (Stamatakis et al., 2005). To mitigate the inorganic scaling on MD membranes, prevention tools including thermal water softening, use of antiscalants, pH control, and magnetic water treatment have been suggested (Warsinger et al., 2015). In addition, one may tailor MD membrane properties with greater hydrophobicity and smaller surface porosity/pore size (Gryta, 2007). Other fouling contributing factors such as temperature, concentration polarization, residence time, and stagnation areas should be carefully examined, when it comes to the membrane module design (Warsinger et al., 2015).

In the context of using the MD discharged brine as the draw solution, robust PRO membranes withstanding high pressures are crucial to maximize the capacity of harvesting the osmotic energy (Han et al., 2013a, 2013b; Straub et al., 2014; Wan et al., 2018; Zhang and Chung, 2013). Up to date, there are no commercially available membranes or membrane modules specifically designed for the PRO application. Best PRO membranes reported in bench-scale were developed by Straub et al. for the flat sheet configuration (Straub et al., 2014) and Wan et al. for the hollow fiber configuration (Wan et al., 2018). They could achieve the peak power densities of 60 W/m^2 at 48 bar using 3 M NaCl and deionized water as the feed pair and 38 W/m^2 at 30 bar using 1.2 M NaCl and deionized water as the feed pair, respectively.

As the discharged brine often carries a certain heat from the MD unit, its temperature effect on PRO membranes needs to be evaluated. A higher temperature benefits water flux and power density mainly owing to the enhanced draw solute diffusivity. Nonetheless, an enhanced diffusivity allows the solute to permeate across the membrane faster than it would be at lower temperatures, which might counteract the positives brought by the temperature effect (Anastasio et al., 2015; She et al., 2012). One should also be aware that increasing the temperature will result in the change of physiochemical properties of both membranes and solutions, directly influencing the ultimate membrane performance. Although fouling on the draw solution side is greatly suppressed, foulants on the feed solution side can be easily brought into the porous membrane substrate by the bulk water flow and accumulate underneath the active layer (Han et al., 2016). To effectively control this, strategies of pretreating the feed solution (Antony et al., 2011; Chen et al., 2015, 2016; Van der Bruggen et al., 2003; Wan and Chung, 2015) and developing antifouling PRO membranes (Han et al., 2015a, 2017, 2018; Li et al., 2016, 2017) as well as membrane processes are normally adopted.

12.4 PROCESS MODELING

Process modeling developed through mass and heat transport fundamentals is an essential pillar to design and optimize the PRO-MD hybrid process. For the MD unit, Chapter 2 gives the detailed mathematical derivation concerning both local and module scenarios across four typical MD processes (i.e., DCMD, VMD, AGMD, and SGMD). For the PRO unit, Cheng and Chung built a universal platform to analytically evaluate the mass transport of various membrane configurations of flat sheet, single- and double-skinned hollow fibers (Cheng and Chung, 2017). Nonetheless, the temperature effect has not been taken into account when formulating the modeling of PRO processes. By measuring the PRO performance of commercial FO membranes supported by woven mesh/spacers, She et al. observed a growth of ~34% in water flux by increasing the testing temperature from 25 °C to 35 °C using 1 M NaCl and 10 mM NaCl as the feed pair (She et al., 2012) while Anastasio et al. reported the potential to double the power density by raising the temperature from 20 °C to 40 °C using 1.5 M NaCl and deionized water as the feed pair (Anastasio et al., 2015). As mentioned earlier, the elevated temperature could enhance both water flux and reverse solute flux at the same time. This trade-off should be optimized through a model coupling simultaneous mass and heat transport across the membrane. Recently, Chowdhury and McCutcheon developed such a model for FO processes and validated it by carrying out experiments at controlled temperature conditions (Chowdhury and McCutcheon, 2018). Since FO and PRO processes share high similarities besides the application of hydraulic pressures on the draw solution side, future studies could extend their work to PRO processes.

On the other hand, the elevated temperature also affects the membrane transport properties (i.e., water permeability (A) and salt permeability (B)). Nonetheless, this aspect has not been fully analyzed for PRO membranes yet. In analogy, the RO membrane system has already established a temperature correction factor (TCF) to calculate A and B at a desired temperature T ($^\circ C$) (Ruiz-Saavedra et. al., 2015). This could serve as a guideline for the PRO application.

$$\text{For } T \leq 25^\circ\text{C} : TCF = \exp\left[3480 \times \left(\frac{1}{298} - \frac{1}{273 + T}\right)\right] \tag{12.1}$$

$$\text{For } T > 25^\circ\text{C} : TCF = \exp\left[2640 \times \left(\frac{1}{298} - \frac{1}{273 + T}\right)\right] \tag{12.2}$$

$$A = A_0 \times TCF \tag{12.3}$$

$$B = B_0 \times TCF \tag{12.4}$$

Where A_0 and B_0 are water and salt permeabilities measured at 25 °C.

Different from traditional PRO processes, another important factor that should be considered is the nonideality of the hypersaline draw solution generated from the high recovery of the upstream MD unit. By assuming relatively dilute draw solution, most PRO studies rely on the van't Hoff equation to calculate its osmotic pressure (π) (Cheng and Chung, 2017).

$$\pi = iCRT \tag{12.5}$$

where i, R, C, and T are the van't Hoff factor, universal gas constant, concentration, and absolute temperature of the draw solution, respectively. Generally speaking, this simple equation gives a reasonable estimation of the osmotic pressure for salinity up to 2 M (Bajraktari et al., 2017; Lee and Kim, 2016). For higher salinities, the osmotic pressure can be calculated from the water activity (γ_w) and molar volume (V) to get more accurate results (Bajraktari et al., 2017; Madsen et al., 2017).

$$\pi = \frac{RT}{V}\ln(\gamma_w) \tag{12.6}$$

Data of γ_w and V could be either obtained using Pitzer correlations or found in the literature (Pitzer et al., 1984). From an engineering point of view, the OLI software is often used to calculate π based on the activity model for quick and precise predictions (Bajraktari et al., 2017; Lee and Kim, 2016).

12.5 ECONOMIC ANALYSIS

The economic analyses of comparing RO, RO-MD, RO-PRO, and RO-MD-PRO were conducted by Choi et al. using theoretical models (Choi et al., 2017). Figure 12.5 shows that the hybrid systems could outperform a RO stand-alone system in terms of reducing water cost and alleviating the disposal and environmental impacts of the RO discharged brine. The electricity cost plays an important role in determining the economic feasibility of hybrid processes (Choi et al., 2017; Wan and Chung, 2016). If the electricity cost exceeds 0.2 US $/kWh, the water costs of hybrid systems (i.e., RO-MD, RO-PRO, and RO-MD-PRO) are lower than that of stand-alone RO. While the heat supply cost for MD is another crucial factor affecting the economic feasibility of MD-hybridized systems (i.e., RO-MD and RO-MD-PRO). Overall, RO-MD-PRO is found to be the most cost-effective configuration at the same level of electricity cost owing to its extra

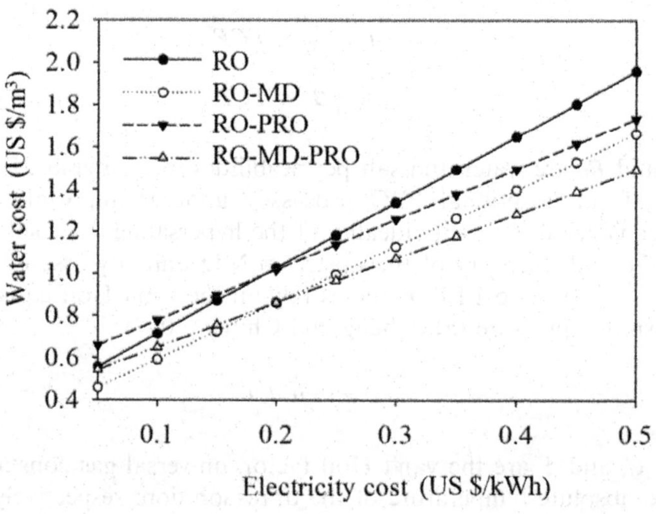

FIGURE 12.5 Effect of electricity cost on water cost for RO, RO-MD, RO-PRO, and RO-MD-PRO (Lee et al., 2019). Copyright (2019) Elsevier.

TABLE 12.2

Comparison of estimated project costs based on various hybrid processes (based on 100 m^3/day plant size) (Lee et al., 2019). Copyright (2019) Elsevier

Processes	RO	RO-MD	RO-PRO	RO-MD-PRO
Initial capital cost (US $/m^3 day)	115.4	115.7	166.6	144.9
Operation cost(US $/m^3)	1.263	1.039	1.233	1.074
Specific energy consumption (kWh/m^3)	3.320	2.809	2.869	2.683

freshwater produced using waste heat via MD and osmotic energy harvested via PRO. By further taking the membrane cost and interest rate into consideration, the estimated project costs with different hybrid processes are presented in Table 12.2.

12.6 CHALLENGES AND FUTURE PROSPECTS

In order to fully maximize the potential of PRO-MD hybrid processes for osmotic power and clean water production, more studies must be dedicated to advance the long-term performance and stability of commercial-scale PRO and MD membrane modules as well as further improve the efficiency of converting low-cost thermal energy to useful works in MD. As restricted by practical losses in an actual process and the thermodynamic limitations, only a fraction of the abundant quantity of waste heat can be feasibly converted. The resultant intrinsically low efficiency suggests that it would be more prudent to utilize the heat resources directly, for instance, in heat pumps and space heating (Lin et al., 2014b). Yet, such direct uses are geographically confined to the locality and availability of the heat sources. In the absence of a matching need for direct use at the source, converting waste heat to electricity for conveyance is the only available option to exploit the low-cost thermal energy. Another method to particularly raise the thermal separation efficiency of MD is to use a solvent (e.g., methanol (Shaulsky et al., 2015)) of high volatility and lower heat capacity and enthalpy of vaporization than water. For the PRO aspect, apart from designing highly robust and permselective membranes, employing novel draw solutes of large molecule sizes (Han et al., 2014; Xia et al., 2018; Zhong et al., 2017) in the close-loop process configuration could minimize the leakage while maintaining the osmotic driving force. By successfully establishing an integrated process model to account for the system nonidealities and economic analysis, further works are required to demonstrate the pilot-scale studies based on the optimized conditions to advance the PRO-MD hybrids toward actual implementation.

REFERENCES

Achilli, A., Prante, J.L., Hancock, N.T., Maxwell, E.B., Childress, A.E., 2014. Experimental results from RO-PRO: A next generation system for low-energy desalination. Environ. Sci. Technol. 48, 6437–6443.

Anastasio, D.D., Arena, J.T., Cole, E.A., McCutcheon, J.R., 2015. Impact of temperature on power density in closed-loop pressure retarded osmosis for grid storage. J. Membr. Sci. 479, 240–245.

Antony, A., Low, J.H., Gray, S., Childress, A.E., Le-Clech, P., Leslie, G., 2011. Scale formation and control in high pressure membrane water treatment systems: A review. J. Membr. Sci. 383, 1–16.

Bajraktari, N., Hélix-Nielsen, C., Madsen, H.T., 2017. Pressure retarded osmosis from hypersaline sources—A review. Desalination 413, 65–85.

Chae, S.H., Seo, J., Kim, J., Kim, Y.M., Kim, J.H., 2018. A simulation study with a new performance index for pressure-retarded osmosis processes hybridized with seawater reverse osmosis and membrane distillation. Desalination 444, 118–128.

Chen, S.C., Amy, G.L., Chung, T.S., 2016. Membrane fouling and anti-fouling strategies using RO retentate from a municipal water recycling plant as the feed for osmotic power generation. Water Res. 88, 144–155.

Chen, S.C., Wan, C.F., Chung, T.S., 2015. Enhanced fouling by inorganic and organic foulants on pressure retarded osmosis (PRO) hollow fiber membranes under high pressures. J. Membr. Sci. 479, 190–203.

Cheng, Z.L., Chung, T.S., 2017. Mass transport of various membrane configurations in pressure retarded osmosis (PRO). J. Membr. Sci. 537, 160–176.

Choi, Y., Kim, S.H., Lee, S., 2017. Comparison of performance and economics of reverse osmosis, membrane distillation, and pressure retarded osmosis hybrid systems. Desalination Water Treat. 77, 19–29.

Chowdhury, M.R., McCutcheon, J.R., 2018. Elucidating the impact of temperature gradients across membranes during forward osmosis: Coupling heat and mass transfer models for better prediction of real osmotic systems. J. Membr. Sci. 553, 189–199.

Chung, T.S., Li, X., Ong, R.C., Ge, Q., Wang, H., Han, G., 2012. Emerging forward osmosis (FO) technologies and challenges ahead for clean water and clean energy applications. Curr. Opin. Chem. Eng. 1, 246–257.

Curcio, E., Drioli, E., 2005. Membrane distillation and related operations—A review. Sep. Purif. Rev. 34, 35–86.

Edwie, F., Chung, T.S., 2013. Development of simultaneous membrane distillation–crystallization (SMDC) technology for treatment of saturated brine. Chem. Eng. Sci. 98, 160–172.

Efraty, A., 2013. Pressure retarded osmosis in closed circuit: a new technology for clean power generation without need of energy recovery. Desalination Water Treat. 51, 7420–7430.

Francis, L., Ghaffour, N., Alsaadi, A.S., Nunes, S.P., Amy, G.L., 2014. Performance evaluation of the DCMD desalination process under bench scale and large scale module operating conditions. J. Membr. Sci. 455, 103–112.

Gerstandt, K., Peinemann, K.V., Skilhagen, S.E., Thorsen, T., Holt, T., 2008. Membrane processes in energy supply for an osmotic power plant. Desalination 224, 64–70.

Ghaffour, N., Lattemann, S., Missimer, T., Ng, K.C., Sinha, S., Amy, G., 2014. Renewable energy-driven innovative energy-efficient desalination technologies. Appl. Energy 136, 1155–1165.

Gryta, M., 2002. Concentration of NaCl solution by membrane distillation integrated with crystallization. Sep. Sci. Technol. 37, 3535–3558.

Gryta, M., 2007. Influence of polypropylene membrane surface porosity on the performance of membrane distillation process. J. Membr. Sci. 287, 67–78.

Gryta, M., Tomaszewska, M., Grzechulska, J., Morawski, A.W., 2001. Membrane distillation of NaCl solution containing natural organic matter. J. Membr. Sci. 181, 279–287.

Han, G., Cheng, Z.L., Chung, T.S., 2017. Thin-film composite (TFC) hollow fiber membrane with double-polyamide active layers for internal concentration polarization and fouling mitigation in osmotic processes. J. Membr. Sci. 523, 497–504.

Han, G., Ge, Q., Chung, T.S., 2014. Conceptual demonstration of novel closed-loop pressure retarded osmosis process for sustainable osmotic energy generation. Appl. Energy 132, 383–393.

Han, G., Liu, J.T., Lu, K.J., Chung, T.S., 2018. Advanced anti-fouling membranes for osmotic power generation from wastewater via pressure retarded osmosis (PRO). Environ. Sci. Technol. 52, 6686–6694.

Han, G., Wang, P., Chung, T.S., 2013a. Highly robust thin-film composite pressure retarded osmosis (PRO) hollow fiber membranes with high power densities

for renewable salinity-gradient energy generation. Environ. Sci. Technol. 47, 8070–8077.
Han, G., Zhang, S., Li, X., Chung, T.S., 2013b. High performance thin film composite pressure retarded osmosis (PRO) membranes for renewable salinity-gradient energy generation. J. Membr. Sci. 440, 108–121.
Han, G., Zhang, S., Li, X., Chung, T.S., 2015a. Progress in pressure retarded osmosis (PRO) membranes for osmotic power generation. Prog. Polym. Sci. 51, 1–27.
Han, G., Zhou, J.L., Wan, C.F., Yang, T.S., Chung, T.S., 2016. Investigations of inorganic and organic fouling behaviors, antifouling and cleaning strategies for pressure retarded osmosis (PRO) membrane using seawater desalination brine and wastewater. Water Res. 103, 264–275.
Han, G., Zuo, J., Wan, C.F., Chung, T.S., 2015b. Hybrid pressure retarded osmosis-membrane distillation (PRO-MD) process for osmotic power and clean water generation. Environ. Sci.: Water Res. Technol. 1, 507–515.
Kim, J., Lee, J., Kim, J.H., 2012. Overview of pressure-retarded osmosis (PRO) process and hybrid application to sea water reverse osmosis process. Desalination Water Treat. 43, 193–200.
Kim, J., Park, M., Shon, H.K., Kim, J.H., 2016. Performance analysis of reverse osmosis, membrane distillation, and pressure-retarded osmosis hybrid processes. Desalination 380, 85–92.
Lee, J., Kim, S., 2016. Predicting power density of pressure retarded osmosis (PRO) membranes using a new characterization method based on a single PRO test. Desalination 389, 224–234.
Lee, J.G., Kim, Y.D., Shim, S.M., Im, B.G., Kim, W.S., 2015. Numerical study of a hybrid multi-stage vacuum membrane distillation and pressure-retarded osmosis system. Desalination 363, 82–91.
Lee, K.L., Baker, R.W., Lonsdale, H.K., 1981. Membranes for power generation by pressure-retarded osmosis. J. Membr. Sci. 8, 141–171.
Lee, S., Choi, J., Park, Y.G., Shon, H., Ahn, C.H., Kim, S.H., 2019. Hybrid desalination processes for beneficial use of reverse osmosis brine: Current status and future prospects. Desalination 454, 104–111.
Li, X., Cai, T., Amy, G.L., Chung, T.S., 2017. Cleaning strategies and membrane flux recovery on anti-fouling membranes for pressure retarded osmosis. J. Membr. Sci. 522, 116–123.
Li, X., Cai, T., Chen, C., Chung, T.S., 2016. Negatively charged hyperbranched polyglycerol grafted membranes for osmotic power generation from municipal wastewater. Water Res. 89, 50–58.
Lin, S., Straub, A.P., Elimelech, M., 2014a. Thermodynamic limits of extractable energy by pressure retarded osmosis. Energy Environ. Sci. 7, 2706–2714.
Lin, S., Yip, N.Y., Cath, T.Y., Osuji, C.O., Elimelech, M., 2014b. Hybrid pressure retarded osmosis-membrane distillation system for power generation from low-grade heat: thermodynamic analysis and energy efficiency. Environ. Sci. Technol. 48, 5306–5313.
Loeb, S., Norman, R.S., 1975. Osmotic power plants. Science 189, 654–655.
Logan, B.E., Elimelech, M., 2012. Membrane-based processes for sustainable power generation using water. Nature 488, 313–319.
Madsen, H.T., Nissen, S.S., Muff, J., Søgaard, E.G., 2017. Pressure retarded osmosis from hypersaline solutions: Investigating commercial FO membranes at high pressures. Desalination 420, 183–190.
McGinnis, R.L., McCutcheon, J.R., Elimelech, M., 2007. A novel ammonia–carbon dioxide osmotic heat engine for power generation. J. Membr. Sci. 305, 13–19.
Morel, F., Hering, J.G., 1993. Principles and applications of aquatic chemistry, Wiley-Interscience, New York.

Pitzer, K.S., Peiper, J.C., Busey, R.H., 1984. Thermodynamic properties of aqueous sodium chloride solutions. J. Phys. Chem. Ref. Data 13, 1.

Ramon, G.Z., Feinberg, B.J., Hoek, E.M.V., 2011. Membrane-based production of salinity-gradient power. Energy Environ. Sci. 4, 4423–4434.

Ruiz-Saavedra, E., Ruiz-García, A., Ramos-Martín, A., 2015. A design method of the RO system in reverse osmosis brackish water desalination plants (calculations and simulations). Desalin. Water Treat. 55, 2562–2572.

Shaulsky, E., Boo, C., Lin, S., Elimelech, M., 2015. Membrane-based osmotic heat engine with organic solvent for enhanced power generation from low-grade heat. Environ. Sci. Technol. 49, 5820–5827.

She, Q., Jin, X., Tang, C.Y., 2012. Osmotic power production from salinity gradient resource by pressure retarded osmosis: Effects of operating conditions and reverse solute diffusion. J. Membr. Sci. 401–402, 262–273.

Sikdar, S.K., 2014. An e-conversation with Prof. Neal Chung. Clean Technol. Environ. Policy 16, 1481–1485.

Stamatakis, E., Stubos, A., Palyvos, J., Chatzichristos, C., Muller, J., 2005. An improved predictive correlation for the induction time of $CaCO_3$ scale formation during flow in porous media. J. Colloid Interface Sci. 286, 7–13.

Straub, A.P., Yip, N.Y., Elimelech, M., 2014. Raising the bar: Increased hydraulic pressure allows unprecedented high power densities in pressure-retarded osmosis. Environ. Sci. Technol. Lett. 1, 55–59.

Thorsen, T., Holt, T., 2009. The potential for power production from salinity gradients by pressure retarded osmosis. J. Membr. Sci. 335, 103–110.

Van der Bruggen, B., Vandecasteele, C., Van Gestel, T., Doyen, W., Leysen, R., 2003. A review of pressure-driven membrane processes in wastewater treatment and drinking water production. Environ. Prog. Sustainable Energy 22, 46–56.

Wan, C.F., Chung, T.S., 2015. Osmotic power generation by pressure retarded osmosis using seawater brine as the draw solution and wastewater retentate as the feed. J. Membr. Sci. 479, 148–158.

Wan, C.F., Chung, T.S., 2016. Maximize the operating profit of a SWRO-PRO integrated process for optimal water production and energy recovery. Renewable Energy 94, 304–313.

Wan, C.F., Yang, T., Gai, W., Lee, Y.D., Chung, T.S., 2018. Thin-film composite hollow fiber membrane with inorganic salt additives for high mechanical strength and high power density for pressure-retarded osmosis. J. Membr. Sci. 555, 388–397.

Warsinger, D.M., Swaminathan, J., Guillen-Burrieza, E., Arafat, H.A., Lienhard, V.J.H., 2015. Scaling and fouling in membrane distillation for desalination applications: A review. Desalination 356, 294–313.

Xia, L., Arena, J.T., Ren, J., Reimund, K.K., Holland, A., Wilson, A.D., McCutcheon, J.R., 2018. A trimethylamine-carbon dioxide draw solution for osmotic engines. AIChE J. 64, 3369–3375.

Zhang, S., Chung, T.S., 2013. Minimizing the instant and accumulative effects of salt permeability to sustain ultrahigh osmotic power density. Environ. Sci. Technol. 47, 10085–10092.

Zhong, Y., Wang, X., Feng, X., Telalovic, S., Gnanou, Y., Huang, K.W., Hu, X., Lai, Z., 2017. Osmotic heat engine using thermally responsive ionic liquids. Environ. Sci. Technol. 51, 9403–9409.

13 Hybrid Forward Osmosis (FO)-MD Processes

Liang Shen, Yan Wang, and Sui Zhang

Another membrane process that is emerging as a potential alternative for desalination and water treatment is forward osmosis (FO). The concept was proposed in the 1970s (Frank 1972; Hough 1973; Kessler and Moody 1976), and has been under rapid development since 2005. By employing a high concentration (and hence a high osmotic pressure) solution as the draw solution at one side of the membrane and a low concentration feed solution at the other side, FO allows the natural permeation of pure water from the feed solution to the draw solution without the need of external hydraulic pressure or heat. Two main advantages are offered by FO: (1) low membrane fouling propensity and (2) high solute rejection. These make FO particularly suitable for applications where fouling and rejection to small solutes are concerns (Ansari et al. 2017). However, unless the draw solution has other uses such as fertilizers, regeneration of the draw solutes is usually required to enable its continuous use. Since MD is insensitive to the salinity of solutions, it emerges as an ideal process to concentrate the diluted draw solution and recover the original high osmotic pressure. The hybrid FO-MD processes have shown strengths such as low fouling tendency and high quality of the product water in many lab-scale demonstrations for water treatment (Ge et al. 2012b; Husnain et al. 2015a; Liu et al. 2016; Nguyen et al. 2016; Xie et al. 2013, 2014; Zhang et al. 2014). In this chapter, we will briefly introduce the concept and recent development of FO, discuss the design and basic properties of FO-MD, and in the last part summarize its applications.

13.1 FORWARD OSMOSIS

13.1.1 INTRODUCTION

The thermodynamics of solutions reveals that all mixtures containing solvents and dissolved solutes are characterized by osmotic pressure (π), which can be estimated using the van't Hoff equation (Yen et al. 2010):

$$\pi = icRT \tag{13.1}$$

where i is the van't Hoff factor, c and T are the solution concentration and temperature, respectively, and R is the gas constant. For most nonelectrolytes dissolved in water, the van't Hoff factor is 1. Clearly, the osmotic pressure of a solution is directly related to its concentration and temperature.

Figure 13.1 plots the osmotic pressure as a function of solute concentration for different aqueous salt solutions at room temperature. A typical seawater consisting of 3.5 wt% NaCl has an osmotic pressure of about 27 bar. Many salts and small organic solutes can dissolve in water at a very high concentration, giving high osmotic pressures of even 1000 atm. Thermodynamics also suggests that when two solutions of different osmotic pressures are separated by a semipermeable membrane, water will naturally transport through the membrane from the low to the high osmotic pressure side in order to balance the chemical potentials. Therefore, it is theoretically possible to use a concentrated solution as the draw solution to extract water from the diluted water sources, such as brackish water or seawater, as illustrated in Figure 13.2. The process is termed as forward osmosis.

Since most practical membranes are asymmetric in structure to maximize the permeation flux, there can be two orientations in the FO process as shown in Figure 13.3, that is, the active layer facing the feed solution (referred to as AL-FS, or FO mode) and the active layer facing the draw solution (referred to as AL-DS, or PRO mode) (Cath et al. 2006). Under the FO mode (Figure 13.3(a)), the diffusion of water dilutes the draw solution in the pores of the support layer. This phenomenon is called as the diluted internal concentration polarization (ICP) that not only sharply declines the effective osmotic pressure difference but also reduce the water flux (Chen et al. 2004; Gray et al. 2006; McCutcheon and Elimelech 2006; Porter 1972). The extent of ICP is closely related to the structure of the support layer, which is usually described by the S parameter (McCutcheon and Elimelech 2006).

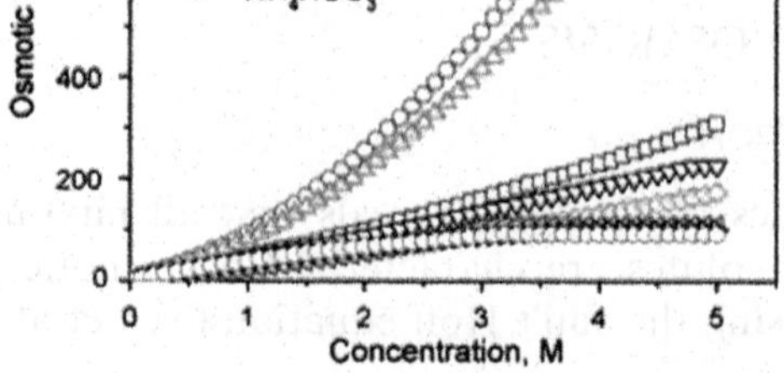

FIGURE 13.1 The osmotic pressure as a function of solution concentration at 25 °C for various potential draw solutions (Cath et al. 2006). Copyright (2006) Elsevier.

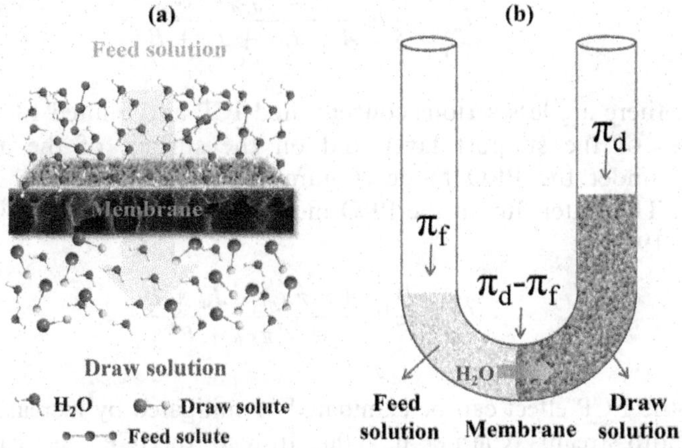

FIGURE 13.2 Schematic illustrations of (a) the FO process and (b) osmotic pressures of the solutions involved in FO.

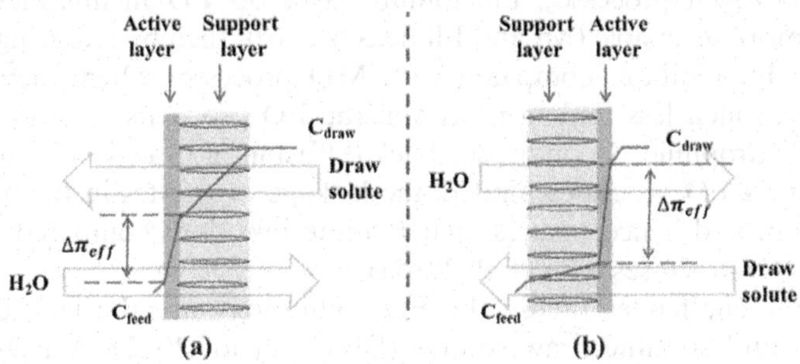

FIGURE 13.3 Orientations in the FO process: (a) active layer facing the feed solution (AL-FS, FO mode); (b) active layer facing the draw solution (AL-DS, PRO mode).

$$S = \frac{t\tau}{\varepsilon} \tag{13.2}$$

where t, τ, and ε are the membrane thickness, tortuosity, and porosity, respectively. Meanwhile, the accumulation of the feed solute on the surface of the active layer would cause the concentrated external concentration polarization (ECP) (Cath et al. 2006; Zhao et al. 2012). Eq. (13.3) gives a clear mathematical model to predict the water flux as a function of S parameter, water permeability (A), salt permeability (B), solute diffusion coefficient (D), osmotic pressure of the bulk feed solution ($\pi_{f,b}$), and effective osmotic pressure of the draw solution ($\pi_{d,m}$) (Loeb et al. 1997).

$$J_w = \frac{D}{S} ln \frac{A \times \pi_{d,m} + B}{A \times \pi_{f,b} + J_w + B} \quad (13.3)$$

In contrast, there are less serious concentrated ICP and diluted ECP occurring in the pores of the support layer and on the surface of the active layer, respectively, under the PRO mode (Figure 13.3(b)) (Cath et al. 2006; Zhao et al. 2012). The water flux in the PRO mode can be determined by Eq. (13.4) (Loeb et al. 1997):

$$J_w = \frac{D}{S} \ln \frac{A \times \pi_{d,m} - J_w + B}{A \times \pi_{f,b} + B} \quad (13.4)$$

In general, the ECP effect can be eliminated or mitigated by increasing the flow velocity of both streams (Cath et al. 2006). Instead, the ICP effect can't be completely eliminated, which can be mitigated by optimizing the structure of the support layer (Cath et al. 2006).

Comparing with other membrane processes, FO offers two major advantages. Firstly, without the hydraulic pressure directly exerted on membrane surfaces in the pressure-driven processes, the fouling layer on FO membranes is usually looser and more reversible (Mi and Elimelech 2010), thereby requiring less efforts in cleaning. In addition, compared with MD processes where membranes are hydrophobic, much less fouling is seen in the FO process since FO membranes are usually hydrophilic. Secondly, the back diffusion of draw solutes into the feed solution during FO processes hinders the leakage of solutes in the feed into the draw solution, and hence ensures higher solute rejections compared to the pressure-driven RO processes (Cui et al. 2014a).

The major challenges facing the FO technology are the lack of desirable membranes and suitable draw solutes (Chung et al. 2012). A FO membrane with high water flux, solute rejection, and fouling resistance is preferred. For the draw solute, appropriate molecular weight, high osmotic pressure, minimal solute leakage, reusability, nontoxicity, and compatibility with FO membranes are required (Chung et al. 2012; Luo et al. 2014). In addition, energy-efficient approaches to recover the draw solutes have been of great interest to the community.

13.1.2 FO Membranes

Developments of FO membranes have been focused on the following aspects: (1) higher water permeability and lower salt permeability of the selective layer; (2) smaller structural parameter; and (3) lower fouling tendency. So far, four different approaches have been reported to prepare FO membranes, including (1) nonsolvent-induced phase separation method (Loeb and Sourirajan 1960), (2) interfacial polymerization method (Cadotte 1981), (3) layer-by-layer self-assembly method (Podsiadlo et al. 2007), and (4) aquaporin incorporated biomimetic membranes (Wang et al. 2012).

The first generation of FO membranes is made of cellulose triacetate (CTA) membranes developed by Hydration Technology Innovations (HTI). It consists of a phase-inversion-induced CTA layer embedded in a woven support, which is much more porous than traditional nonwoven supports used for RO membranes. In addition, the total thickness of the membrane is only around 50 μm so as to reduce the *S* parameter and hence minimize the ICP effect in FO.

Most FO membranes are polyamide-based thin film composite (TFC) membranes, due to their ease of fabrication, good temperature and pH stability, and high separation performances (Lee et al. 2011). Typical TFC membranes are formed by the interfacial polymerization of aromatic amine and acyl chloride monomers on a porous support layer (Lau et al. 2012). Various methods have been reported to enhance the performance of TFC membranes in FO processes, mainly in terms of modifications of the polyamide layer and design of the support layer.

Modifications of the polyamide layer can be conducted by involving additives during the processes of interfacial polymerization, surface modification, and post-treatment. The addition of additives into the monomer solutions, such as catalyst, surfactant, and cosolvent, impacts on the reaction rate or degree, thus optimizing the microstructure and morphology of the resulting membranes. Surface modifications with the aid of hydrophilic functional materials tune the surface properties of the selective layer, thus improving the separation performance and antifouling properties. Post-treatments are generally conducted by immersing membranes into certain solutions or solvents to swell up polyamide chains, dissolve small polymer fragments, or remove unreacted monomers, including alcohols (Han et al. 2013; Zhang et al. 2013), N,N-dimethyl formamide (Cui et al. 2014b), and SDS/glycerol (Ong et al. 2015).

On the other hand, changing the phase inversion process to tune the microstructure of the support layer can effectively mitigate the ICP effect and facilitate the uniform formation of the polyamide selective layer, thus improving the membrane separation performance. It can be realized by incorporating nanomaterials, blending with specific polymers (Sun et al. 2014), taking advantages of hydrophilic polymers (Widjojo et al. 2013), and adding additives or cosolvents in dope solutions (Emadzadeh et al. 2014). Additionally, electrospun nanofiber membranes are characterized by low tortuosity, high porosity, and interconnected pores, and have been employed to prepare FO membranes with low ICP and high permeability (Song et al. 2011).

13.1.3 Draw Solutes and Recovery Methods

Generally, the appropriate selection of suitable draw solutes impacts significantly on the efficacy and sustainability of the FO process. Characteristics of draw solutions are closely related to the separation performance, including osmotic pressure, solubility, viscosity, molecular weight, concentration, and temperature. For example, the water flux is directly related to the viscosity and osmotic pressure of the draw solution. To generate a high osmotic pressure, the solubility of the solute needs to be high to allow a higher concentration.

Typical draw solutes can be classified into the following types as summarized in Table 13.1, based on their material properties. Inorganic salts are commonly used draw solutes, which generally feature high solubility, low viscosity, good diffusivity, and hence produce high water flux. However, due to their relatively small size, salt leakage is a concern. Nanofiltration (NF), reverse osmosis (RO), or MD is usually needed to recover the draw solutes. To minimize the solute leakage and facilitate the regeneration process, a series of novel draw solutes have been developed, such as organic compounds (Yen et al. 2010), polymers (Ge et al. 2012a),

TABLE 13.1

Overview of various draw solutes that have been reported in literature

Draw solution type	Examples	Advantages	Disadvantages	Recovery method
Inorganic salts	$NaCl_2$, $MgCl_2$, Na_2SO_4	High solubility, high osmotic pressure, high water flux	Salt leakage	RO, NF, MD
Organic compounds	Zwitterions, sodium salts of organic salts	High osmotic pressure, high water flux, low solute leakage	Complex synthesis procedures, limited solubility	NF, UF, MD
Polymers	Copolymers, polyacrylic acid	Low solute leakage	High viscosity, limited solubility, low water flux	UF, MD
Nanoparticles	Magnetic nanoparticles, carbon quantum dot	Easy recovery, low solute leakage	Low water flux, easy aggregation	UF, MD
Thermal-responsive hydrogels	Poly(N-isopropylacrylamide)	Easy recovery, no leakage	Low water flux	Heating
Metathesis precipitable salts	$Al_2(SO_4)_3$ + CaO	Easy recovery	Chemicals needed in recovery, toxicity issue	Precipitation
Dissolved gas or volatiles	SO_2, NH_3, CO_2	High osmotic pressure	Severe solute leakage, toxicity issue	Heating
NH_3-CO_2 combination	NH_4HCO_3	High osmotic pressure	Severe solute leakage, bad smell	Heating
Switchable polarity solvents	Tertiary amines-CO_2	High osmotic pressure, high water flux, low solute leakage, easy recovery	Toxicity	Heating
Direct use	Glucose, fertilizers	Without recovery	Not pure product water	None

nanoparticles, stimuli-responsive materials (Cai et al. 2013), etc. (Ling et al. 2011). Additionally, some draw solutions, after being diluted by the permeation flux in FO processes, can be directly used for subsequent applications such as drinks, irrigation, or desert restoration (Duan et al. 2014; Phuntsho et al. 2012). Such draw solutes include glucose, fructose, edible saccharide, and fertilizers.

According to their physicochemical properties, diluted draw solutions can be regenerated by various methods as summarized in Table 13.1, including thermal separation, membrane-based separation (RO, NF, ultrafiltration, MD), precipitation, and stimuli-response (heat, magnetic field). Generally, the direct thermal separation process is energy intensive. Precipitation can be more energy-efficient, but costs additional chemicals and may generate toxic by-products. Pressure-driven membrane processes may be relatively energy saving, but the high operating pressure required to regenerate draw solutes is a concern. Comparing with other methods, MD has two unique features: (1) its insensitivity to the solute concentration is particularly useful in the regeneration of high salinity draw solutions, and (2) its possibility to utilize low-grade waste heat from many industrial processes as the energy source lowers the overall energy cost. The integrated FO-MD process has hence received broad interests in many different applications.

13.2 FO-MD HYBRID PROCESS

13.2.1 INTRODUCTION

Figure 13.4 shows a schematic illustration of the lab-scale FO-MD hybrid system. When the feed solution is flown through the FO membrane, water is drawn automatically to the draw solution side due to the transmembrane

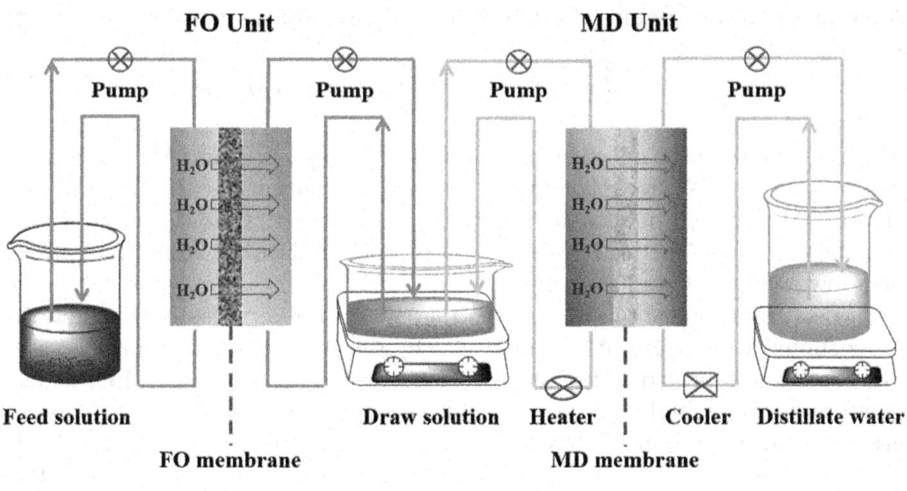

FIGURE 13.4 A schematic diagram of the FO-MD hybrid system.

difference in osmotic pressure, increasing the volume of the draw solution and diluting its concentration. In the meantime, the draw solution is continuously concentrated by flowing through the MD membrane where it is heated to a temperature of 50–80 °C. Water permeates through the membrane due to the vapor pressure difference, and is cooled and collected as the pure water product at the MD side. Clearly, the integrated FO-MD system enables simultaneous recovery of draw solutions and production of clean water, and forms a complete and sustainable cycle for water treatment.

Comparing with MD, FO-MD avoids the direct physical contacts between feed water sources and MD membranes. It has been discussed in the earlier chapters that due to the hydrophobic nature of MD membranes, severe fouling may occur in the MD process, especially when wastewaters contain high concentrations of foulants. Fouling leads to frequent physical and chemical cleaning, and may cause damage to the membranes. Since FO is an intrinsically low-fouling process, the hybrid FO-MD system can effectively suppress fouling and reduce the operational cost.

In addition, the main benefit of FO-MD in comparison with other processes, such as FO-NF and FO-RO, is its ability to recover high-concentration draw solutions, which makes it possible to treat high-salinity wastewater, and increase the productivity of the FO process.

13.2.2 Process Design

To ensure continuous, sustainable operations of the FO-MD process, engineering design of various main parameters such as membrane area, permeability, draw solution concentration, flow rate, and temperature is needed. The aim of engineering design is to maintain constant volume and concentration of the draw solution, so that steady state can be reached in long-term operations. A simple equation should be established in the steady state as described next:

$$J_{W,FO} \times A_{S,FO} = J_{W,MD} \times A_{S,MD} \tag{13.5}$$

where A_S refers to the effective area, and the subscripts FO and MD refer to the two processes, respectively. The water flux in FO is given by either Eq. (13.3) or (13.4), and is closely related to the membrane properties, flow conditions, and osmotic pressures. The water flux in MD is mainly related to membrane properties, flow conditions, and temperature.

In particular, one should take note that the osmotic pressure of solutions is proportional to temperature, as indicated by the van't Hoff equation. Since the draw solution is heated in the MD process and is recycled back to FO, its temperature becomes higher than its original temperature and hence generates a greater osmotic pressure difference across the membrane. Meanwhile, solution viscosity decreases while solute diffusivity increases at a higher temperature, which are beneficial for water transport. In addition, because of thermal conduction from the hot draw solution, it increases the feed solution temperature.

Though such effect compromises the water production by MD due to the heat loss from the draw solution to the feed solution, the overall higher flux achieved in FO is increased because of the higher draw solution temperature. Accordingly, the temperature effects have to be taken into consideration in the calculation of osmotic pressure and water flux. A detailed analysis of temperature profile along the FO membrane module in the steady state will be necessary and should be carefully evaluated during the process design.

However, it has to be noted that a high feed temperature may not be suitable for certain applications where the solutes are sensitive to temperature. In such cases, additional operational measures need to be taken to ensure a reasonable feed temperature.

13.3 APPLICATIONS OF FO-MD PROCESSES

The hybrid FO-MD process has been widely applied in various fields in the past decade, mostly in lab scale, due to its low membrane fouling tendency and high quality product water. As will be introduced in the following, the most popular application is in the field of wastewater treatment. A variety of water sources have been studied, such as sewage wastewater, sludge wastewater, domestic wastewater, human urine, dye-containing wastewater, and oily wastewater. Other applications, for example, desalination and concentration of proteins will also be summarized.

13.3.1 Sewage Treatment

Sewage refers to the wastewater in the sewer system that is discharged from our daily life. Recovering clean water from the sewage has been practiced in modern urban cities by the use of numerous capital and energy intensive systems (Xie et al. 2013). The major obstacle for the implementation of sewer mining is the development of a treatment system which can produce high-quality water from raw sewage and is simple and robust enough for decentralized applications. Recent researchers have found that the integrated FO-MD system can be a promising candidate for treating sewage.

Xie et al. (Xie et al. 2013) utilized a FO-MD hybrid process for direct sewer mining. It can be found in Figure 13.5 that the trace organic contaminants (TrOC) rejection was substantially enhanced in the FO-MD hybrid system, due to the double rejecting effects of the FO and MD processes. The effect was especially significant for the removal of hydrophobic or semivolatile TrOCs, which can only be moderately removed by either process individually. As a result, TrOC concentrations in the distillate water in the FO-MD hybrid system were much less than those in the single MD system.

13.3.2 Sludge Treatment

Sludge is mainly produced in wastewater treatment plants, and its dewatering treatment accounts for nearly 50–60% of the total operating cost of the entire

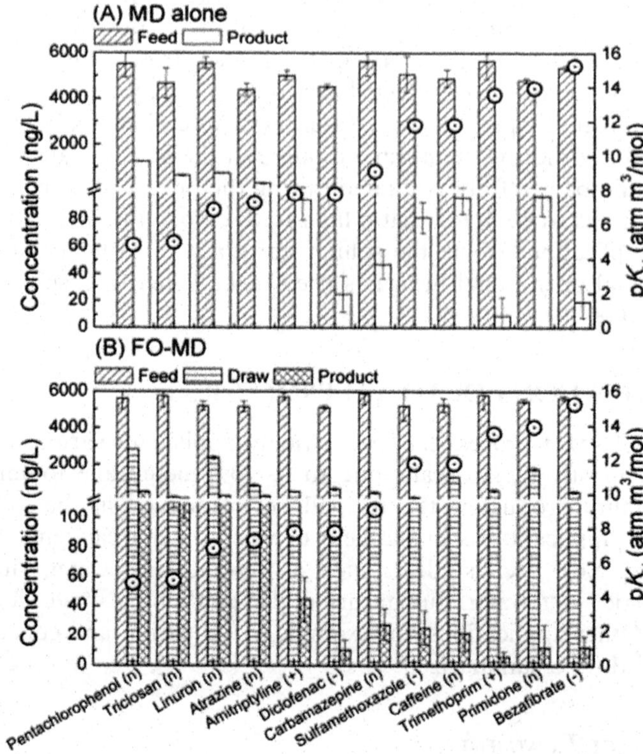

FIGURE 13.5 Concentrations of trace organic contaminants in (A) feed and product water of the MD alone system and (B) feed, draw, and product water of the FO-MD system (Xie et al. 2013). Copyright (2013) American Chemical Society.

wastewater treatment process (Nguyen et al. 2016). Traditional sludge dewatering techniques, such as clarification and gravity thickening, can hardly meet the strict regulations for effluent water quality (Kim et al. 2010). Additionally, the stabilization process of the digested sludge produces a high-nutrient-rich concentrate which poses challenges in lowering the N and P content to meet the target discharge limits (Wild et al. 1997).

When traditional membrane processes are applied to solve the problems, significant fouling occurs (Wang et al. 2008). Recently, Nguyen et al. (Nguyen et al. 2016) employed the highly charged phosphate as the draw solute in FO-MD integrated systems for dewatering sludge. After dewatering by the FO process, the sludge became rich in nutrients and could be further used for agriculture. In addition, after 15 h of dewatering operation, the mixed liquor suspended solid (MLSS) concentration with an initial concentration of 3500 mg/L reached to 19,800 and 22,000 mg/L for the PRO and FO modes, respectively. Moreover, a PTFE MD membrane with a pore size of 0.45 μm was effective for recovering the phosphate (100% rejection).

As mentioned previously, there are rich N and P elements in the sludge concentrate. It is known that the release of phosphorous in the aquatic environment leads to the algal bloom and other severe ecological influences (Conley and Likens 2009). Xie et al. (Xie et al. 2014) utilized a hybrid FO-MD system to recover phosphorus from the digested sludge by forming struvites ($MgNH_4PO_4{\cdot}6H_2O$), as illustrated in Figure 13.6. Due to the bidirectional diffusion in the FO system, that is, the diffusion of protons from the feed solution into the draw solution elevated the feed solution pH, and the diffusion of Mg^{2+} ($MgCl_2$ as the draw solute) from the draw solution into the feed solution, the phosphorous could be deposited by the formation of struvite crystals, which could be further used as an excellent slow-release fertilizer (Debashan and Bashan 2004). The MD process was applied to recover the diluted draw solution and extract clean water from the digested sludge concentrate.

Similarly, nitrogen, mainly in forms of urea and ammonium, can deteriorate the water quality, and its discharge is also strictly restricted by regulations (Wang and Wang 2009). Husnain et al. (Husnain et al. 2015b) employed the combined system of FO-MD to treat the high-nutrient side stream from anaerobic digester effluents and dewatering flows. The NH_3-N rejection in the FO-MD hybrid system was more than 98%, while in the separate MD system, the rejection only ranged from 62% to 72%. The poor rejection in the MD system was ascribed to the high content of organic solids in the feed stream, which led to the wetting problem of the MD membrane and thus the passage of contaminants. Apparently, the hybrid FO-MD system helps maintain a desirable high rejection to nitrogen.

13.3.3 Domestic Wastewater Treatment

Various contaminants generated by our daily life cause serious pollution to the water resources, such as P, N, heavy metal, microorganisms, and TrOC. To

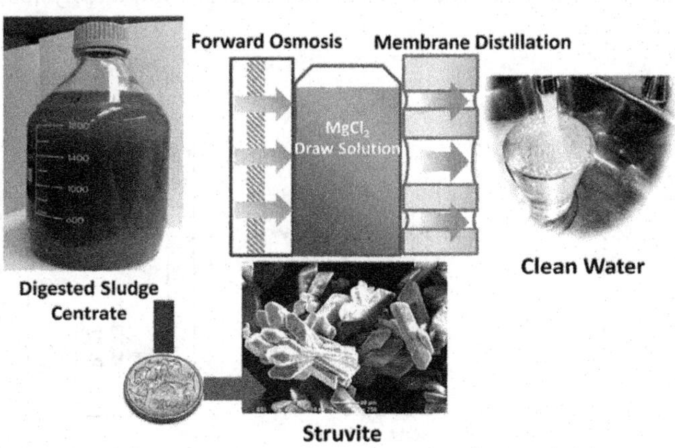

FIGURE 13.6 Schematic illustration of a FO-MD hybrid system for the extraction of phosphorous from the digested sludge (Xie et al. 2014). Copyright (2014) American Chemical Society.

minimize their environmental impacts, different separation and purification technologies have been applied to separate contaminants from the water sources. Among them, the hybrid process of FO-MD shows great potentials in the reclamation of domestic wastewater.

Su et al. (Su et al. 2013) employed the FO-MD hybrid system for the reclamation of wastewater containing heavy metal ions. They found that equal or similar water flux for both processes favored the sustainable production of clean water from the wastewater. In addition, Husnain et al. (Husnain et al. 2015a) also applied the FO-MD hybrid system in the treatment of wastewater. After treating by the hybrid system, the ammonium concentration (~0.18 mg/L) could not be detected in the distillate solution (300–495 mg/L in the feed solution) as exhibited in Figure 13.7, indicating the almost 100% rejection. Additionally, the chemical oxygen demand (COD) was also completely removed by this integrated system. Moreover, the removal of natural As (III) by the hybrid system was more efficient than that of the single FO system. For the removal of As (V), both the FO and FO-MD systems were effective with a high rejection. It was also found that regardless of the feed composition (synthetic wastewater or real wastewater secondary effluent), the permeate conductivity was even lower than that of deionized water, demonstrating the extremely low passage of solutes across the MD membrane.

13.3.4 Dye Wastewater Treatment

Dyes are extensively used as coloring agents in manufacturing industries of papers, plastics, and textiles. Since many organic dyes are toxic or even carcinogenic and non-biodegradable, their discharge may lead to severe environmental contamination and fearful threats to human health (Han et al. 2009; Minear 1996). Therefore, it is urgent to pretreat dye-containing wastewaters before releasing them into the environment. Traditional physical, chemical,

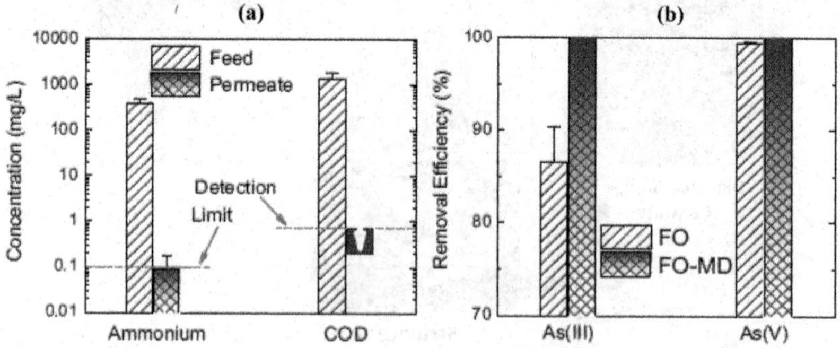

FIGURE 13.7 (a) Ammonium and COD, and (b) As removal by FO-only and FO-MD systems (Husnain et al. 2015a). Copyright (2015) Elsevier.

and electrochemical technologies are difficult to thoroughly remove dyes from wastewater (Daneshvar et al. 2002; Río et al. 2011). Alternatively, membrane-based separation technologies have proven to be practical approaches for the removal of dyes and recovery of reusable water because of their high rejection, low energy consumption, and ease of scale-up.

Ge et al. (Ge et al. 2012b) applied the FO-MD hybrid system into the dye wastewater treatment, using a poly(acrylic acid) sodium (PAA-Na) salt as the draw solute. It can be seen from Figure 13.8 that, as compared to the individual FO system, the acid orange 8 concentration in the feed solution was increased by 2–3 times more by the FO-MD hybrid system after the 2-h treatment, which was attributed to the simultaneous concentration of the draw solution by MD that maintained the effective osmotic pressure gradient across the membrane.

13.3.5 Oily Wastewater Treatment

Produced water which usually contains dispersed oils, suspended particles, and dissolved solutes is a typical by-product in the oil and gas manufacturing industries, and is one type of the major oily wastewaters (Daiminger et al. 1995). The concentration of minerals and/or salts in oily wastewater is varied from several to tens of thousands ppm (Daiminger et al. 1995). The oil content may vary in a wide range, but mostly between 100 and 1000 ppm, and is usually emulsified by natural or added surfactants. Additionally, other components are also found in the produced water, such as formic acid, acetic acid, and alcohols (Islam and Chakma 1989, Us Epa 2004). With the increasing demand on fossil fuels, the generation of oily wastewater would worsen the environment and government regulations have started to impose discharge standards on the maximum concentrations of oil and other contaminants to the industries.

Overall, current technologies aim at further reuse of the cleaned water (Gupta and Hlidek 2009) or the safe discharge of wastewaters into the

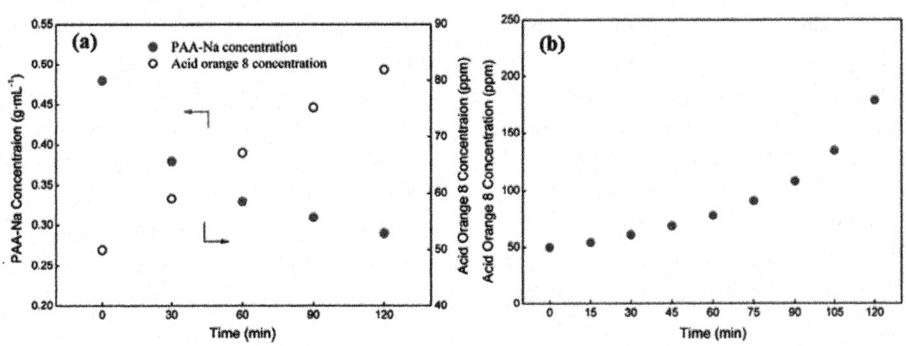

FIGURE 13.8 (a) Variations of PAA-Na and acid orange 8 concentrations with time in the FO process; (b) variations of acid orange 8 concentration in the FO-MD hybrid system (Ge et al. 2012b). Copyright (2012) American Chemical Society.

freshwater cycle (Chakrabarty et al. 2010). However, conventional technologies mostly failed to produce low hydrocarbon content in the treated water, particularly for the finely dispersed oil/water emulsions with droplet sizes less than 10 μm (Chakrabarty et al. 2010). Moreover, the water recovery rate can only reach to 50–60%, the rest of which is still polluted by oil and other components (Gupta and Hlidek 2009). Alternatively, membrane-based separation technologies are possibly promising solutions for treating micrometer-sized oily wastewater with a rejection ranging from 80% to 99% (Ohya et al. 1998). Typically, UF membranes with pores between 0.001 and 0.1 μm have been widely applied in treating emulsified oil and suspended solids (Li et al. 2006), but severe membrane fouling easily occurs (Xu et al. 1999). Therefore, the FO process with low fouling tendency and high rejections to salts and most contaminants becomes a promising candidate for separation oil/water systems.

Zhang et al. (Zhang et al. 2014) employed the FO-MD hybrid system for sustainable recovery of water from the oily wastewater. After 42–45 h of treatment, the water recovery ratio reached to 90%. As shown in Table 13.2, the concentration of NaCl in the final permeate was sharply reduced. Moreover, the oil content in both the FO and MD permeates is negligible, demonstrating the efficient water recovery from the oily wastewater by the FO-MD hybrid system.

TABLE 13.2
The initial and final compositions of the feed, permeate, and draw solutions for FO and MD of the hybrid FO-MD system at 60 °C (Zhang et al. 2014). Copyright (2014) Elsevier

	2 M NaCl as the draw solution (Series #1)			5 M NaCl as the draw solution (Series #2)		
Composition	**Oil (ppm)**	**Acetic acid (ppm)**	**NaCl**	**Oil (ppm)**	**Acetic acid (ppm)**	**NaCl**
Initial feed solution (FO)	4000	1000	0.04	4000	1000	0.2
Permeate (FO)	Trace	237	–	Trace	219	–
Final feed solution (FO, calculated)	47,000	9213	0.47 M	42,200	8440	2.11 M
Draw solution (FO) at the end of FO operation (the initial composition for the MD)	Trace	54	1.55 M	Trace	49	3.87 M
Permeate (MD)	–	29	1 ppm	–	28	3 ppm
Final draw solution (MD, calculated)	Trace	61	2.0 M	Trace	55	5.0 M

13.3.6 Human Urine Treatment

Human urine contains 95% water, 3.5% organics, and 1.5 % inorganic salt (Lind et al. 2001), and contributes to 80% of the total nitrogen and 50% of the total phosphate content in domestic wastewater (Larsen* et al. 2009). Urea and ammonium are the main forms of the N element, which can lead to algae bloom and eutrophication in water (Wang and Wang 2009).

To meet the effluent quality standard, long hydraulic and sludge retention times are required during the traditional wastewater treatment process for the safe discharge of urine (Tice and Kim 2014). In addition, the rich nutriments in human urine can be a promising resource. Therefore, the concentration and recovery of nutrients resources from the urine are of great interest (Etter et al. 2011). Several approaches have been reported, including RO, electrodialysis, distillation, and freeze/thaw concentration (Maurer et al. 2006). Meanwhile, the reclamation of potable water from urine is significant for living in some special environments, such as the space station (Zhao et al. 2013). A process that is reliable, durable, lightweight, efficient, and easy for operation is required. An apparatus combining biological purification and UV photocatalytic oxidation was reported previously to treat the urine solution to produce clean water that meets the drinking water quality.

Currently, membrane-based technologies have been widely applied to the human urine treatment, which are safer and more stable as compared to the oxidation and biological processes (Cath et al. 2005). Among them, the FO process is effective for the removal of phosphate and ammonium with high rejections (McCutcheon and Elimelech 2006).

Liu et al. (Liu et al. 2016) employed the hybrid system of FO-MD to remove the contaminants in human urine. As shown in Figure 13.9, in a FO system alone, rejections for total organic carbon (TOC), total nitrogen (TN), and NH_4^+-N reached to 97%, 95%, and 99%, respectively. By coupling with MD, the rejections went up to approximately 100%. In addition, regardless of the draw solution concentration and temperature, concentrations of TOC, TN, and NH_4^+-N in the distillate were much lower than those in the feed and draw solutions. Additionally, concentrations of TOC, TN, and NH_4^+-N in the product water treated by the FO-MD process were much lower than those of product water generated by MD alone.

13.3.7 Desalination

RO is now the dominant technology for seawater desalination in the market. However, numerous foulants are present in seawater, which lead to intensively chemical treatments and high energy consumption. With its low fouling propensity, the FO process provides an alternative approach for desalination. Nonetheless, since the typical concentration of seawater is 3.5 wt%, corresponding to an osmotic pressure of around 27 bar, seawater desalination usually involves high salinity and high osmotic pressure. Two major challenges need to be addressed before FO can be practically used: Firstly, the preparation of draw solutions

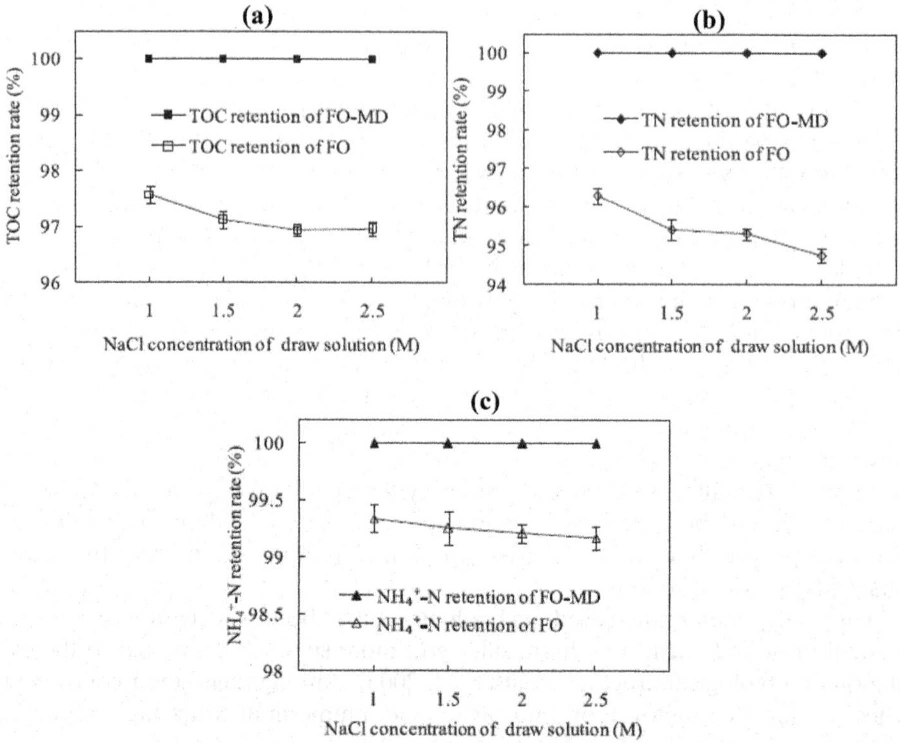

FIGURE 13.9 Retention rates of different contaminants as a function of draw solution concentration in the FO-MD hybrid system: (a) TOC, (b) TN, and (c) NH_4^+N (Liu et al. 2016). Copyright (2016) Elsevier.

that offer a high osmotic pressure, and secondly, the regeneration of high-salinity draw solutions.

Many research efforts have been devoted to the design of new draw solutes, as is discussed in the earlier section. Meanwhile, MD serves as a good platform for the recovery of such solutes, thanks to its following advantages: (1) relatively stable performance at high feed salinity; (2) high-quality product water; (3) 100% theoretical rejections for nonvolatile feed solutes; (4) potentially energy saving if a low-grade waste heat is utilized.

Zhao et al. (Zhao et al. 2014) synthesized a thermoresponsive copolymer of poly(sodiumstyrene-4-sulfonate-*co*-N-isopropylacrylamide) (PSSS-PNIPAM) as the draw solutes for seawater desalination, and the diluted draw solution was then regenerated by the MD process (Figure 13.10). Generally, below the low critical solution temperature (LCST), the copolymer was soluble in water; and above LCST, phase separation occurred. Therefore, the FO process was performed at ambient temperature (<LCST), allowing the maximum osmotic pressure in the draw solution by the fully expanded polymer chains. Meanwhile, the MD process

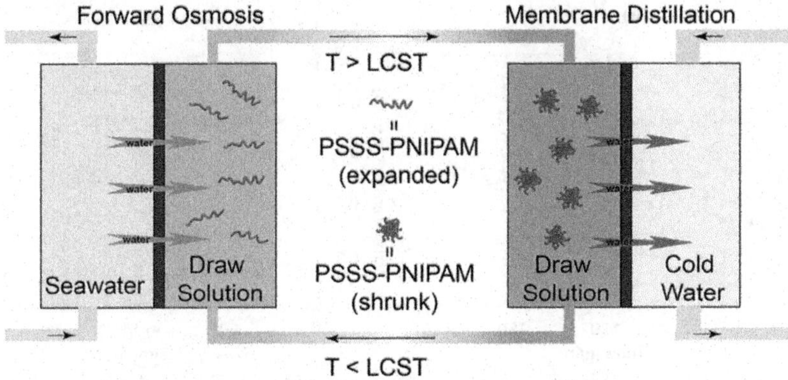

FIGURE 13.10 Schematic illustration of integrated FO-MD process for seawater desalination (Zhao et al. 2014). Copyright (2014) Elsevier.

was conducted at 50 °C (>LCST), where the agglomerated chains produced a lower osmotic pressure and a higher effective water vapor pressure, facilitating the separation of water from the diluted draw solution.

Ge et al. (Ge et al. 2016) utilized the multicharged hydroacid complex of Na-Cr-OA as the draw solute to remove As (III) from water. The FO-MD integrated system not only made As (III) removal sustainable by reconcentrating the diluted draw solution, but also further reduced the As (III) concentration (<10 μg/L) in the product water (5×10^4 to 1×10^6 μg/L and 30–3000 μg/L in the feed solution and FO permeate, respectively). Additionally, Wang et al. (Wang et al. 2015) employed a draw solution Na_5Fe-CA with a high solubility, low viscosity, and excellent thermal stability in the hybrid FO-MD process for desalination, achieving a minimal reverse solute flux and a high seawater desalination flux of 6/32 LMH (FO/MD).

13.3.8 Concentration of Proteins

Pure and concentrated protein products are important for comprehensive understanding of protein function and manufacture (van Reis and Zydney 2007), and are also critical for many pharmaceutical and medical applications. Since most protein products are labile and heat sensitive, athermal separation processes are favored, such as membrane processes. In consideration of the tendency of proteins to foul on membrane surfaces and the potential denaturation of proteins under high hydraulic pressure, FO-MD may be a good option for protein enrichment. However, one problem concerning FO-MD is the elevation of feed temperature due to the heat transfer from the heated draw solution. A temperature controller needs to be installed in the design of such a process.

Wang et al. (Wang et al. 2011) applied an integrated FO-MD system for the enrichment of proteins. The BSA concentration in the feed solution was enriched from 1 g/L to roughly 2.1 g/L over 4 h (as shown in Figure 13.11(a)).

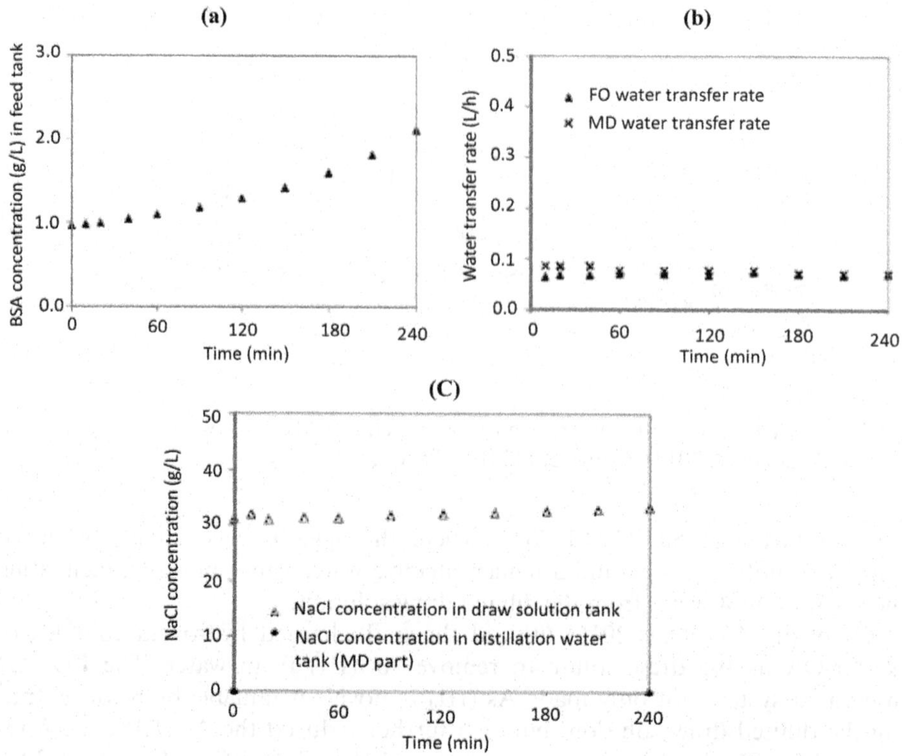

FIGURE 13.11 BSA concentration in the feed tank, water transfer rate, NaCl concentrations in draw solution tank and distillate water tank as a function of time (Wang et al. 2011). Copyright (2011) Elsevier.

Besides, Figure 13.11(b) shows a stable FO water transfer rate was kept during the entire test, which matched well with that of the MD process. Furthermore, the NaCl concentration in the draw solution tank changed slightly, indicating the continuous regeneration of draw solution during the whole test (Figure 13.11(c)). Figure 13.11(c) also displays that the NaCl concentration in the distillate tank was almost negligible during the whole test, demonstrating no significant leakage of NaCl from the draw solution to the distillate.

13.4 SUMMARY

In summary, tremendous research efforts have been made in applying the FO-MD hybrid process for treating various feed solutions, such as seawater, domestic wastewater, protein solutions, etc. Table 13.3 briefly summarizes the different compositions and separation performances in correspondence to the wide variety of feed solutions in literature studies. In general, high rejections to both the draw and feed solutes have been achieved, which demonstrates the potential of

TABLE 13.3
A summary of the applications of the FO-MD hybrid system

Feed solution type	Feed compositions	Draw solution	Water flux (FO)/ LMH	Water flux (MD)/ LMH	Solute rejection/%	Ref.
Sewage wastewater	Raw sewage, 0.5 μg/L TrOC	1.5 M NaCl	8.0	8.0	*Rs*(TrOC) > 99.5%	(Xie et al. 2013)
High-nutrient sludge wastewater	3500 mg/L MLSS, 300 ± 5 mg/L DOC, 100 ± 2 mg/L NH_4^+-N, 100 ± 3 mg/L PO_4^{3-}-P	0.05–0.4 M Na_3PO_4 $\cdot 12H_2O$	8.57–16.20	10.28	*Rs*(Na_3PO_4) =100%	(Nguyen et al. 2016)
Human urine wastewater	5298 ± 792 mg/L TOC, 7523 ± 1097 mg/L TN, 1125 ± 147 mg/L NH_4^+	2.0 M NaCl	~6.0	~8.0	*Rs*(TN/TOC/ N-NH_4^+) > 98%	(Liu et al. 2016)
Domestic wastewater	300–495 mg/L NH_4Cl, 985–1800 mg/L COD, 100–200 mg/L $NaAsO_2$	1.0 M NaCl	~10.0	~15.0	*Rs*(NH_3/ COD/As) > 98%	(Husnain et al. 2015a)
Oily wastewater	0.04/0.2 M NaCl, 4000 ppm petroleum, 444 ppm Tween 80, 1000 ppm acetic acid	2/5 M NaCl	26.7	5.8	*Rs*(oil/NaCl) ≈ 100%	(Zhang et al. 2014)
Dye-containing wastewater	50 ppm acid orange 8	0.48 g/mL PAA-Na	~25.0	~8.0	/	(Ge et al. 2012b)
Digested sludge wastewater	NH_4^+, PO_4^{3-}, Mg^{2+}	1.5 M $MgCl_2$	9.0	9.0	*Rs*(NH_4^+) = 90% *Rs*(PO_4^{3-}) = 97%	(Xie et al. 2014)
Seawater	0.6 M NaCl	33.3 wt% PSSS-PNIPAM	4.0	2.7	/	(Zhao et al. 2014)
Protein-containing solution	BSA	2.0 M NaCl	~5.0	~15.0	*Rs*(NaCl) ≈ 100%	(Wang et al. 2011)

* *Rs* means "Solute rejection"

the hybrid FO-MD system as a complete, sustainable cycle for draw solution regeneration and water recovery/solute concentration.

13.5 CONCLUSION AND OUTLOOK

The FO-MD hybrid system possesses the strengths of both processes, that is, low membrane fouling propensity and high quality of produced water. The

FO process endows an effective pretreatment for the MD process to reduce membrane fouling, and the MD process is an excellent complementation for the FO process to regenerate the diluted draw solution. So far, most studies are conducted in the lab scale and have demonstrated the technical feasibility of FO-MD for various applications, such as wastewater treatment, seawater desalination, and concentration of protein. However, more studies are needed in the following aspects in the future to fully explore its potential and assess its economical, energy, and practical feasibility: (1) Process design that takes into account more comprehensive factors, for example, temperature effects; (2) accumulation of permeate in the FO process (e.g., trace organics) in the draw solution and its long-term impact; (3) fouling that might be severe in the MD process for recovery of new types of draw solutes, if the latter contains lots of organic components; (4) integration of the process into the waste heat network; (5) analyses on the energy efficiency and cost effectiveness of the FO-MD system; and (6) pilot-scale studies.

LIST OF ABBREVIATIONS AND NOMENCLATURES

Abbreviations

COD	Chemical oxygen demand
CTA	Cellulose triacetate
ECP	External concentration polarization
FO	Forward osmosis
HTI	Hydration Technology Innovations
ICP	Internal concentration polarization
LCST	Low critical solution temperature
MD	Membrane distillation
MLSS	Mixed liquor suspended solid
NF	Nanofiltration
PAA-Na	Poly(acrylic acid) sodium
PRO	Pressure retarded osmosis
PSSS-PNIPAM	Poly(sodiumstyrene-4-sulfonate-*co*-N-isopropylacrylamide)
RO	Reverse osmosis
TFC	Thin-film composite
TN	Total nitrogen
TrOC	Trace organic contaminants
TOC	Total organic carbon

Nomenclatures

A	Water permeability
AL-DS	Active layer facing draw solution
AL-FS	Active layer facing feed solution
$A_{S,\ FO}$	Effective area of FO Membrane

$A_{S,\ MD}$	Effective area of MD Membrane
B	Salt permeability
c	Concentration
D	Diffusion coefficient
i	Van't Hoff factor
J_w	Water flux
$J_{w,\ FO}$	Water flux in FO process
$J_{w,\ MD}$	Water flux in MD process
R	Gas constant
Rs	Solute rejection
S	Membrane structural parameter
T	Temperature
t	Thickness
π	Osmotic pressure
$\pi_{d,m}$	Effective osmotic pressure of the draw solution
$\pi_{f,b}$	Osmotic pressure of the bulk feed solution
τ	Tortuosity
ε	Porosity

REFERENCES

Ansari, A.J., Hai, F.I., Price, W.E., Drewes, J.E. and Nghiem, L.D. (2017) Forward osmosis as a platform for resource recovery from municipal wastewater - A critical assessment of the literature. Journal of Membrane Science 529, 195–206.

Cadotte, J.E. (1981) Interfacially synthesized reverse osmosis membrane. U.S. Patent 4,277,344. 1981-7-7.

Cai, Y., Shen, W., Wang, R., Krantz, W.B., Fane, A.G. and Hu, X. (2013) CO_2 switchable dual responsive polymers as draw solutes for forward osmosis desalination. Chemical Communications (Cambridge) 49(75), 8377–8379.

Cath, T.Y., Childress, A.E. and Elimelech, M. (2006) Forward osmosis: principles, applications, and recent developments. Journal of Membrane Science 281(1–2), 70–87.

Cath, T.Y., Gormly, S., Beaudry, E.G., Flynn, M.T., Adams, V.D. and Childress, A.E. (2005) Membrane contactor processes for wastewater reclamation in space: part I. Direct osmotic concentration as pretreatment for reverse osmosis. Journal of Membrane Science 257(1), 85–98.

Chakrabarty, B., Ghoshal, A.K. and Purkait, M.K. (2010) Cross-flow ultrafiltration of stable oil-in-water emulsion using polysulfone membranes. Chemical Engineering Journal 165(2), 447–456.

Chen, J.C., Li, Q. and Elimelech, M. (2004) In situ monitoring techniques for concentration polarization and fouling phenomena in membrane filtration. Advances in Colloid and Interface Science 107(2–3), 83–108.

Chung, T.-S., Li, X., Ong, R.C., Ge, Q., Wang, H. and Han, G. (2012) Emerging forward osmosis (FO) technologies and challenges ahead for clean water and clean energy applications. Current Opinion in Chemical Engineering 1(3), 246–257.

Conley, D.J. and Likens, G.E. (2009) Ecology. Controlling eutrophication: nitrogen and phosphorus. Science 323(5917), 1014.

Cui, Y., Ge, Q., Liu, X.-Y. and Chung, T.-S. (2014a) Novel forward osmosis process to effectively remove heavy metal ions. Journal of Membrane Science 467, 188–194.

Cui, Y., Liu, X.-Y. and Chung, T.-S. (2014b) Enhanced osmotic energy generation from salinity gradients by modifying thin film composite membranes. Chemical Engineering Journal 242, 195–203.

Daiminger, U., Nitsch, W., Plucinski, P. and Hoffmann, S. (1995) Novel techniques for oil/water separation. Journal of Membrane Science 99(2), 197–203.

Daneshvar, N., Salari, D. and Aber, S. (2002) Chromium adsorption and Cr(VI) reduction to trivalent chromium in aqueous solutions by soya cake. Journal of Hazardous Materials 94(1), 49–61. .

Debashan, L.E. and Bashan, Y. (2004) Recent advances in removing phosphorus from wastewater and its future use as fertilizer (1997–2003). Water Research 38(19), 4222–4246.

Duan, J., Litwiller, E., Choi, S.H. and Pinnau, I. (2014) Evaluation of sodium lignin sulfonate as draw solute in forward osmosis for desert restoration. Journal of Membrane Science 453(2), 463–470.

Emadzadeh, D., Lau, W.J., Matsuura, T., Ismail, A.F. and Rahbari-Sisakht, M. (2014) Synthesis and characterization of thin film nanocomposite forward osmosis membrane with hydrophilic nanocomposite support to reduce internal concentration polarization. Journal of Membrane Science 449, 74–85.

Etter, B., Tilley, E., Khadka, R. and Udert, K.M. (2011) Low-cost struvite production using source-separated urine in Nepal. Water Research 45(2), 852–862.

Frank, B.S. (1972) Desalination of sea water. U.S. Patent 3,670,897. 1972-6-20.

Ge, Q., Han, G. and Chung, T.-S. (2016) Effective as(III) removal by a multi-charged hydroacid complex draw solute facilitated forward osmosis-membrane distillation (FO-MD) processes. Environmental Science & Technology 50(5), 2363–2370.

Ge, Q., Su, J., Amy, G.L. and Chung, T.S. (2012a) Exploration of polyelectrolytes as draw solutes in forward osmosis processes. Water Research 46(4), 1318–1326.

Ge, Q., Wang, P., Wan, C. and Chung, T.S. (2012b) Polyelectrolyte-promoted forward osmosis-membrane distillation (FO-MD) hybrid process for dye wastewater treatment. Environmental Science & Technology 46(11), 6236–6243.

Gray, G.T., McCutcheon, J.R. and Elimelech, M. (2006) Internal concentration polarization in forward osmosis: role of membrane orientation. Desalination 197(1), 1–8.

Gupta, D.V.S. and Hlidek, B.T. (2009) Frac fluid recycling and water conservation: a case history. Spe Production & Operations 25(1), 65–69.

Han, F., Kambala, V.S.R., Srinivasan, M., Rajarathnam, D. and Naidu, R. (2009) Tailored titanium dioxide photocatalysts for the degradation of organic dyes in wastewater treatment: a review. Applied Catalysis A General 359(1), 25–40.

Han, G., Zhang, S., Li, X. and Chung, T.-S. (2013) High performance thin film composite pressure retarded osmosis (PRO) membranes for renewable salinity-gradient energy generation. Journal of Membrane Science 440, 108–121.

Hough, W. (1973) Forward-osmosis solvent extraction. U.S. Patent 3,721,621. 1973-3-20.

Husnain, T., Liu, Y., Riffat, R. and Mi, B. (2015a) Integration of forward osmosis and membrane distillation for sustainable wastewater reuse. Separation and Purification Technology 156, 424–431.

Husnain, T., Mi, B. and Riffat, R. (2015b) A combined forward osmosis and membrane distillation system for sidestream treatment. Journal of Water Resource and Protection 07(14), 1111–1120.

Islam, M.R. and Chakma, A. (1989) Mechanisms and process selection criteria for waste water treatment in petroleum production. Annual Technical Meeting, Petroleum Society of Canada.

Kessler, J. and Moody, C. (1976) Drinking water from sea water by forward osmosis. Desalination 18(3), 297–306.

Kim, H.-G., Jang, H.-N., Kim, H.-M., Lee, D.-S. and Chung, T.-H. (2010) Effects of the sludge reduction system in MBR on the membrane permeability. Desalination 250(2), 601–604.

Larsen*, T.A., Alder, A.C., Eggen, R.I.L., Maurer, M. and Lienert, J. (2009) Source separation: will we see a paradigm shift in wastewater handling?. Environmental Science & Technology 43(16), 6121–6125.

Lau, W.J., Ismail, A.F., Misdan, N. and Kassim, M.A. (2012) A recent progress in thin film composite membrane: a review. Desalination 287, 190–199.

Lee, K.P., Arnot, T.C. and Mattia, D. (2011) A review of reverse osmosis membrane materials for desalination – development to date and future potential. Journal of Membrane Science 370(1–2), 1–22.

Li, Y.S., Yan, L., Xiang, C.B. and Hong, L.J. (2006) Treatment of oily wastewater by organic–inorganic composite tubular ultrafiltration (UF) membranes. Desalination 196(1), 76–83.

Lind, B.-B., Ban, Z. and Bydén, S. (2001) Volume reduction and concentration of nutrients in human urine. Ecological Engineering 16(4), 561–566.

Ling, M.M., Chung, T.S. and Lu, X. (2011) Facile synthesis of thermosensitive magnetic nanoparticles as "smart" draw solutes in forward osmosis. Chemical communications (Cambridge) 47(38), 10788–10790.

Liu, Q., Liu, C., Zhao, L., Ma, W., Liu, H. and Ma, J. (2016) Integrated forward osmosis-membrane distillation process for human urine treatment. Water Research 91, 45–54.

Loeb, S. and Sourirajan, S. (1960) Sea water demineralization by means of an osmotic membrane, University of California, Los Angeles, Department of Engineering Report.

Loeb, S., Titelman, L., Korngold, E. and Freiman, J. (1997) Effect of porous support fabric on osmosis through a Loeb-Sourirajan type asymmetric membrane. Journal of Membrane Science 129(2), 243–249.

Luo, H., Wang, Q., Zhang, T.C., Tao, T., Zhou, A., Chen, L. and Bie, X. (2014) A review on the recovery methods of draw solutes in forward osmosis. Journal of Water Process Engineering 4, 212–223.

Maurer, M., Pronk, W. and Larsen, T.A. (2006) Treatment processes for source-separated urine. Water Research 40(17), 3151–3166.

McCutcheon, J.R. and Elimelech, M. (2006) Influence of concentrative and dilutive internal concentration polarization on flux behavior in forward osmosis. Journal of Membrane Science 284(1–2), 237–247.

Mi, B. and Elimelech, M. (2010) Organic fouling of forward osmosis membranes: fouling reversibility and cleaning without chemical reagents. Journal of Membrane Science 348(1), 337–345.

Minear, R.A. (1996) Water disinfection and natural organic matter: history and overview. Acs Symposium 649, 1–9.

Nguyen, N.C., Nguyen, H.T., Ho, S.T., Chen, S.S., Ngo, H.H., Guo, W., Ray, S.S. and Hsu, H.T. (2016) Exploring high charge of phosphate as new draw solute in a forward osmosis-membrane distillation hybrid system for concentrating high-nutrient sludge. Science of the Total Environment 557–558, 44–50.

Ohya, H., Kim, J.J., Chinen, A., Aihara, M., Semenova, S.I., Negishi, Y., Mori, O. and Yasuda, M. (1998) Effects of pore size on separation mechanisms of microfiltration of oily water, using porous glass tubular membrane. Journal of Membrane Science 145(1), 1–14.

Ong, R.C., Chung, T.-S., de Wit, J.S. and Helmer, B.J. (2015) Novel cellulose ester substrates for high performance flat-sheet thin-film composite (TFC) forward osmosis (FO) membranes. Journal of Membrane Science 473, 63–71.

Phuntsho, S., Shon, H.K., Majeed, T., El, S.I., Vigneswaran, S., Kandasamy, J., Hong, S. and Lee, S. (2012) Blended fertilizers as draw solutions for fertilizer-drawn forward osmosis desalination. Environmental Science & Technology 46(8), 4567–4575.

Podsiadlo, P., Kaushik, A.K., Arruda, E.M., Waas, A.M., Shim, B.S., Xu, J., Nandivada, H., Pumplin, B.G., Lahann, J. and Ramamoorthy, A. (2007) Ultrastrong and stiff layered polymer nanocomposites. Science 318(5847), 80–83.

Porter, M.C. (1972) Concentration polarization with membrane ultrafiltration. Industrial & Engineering Chemistry Product Research and Development 11(3), 234–248.

Río, A.I.D., Fernández, J., Molina, J., Bonastre, J. and Cases, F. (2011) Electrochemical treatment of a synthetic wastewater containing a sulphonated azo dye. Determination of naphthalenesulphonic compounds produced as main by-products. Desalination 273(2–3), 428–435.

Song, X., Liu, Z. and Sun, D.D. (2011) Nano gives the answer: breaking the bottleneck of internal concentration polarization with a nanofiber composite forward osmosis membrane for a high water production rate. Advanced Materials 23(29), 3256–3260.

Su, J., Ong, R.C., Wang, P., Chung, T.-S., Helmer, B.J. and de Wit, J.S. (2013) Advanced FO membranes from newly synthesized CAP polymer for wastewater reclamation through an integrated FO-MD hybrid system. AIChE Journal 59(4), 1245–1254.

Sun, Y., Xue, L., Zhang, Y., Zhao, X., Huang, Y. and Du, X. (2014) High flux polyamide thin film composite forward osmosis membranes prepared from porous substrates made of polysulfone and polyethersulfone blends. Desalination 336, 72–79.

Tice, R.C. and Kim, Y. (2014) Energy efficient reconcentration of diluted human urine using ion exchange membranes in bioelectrochemical systems. Water Research 64, 61–72.

Us Epa, O. (2004) Evaluation of impacts to underground sources of drinking water by hydraulic fracturing of coalbed methane reservoirs study.

van Reis, R. and Zydney, A. (2007) Bioprocess membrane technology. Journal of Membrane Science 297(1–2), 16–50.

Wang, H., Chung, T.S., Tong, Y.W., Jeyaseelan, K., Armugam, A., Chen, Z., Hong, M. and Meier, W. (2012) Highly permeable and selective pore-spanning biomimetic membrane embedded with Aquaporin Z.

Wang, H. and Wang, H. (2009) Mitigation of lake eutrophication: loosen nitrogen control and focus on phosphorus abatement. Progress in Natural Science 19(10), 1445–1451.

Wang, K.Y., Teoh, M.M., Nugroho, A. and Chung, T.-S. (2011) Integrated forward osmosis–membrane distillation (FO–MD) hybrid system for the concentration of protein solutions. Chemical Engineering Science 66(11), 2421–2430.

Wang, P., Cui, Y., Ge, Q., Fern Tew, T. and Chung, T.-S. (2015) Evaluation of hydroacid complex in the forward osmosis–membrane distillation (FO–MD) system for desalination. Journal of Membrane Science 494, 1–7.

Wang, Z., Wu, Z., Hua, J., Wang, X., Du, X. and Hua, H. (2008) Application of flat-sheet membrane to thickening and digestion of waste activated sludge (WAS). Journal of Hazardous Materials 154(1), 535–542.

Widjojo, N., Chung, T.-S., Weber, M., Maletzko, C. and Warzelhan, V. (2013) A sulfonated polyphenylenesulfone (sPPSU) as the supporting substrate in thin film composite (TFC) membranes with enhanced performance for forward osmosis (FO). Chemical Engineering Journal 220, 15–23.

Wild, D., Kisliakova, A. and Siegrist, H. (1997) Prediction of recycle phosphorus loads from anaerobic digestion. Water Research 31(9), 2300–2308.

Xie, M., Nghiem, L.D., Price, W.E. and Elimelech, M. (2013) A forward osmosis-membrane distillation hybrid process for direct sewer mining: system performance and limitations. Environmental Science & Technology 47(23), 13486–13493.

Xie, M., Nghiem, L.D., Price, W.E. and Elimelech, M. (2014) Toward resource recovery from wastewater: extraction of phosphorus from digested sludge using a hybrid forward osmosis–membrane distillation process. Environmental Science & Technology Letters 1(2), 191–195.

Xu, Z.-L., Chung, T.-S., Loh, K.-C. and Lim, B.C. (1999) Polymeric asymmetric membranes made from polyetherimide/polybenzimidazole/poly(ethylene glycol) (PEI/PBI/PEG) for oil–surfactant–water separation. Journal of Membrane Science 158(1), 41–53.

Yen, S.K., Mehnas Haja N.F., Su, M., Wang, K.Y. and Chung, T.-S. (2010) Study of draw solutes using 2-methylimidazole-based compounds in forward osmosis. Journal of Membrane Science 364(1–2), 242–252.

Zhang, S., Fu, F. and Chung, T.-S. (2013) Substrate modifications and alcohol treatment on thin film composite membranes for osmotic power. Chemical Engineering Science 87, 40–50.

Zhang, S., Wang, P., Fu, X. and Chung, T.S. (2014) Sustainable water recovery from oily wastewater via forward osmosis-membrane distillation (FO-MD). Water Research 52, 112–121.

Zhao, D., Wang, P., Zhao, Q., Chen, N. and Lu, X. (2014) Thermoresponsive copolymer-based draw solution for seawater desalination in a combined process of forward osmosis and membrane distillation. Desalination 348, 26–32.

Zhao, S., Zou, L., Tang, C.Y. and Mulcahy, D. (2012) Recent developments in forward osmosis: opportunities and challenges. Journal of Membrane Science 396, 1–21.

Zhao, Z.-P., Xu, L., Shang, X. and Chen, K. (2013) Water regeneration from human urine by vacuum membrane distillation and analysis of membrane fouling characteristics. Separation and Purification Technology 118, 369–376.

14 Freeze Desalination-Membrane Distillation-Crystallization (FD-MD-C)

Kang-Jia Lu, Zhen Lei Cheng, Jian Chang, Lin Luo, and Tai-Shung Chung

14.1 INTRODUCTION

In order to mitigate the water scarcity, various desalination technologies have been developed in the past few decades. Among them, desalination based on membrane technologies contributes to nearly 68% of the total desalination capacity, and this figure is constantly increasing as the membrane market grows rapidly (Nakoa et al., 2016). Nevertheless, typical membrane processes such as reverse osmosis (RO) have a hydraulic recovery limit of 75%–85% due to the economic concern. Thus, at least 15%–25% of saline water must be discarded as brine (Cappelle et al., 2017). Unfortunately, direct discharge to ocean or surface water is still the most common method for brine disposal, and its detrimental impact on ecosystems has been widely reported (Ahmed et al., 2001; El-Naas, 2011; Missimer and Maliva, 2018). One of the sustainable solutions is to develop zero liquid discharge desalination (ZLDD) systems that convert seawater into potable water and valuable salts without generating a waste stream.

The combination of membrane distillation and crystallization (MD-C) has been proposed to achieve zero liquid discharge (ZLD) (Creusen et al., 2013; Edwie and Chung, 2013; Guan et al., 2012; Julian et al., 2016; Quist-Jensen et al., 2016; Suk and Matsuura, 2006). In an MD-C system, MD is mainly used for water recovery. The concentrate from MD is then sent to a crystallizer where value-added salt crystals are extracted (Chan et al., 2005; Julian et al., 2016). By returning the crystallizer brine to MD where it combines with the feed solution for the next cycle of water and salt recovery, a zero-discharge status is achieved.

However, the MD-C process suffers from the intensive energy consumption (Al-Obaidani et al., 2008). One of the solutions is to couple it with renewable energy or low-grade waste energy.

Solar energy is the most commonly used green energy to power the system, demonstrating a great potential to lower the energy cost of MD (Cipollina et al., 2012; Khayet, 2013; Selvi and Baskaran, 2015; Shim et al., 2015). Another widely used method is to integrate MD-C with other water recovery processes in order to reduce the specific energy consumption. One of the low-cost water recovery technologies is freeze desalination (FD). As mentioned in Chapter 11, FD refers to the process that extracts fresh water by harvesting ice crystals from the chilled saline water (Kadi and Janajreh, 2017; Wang and Chung, 2012). Since the latent heat of ice fusion is only one-seventh of that of water vaporization, it is more energy saving to produce water by FD instead of distillation (Chang et al., 2016). In addition, the energy cost can be further reduced if the cryogenic energy for FD is provided by waste energy. For example, in Singapore, more than 6 million tons per annum (Mtpa) of natural gas (NG) is imported from overseas (Singapore, 2017). For the ease of transportation, NG is stored in a liquid form at around −162 °C. Thus, it must be regasified in LNG terminals before being supplied to end users. Because of the low temperature and high energy density of LNG, an enormous amount of high-quality cold energy is reserved in the regasification process, making LNG a perfect cooling source for FD (Chang et al., 2016; Wang and Chung, 2012). On the other hand, similar to many conventional water recovery technologies, FD is vulnerable to high salt concentrations (Chang et al., 2016). Ice crystals generated from FD have a high surface energy and an enormous specific surface area due to their small sizes. As a result, salts tend to adhere on the surfaces of ice crystals. Pure water is needed to wash these salts off from the ice crystals (Chang et al., 2016). When using a high-salinity feed, a large number of salt crystals will be adsorbed on ice surfaces. Therefore, a great amount of washing water must be consumed. This will offset the amount of pure water generated from ice and makes the process uneconomical. By contrast, the performance of MD-C is less affected by the feed salinity. One can integrate FD and MD-C together so that they are supplementary to each other in order to achieve an improved water recovery with less energy consumption.

In this chapter, an FD-MD-C hybrid system with an ambitious goal to achieve 100% recovery of water and salts will be illustrated and discussed. Effects of important operating parameters such as (1) the feed temperature, concentration, and distillate temperature of MD and (2) the recovery ratio of FD on system operating curves and energy consumptions will be systematically discussed. Eventually, a case study where an optimized system is supported by solar energy and cold energy from regasification of LNG will be discussed. It is believed that this chapter may provide valuable insights to design and evaluate the energy consumption of ZLDD systems consisting of FD and MD-C.

14.2 WORKING PRINCIPLES OF AN FD-MD-C HYBRID SYSTEM FOR ZLDD

A schematic diagram of the FD-MD-C hybrid system designed by Lu et al. is shown in Figure 14.1 (Lu et al., 2019). Seawater is fed to an FD unit where nearly pure ice crystals are generated and separated from the concentrated brine. The concentrated brine is then sent to the feed tank of a hollow fiber MD unit. Subsequently, a designated amount of the feed is drawn from the feed tank, heated to a set temperature, and sent to the shell side of the MD module. Inside

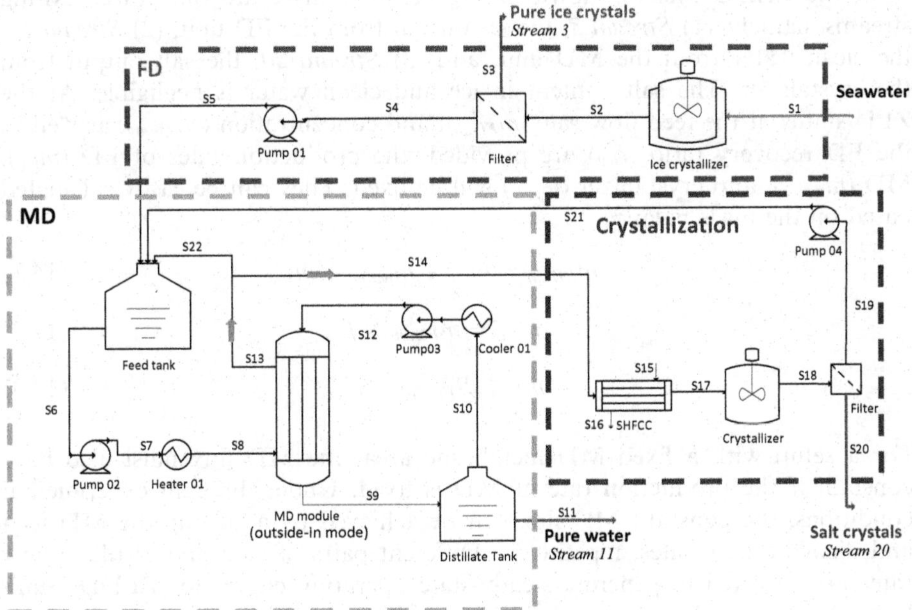

FIGURE 14.1 Schematic diagram of the freeze desalination (FD)-membrane distillation and crystallization (MD-C) hybrid system with zero liquid discharge (ZLD).

the MD unit, water vapor transports across the membrane and is condensed by the cold stream in the lumen side. At the steady state, the rate of vapor condensation is equivalent to the water production rate at *Stream 11*. Due to the loss of water, the feed is concentrated in MD. Part of the feed outlet stream is sent to a solid hollow fiber cooling crystallizer (SHFCC) where it will be cooled for salt crystallization (Luo et al., 2017; Zarkadas and Sirkar, 2004). The remaining feed outlet is recycled to the feed tank. The chilled stream from SHFCC enters a crystallizer where salt precipitation takes place. Subsequently, the salt is separated from the stream by a filter and the brine is recycled to the feed tank, as indicated by *Streams 20* and *19*, respectively.

For the sake of simplicity, several streams and parameters were predetermined. A PVDF membrane with an outer diameter of 1017 μm, a thickness of 109.7 μm, a porosity of 85%, an average pore size of 150 nm, and a membrane tortuosity of 2 was used in DCMD. For the illustration purpose, the DCMD module was set to be 1 m long, with a diameter of 0.5 inch and a packing density of 50%. The total membrane area was 0.14 m^2. In real applications, the feed composition and flow rate for the FD-MD-C hybrid plant are usually fixed. Therefore, we set the feed stream to be a model seawater (3.5 wt% of NaCl, 25 °C) with a flow rate of 3 kg/h. The feed flow rate was set so that a wide range of operating curves can be explored. The stream temperature at the outlet of SHFCC unit (*Stream* 17) was set to room temperature (25 °C).

In the ZLDD system shown in Figure 14.1, there are only three exiting streams, namely, (1) *Stream 3*, the ice output from the FD unit, (2) *Stream 11*, the clean water from the MD unit, and (3) *Stream 20*, the salt output from the crystallizer. The salt content in ice and clean water is negligible. At the ZLD status, if the feed flow rate (m_{FD_in}) and concentration (S_{FD_in}), as well as the FD recovery ratio (f_{ice}) are provided, the production rates of FD (m_{ice}), MD (m_{water}), and crystallizer (m_{salt}) will be fixed. They can be easily calculated based on the mass balance:

$$m_{FD\_in} = m_{ice} + m_{water} + m_{salt} \tag{14.1}$$

$$m_{ice} = m_{FD\_in} \times f_{ice} \tag{14.2}$$

$$m_{FD\_in} S_{FD\_in} = m_{salt} \tag{14.3}$$

For a setup with a fixed MD membrane area, the MD flux must also be a constant if the production rate of MD is fixed. Under different experimental conditions, the constant MD flux can be achieved by adjusting the MD feed and distillate flow rates, repetitively. Different pairs of feed and distillate flow rates are depicted to generate steady-state operating curves for ZLDD. Similarly, the salt production rate of the crystallizer is also constant at the ZLD status. It is realized by controlling the flow rate of the crystallizer feed (i.e., *Stream 14*) depending on the outlet concentration of the MD feed stream (i.e., *Stream 13*).

Energy transfer of the hybrid system takes place in all three units. In the FD unit, cooling energy Q_{FD_Cool} is required to bring down the feed temperature to the freezing point and to form ice crystals. In the crystallization unit, cooling energy Q_{MC_Cool} is needed to cool the hot brine from MD and to form salt crystals. In the MD unit, Q_{MD_Cool} is equivalent to the total energy transferred cross the MD membrane; Q_{MD_Heat} is the total energy required to heat (1) the FD brine before it entering MD (*Stream 5*), (2) the crystallizer brine back to the feed tank (*Stream 21*), and (3) the recycled MD feed (*Stream 22*) to the designated feed inlet temperature. The total pumping energy Q_{pump} is evaluated based on the pressure drop (ΔP) and volumetric flow velocity (v) by the following equation:

$$Q_{pump} = \sum_i \Delta P_i v_i \tag{14.4}$$

In a ZLDD system, both water and salt are valuable products. Therefore, the specific energy consumption (SEC) in this work is not defined based on water production but the sum of both water and salt productions (i.e., the total amount of feed processed). Different types of specific energy consumptions are calculated based on the following equations:

$$SEC_i = \frac{|Q_i|}{m_{FD\_in}} \tag{14.5}$$

where the subscript i denotes the type of energy. The total specific energy consumption (SEC_{tot}) is the sum of all different types of specific energy consumptions, namely, SEC of (1) the pumping energy (SEC_{pump}), (2) the cooling for FD (SEC_{FD_Cool}), (3) the cooling for MD (SEC_{MD_Cool}), (4) the cooling for the crystallizer (SEC_{MC_Cool}), and (5) the heating for MD (SEC_{MD_Heat}).

$$SEC_{tot} = SEC_{pump} + SEC_{FD\_Cool} + SEC_{MD\_Cool} + SEC_{MC\_Cool} + SEC_{MD\_Heat} \tag{14.6}$$

14.3 EFFECTS OF DIFFERENT PARAMETERS ON SYSTEM OPERATING CURVES AND ENERGY CONSUMPTIONS

14.3.1 EFFECTS OF MD FEED INLET TEMPERATURE

Many studies have investigated the effects of feed inlet temperature in an MD unit. However, very few works study this parameter in a hybrid system that considers both feed recycling and combined effects from different units. In this section, feed temperature is varied from 65 to 80 °C. For the illustration purpose, the recovery ratio of FD, distillate temperature, and feed inlet concentration of MD are set to be 30%, 5 °C, and 27%, respectively. Figure 14.2A shows the operating curves of MD in the hybrid system with the goal to achieve the ZLD status. This figure is obtained by varying the feed and distillate flow rates via iteration until ZLD status is reached. As mentioned previously, the overall water production rate of the hybrid system is fixed at the ZLD status, thus the MD flux across the membrane, which equals to the ratio of MD production rate to membrane area, must be fixed. Therefore, if the heat and mass transfer coefficients are increased on the feed side by increasing the feed flow rate, they should be reduced on the other side by decreasing the distillate flow rate in order to keep a constant

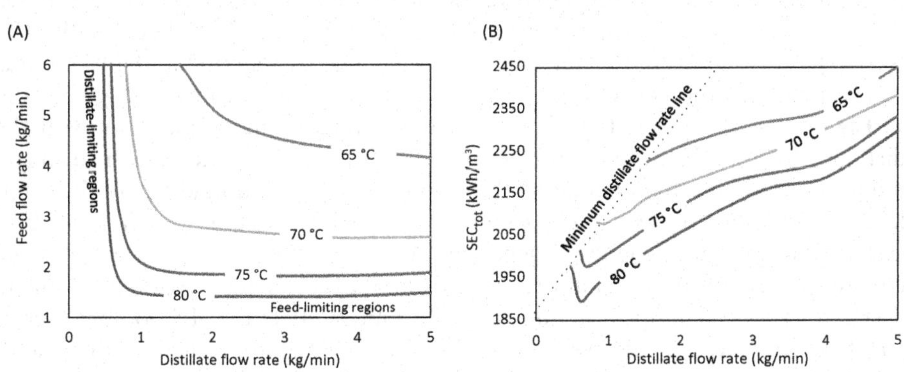

FIGURE 14.2 (A) Operating curves. (B) Total specific energy consumption at different feed temperatures and distillate flow rates.

flux. As a result, the feed and distillate flow rates have a convex relationship; a higher feed flow rate corresponds to a lower distillate flow rate and vice versa. This phenomenon becomes more severe at high MD feed temperatures, the sharp convex relationship transforms the operating curves into two interesting regimes, namely, feed-limiting and distillate-limiting regions.

As shown in Figure 14.2A, the feed-limiting region takes place near the lower boundary of the feed flow rate where the adjustment of distillate flow rate has a minimal effect on MD performance while a slight change in feed flow rate causes a drastic impact. This phenomenon arises from the fact that the exponent of the Reynolds number in calculating the heat transfer coefficient is smaller than one, indicating that the heat and mass transfer coefficients increase sublinearly with the Reynolds number. Thus, the enhancive effect of the feed flow rate on heat and mass transfer is significant only at the beginning but it will eventually reach a plateau. In the feed-limiting region, the distillate flow rate is high while the feed flow rate is low. Thus, the adjustment of the feed flow rate results in a much more obvious effect than that of the distillate one. Similarly, the operating curves enter the distillate-limiting region near the lower boundary of the distillate flow rate. In MD, the driving force for water transport is proportional to the feed temperature. A high feed temperature shifts the operating curves to the lower left corner of Figure 14.2A. In other words, since the system can tolerate more temperature and concentration polarizations at high temperatures, lower feed flow rates must be chosen for the MD unit in order to keep a constant water flux.

The effects of feed inlet temperature and distillate flow rates on total specific energy consumption are depicted in Figure 14.2B. Since every distillate flow rate corresponds to a specific feed flow rate at a given feed temperature as shown in Figure 14.2A, only distillate flow rate is plotted as the x-axis in Figure 14.2B to avoid redundancy. Figure 14.2B suggests that an elevated feed temperature in MD lowers the total specific energy consumption of the hybrid system. In contrast, generally more energy is consumed if MD is operated at a higher distillate flow rate (or a lower feed flow rate). In order to explore the physical meaning of the trends, the total specific energy consumption has been decomposed and the trends of each individual energy component are shown in Figure 14.3. Because the FD recovery ratio is fixed in this section, the cooling energy required in the FD unit is constant for all the cases. Therefore, SEC_{FD_Cool} is not shown in Figure 14.3.

Figure 14.3A depicts that in general the pumping energy increases with the distillate flow rate because the pressure drop along the lumen side for the distillate fluid is more severe than that along the shell side for the feed fluid due to the smaller flow channels. However, comparing to other energy consumptions in the hybrid system, the pumping energy is generally negligible. Figure 14.3B illustrates the cooling energy consumption in MD as a function of feed temperature and distillate flow rate. The feed temperature does not have much effect on MD cooling energy, which is attributed to the constant flux at the ZLD status. On the other hand, more cooling energy is needed at a larger distillate flow rate because of the lower thermal efficiency (i.e., a higher conductive energy loss across the membrane).

Figure 14.3C shows that the required cooling energy for the crystallizer decreases with an increase in MD distillate flow rate. Since, under the ZLD

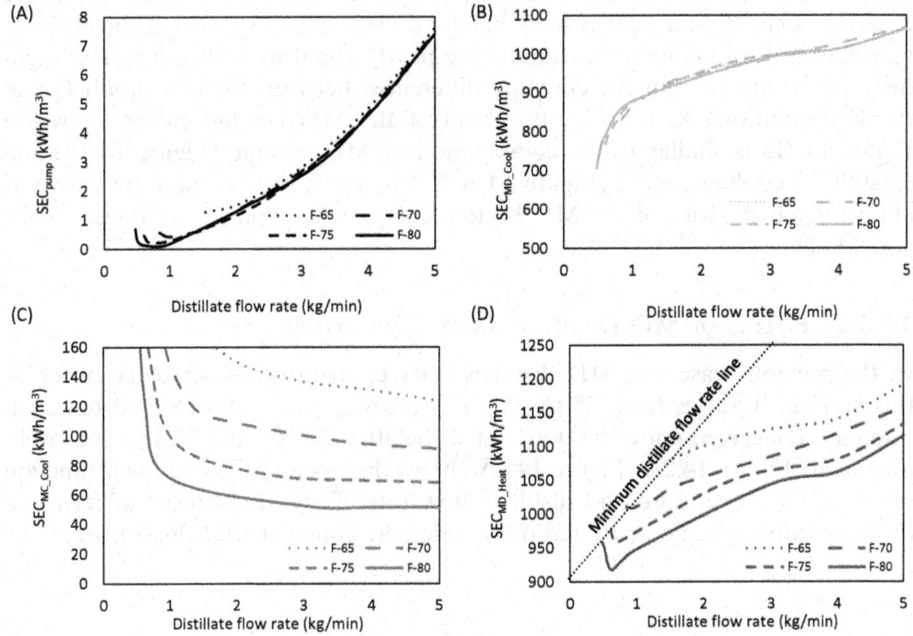

FIGURE 14.3 Different specific energy consumptions versus distillate flow rates: (A) system pumping energy, (B) MD cooling energy, (C) crystallization cooling energy, and (D) MD heating energy.

status, the feed flow rate must decrease when the distillate flow rate increases in order to keep the same MD flux, this results in a higher feed outlet concentration and a lower feed outlet temperature. As mentioned previously, a portion of the MD feed outlet stream will be sent to the crystallizer for salt production. At the ZLD status, the salt production rate is fixed. If the MD feed outlet concentration increases, a smaller portion of it would be transported to the crystallizer; thus, less cooling energy is required. In addition, a lower MD feed outlet temperature facilitates the salt precipitation, which also results in a lower cooling energy required in the crystallizer. Similarly, the required cooling energy in the crystallizer also decreases if the MD feed inlet temperature increases. This is because a higher feed inlet temperature results in a lower feed flow rate as displayed in Figure 14.2A. At a constant MD flux, it leads to a more concentrated MD feed outlet, thus less cooling energy is required in the crystallizer to induce crystallization. In a nutshell, the required cooling energy for the crystallizer is strongly proportional to the feed flow rate of MD. A higher feed flow rate causes a larger SEC_{MD_Cool} and vice versa. Therefore, the trends of MD feed flow rate and SEC_{MD_Cool} versus distillate flow rate (i.e., Figure 14.2A vs. 14.3C) are quite similar.

Figure 14.3D shows the total heating energy of MD as a function of distillate flow rate, which is equal to the sum of the energy to heat the (1) FD brine (*stream 5*), (2) brine from the crystallizer (*stream 21*), and (3) recycled MD feed (*stream 22*) to the designated feed inlet temperature. At steady state, the entire ZLDD system should be thermally balanced. The results show that the absolute values of hot and cold energy inputs are nearly equal to each other. The slight discrepancy arises from the enthalpy differences between the end products and the feed solution. As a result, the trend of the MD heating energy shown in Figure 14.3D is similar to the combination of MD cooling (Figure 14.3B) and crystallizer cooling energy (Figure 14.3C). This leads to that the total energy is almost equal to twice of the MD heating energy when comparing Figure 14.2B and 14.3D.

14.3.2 Effects of MD Distillate Inlet Temperature

In the previous case, the MD distillate inlet temperature is set to be 5 °C. In this section, it varies from 5° to 45°. The recovery ratio of FD, feed temperature, and concentration of MD are set to be 30%, 80 °C, and 27%, respectively. Similar to Figure 14.2A, Figure 14.4A shows the operating curves as a function of distillate temperature and distillate flow rate. They also consist of feed and distillate-limiting regions. In addition, under the constant MD flux set by ZLD,

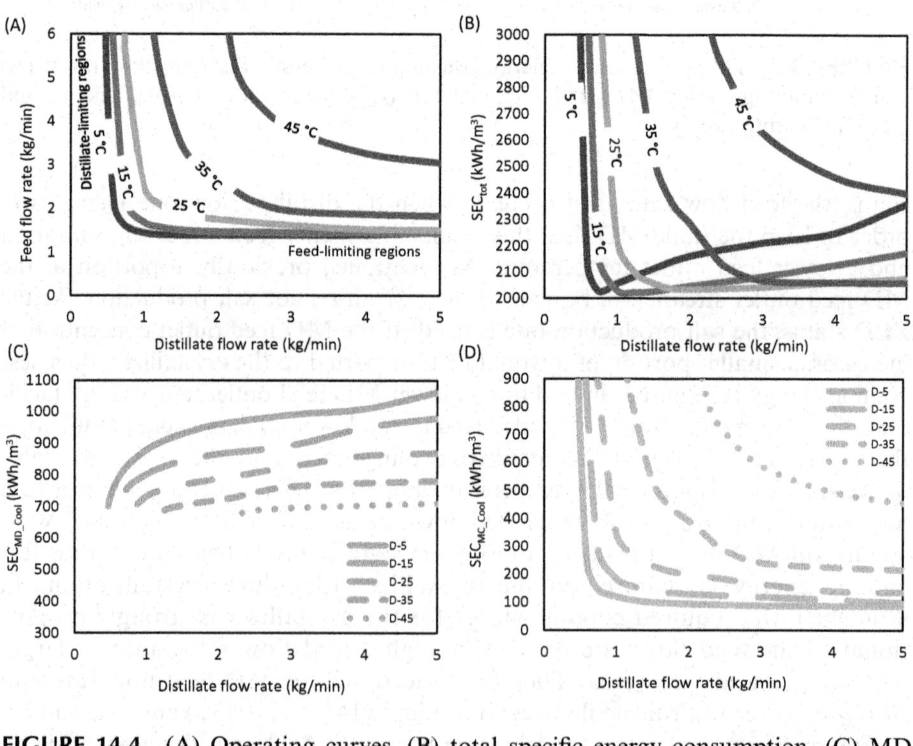

FIGURE 14.4 (A) Operating curves, (B) total specific energy consumption, (C) MD cooling energy, and (D) crystallization cooling energy at different distillate temperatures and distillate flow rates.

a lower distillate temperature corresponds to a smaller feed flow rate due to the greater driving force for water transport across the membrane.

The trend of the total energy consumption for the hybrid system is depicted in Figure 14.4B. In the distillate-limiting region, an increase in distillate flow rate leads to a drastic drop in total energy consumption, and a lower distillate temperature is beneficial for reducing the total energy consumption. However, as distillate flow rate increases, the decrease of the total energy slows down or even inverts to increase. To investigate this trend, the constituent components of total energy consumption are analyzed. As mentioned in the previous section, (1) the cooling energy required in FD is fixed, (2) the pumping energy is negligible, and (3) the heating energy consumption in MD has the same trend as the total energy consumption; therefore, only cooling energy components in MD and crystallization units are displayed. Figure 14.4C shows that the cooling energy consumption in MD increases gradually with a larger distillate flow rate and a lower distillate temperature, which can be attributed to the lower energy efficiency. Comparing Figure 14.4A and 14.4D, a great similarity can be observed. This can be ascribed to the strong proportional relationship between the crystallization cooling energy and the feed flow rate as elaborated in the previous section. By combining Figure 14.4C and 14.4D, the trend of the total energy consumption shown in Figure 14.4B can therefore be easily understood.

14.3.3 Effect of MD Feed Inlet Concentration

Figure 14.5A and 14.5B displays the operating curves and total specific energy consumptions as a function of MD feed inlet concentration because it is another important parameter of the hybrid system. The feed inlet concentration in this section is chosen in the range of 26% to 27.5% based on the reason stated elsewhere (Lu et al., 2019), while the recovery ratio of FD, feed temperature, and distillate temperature of MD are set at 30%, 80 °C, and 5 °C, respectively. Even though both temperature and concentration polarizations are considered in the modeling, there is no obvious difference among the operating curves. Figure 14.5C also shows that the required cooling energy for MD does not vary much with the feed concentration. The results can be ascribed to (1) the weak dependence of MD driving force on feed concentration and (2) the constant MD flux required in the ZLDD system. However, previous works have shown that feed concentration does affect the membrane performance due to inorganic scaling; this issue and the mitigating measures will be discussed later. In contrast, Figure 14.5D displays that the required cooling energy for the crystallizer is reduced with an increase in feed concentration. This is because a less amount of MD feed outlet is needed in the crystallizer as discussed in the preceding sections. However, the cooling energy curves are getting close to one another, indicating that the amount of energy reduction is gradually reduced as the feed inlet concentration increases to about 27.5%. A feed inlet concentration of 26.5% is therefore selected for the following study because it offers the advantages of less cooling energy for salt crystallization and a lower tendency for inorganic scaling.

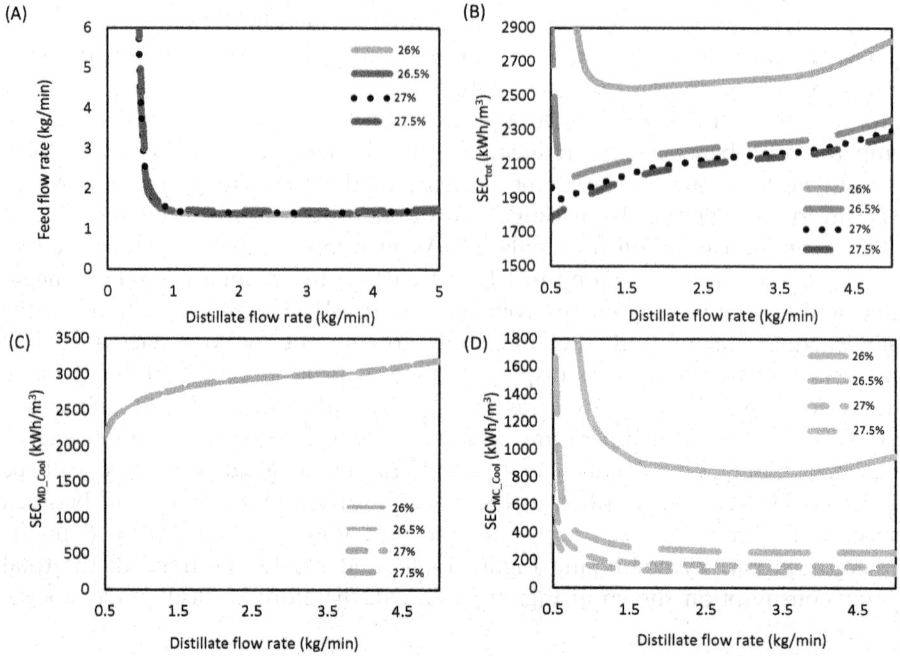

FIGURE 14.5 (A) Operating curves, (B) total specific energy consumption, (C) MD cooling energy, and (D) crystallization cooling energy at different feed concentrations and distillate flow rates.

14.3.4 Potential Scaling Issues at High Feed Concentrations

Scaling cannot be neglected in the MD-C hybrid system where a high salinity is indispensable. Salt scales can block membrane pores, induce membrane wetting, or even cause permanent damages to the membranes (Rezaei et al., 2018; Tijing et al., 2015; Warsinger et al., 2015); thus, measures must be taken to minimize the crystallization inside the membrane modules. Figure 14.6A displays the bulk and interfacial salt concentrations (referred to as S_b and S_{fm}, respectively) along the membrane module if the feed flow rate is low and about 1.38 kg/min. The salt concentration at the membrane surface deviates greatly from the bulk concentration due to the effect of concentration polarization. The degree of concentration polarization is exponentially proportional to the flux as shown in the following equation:

$$\text{CPC} = \exp\left(\frac{J}{\rho_w k}\right) \tag{14.7}$$

where ρ_w is the density of the feed solution and k is the mass transfer coefficient. J is the membrane flux. The flux and concentration polarization coefficient (CPC) along the module are also shown in Figure 14.6B. To investigate the possibility of scaling, the metastable concentration limit for the nucleation of

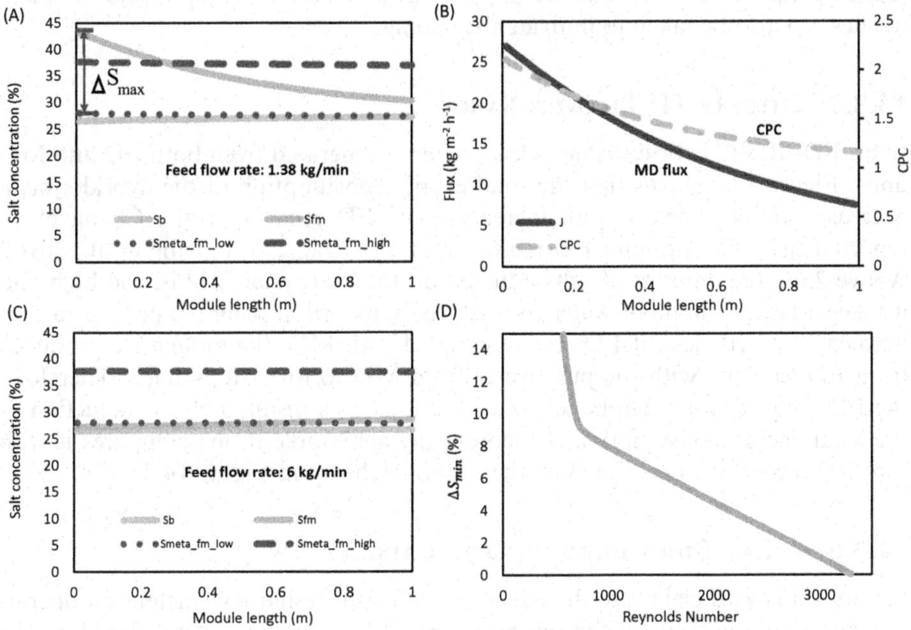

FIGURE 14.6 (A) Salt concentration along the membrane module (feed flow rate: 1.38 kg/min; Re). (B) MD flux and concentration polarization coefficient (CPC) along the module (feed flow rate: 1.38 kg/min). (C) Salt concentration along the membrane module (feed flow rate: 6 kg/min). (D) ΔS_{max} versus Reynolds number.

NaCl crystals (S_{meta_fm}) is evaluated on the feed temperature at the membrane surface $\left(T_{fm}\right)$. The metastable concentration limit is calculated based on the maximum allowable supersaturation. Usually, the feed reaches the metastable limit before crystallization takes place and the difference between S_{fm} and S_{meta_fm} is considered as the driving force for salt crystallization. However, measurements of the metastable concentration limit depend on experimental conditions and its value varies vastly in the literature (Chlanese et al., 1986; Mullin, 2001). Therefore, both the lowest ($S_{meta\_fm\_low}$) and highest ($S_{meta\_fm\_high}$) reported values, which correspond to degrees of supersaturation of 1.01 and 1.6, respectively, are plotted and the true metastable limit is believed to be in-between.

Figure 14.6A shows that the salt concentration on the membrane surface at the low feed flow can be even higher than the highest reported metastable limit. Salt crystallization on the membrane surface is almost inevitable. To mitigate the scaling, the mass transfer rate is enhanced by improving the feed flow rate. As shown in Figure 14.6C, the interfacial salt concentration drops drastically to near the lowest reported metastable limit at the feed flow rate of 6 kg/min. Figure 14.6D depicts the largest difference (ΔS_{max}) between S_{fm} and $S_{meta\_fm\_low}$ along the module as a function of Reynolds number. ΔS_{max} vanishes when the Reynolds number reaches 3300. Besides increasing the Reynolds number, measures such as reducing the membrane flux by increasing the membrane area and frequent washes can also be taken to mitigate the scaling.

14.3.5 Effect of FD Recovery Ratio

In an FD-MD-C hybrid system, clean water is generated from both FD and MD units. Figure 14.7 shows that the total energy consumption of the hybrid system decreases almost linearly with increasing the FD recovery ratio. Chang et al. reported that the optimum FD recovery ratio is near 25% (Chang et al., 2016). Above 25%, the amount of salts adhered on the ice crystals will be too high and it takes a large volume of water to wash the salts off, making the process uneconomical. Nevertheless, if FD can be coupled with MD, the molten ice produced from FD can mix with the pure water from MD to lower its salinity. Therefore, the FD recovery ratio can be increased to 30%. As a result, a 29.2% reduction of the total energy consumption is theoretically achievable if one compares it with zero FD recovery, as predicted by the equation shown in Figure 14.7.

14.3.6 A Case Study Based on Optimizations

Previous analyses elaborate the effects of different design parameters on operating curves and energy consumptions of the ZLDD hybrid system. Based on the analyses, an MD feed temperature of 80 °C, a distillate temperature of 5 °C, and an FD recovery of 30% are chosen in order to minimize the total specific energy consumption. An MD feed concentration of 26.5% is also selected in consideration of both energy consumption and scaling possibility. Without any heat recovery, the hot and cold energy requirements are 990 and 1010 kWh/m^3, respectively. Since heat integrations between hot and cold streams are possible,

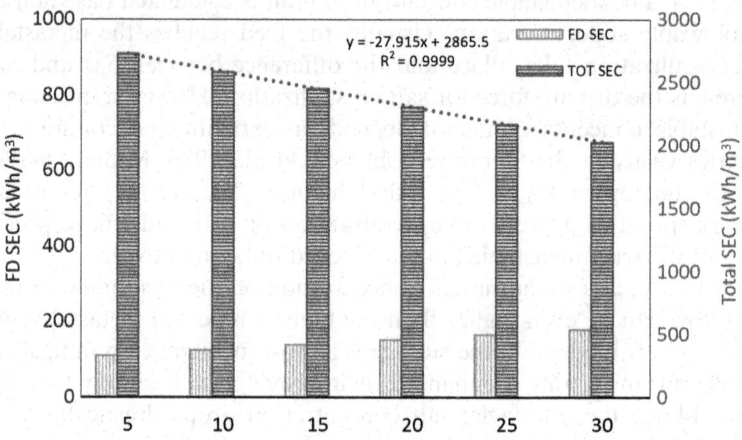

FIGURE 14.7 FD cooling energy and total energy consumption versus the FD recovery rate.

one can further reduce the energy consumption. For example, the recycling stream from the crystallizer (i.e., *Stream 21*) has to be heated from 25 to 80 °C, while the feed to the crystallizer (i.e., *Stream 14*) must be cooled from 53.4 to 25 °C. By heat exchanging between these two streams, some of their heating and cooling energies could be saved. Table 14.1 summarizes all the streams that either absorb or release heat, their supply and target temperatures, as well as heat flow rates. A widely applied technology, Pinch analysis, can be employed to evaluate the minimum energy requirement via heat integration. If a minimum temperature difference of 5 °C between the hot and cold streams is adopted for the Pinch analysis, the minimum specific hot and cold energies can drop to 810 and 830 kWh/m^3, respectively, indicating that 18% of the total energy can be saved after the heat integration of hot and cold streams. The values agree with the literature data for single-stage DCMD desalination (Khayet et al., 2006; Shim et al., 2015; Zakrzewska-Trznadel et al., 1999).

Though the energy consumption seems high compared to that of RO, it is important to take note that ZLD is achieved, which is impossible in RO. On the other hand, to make ZLD come true in the proposed hybrid system, one must utilize renewable energy or industrial waste energy in order to lower the energy cost. A semiconceptual case study is elucidated below by using solar energy and cold energy from regasification of LNG to power the hybrid system with the heat integrations.

If we aim to set up a hybrid pilot system that processes 72 kg of seawater per day with a goal to meet ZLDD, the pilot would have a salt production of 2.52 kg/day and a water production of 69.48 kg/day. 30% of the water recovery is achieved by the FD unit while the rest is accomplished by the DCMD unit. The daily hot and cold energy consumptions are calculated to be 58.3 and

TABLE 14.1

Streams involved in the pinch analysis

Stream name	Supply temperature [°C]	Target temperature [°C]	Heat capacity flowrate [kW/°C]	Heat flow [kW]	Energy type	Stream type
S3-1	−2.3	−2.3	–	0.0028	Latent	Cold
S3-2	−2.3	25	0.00035	0.0095	Sensible	Cold
S5	−2.3	80	0.00077	0.0632	Sensible	Cold
S21	25	80	0.00567	0.3121	Sensible	Cold
S22	53.38	80	0.02277	0.6061	Sensible	Cold
S1-1	25	−2.3	0.00111	0.0302	Sensible	Hot
S1-2	−2.3	−2.3	–	0.0093	Latent	Hot
S10	50.31	5	0.01742	0.7894	Sensible	Hot
S14	53.38	25	0.00568	0.1613	Sensible	Hot
S11	50.31	25	0.00081	0.0206	Sensible	Hot

S3-1, S3-2 and S1-1, S1-2 are streams 3 and 1 at different phases.

59.8 kWh, respectively. The hot energy can be compensated by solar energy in the form of either thermal energy or electricity. The former has a larger energy efficiency while the latter can provide a more stable energy input by being integrated into the power grid (Khayet, 2013; Lee et al., 2018; Selvi and Baskaran, 2015; Shim et al., 2015). Take a solar panel installed in Singapore which produces electricity as an example. It could generate a stable electricity output of 17.32 ± 1.11 kWh month^{-1} m^{-2} throughout the entire year of 2017 (Lu et al., 2019). Therefore, solar panels with an effective area of 50.5 m^2 can provide 50% of the heat energy for the hybrid system. The cold energy consumed in the pilot can be 100% supplied from the regasification of LNG because of its abundance in Singapore and it adsorbs 1040 kJ/kg during the regasification (Dorosz et al., 2018). Therefore, LNG of 207 kg is needed to provide the daily cold energy consumption for the hybrid system.

14.4 CONCLUSIONS

This chapter illustrates a novel FD-MD-C hybrid system for ZLDD. With the help of mathematical modeling, the effects of MD feed and distillate temperature, feed concentration, and FD recovery ratio on system operating curves, and energy consumption have been analyzed. In general, a high MD feed temperature and concentration, a low distillate temperature, and a large FD recovery ratio are preferred to reduce the overall energy consumption. Nevertheless, many other factors such as boundary conditions of the operating parameters and scaling effects should also be considered in real applications. In addition, the selection of feed and distillate flow rates significantly affects the energy consumption. For an optimized lab-scale ZLDD pilot with a daily output of 2.52 kg of salt and 69.48 kg of water, 50% of its heating energy can be supplied by solar panels with an effective area of 50.5 m^2 and 100% of its cooling energy can be provided by the regasification of 207 kg LNG. We hope this chapter provides a valuable guidance for designing a low-coast FD-MD-C hybrid desalination system with zero liquid discharge.

ACKNOWLEDGEMENT

This chapter was reprinted from K. J. Lu, Z. L. Cheng, J. Chang, L. Luo, T. S. Chung. Design of zero liquid discharge desalination (ZLDD) systems consisting of freeze desalination, membrane distillation, and crystallization powered by green energies. Desalination 2019, 458, 66–75. Copyright (2019), with permission from Elsevier.

NOMENCLATURES

Abbreviations

CPC concentration polarization coefficient
DCMD direct contact membrane distillation

FD	freeze desalination
LEP	liquid entry pressure [kPa]
LNG	liquefied natural gas
MC	membrane crystallization
MD	membrane distillation
Re	Reynolds number
SEC	specific energy consumption [kWh/m^3]
SHFCC	solid hollow fiber cooling crystallizer
TPC	temperature polarization coefficient
ZLD	zero liquid discharge
ZLDD	zero liquid discharge desalination

English letters

f_{ice}	FD recovery ratio
m	mass flow rate [kg/s]
P	pressure drop [Pa]
Q	heat transfer rate [W]
S	salt concentration [wt%]
v	volumetric flow rate [m^3/s]

Subscripts

FD_Cool	cooling energy required in FD unit
FD_in	FD feed inlet
ice	ice generated in FD unit
MC_Cool	cooling energy required in the crystallizer
MD_Cool	cooling energy required in MD unit
MD_Heat	heat energy required in MD unit
pump	pumping energy
salt	salt crystals
tot	total
w, water	water

REFERENCES

Ahmed, M., Shayya, W.H., Hoey, D., Al-Handaly, J., 2001. Brine disposal from reverse osmosis desalination plants in Oman and the United Arab Emirates. Desalination 133, 135–147. doi: 10.1016/S0011-9164(01)80004-7

Al-Obaidani, S., Curcio, E., Macedonio, F., Di Profio, G., Al-Hinai, H., Drioli, E., 2008. Potential of membrane distillation in seawater desalination: thermal efficiency, sensitivity study and cost estimation. J. Memb. Sci. 323, 85–98. doi: 10.1016/j.memsci.2008.06.006

Cappelle, M., Walker, W.S., Davis, T.A., 2017. Improving desalination recovery using zero discharge desalination (ZDD): a process model for evaluating technical feasibility. Ind. Eng. Chem. Res. 56, 10448–10460. doi: 10.1021/acs.iecr.7b02472

Chan, M.T., Fane, A.G., Matheickal, J.T., Sheikholeslami, R., 2005. Membrane distillation crystallization of concentrated salts – flux and crystal formation. J. Memb. Sci. 257, 144–155. doi: 10.1016/j.memsci.2004.09.051

Chang, J., Zuo, J., Lu, K.J., Chung, T.S., 2016. Freeze desalination of seawater using LNG cold energy. Water Res. 102, 282–293. doi: 10.1016/j.watres.2016.06.046

Chlanese, A., Cave, S.D., Mazzarotta, B., 1986. Solubility and metastable zone width of sodium chloride in water-diethylene glycol mixtures. J. Chem. Eng. Data 31, 329–332. doi: 10.1021/je00045a020

Cipollina, A., Di Sparti, M.G., Tamburini, A., Micale, G., 2012. Development of a membrane distillation module for solar energy seawater desalination. Chem. Eng. Res. Des. 90, 2101–2121. doi: 10.1016/j.cherd.2012.05.021

Creusen, R., van Medevoort, J., Roelands, M., van Renesse van Duivenbode A., Hanemaaijer, J.H., van Leerdam, R, 2013. Integrated membrane distillation-crystallization: process design and cost estimations for seawater treatment and fluxes of single salt solutions. Desalination 323, 8–16. doi: 10.1016/j.desal.2013.02.013

Dorosz, P., Wojcieszak, P., Malecha, Z., 2018. Exergetic analysis, optimization and comparison of LNG cold exergy recovery systems for transportation. Entropy 20. doi: 10.3390/e20010059

Edwie, F., Chung, T.S., 2013. Development of simultaneous membrane distillation-crystallization (SMDC) technology for treatment of saturated brine. Chem. Eng. Sci. 98, 160–172. doi: 10.1016/j.ces.2013.05.008

Kadi K.E., Janajreh, I., 2017. Desalination by freeze crystallization: an overview. Int. J. Therm. Environ. Eng. 15, 103–110. doi: 10.5383/ijtee.15.02.004

El-Naas, M.H., 2011. Reject brine management. Desalation, Trends Technol 237–252. doi:10.5772/60142

Guan, G., Wang, R., Wicaksana, F., Yang, X., Fane, A.G., 2012. Analysis of membrane distillation crystallization system for high salinity brine treatment with zero discharge using Aspen flowsheet simulation. Ind. Eng. Chem. Res. 51, 13405–13413. doi: 10.1021/ie3002183

Julian, H., Meng, S., Li, H., Ye, Y., Chen, V., 2016. Effect of operation parameters on the mass transfer and fouling in submerged vacuum membrane distillation crystallization (VMDC) for inland brine water treatment. J. Memb. Sci. 520, 679–692. doi: 10.1016/j.memsci.2016.08.032

Khayet, M., 2013. Solar desalination by membrane distillation: dispersion in energy consumption analysis and water production costs (a review). Desalination 308, 89–101. doi: 10.1016/j.desal.2012.07.010

Khayet, M., Mengual, J.I., Zakrzewska-Trznadel, G., 2006. Direct contact membrane distillation for nuclear desalination, part II: experiments with radioactive solutions. Int. J. Nucl. Desalin. 2, 56–73. doi:10.1504/IJND.2006.009505

Lee, J.G., Kim, W.S., Choi, J.S., Ghaffour, N., Kim, Y.D., 2018. Dynamic solar-powered multi-stage direct contact membrane distillation system: concept design, modeling and simulation. Desalination 435, 278–292. doi: 10.1016/j.desal.2017.04.008

Lu, K.J., Cheng, Z.L., Chang, J., Luo, L., Chung, T., 2019. Design of zero liquid discharge desalination (ZLDD) systems consisting of freeze desalination, membrane distillation, and crystallization powered by green energies. Desalination 458, 66–75. doi: 10.1016/j.desal.2019.02.001

Luo, L., Chang, J., Chung, T.S., 2017. Cooling crystallization of sodium chloride via hollow fiber devices to convert waste concentrated brines to useful products. Ind. Eng. Chem. Res. 56, 10183–10192. doi: 10.1021/acs.iecr.7b02818

Missimer, T.M., Maliva, R.G., 2018. Environmental issues in seawater reverse osmosis desalination: intakes and outfalls. Desalination 434, 198–215. doi: 10.1016/j.desal.2017.07.012

Mullin, J.W., 2001. Crystallization, 4th edition. Butterworth-Heinemann, Oxford. doi:10.1002/0471238961.0318251918152119.a01.pub3

Nakoa, K., Rahaoui, K., Date, A., Akbarzadeh, A., 2016. Sustainable zero liquid discharge desalination (SZLDD). Sol. Energy 135, 337–347. doi: 10.1016/j.solener.2016.05.047

Quist-Jensen, C.A., Macedonio, F., Drioli, E., 2016. Membrane crystallization for salts recovery from brine – an experimental and theoretical analysis. Desalin. Water Treat. 57, 7593–7603. doi: 10.1080/19443994.2015.1030110

Rezaei, M., Warsinger, D.M., Lienhard V. J.H., Duke M.C., Matsuura, T., Samhaber, W. M., 2018. Wetting phenomena in membrane distillation: mechanisms, reversal, and prevention. Water Res. 139, 329–352. doi: 10.1016/j.watres.2018.03.058

Selvi, S.R., Baskaran, R., 2015. Solar photovoltaic-powered membrane distillation as sustainable clean energy technology in desalination. Curr. Sci. 109, 1247–1254.

Shim, W.G., He, K., Gray, S., Moon, I.S., 2015. Solar energy assisted direct contact membrane distillation (DCMD) process for seawater desalination. Sep. Purif. Technol. 143, 94–104. doi: 10.1016/j.seppur.2015.01.028

Singapore, E.M.A., 2017. Overview of gas market [WWW Document]. URL www.ema.gov.sg/Gas_Market_Overview.aspx

Suk, D.E., Matsuura, T., 2006. Membrane-based hybrid processes: a review. Sep. Sci. Technol. 41, 595–626. doi: 10.1080/01496390600552347

Tijing, L.D., Woo, Y.C., Choi, J.S., Lee, S., Kim, S.H., Shon, H.K., 2015. Fouling and its control in membrane distillation-a review. J. Memb. Sci. 475, 215–244. doi: 10.1016/j.memsci.2014.09.042

Wang, P., Chung, T.S., 2012. A conceptual demonstration of freeze desalination-membrane distillation (FD-MD) hybrid desalination process utilizing liquefied natural gas (LNG) cold energy. Water Res. 46, 4037–4052. doi: 10.1016/j.watres.2012.04.042

Warsinger, D.M., Swaminathan, J., Guillen-Burrieza, E., Arafat, H.A., Lienhard V, J.H., 2015. Scaling and fouling in membrane distillation for desalination applications: a review. Desalination 356, 294–313. doi: 10.1016/j.desal.2014.06.031

Zakrzewska-Trznadel, G., Harasimowicz, M., Chmielewski, A.G., 1999. Concentration of radioactive components in liquid low-level radioactive waste by membrane distillation. J. Memb. Sci. 163, 257–264. doi: 10.1016/S0376-7388(99)00171-4

Zarkadas, D.M., Sirkar, K.K., 2004. Solid hollow fiber cooling crystallization. Ind. Eng. Chem. Res. 43, 7163–7180. doi: 10.1021/ie0401004

Section C

Pilots & Economic Studies

15 Economic Evaluation of Membrane Distillation Systems

Yong-jun Choi and Sangho Lee
School of Civil and Environmental Engineering, Kookmin University, Jeongneung-Dong, Seongbuk-Gu, Seoul, Republic of Korea

15.1 INTRODUCTION

Fresh water is an inevitable resource for human and life on earth and for economic and industrial development. Unfortunately, it is being depleted due to enormous water demands, resulting in serious problems associated with water shortage (Jones et al., 2019; Richter et al., 2013; Teow and Mohammad, 2019). Approximately 41% of the world population live in the arid regions, thus fresh water shortage is becoming a worldwide problem (Parfit, 1993; Service, 2006). Accordingly, many countries have explored sustainable sources of fresh water, which is also less affected by the impact of climate changes (Djuma et al., 2016). Among them, seawater desalination has been considered as an ongoing source of water supply (Cipollina et al., 2009; Darre and Toor, 2018). Over the past few decades, a number of technologies have been developed to produce fresh water from seawater by separating salts, including thermal distillation (multistage flash distillation, MSF; multieffect distillation, MED; mechanical vapor pressure compression distillation, MVC), membrane separation (reverse osmosis, RO; nanofiltration; NF), freezing, and electrodialysis (Fritzmann et al., 2007; Teow and Mohammad, 2019). Recently, RO has become a dominant technology (Greenlee et al., 2009; Rautenbach et al., 1997; Sukitpaneenit and Chung, 2012). Nevertheless, RO technology also has drawbacks such as high electricity consumption and low recovery ratio of product water (Elimelech and Phillip, 2011; McCutcheon et al., 2006).

Recently, membrane distillation (MD) has drawn attention as a novel technology to overcome the disadvantages of RO technology (Alkhudhiri et al., 2012; Drioli et al., 2015; Hitsov et al., 2018). MD is a separation process using a vapor pressure, which results from the temperature difference between the feed and permeate water (Lawson and Lloyd, 1997). Hydrophobic microporous membranes enable the transport of water vapor through their pores, while maintaining vapor–liquid interfaces at the pore entrance, but it does not participate in the actual separation process. MD has several advantages over RO and other desalination processes for the treatment of saline water and wastewater (Alklaibi and Lior, 2005; Mariah et al., 2006; Martinetti et al., 2009). Because water is transported through the membrane only in a vapor phase, MD can offer complete rejections of all nonvolatile constituents in the feed solution; thus, almost 100% rejections of ions, dissolved nonvolatile organics, colloids, and pathogenic microorganisms can be achieved via the MD process (Drioli et al., 2015). But more importantly, due to the discontinuity of the liquid phase across the membrane, water flux in MD is not influenced by the osmotic pressure gradient across the membrane (Tijing et al., 2015). Consequently, MD has potential for the treatment of highly saline solutions that cannot be treated by RO due to their high osmotic pressures (Alklaibi and Lior, 2005; Ruiz Salmón and Luis, 2018).

Nevertheless, MD is still in its early stage in terms of commercial applications (Wang and Chung, 2015). Therefore, further work is still required to bring MD technologies into practice, including development of new membrane materials (Eykens et al., 2017), fouling and wetting control (Rezaei et al., 2018; Tijing et al., 2015), design of optimization systems (Hitsov et al., 2015), and development of heat exchange systems (Zhang et al., 2015). Moreover, there is only limited information on economics of MD compared with other desalination technologies (Hitsov et al., 2018; Kesieme et al., 2013; Moore et al., 2018). In this context, this study aims to evaluate the economic feasibility of the MD and RO-MD hybrid systems and to explore conditions where MD processes are able to be more competitive than a RO stand-alone system. To do this, a theoretical analysis was carried out using a simple cost model to investigate the effects of flux, recovery, membrane, and energy price on the cost of water production by different desalination technologies.

15.2 COST MODEL

In order to analyze the effects of major parameters such as flux, recovery, membrane, and energy cost on RO, MD, and RO-MD hybrid systems, a set of cost functions were used (Choi et al., 2016; Darwish et al., 1989; Dreyzin, 2006; Malek et al., 1996; Marcovecchio et al., 2005; Marechal et al., 2008; Sommariva and Mott MacDonald International Ltd., 2004). Based on these cost functions, a theoretical model was developed to evaluate the economics of RO, MD, and RO-MD hybrid systems (Choi et al., 2016).

15.2.1 RO Cost Model

The RO system is made of five major parts: The seawater intake and pretreatment process, high-pressure pump, booster pump, RO membrane module, and energy recovery device. The capital and operating costs of the intake and pretreatment are expressed as follows (Malek et al., 1996; Marcovecchio et al., 2005; Marechal et al., 2008):

$$CC_{IP}[\$] = 997 \times (Q_{IP}[m^3/d])^{0.8} \tag{15.1}$$

$$OC_{IP}[\$/d] = \frac{0.028^{*} P_{IP}[bar] * Q_{IP}[m^3/d]^{*} D_{EP}[\$/kwh]}{\eta_{IP}} * PLF \tag{15.2}$$

where CC and OC denote the capital cost and operating cost, respectively. The subscript IP denotes the intake and pretreatment. Q_{IP} is the intake and pretreatment flow rate, P_{IP} is the pressure of intake and pretreatment, D_{EP} is the unit electricity price, η_{IP} is the efficiency of the intake and pretreatment pump, and PLF is the plant load factor.

The capital and operating costs of the high-pressure pump, booster pump, and energy recovery device are expressed as follows (Malek et al., 1996; Marcovecchio et al., 2005; Marechal et al., 2008):

$$CC_{HP}[\$] = \frac{Q_f[m^3/d] \times (393,000 + 10,710 \times P_{f,in}[bar])}{30,000} \tag{15.3}$$

$$OC_{HP}[\$/d] = \frac{0.028 \times P_{f,in}[bar] \times Q_f[m^3/d] \times D_{EP}[\$/kWh]}{\eta_{P\_HP}} \times PLF \tag{15.4}$$

$$CC_{BP}[\$] = \frac{(Q_f[m^3/d] - Q_p[m^3/d]) \times (393,000 + 10,710 \times P_{f,in}[bar])}{60,000} \tag{15.5}$$

$$OC_{BP}[\$/d] = \frac{0.028 \times (P_{f,in}[bar] - P_{f,out}[bar]) \times \eta_{ERD} \times (Q_f - Q_p)[m^3/d] \times D_{EP}}{\eta_{P\_BP}} \times PLF \tag{15.6}$$

$$CC_{ERD}[\$] = \frac{(Q_f[m^3/d] - Q_p[m^3/d]) \times (393,000 + 10,710 \times P_{f,in}[bar])}{40,000} \tag{15.7}$$

where subscripts, HP, BP, and ERD denote the high-pressure pump, booster pump, and energy recovery device, respectively. Q_f is the feed flow rate, Q_p is the permeate flow rate, $P_{f,in}$ is the inlet pressure of RO vessel, and $P_{f,out}$ is the outlet pressure of RO vessel.

Assuming that the capital cost of the membrane module is linear to the membrane area, the annualized capital cost of the membrane is calculated by the following (Maskan et al., 2004):

$$CC_{Mem}[\$] = Area_{Mem} \times C_{mem}[\$/m^2] \tag{15.8}$$

where subscript Mem denotes the membrane, $Area_{Mem}$ is the total membrane area, and C_{Mem} is the unit membrane cost.

The total capital cost is composed of the direct capital cost and the indirect capital cost. The direct capital cost is the sum of the cost for plant equipment and the cost for site development. The cost for site development is set at 20% of the cost for plant equipment (Malek et al., 1996). The indirect capital cost is set at 30% of the direct capital cost (Malek et al., 1996). The annualized capital cost is calculated based on a previous study (Lu et al., 2007). The total and annual capital costs of RO process are expressed as follows (Malek et al., 1996; Marcovecchio et al., 2005; Marechal et al., 2008; Lu et al., 2007):

$$CC_{Equipment}[\$] = CC_{IP} + CC_{HP} + CC_{BP} + CC_{ERD} + CC_{Mem\_RO}CC_{Mem} \quad (15.9)$$

$$CC_{Site}[\$] = CC_{Equipment} \times 0.2 \quad (15.10)$$

$$DCC_{RO}[\$] = CC_{Equipment} + CC_{Site} \quad (15.11)$$

$$ICC_{RO}[\$] = DCC_{RO} \times 0.3 \quad (15.12)$$

$$TCC_{RO}[\$] = DCC_{RO} + ICC_{RO} \quad (15.13)$$

$$ACC_{RO}[\$/y] = TCC_{RO} \times \frac{i \times (1+i)^n}{(1+i)^n - 1} \quad (15.14)$$

where DCC_{RO} is the direct capital cost, ICC_{RO} is the indirect capital cost, TCC_{RO} is the total capital cost, ACC_{RO} is the annual capital cost, i is the interest rate, and n is the plant lifetime.

The annual operating cost (AOC) is composed of the annual power cost, annual membrane replacement cost, and other costs (e.g., labor, chemicals, and maintenance). The annual operating costs of RO process are expressed as follows (Malek et al., 1996; Marcovecchio et al., 2005; Marechal et al., 2008):

$$OC_{Power}[\$/y] = (OC_{IP} + OC_{HP} + OC_{BP}) \times 365 \quad (15.15)$$

$$OC_{MR}[\$/y] = CC_{Mem} \times 0.2 \quad (15.16)$$

$$OC_{etc}[\$/y] = AOC_{RO} \times 0.3 \quad (15.17)$$

$$AOC_{RO}[\$/y] = OC_{power} + OC_{MR} + OC_{etc} \quad (15.18)$$

where OC_{Power} is the annual power cost, OC_{MR} is the annual membrane replacement cost, and OC_{etc} is the sum of other costs.

Finally, the water cost of RO process is calculated as follows:

$$WC_{RO}[\$/m^3] = \frac{(ACC_{RO} + AOC_{RO})}{(365 \times Q_p \times PLF)} \quad (15.19)$$

15.2.2 MD Cost Model

MD process has four different configurations, including direct contact MD, air gap MD, sweeping gas MD, and vacuum MD (VMD). Among them, the VMD configuration was selected. In VMD, vacuum is applied in the permeate side of the membrane module by means of a vacuum pump. As vacuum is applied on the permeate side, the distillate production rate increased and thermal energy loss is reduced (Alklaibi and Lior, 2005).

The VMD system is made of five major parts: The seawater intake and pretreatment process, vacuum pressure pump, feed pump, VMD membrane module, and heat exchanger. The capital and operating costs of the intake and pretreatment are expressed as follows (Malek et al., 1996; Marcovecchio et al., 2005; Marechal et al., 2008):

$$CC_{IP}[\$] = 997 \times (Q_{IP}[m^3/d])^{0.8} \tag{15.20}$$

$$OC_{IP}[\$/d] = \frac{0.028 \times P_{IP}[bar] \times Q_{IP}[m^3/d] \times D_{EP}[\$/kWh]}{\eta_{IT\_S}} \times PLF \tag{15.21}$$

The capital and operating costs of the vacuum pressure pump, feed pump, and heat exchanger are expressed as follows (Al-Obaidania et al., 2008; Malek et al., 1996; Marcovecchio et al., 2005; Marechal et al., 2008):

$$CC_{VP}[\$] = \frac{Q_f[m^3/d] \times (393,000 + 10,710))}{12,000 \times P_{VP}[mbar]} \tag{15.22}$$

$$OC_{VP}[\$/d] = 1.0 \times \left[\frac{kWh}{m^3}\right] Q_p[m^3/d] \times D_{EP}[\$/kWh] \times PLF \tag{15.23}$$

$$CC_{FP}[\$] = 4.78 \times 10^{-6} \times Q_f[m^3/d] \times 120,000 \tag{15.24}$$

$$OC_{FP}[\$/d] = \frac{0.028 \times (P_{f,in}[bar]) \times (Q_f[m^3/d]) \times D_{EP}}{\eta_{FP}} \times PLF \tag{15.25}$$

$$CC_{HX}[\$] = 1000 \times Area_{HX} \tag{15.26}$$

where subscripts, *VP*, *FP*, and *HX* denote the vacuum pressure pump, feed pump, and heat exchanger, respectively, P_{VP} is the vacuum pressure of VMD vessel and $Area_{HX}$ is the total heat exchanger area.

Assuming that the capital cost of the membrane module is linear to the membrane area, the annualized capital cost of the membrane is calculated by the following same formulation as RO membrane module (Maskan et al., 2004):

$$CC_{Mem}[\$] = Area_{Mem} \times C_{mem}[\$/m^2] \tag{15.27}$$

The total and annual capital costs of VMD system are expressed as follows:

$$CC_{Equipment}[\$] = CC_{IP} + CC_{VP} + CC_{FP} + CC_{HX} + CC_{Mem} \tag{15.28}$$

$$CC_{Site}[\$] = CC_{Equipment} \times 0.2 \tag{15.29}$$

$$DCC_{MD}[\$] = CC_{Equipment} + CC_{Site} \tag{15.30}$$

$$ICC_{MD}[\$] = DCC_{RO} \times 0.3 \tag{15.31}$$

$$TCC_{MD}[\$] = DCC_{MD} + ICC_{MD} \tag{15.32}$$

$$ACC_{MD}[\$/y] = TCC_{MD} \times \frac{i \times (1+i)^n}{(1+i)^n - 1} \tag{15.33}$$

The annual operating costs of VMD system are expressed as follows:

$$OC_{Steam}[\$/d] = (Steam[kg/day] \times C_{Steam}[\$/kg]) \tag{15.34}$$

$$OC_{Power}[\$/y] = (OC_{IP} + OC_{VP} + OC_{FP} + OC_{Steam}) \times 365 \tag{15.35}$$

$$OC_{MR}[\$/y] = CC_{Mem} \times 0.2 \tag{15.36}$$

$$OC_{etc}[\$/y] = AOC_{MD} \times 0.3 \tag{15.37}$$

$$AOC_{MD}[\$/y] = OC_{power} + OC_{MR} + OC_{etc} \tag{15.38}$$

where OC_{Steam} is the daily steam cost and C_{Steam} is the unit steam cost.

Finally, the water cost of VMD system is as follows:

$$WC_{MD}[\$/m^3] = \frac{(ACC_{MD} + AOC_{MD})}{(365 \times Q_P \times PLF)} \tag{15.39}$$

15.3 RESULTS AND DISCUSSIONS

The economics of seawater desalination were theoretically assessed for these systems:

(1) RO stand-alone system (water production capacity of 10,000 m^3/d)
(2) MD stand-alone system (water production capacity of 10,000 m^3/d)
(3) RO-MD hybrid system (water production capacity of 10,000 m^3/d).

In the RO-MD hybrid system, the RO brine is treated by the MD system to increase overall system recovery ratio. Model parameters and operating conditions used in this analysis are summarized in Table 15.1.

15.3.1 Economic Analysis for RO System

Table 15.2 shows calculation results for a RO plant using the parameters and operating conditions in Table 15.1. For this calculation, the permeate flux, feed pressure, and recovery were set to 12 LMH, 55.0 bar, and 40%,

TABLE 15.1
Process parameters and operating conditions

Parameter	Value	Parameter	Value
RO		**MD**	
Membrane area	40 m^2	Membrane area	20 m^2
Permeate flux	12 LMH	Permeate flux	1–20 LMH
Feed pressure	55.0 bar	Feed pressure	1.0 bar
Recovery	40%	Recovery	20–60%
Feed TDS	35,000 mg/L	Feed TDS	35,000 mg/L
Efficiency		**Efficiency**	
Intake and pretreatment pump	80%	Intake and pretreatment pump	80%
High-pressure pump	75%	Vacuum pump	–
Booster pump	80%	Feed pump	80%
Energy recovery device	90%	Heat exchanger	90%
Cost		**Cost**	
Electricity bill	0.08 \$/kWh	Electricity bill	0.08 \$/kWh
Membrane cost	50 \$/m^2	Membrane cost	30–100 \$/m^2
Plant load factor	0.91	Steam cost	0–5 \$/ton
		Plant load factor	0.91
Interest rate	0.08	Interest rate	0.08
Plant life	20 years	Plant life	20 years

respectively, as shown in Table 15.1. The recovery and pressure of intake and pretreatment were set to 90% and 5.0 bar, respectively.

As a result of the calculations, the water cost of RO system was found to be 0.83 \$/m^3. The cost of the RO plant construction was 49.5% from the total water cost shown in Figure 15.1. The cost of energy, membrane replacement, and others (including labor, chemicals, filters, and miscellaneous items) of the RO plant construction was 23.8, 11.5, and 11.5%, respectively, from the total water cost.

15.3.2 Economic Analysis for MD System

The two methods were used to perform economic evaluation for 10,000 m^3/d MD and RO-MD hybrid desalination systems. The recovery, flux, steam cost, and membrane cost of MD system were determined as most important factors to influence the water cost. In the first method, the recovery and flux of MD system were adjusted, while the other factors including steam cost and membrane cost of MD system were fixed. The recovery changed from 20 to 60% and the permeate flux changed from 10 to 30 LMH. The steam cost and membrane cost were set to 1 \$/ton and 50 \$/m^2, respectively. The other model parameters and operating conditions are listed in Table 15.1.

TABLE 15.2
Calculation results for a 10,000 m^3/d RO system

Energy consumption	
Intake and Pretreatment	201 kW
High-pressure pump	848 kW
Booster pump	94 kW
Specific energy	2.74 kWh/m^3
Capital cost	
Intake and pretreatment	2,913,000 $
High-pressure pump	2,946,150 $
Booster pump	878,480 $
Energy recovery device	964,380 $
Membrane	1,750,000 $
Site	1,890,400 $
Direct capital cost	11,342,410 $
Indirect capital cost	3,402,700 $
Capital cost	14,742,680 $
Annual capital cost	1,501,800 $/year
Operating cost	7,209,100 $/year
Power cost	
Membrane replacement	350,000 $/year
Labor, chemicals, filters, and other items	458,961 $/year
Annual operating cost	1,529,871 $/year
Water cost	0.83 $/$m^3$

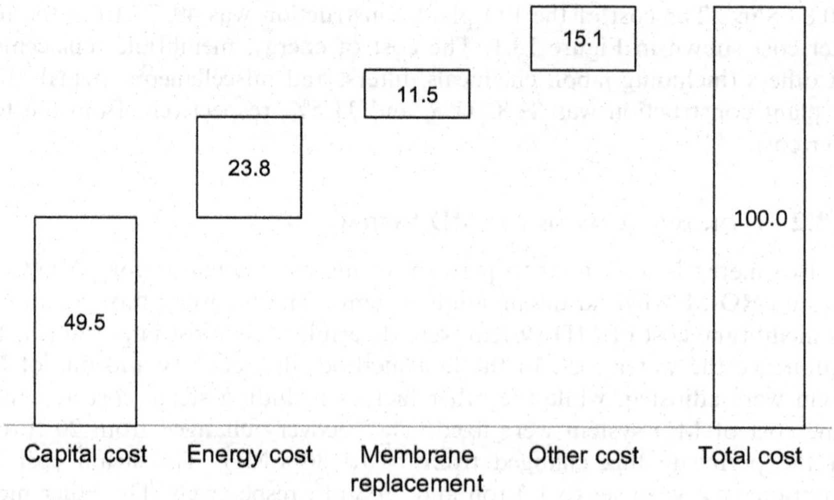

FIGURE 15.1 Percent of water cost of 10,000 m^3/d RO plant.

On the other hand, in the second method, the steam cost and membrane cost were adjusted and the other factors including recovery and permeate flux were fixed. The steam cost changed from 1 to 5 \$/ton and the membrane cost changed from 30 to 100 \$/m^2. The recovery and permeate flux were set to 40% and 12 LMH, respectively, the same as RO system. In both cases, the permeate flux was not coupled with operating temperature in MD system and depended on the membrane properties.

In MD system, flux and recovery are two main operating parameters that determine the process performance. Accordingly, the effect of these parameters on water cost of MD was investigated as shown in Figure 15.2. The water cost of MD system was estimated to vary from 0.67 to 4.16 \$/m^3, by the changes in the permeate flux and recovery. The water cost of MD system decreases with increasing recovery and permeate flux due to the decreasing requirements of large capacity intake and pretreatment, thermal energy, and the required membrane area. Especially, the water cost is significantly affected by the permeate flux, suggesting that the flux should be higher to achieve a lower water cost.

In addition to water cost, the annual capital and operating cost were considered at different permeate fluxes and recovery in Figures 15.3 and 15.4. The overall trends of the annual capital cost and operating cost are the same as that of water cost. The recovery is inversely proportional to capital and operating costs of intake and pretreatment process, vacuum pressure pump, feed pump, and steam cost. The permeate flux is also inversely proportional to membrane cost and membrane replacement cost. At the same recovery (40%) and permeate flux (12 LMH) conditions to RO, the capital cost of MD system decreased down to approximately 31% since the sum of the capital

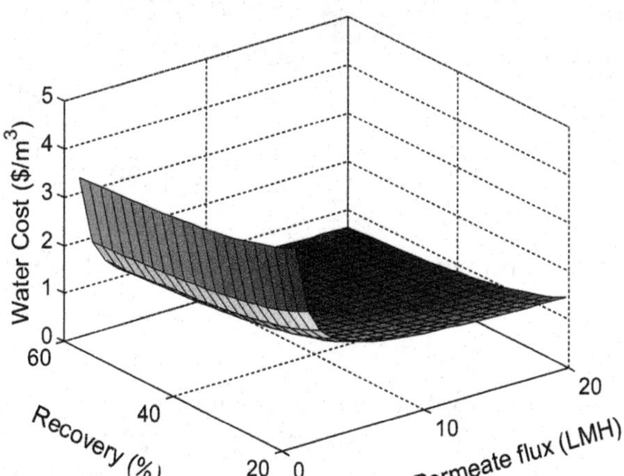

FIGURE 15.2 Effect of permeate flux and recovery on water cost for MD system.

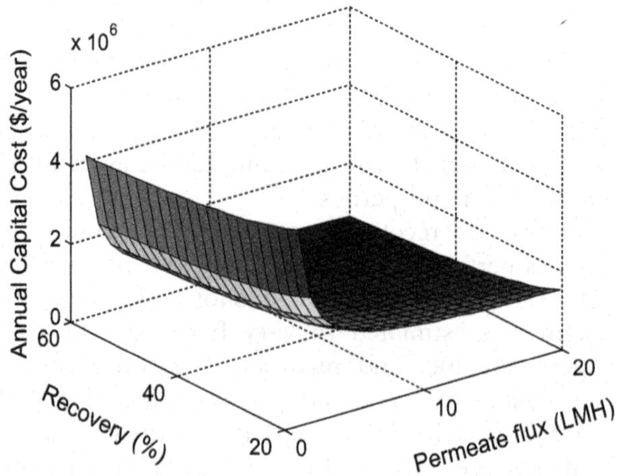

FIGURE 15.3 Effect of permeate flux and recovery on annual capital cost for MD system.

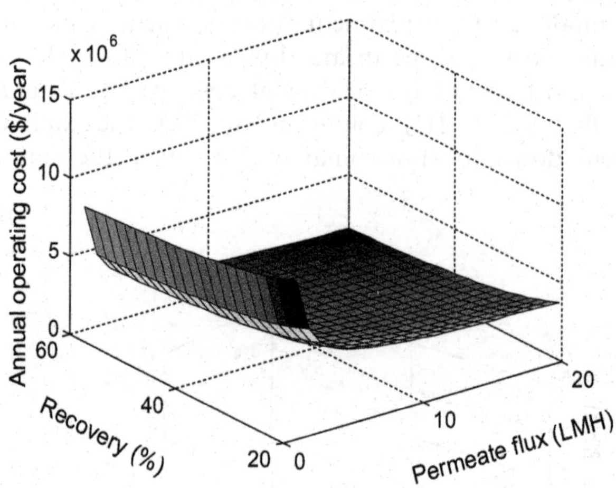

FIGURE 15.4 Effect of permeate flux and recovery on annual operating cost for MD system.

cost of high-pressure pump, booster pump, and ERD is higher than the sum of the capital cost of vacuum pressure pump, feed pump, and heat exchanger under these simulation conditions. However, the operating cost increased to approximately 60%, since the energy cost of MD system is much higher than that of RO process.

Figure 15.5 demonstrates the ratio of AOC to total cost (AOC+ACC) for MD system as a function of permeate flux and recovery. This ratio indicates the relative importance of operating cost for the MD system. Although the annual operating cost and the annual capital cost increase with a decrease in permeate flux, it seems that the annual capital cost is more sensitive. Accordingly, the ratio decreases with decreasing the permeate flux. This suggests that the operating cost is the main factor affecting the water cost at a high flux condition.

Figure 15.6 illustrates the performance ratio (PR) of MD system as a function of the permeate flux and recovery. The PR (also called as grained output ratio, or GOR) is a measure of how much thermal energy is consumed in a desalination system (Al-Obaidania et al., 2008). The simplest definition is as follows: amount of product water/amount of used stream. Typically, the value of PR ranges from 1.0 to 10. Higher PR systems consume less energy, therefore, have lower operating costs. The PR of MD system was estimated to range from 2.0 to 5.5, by changing the permeate flux and recovery. In addition, the PR of MD system increases with increasing the recovery of the product water. According to the simulation results, the PR was found to be related to the recovery.

In VMD system, thermal energy is provided in the form of steam, which has the latent heat. Accordingly, the steam cost is one of the important factors affecting the water cost of MD system. Figure 15.7 shows how membrane and steam costs affect the water cost of MD system. As expected, the water cost of MD system is sensitive to the steam cost. The water cost of MD plant was found to range from 0.41 to 2.7 $/m^3$, with the changes in the steam cost and

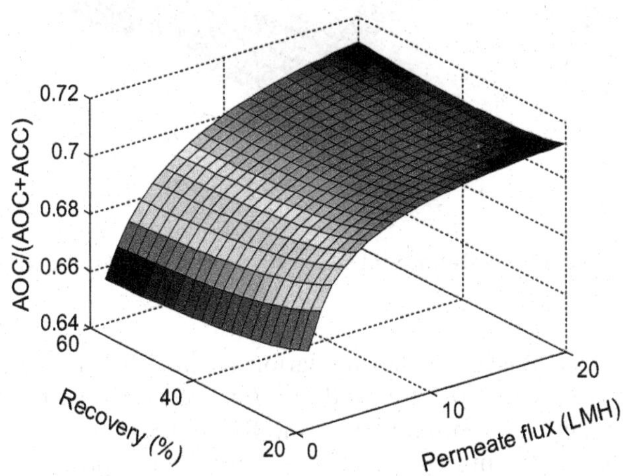

FIGURE 15.5 Effect of permeate flux and recovery on the ratio of annual operating cost (AOC) to total cost (AOC+ACC) for MD system.

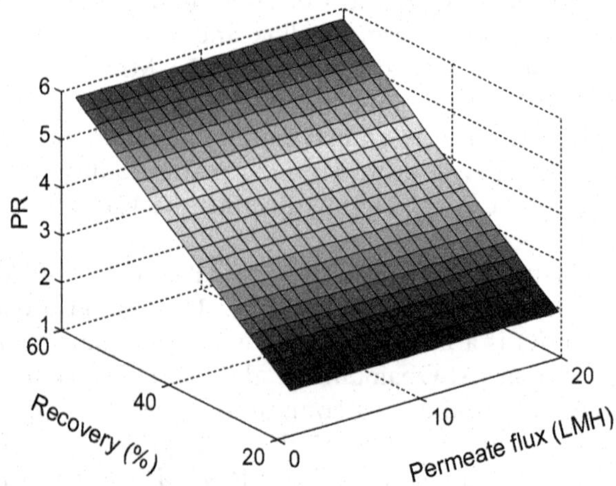

FIGURE 15.6 Effect of permeate flux and recovery on PR for MD system.

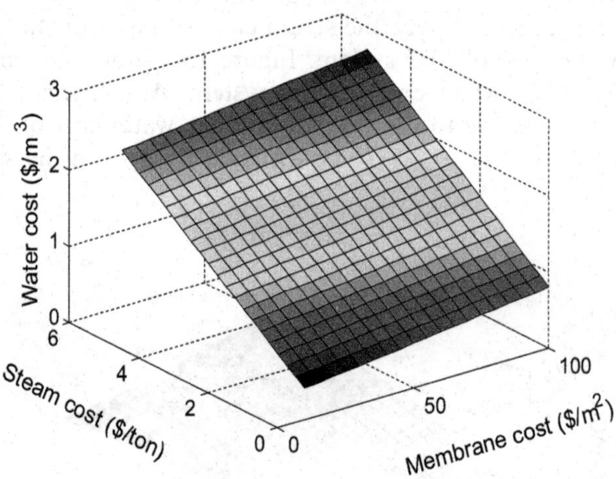

FIGURE 15.7 Effect of membrane and steam costs on water cost for MD system.

the membrane cost. Based on the simulation results, the membrane and steam costs should not exceed 1.0 \$/ton and 60 \$/m^2, respectively, to make the MD system cost-competitive compared with the RO system.

The dependences of annual capital cost and annual operating cost of MD system on membrane and steam costs are demonstrated as 3D plots in Figures 15.8 and 15.9. The annual capital cost is directly proportional to the membrane cost, but the steam cost has no direct effect on the capital cost. The

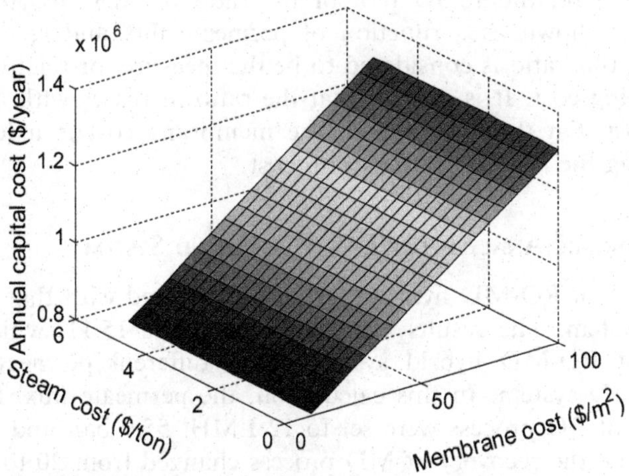

FIGURE 15.8 Effect of membrane and steam costs on annual capital cost for MD system.

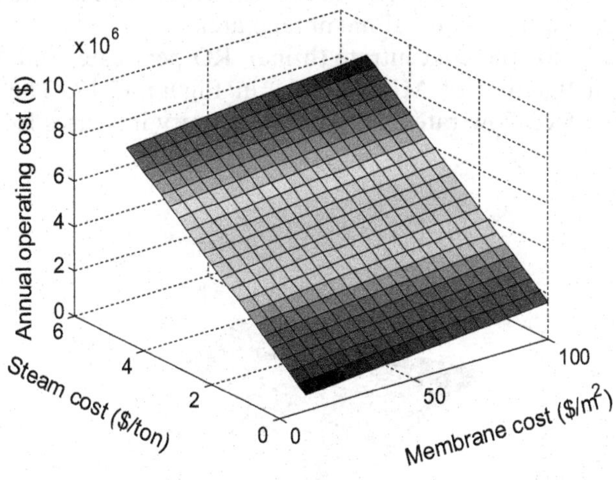

FIGURE 15.9 Effect of membrane and steam costs on annual operating cost for MD system.

annual operating cost shows similar behaviors to the water cost. The annual capital cost is inversely proportional to the membrane and steam costs. Under the same conditions, the operating cost is higher than the capital cost due to the consumption of a significant amount of thermal energy.

In Figure 15.10, the 15.3D plot of the ratio of AOC to AOC+ACC for MD system is shown as a function of permeate flux and recover. As mentioned above, this ratio is considered to be the measure for relative importance of the operating cost. It is evident that the ratio increases with an increase in the steam cost. On the other hand, the membrane cost is a less influential factor affecting the ratio than the steam cost.

15.3.3 Economic Analysis for RO-MD Hybrid System

The water cost of RO-MD hybrid system is compared with that of the standalone MD system. The results are shown in Figure 15.11, which shows the water cost of RO-MD hybrid system under different permeate fluxes and recovery of MD system. In this calculation, the permeate flux, feed pressure, and recovery of RO process were set to 12 LMH, 55.0 bar, and 40%, respectively. Moreover, the recovery of MD process changed from 20 to 60% and the permeate flux changed from 1 to 20 LMH. The steam cost and MD membrane costs were set to 1 \$/ton and 50 $\$/m^2$, respectively. The simulation results showed that the water cost of RO-MD hybrid system ranges from 0.7 to 1.85 $\$/m^3$ with the changes in the permeate flux and recovery of MD system. The water cost decreases with increasing recovery and permeate flux of MD system due to the reduction in the capacity of intake and pretreatment, thermal energy, and the required membrane area.

The flow rates for the concentrate (brine), RO permeate, and MD permeate are shown as a function of MD recovery in Figure 15.12. The sum of three flow rates is the feed flow rate. As the MD recovery increases from 20 to 60%,

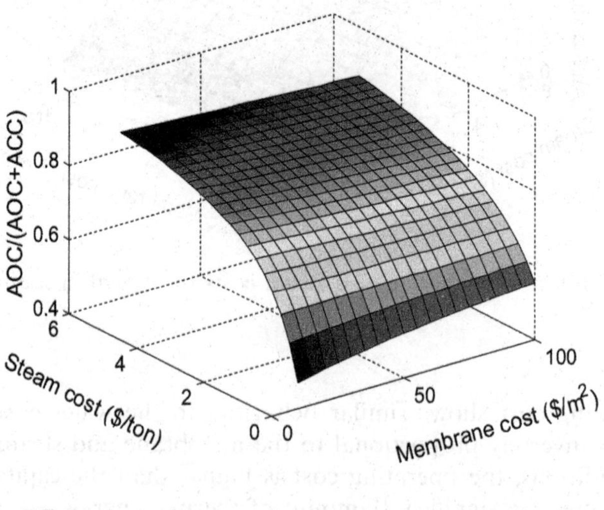

FIGURE 15.10 Effect of membrane and steam costs on the ratio of annual operating cost (AOC) to total cost (AOC+ACC) for MD system.

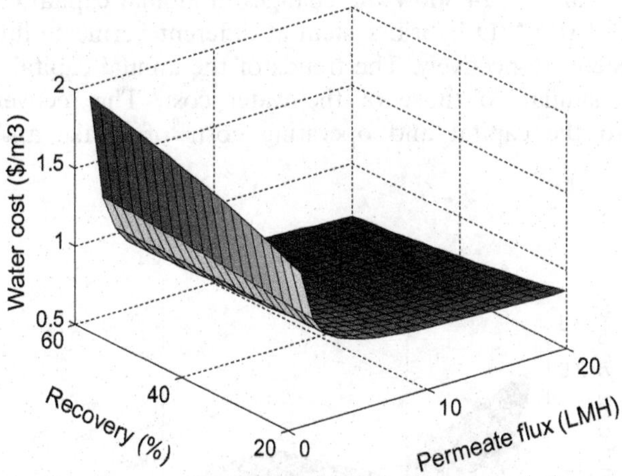

FIGURE 15.11 Effect of permeate flux and recovery on water cost for RO-MD hybrid system.

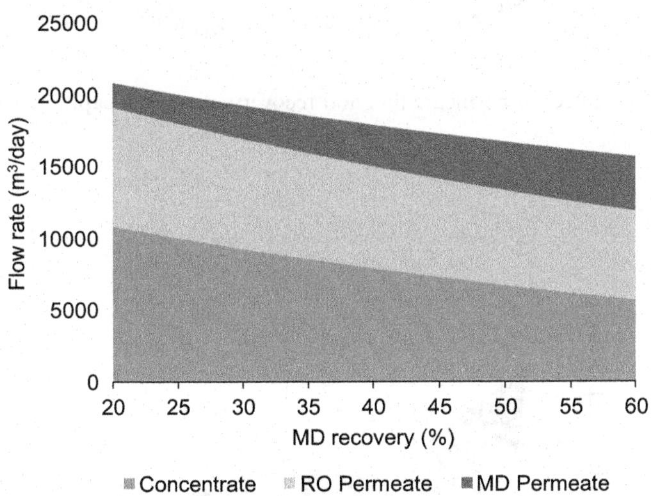

FIGURE 15.12 Dependence of flow rates of concentrate, RO permeate, and MD permeate on MD recovery in RO-MD hybrid system.

the overall recovery increases from 52 to 76%. This leads to a reduction in both the feed and the concentrate. Since the discharge of the concentrate of the desalination plants becomes an issue, the RO-MD hybrid system with a high MD recovery can provide an additional benefit.

Figures 15.13 and 15.14 show the changes in annual capital cost and annual operating cost of RO-MD hybrid system at different permeate fluxes and recovery of MD system, respectively. The trends of the annual capital cost and operating cost are similar to those of the water cost. The recovery is inversely proportional to the capital and operating costs of intake and pretreatment

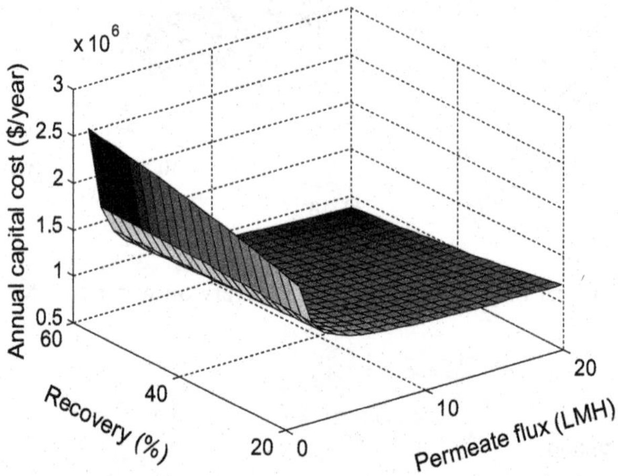

FIGURE 15.13 Effect of permeate flux and recovery on annual capital cost for RO-MD hybrid system.

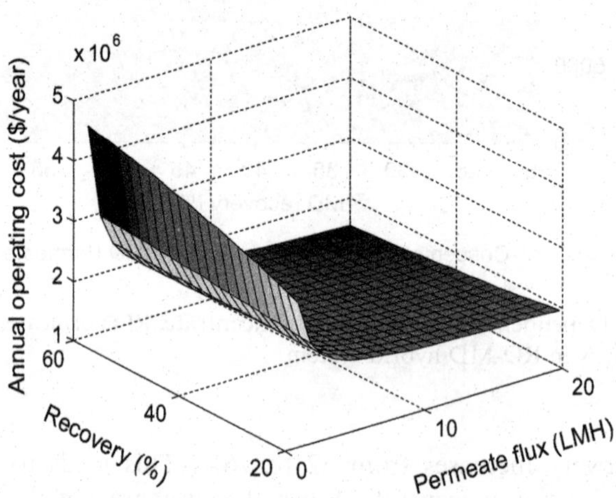

FIGURE 15.14 Effect of permeate flux and recovery on annual operating cost for RO-MD hybrid system.

process, vacuum pressure pump, and feed pump. Moreover, the permeate flux is inversely proportional to the membrane cost and membrane replacement cost. At the same recovery (40%) and permeate flux (12 LMH) conditions to RO and MD, the capital cost of RO-MD hybrid plant decreases by approximately 30% because the recovery of the hybrid plant is 64% under the given conditions. But the operating cost increases to approximately 18% since the power cost of MD system is higher than that of RO system.

As shown in Figure 15.15, the ratio of AOC to AOC+ACC of RO-MD hybrid system is shown at different permeate fluxes and recovery. Compared with the results in Figure 15.7, which is for the stand-alone MD system, the dependence of the ratio on the permeate flux is less significant in RO-MD hybrid system. Moreover, the ratios are less than 0.64 in RO-MD hybrid system, which are significantly lower than those in MD system (0.66–0.72).

Figure 15.16 shows the effect of membrane and steam costs on the water cost of RO-MD hybrid system. As expected, the water cost of RO-MD hybrid system is sensitive to the membrane and steam costs. The water cost of RO-MD hybrid system is estimated between 0.59 and 1.57 $\$/m^3$, with the changes in the steam cost and membrane cost. In addition, the water cost of RO-MD hybrid system is lower than that of stand-alone MD system in the same simulation conditions. Moreover, the steam cost has a less significant effect on the hybrid system than stand-alone MD system. To make the water cost of RO-MD hybrid system competitive, the membrane and steam cost should not exceed 1.5 \$/ton and 60 $\$/m^2$, respectively.

The annual capital cost and annual operating cost of RO-MD hybrid system are analyzed at different membrane and steam costs. It is evident from

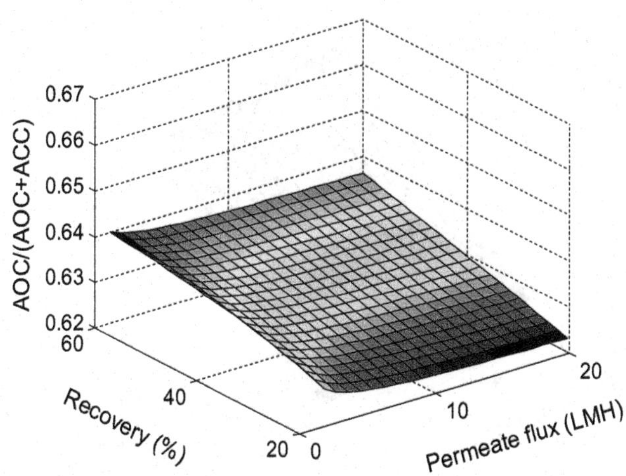

FIGURE 15.15 Effect of permeate flux and recovery on the ratio of annual operating cost (AOC) to total cost (AOC+ACC) for RO-MD hybrid system.

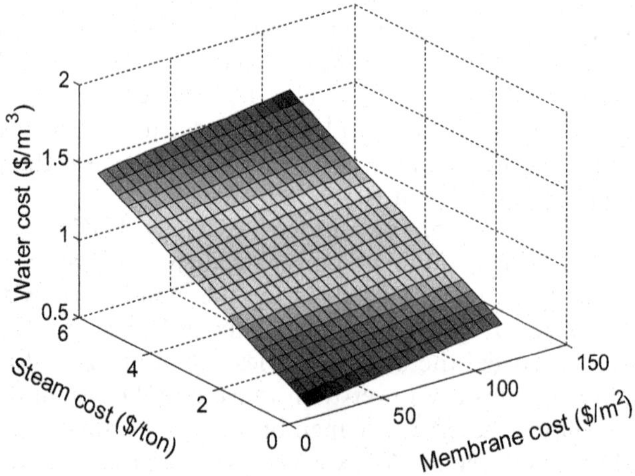

FIGURE 15.16 Effect of membrane and steam costs on water cost for RO-MD hybrid system.

Figure 15.17 that the annual capital cost only depends on the membrane cost. On the other hand, the annual operating cost affects both membrane cost and steam cost as shown in Figure 15.18. Under the same conditions, the operating cost is much higher than the capital cost due to the consumption of a large amount of thermal energy.

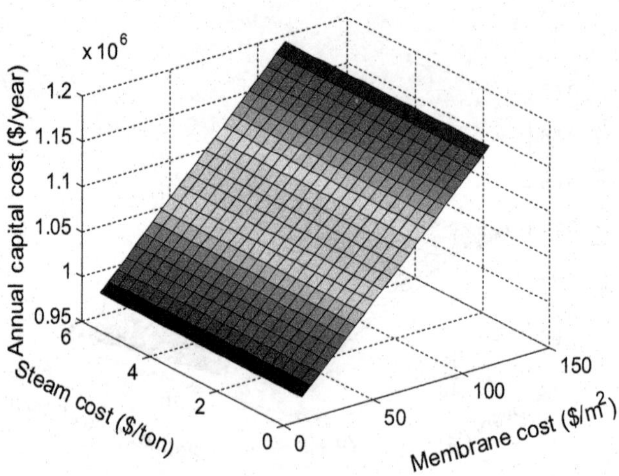

FIGURE 15.17 Effect of membrane and steam costs on annual capital cost for RO-MD hybrid system.

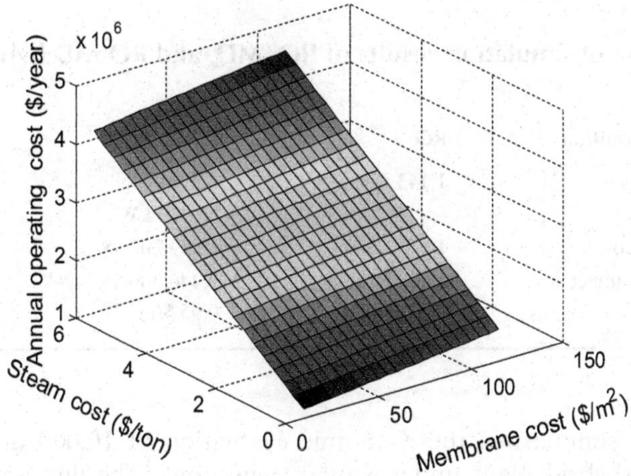

FIGURE 15.18 Effect of membrane and steam costs on annual operating cost for RO-MD hybrid system.

Figure 15.19 shows the ratio of AOC to AOC+ACC of RO-MD hybrid system at different membrane and steam costs. The ratio increases significantly with increasing the steam cost, indicating that the operating cost becomes more significant. On the other hand, the membrane cost shows negligible effects on the ratio. Under the simulation condition, the ratio ranges from 0.54 to 0.8.

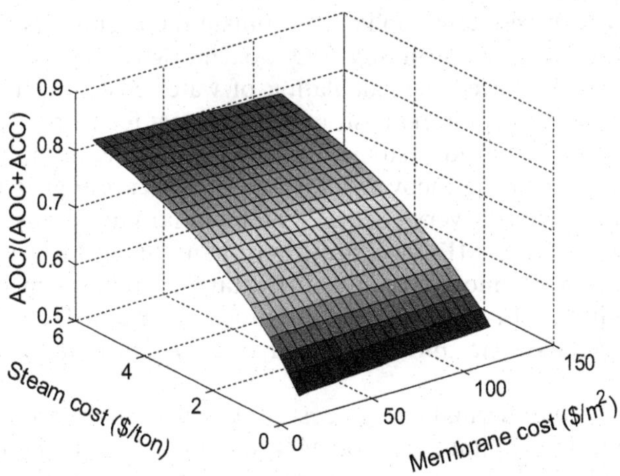

FIGURE 15.19 Effect of membrane and steam costs on the ratio of annual operating cost (AOC) to total cost (AOC+ACC) for RO-MD hybrid system.

TABLE 15.3
Comparison of simulation results of RO, MD, and RO-MD hybrid systems

Energy consumption	RO	MD	RO-MD
Electric energy	1,143 kW	649 kW	883 kW
Thermal energy	–	66,332 kW	24,871 kW
Total capital cost	14,742,680 \$	10,093,000 \$	10,398,000 \$
Annual operating cost	1,529,871 \$/y	2,446,300 \$/y	1,809,700 \$/y
Water cost	0.83 \$/m^3	0.95 \$/m^3	0.79 \$/m^3

Table 15.3 summarizes the economic evaluation of 10,000 m^3/d RO, MD, and RO-MD hybrid plant under similar conditions. The flux and recovery of RO and MD were set to 12 LMH and 40% equally. The membrane cost of RO and MD was set to 50 \$/m^2 and the steam cost was set to 1.0 \$/ton. In this case, the water cost of MD is higher than the water cost of RO system. On the other hand, the water cost of RO-MD system is lower than that of RO system. Only RO-MD hybrid system may have cost-competiveness when the steam cost is 1.0 \$/ton under these simulation conditions. This means that MD and RO-MD hybrid system should use free waste heat to gain price competitive advantage.

15.4 CONCLUSIONS

This study examined the economic feasibility of RO, MD, and RO-MD hybrid systems to explore conditions where MD has economic feasibility. The theoretical model was successfully applied to estimate the economics of three different desalination systems including RO system, MD system, and RO-MD system. The model allowed the calculation of water cost as a function of permeate flux, recovery, membrane cost, and steam cost for these systems.

MD and RO-MD hybrid systems may have economic feasibility compared with RO system when the recovery and flux of MD are higher than those of RO and the steam cost is very cheap. Under similar conditions in Table 15.3, the water costs of RO-MD can be lower than that of RO system. If the steam cost becomes almost free, the MD stand-alone system can also have economic feasibility. In other words, the most important factor affecting the economics of MD system and RO-MD hybrid system is the cost of thermal energy source.

In addition to the thermal energy cost, flux and recovery are also important factors affecting the capital costs for MD and RO-MD hybrid systems. More research should be done to improve MD flux and recovery by developing novel membrane materials, designing new modules, and optimizing the process configurations.

NOMENCLATURE

Abbreviations

ACC	annual capital cost
AOC	annual operating cost
CC	capital cost
DCC	direct capital cost
ICC	indirect capital cost
MD	membrane distillation
OC	operating cost
PLF	plant load factor
Q	flow rate
RO	reverse osmosis
TCC	total capital cost
WC	water cost

SUBSCRIPTIONS

BP	booster pump
ERD	energy recovery device
etc	sum of other costs
f	feed flow rate
FP	feed pump
HP	high-pressure pump
HX	heat exchanger
IP	intake and pretreatment
Mem	membrane
MP	membrane replacement cost
p	permeate flow rate
Power	power cost
Stream	steam cost
VP	vacuum pressure pump

REFERENCES

Alkhudhiri, A., Darwish, N. and Hilal, N. (2012) Membrane distillation: A comprehensive review. Desalination 287, 2–18.

Alklaibi, A.M. and Lior, N. (2005) Membrane-distillation desalination: Status and potential. Desalination 171(2), 111–131.

Al-Obaidania, S., Curcio, E., Macedonio, F., Profio, G.D., Al-Hinai, H. and Drioli, E. (2008) Potential of membrane distillation in seawater desalination: Thermal efficiency, sensitivity study and cost estimation. J. Membr. Sci. 323, 85–98.

Choi, Y.J., Lee, S., Koo, J. and Kim, S.-H. (2016) Evaluation of economic feasibility of reverse osmosis and membrane distillation hybrid system for desalination. Desalination Water Treat. 57(51), 24662–24673.

Cipollina, A., Micale, G. and Rizzuti, L. (2009) Seawater Desalination: Conventional and Renewable Energy Processes (Green Energy and Technology), Springer Science & Business Media, New York, NY.

Darre, N.C. and Toor, G.S. (2018) Desalination of Water: A Review. Current Pollution Reports 4(2), 104–111.

Darwish, M.A., Abdel-Jawad, M. and Aly, G.S. (1989) Technical and economical comparison between large capacity multi stage flash and reverse osmosis desalting plants. Desalination 72, 367–379.

Djuma, H., Bruggeman, A., Eliades, M. and Lange, M.A. (2016) Non-conventional water resources research in semi-arid countries of the Middle East. Desalination Water Treat. 57(5), 2290–2303.

Dreyzin, Y. (2006) Ashkelon seawater desalination project – Off-taker's self costs, supplied water costs and benefits. Desalination 190, 104–116.

Drioli, E., Ali, A. and Macedonio, F. (2015) Membrane distillation: Recent developments and perspectives. Desalination 356, 56–84.

Elimelech, M. and Phillip, W.A. (2011) The future of seawater desalination: Energy, technology, and the environment. Science 333(6043), 712–717.

Eykens, L., De Sitter, K., Dotremont, C., Pinoy, L. and Van der Bruggen, B. (2017) Membrane synthesis for membrane distillation: A review. Sep. Purif. Technol. 182, 36–51.

Fritzmann, C., Löwenberg, J., Wintgens, T. and Melin, T. (2007) State-of-the-art of reverse osmosis. Desalination 216(2007), 1–76.

Greenlee, L.F., Lawler, D.F., Freeman, B.D., Marrot, B. and Moulin, P. (2009) Reverse osmosis desalination: Water sources, technology, and today's challenges. Water Res 43, 2317–2348.

Hitsov, I., Maere, T., De Sitter, K., Dotremont, C. and Nopens, I. (2015) Modelling approaches in membrane distillation: A critical review. Sep.Purif. Technol. 142, 48–64.

Hitsov, I., Sitter, K.D., Dotremont, C. and Nopens, I. (2018) Economic modelling and model-based process optimization of membrane distillation. Desalination 436, 125–143.

Jones, E., Qadir, M., van Vliet, M.T.H., Smakhtin, V. and Kang, S.M. (2019) The state of desalination and brine production: A global outlook. Sci. Environ. 657, 1343–1356.

Kesieme, U.K., Milne, N., Aral, H., Cheng, C.Y. and Duke, M. (2013) Economic analysis of desalination technologies in the context of carbon pricing, and opportunities for membrane distillation. Desalination 323, 66–74.

Lawson, K.W. and Lloyd, D.R. (1997) Membrane distillation. J. Membr. Sci. 124 (1), 1–25.

Lu, Y.Y., Hu, Y.D., Zhang, X.L., Wu, L.Y., and Liu, Q.Z. (2007) Optimum design of reverse osmosis system under different feed concentration and product specification. J. Membr. Sci. 287, 219–229.

Malek, A., Hawlader, M.N.A. and Ho, J.C. (1996) Design and economics of RO seawater desalination. Desalination 105, 245–261.

Marcovecchio, M.G., Aguirre, P.A. and Scenna, N.J. (2005) Global optimal design of reverse osmosis networks for seawater desalination: Modeling and algorithm. Desalination 184, 259–271.

Marechal, F., Aoustin, E. and Bréant, P. (2008) Multi-objective optimization of RO desalination plants. Desalination 222, 96–118.

Mariah, L., Buckley, C.A., Brouckaert, C.J., Curcio, E., Drioli, E., Jaganyi, D. and Ramjugernath, D. (2006) Membrane distillation of concentrated brines – Role of water activities in the evaluation of driving force. J. Membr. Sci. 280, 937–947.

Martinetti, C.R., Childress, A.E. and Cath, T.Y. (2009) High recovery of concentrated RO brines using forward osmosis and membrane distillation. J. Membr. Sci. 331 (1–2), 31–39.

Maskan, F., Wiley, D.E., Johnston, L.P.M. and Clements, D.J. (2004) Optimal design of reverse osmosis module networks. AIChE J. 46(6), 946–954.

McCutcheon, J.R., McGinnis, R.L. and Elimelech, M. (2006) Desalination by ammonia–carbon dioxide forward osmosis: Influence of draw and feed solution concentrations on process performance. J. Membr. Sci. 278, 114–123.

Moore, S.E., Mirchandani, S.D., Karanikola, V., Nenoff, T.M., Arnold, R.G. and Eduardo Sáez, A. (2018) Process modeling for economic optimization of a solar driven sweeping gas membrane distillation desalination system. Desalination 437, 108–120.

Parfit, M., Conniff, M., Mitchell J.G. and Ellis, W.S. (1993), Water, The Power, Promise, and Turmoil of North America's Fresh Water, National Geographic Special Edition.

Rautenbach, R., Linn, T. and Al-Gobaisi, D.M.K. (1997) Present and future pretreatment concepts – Strategies for reliable and low-maintenance reverse osmosis seawater desalination. Desalination 110, 97–106.

Rezaei, M., Warsinger, D.M., Lienhard V, J.H., Duke, M.C., Matsuura, T. and Samhaber, W.M. (2018) Wetting phenomena in membrane distillation: Mechanisms, reversal, and prevention. Water Res. 139, 329–352.

Richter, B.D., Abell, D., Bacha, E., Brauman, K., Calos, S., Cohn, A., Disla, C., O'Brien, S.F., Hodges, D., Kaiser, S., Loughran, M., Mestre, C., Reardon, M. and Siegfried, E. (2013) Tapped out: how can cities secure their water future? Water Policy 15(3), 335–363.

Ruiz Salmón, I. and Luis, P. (2018) Membrane crystallization via membrane distillation. Chem. Eng. Process. Process Intensif. 123, 258–271.

Service, R.F. (2006) Desalination freshens up. Science 313, 1088–1090.

Sommariva, C. and Mott MacDonald International Ltd. (2004) Desalination Management and Economics, Faversham House Group, Surrey.

Sukitpaneenit, P. and Chung, T.S. (2012) High performance thin-film composite forward osmosis hollow fiber membranes with macrovoid-free and highly porous structure for sustainable water production. Environ. Sci. Technol. 46, 7358–7365.

Teow, Y.H. and Mohammad, A.W. (2019) New generation nanomaterials for water desalination: A review. Desalination 451, 2–17.

Tijing, L.D., Woo, Y.C., Choi, J.-S., Lee, S., Kim, S.H. and Shon, H.K. (2015) Fouling and its control in membrane distillation – A review. J. Membr. Sci. 475, 215–244.

Wang, P. and Chung, T.S. (2015) Recent advances in membrane distillation processes: Membrane development, configuration design and application exploring. J. Membr. Sci. 474, 39–56.

Zhang, Y., Peng, Y., Ji, S., Li, Z. and Chen, P. (2015) Review of thermal efficiency and heat recycling in membrane distillation processes. Desalination 367, 223–239.

16 Membrane Distillation for Industrial Water Treatment

Experiences from Pilot Trials

Mikel C. Duke and Noel Dow
Institute for Sustainable Industries and Liveable Cities, Victoria University, Melbourne, Victoria, Australia

16.1 INTRODUCTION

16.1.1 CURRENT STATE OF MEMBRANE DISTILLATION RESEARCH AND DEMONSTRATIONS

Membrane distillation (MD) technology has great potential as a water treatment technology. It principally works much like a thermal water treatment process, but offers additional advantages stemming from the use of the membrane to make it more compact, and restrict the surface of the fluid to be evaporated thus relaxing the needs to have controlled surface evaporation from either planar or round (e.g., droplets) geometries which can be subject to the effects of gravity. With numerous studies on the membrane materials, fouling effects (including the unique MD fouling known as 'wetting'), performance modeling and optimization, and bench scale performance optimization, it appears there is an abundance of both fundamental and practical knowledge for numerous applications. However, due to the small scale of these tests, the question remains on what will happen when installed in a real context. It has been reported recently that MD adoption by industry is low (Cassard and Park 2018) and indeed no available references to plants adopted by industry for water treatment can be found. Practical benefit is seen by demonstrating it to operate over substantially long terms (months to years) while meeting the treatment and thermal energy constraints of a given site. Broadly, MD research is directed toward production of drinking water from saline sources such as seawater and groundwater and also for water (and other product) recovery from industrial wastewater. Each of these cases presents a unique challenge for MD technology, where even one successful example here may not be applicable in another situation. Every case of performance and integration is unique and presents an opportunity for researchers to explore a range of cases that have benefit to the community and overcome challenges in every situation to provide a range of working examples of how MD technology has resolved a water treatment challenge.

Researchers have identified important differences to performance in comparing bench to pilot studies, even using idealised feed solutions made from NaCl (Eykens et al. 2017). For example, it was found under certain test conditions using Aquastill spiral wound modules, that the higher DCMD fluxes compared to AGMD typically seen at lab scale were actually reversed, showing lower DCMD flux than AGMD using 7.2 m^2 pilot modules. This example demonstrates the value of pilot studies to show performance effects in certain practical situations. It is these types of works that will instil confidence to move the technology into commercial reality.

16.1.2 Purpose of This Review of Demonstrations

This chapter presents the current status of MD pilot trials published in the literature, describing how the motivation was derived and the outcomes. Although there are various suppliers of MD systems which could be used in industrial applications (Cassard and Park 2018), the purpose here is to address lack of widely available knowledge on the experiences found in operating MD pilot trials by reviewing and documenting the common issues found in the small number of published isolated works, as well as to highlight their resolution and/or analysis of these issues and needs for future work. The term 'pilot trial' relates to cases where a system was tested on an actual feed water for a sufficient period of time (e.g., more than 1 week), in particular was installed on site with convenient access to the feed water (but cases where the water was brought to the plant have also been considered). Further, literature captured here relates to use of actual pilot equipment that is site-appropriate, as opposed to laboratory-oriented and scaled systems (i.e., single layer or several fiber modules), whereby issues associated with handling commercially oriented module designs are the most useful to represent for potential technology adopters. Specifically, experiences beyond the measurement of preliminary flux and rejection data are of interest in this chapter. Plant management experiences such as start-up and shut-down, membrane cleaning, supporting equipment performance (e.g., heat exchangers), and replacement are particularly emphasized in order to highlight important experiences that are relevant to industry.

16.2 APPLICATION TO SEAWATER AND GROUNDWATER DESALINATION

Earlier pilot trials of MD focused on the application of desalination, especially seawater. A history of pilot trials as of 2013 was conducted in a review by Camacho et al. in 2013 (Camacho et al. 2013). At this time, the key MD technology developers identified were Fraunhofer ISE, Memstill, Aquastill,

Scarab and Memsys. These earlier installations in pilot trials starting in the early 2000s mark the beginning of the wider exploration into the practical application of MD, but more specifically the experiences published in just the last 3–4 years must be captured. Such experiences include cases of integration into industries, plant economics, sustained performance over longer term (months to years), repeated membrane cleaning operations, and autopsy of membranes after long-term operations.

16.2.1 Managing Scaling and Membrane Deterioration

The most intuitive issue for MD on seawater and groundwater is fouling from scaling. Although this review is focused on pilot studies, bench tests on DCMD on seawater RO brine found high recovery limitations due to calcium scale formation (Zhang et al. 2015). Early experiences with scaling on a 0.66 m^2 membrane area hollow fiber pilot plant were documented in 2008 (Song et al. 2008). The pilot plant was used under various conditions over 3 months, for example, concentrating seawater from 2.9 wt% to 19 wt%. Declining flux was attributed to calcium sulfate scale. Another of the earlier pilot experiences documenting cleaning was presented in 2010 (Dow et al. 2010). This MD trial included results of a field test lasting 8 days using DCMD with a custom-made multilayer module with 1.4 m^2 active membrane area and heat obtained from a solar source. The feed was brine from an RO plant that was used to produce drinking water from saline groundwater. Membrane fouling from inorganic scale was removed by using a mineral acid cleaning cycle. The experience from this trial also revealed that water produced at half the design capacity of the plant because of the intermittency of solar collection and fouling from scaling.

More recently, pilot experiences were reported showing wider equipment practicalities. Pilot trials conducted at the Plataforma Solar de Almería (PSA) in Spain have been published and provide critical experiences associated with professional setups and associated management of solar energy as well as operation on seawater. For example in 2018, J.A. Andrés-Mañas et al. trialled a 6.4 m^2 Aquaver BV with MEMSYS module for desalination of seawater (Andrés-Mañas et al. 2018) as shown in Figure 16.1. They found issues related to scaling on the heat exchanger as shown in Figure 16.2. Consequently, important features presented in this work were trials of acid cleaning and use of antiscalants, which are common practices used in RO plants to maintain long-term operation.

Pilot trials integrated with solar collectors on the Scarab AB module at 2.8 m^2 membrane area showed longer term operation issues with scale formation from steel corrosion by-products, particularly iron oxide (Guillen-Burrieza et al. 2014). This study presented important findings toward the practical experience of MD application that would not have been identified in ideal lab conditions, or even short pilot scale studies. Intermittent operation of a model seawater solution over 3 years found a predominantly iron scale build-up. Figure 16.3 shows the membranes after pilot testing, while Figure

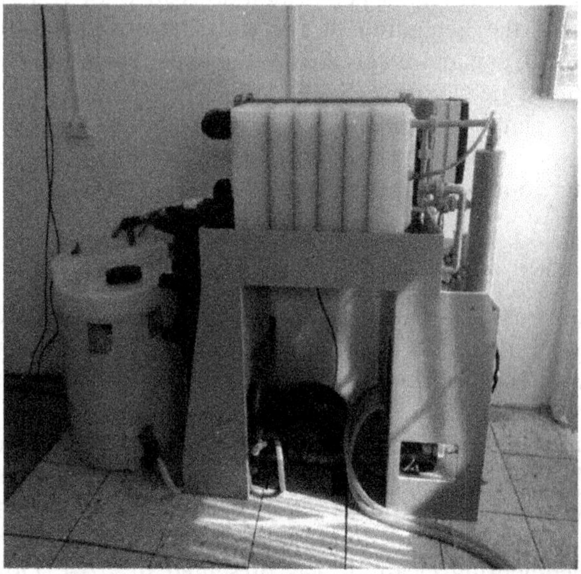

FIGURE 16.1 MEMSYS pilot plant trialled for seawater desalination (Andrés-Mañas et al. 2018). Original color image available from source reference. Copyright (2018) Elsevier.

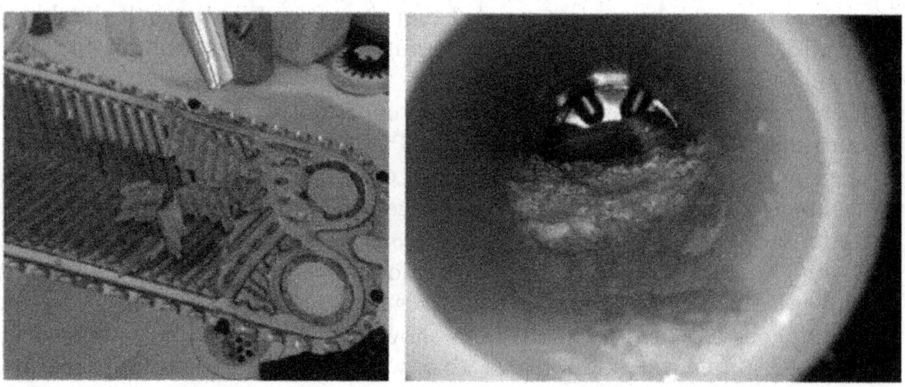

FIGURE 16.2 Scaling over plate heat exchanger (left) and heat exchanger inlet (right) observed after processing 36 m^3 of Mediterranean seawater by MD without use of anti-scalant (Andrés-Mañas et al. 2018). Original color image available from source reference. Copyright (2018) Elsevier.

16.4 shows the measure of key MD membrane material properties and how they changed as a result of the long-term, intermittent desalination of the model seawater solution. Contact angle (CA) which indicates the membrane's apparent hydrophobicity was reported to decline to 84% of its initial value of

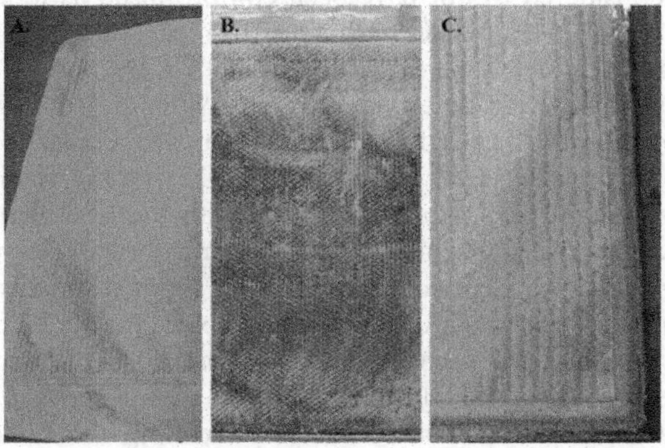

FIGURE 16.3 Images taken of membranes used in pilot trial of a Scarab MD module, showing unused to surface (A), fouled top surface (B), and fouled back surface (C) (Guillen-Burrieza et al. 2014). Original color image available from source reference. Copyright (2014) Elsevier.

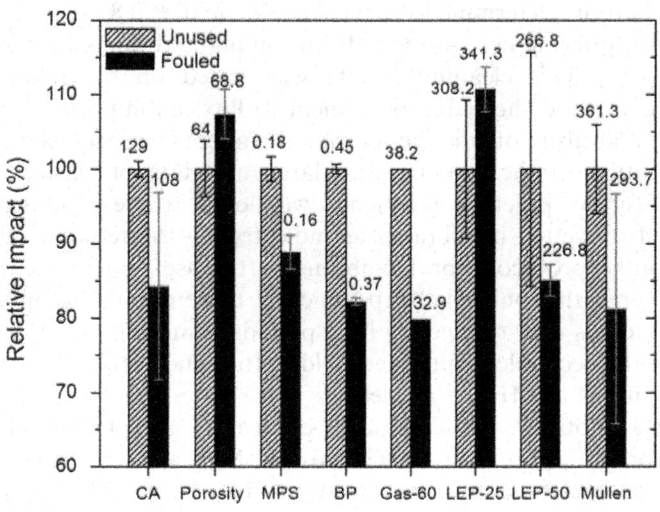

FIGURE 16.4 Assessment of the change in membrane properties as 'relative impact' after long-term operation of the membranes in the Scarab MD modules used at the PSA in Spain. The impact value determined by the change in their principle measurement units for contact angle (CA), porosity, mean pore size (MPS), bubble point (BP), gas permeability at 60 psi, liquid entry pressure at 25 °C (LEP-25) and 50 °C (LEP-50), and Mullen burst test (Guillen-Burrieza et al. 2014). Copyright (2014) Elsevier.

128° and was considered to not have been compromised as it stayed above an average value of 90°. Meanwhile, reductions in mean pore size (MPS), bubble point (BP), and 60 PSI gas permeability (gas-60) were observed above measurement errors which were determined to be related to fouling of the upper layers of the membrane. The reduction in BP was also correlated to the rise in liquid entry pressure measured at 25 °C (LEP-25) of the fouling membrane. However, an important observation was made with the LEP measured at 50 °C (LEP-50), showing a sharp decline in the fouled membrane. No specific material reason directly related to LEP could be given as the measurement is influenced by several properties, but was attributed to the defects in the PTFE membrane's fibril structure and it was thus concluded that there is a destruction of the membrane structure as a result of field testing.

Further investigations were conducted by cleaning the fouled membranes with deionized (DI) water, a range of organic acids, metal chelating agents (EDTA) and sodium tripolyphosphate (STPP), first on membrane samples tested in the lab then on the trial fouled membranes (Guillen-Burrieza et al. 2014). The range of measurements listed previously as well as Mullen burst tests and scanning electron microscopy (SEM) on membranes was conducted before and after cleaning with the following range of agents: DI water, citric acid, sulfuric acid, formic acid, and a combination of oxalic, EDTA, and STPP. The authors determined 0.1 wt% oxalic acid + 0.8 wt% citric acid was most suitable (offering best recovery from fouling and minimal degradation of the membrane). This cleaning regime was tested on the pilot plant, and although the method showed improvement to flux, an important experience in moving from lab to pilot was the inferior cleaning effect was related to poorer flow distribution in the module, and large amounts of rinsing water were required. A further practical experience was noted, where periods of inactivity 'dry out' led to wetted membranes as indicated by the reduced salt rejection. Correct storage protocols for membranes after use in practice was recommended. Overall this published experience of a membrane autopsy and practical pilot findings over sufficiently long periods (~months) are highly valuable for moving the technology into the field. More such studies on similar and other applications of MD are needed.

Most MD pilot tests are conducted on readily available membrane types, while membranes specifically developed for MD application is a topic for researchers as presented in a review from 2015 by García-Fernández et al. (García-Fernández et al. 2015). Example custom development approaches include increased hydrophobicity (Lu et al. 2016) and reduced fouling chemistries (Zuo and Wang 2013). However, a trial on a custom modified membrane was recently presented in 2018 to directly deal with the issue of fouling in the application of seawater treatment (Mansour et al. 2018). The modification featured graphene nanoplatelet (GNP) coatings on membranes in a direct contact MD (DCMD) and the pilot system held 0.05 m^2 of active membrane area. Polyethylene membranes from Aquastill were used and an important feature of this test was the use of real brine of 55,000 mg/L concentration from a seawater reverse osmosis plant operating in the UAE. This enabled studies not only on the issues related to inorganics but also organics which were present in actual seawater. Individual experiments lasted 4 h each with flux of 16.7 L/m^2/h and specific thermal energy consumption (STEC) of 152 kWh/m^3 being recorded, then a 77 h period showed severe flux decline of 69% as a result of fouling where the main fouling components identified were inorganic precipitates, inorganic cleaning chemicals, and antiscalants which came from the usual operation of seawater RO plants. This fouling was mitigated with the novel coating of the commercial polyethylene membrane with GNP. With a set of 10 h tests on the same RO brine as their long-term tests are shown in Figure 16.5, compared to the commercial polyethylene membrane the GNP-coated membrane flux decline was reduced to as low as 6.7%. Figure 16.5 also shows performance with different loadings of GNPs in the coating were tested and compared to the commercial polyethylene membranes. Membrane characterization features associated with the GNP included increased hydrophobicity measured by contact angle, increase in the liquid entry pressure, and higher tensile strength. Despite the reduced porosity from nanofilling with GNPs, the improved properties led to higher vapor fluxes.

16.2.2 Assessment of Thermal Energy Requirement: Optimizing Conditions, Impact from Fouling, and Module Design

Numerous desktop studies have already been conducted to represent the energy requirement for MD, but only recently pilot trials have been validating the estimations with real data on commercially available modules. Duong et al. utilized an Aquastill AGMD system with 7.2 m^2 membrane area in 2016 to explore the energy efficiency in a seawater desalination application (Duong et al. 2016). The thermal efficiency in operating real pilot systems with actual feed streams is

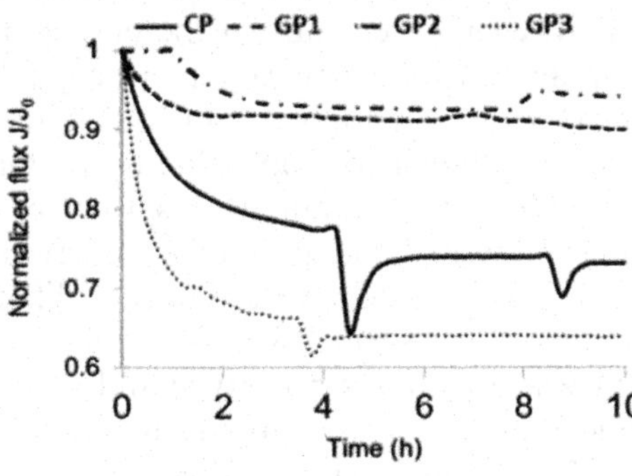

FIGURE 16.5 Normalized flux results for a 10 h test by MD on a seawater RO brine using a commercial polyethylene (CP) membrane used for MD, as well the novel graphene nanoplatelet (GNP) coated CP membranes with 0.08 wt% loading (GP1), 0.16 wt% loading (GP2), and 0.2 wt% loading GP3 (Mansour et al. 2018). Copyright (2018) Elsevier.

critical to developing confident values as references to future practical applications. Operating the MD system with 70 °C hot side, they found the STEC as low as 90 kWh/m^3 water treated could be achieved, with an associated specific electrical energy consumption (SEEC) of 0.13 kWh/m^3 water treated. The operation of the system with a water recovery of 5% ensured issues such as calcium scaling were not noticed over the 9 h run period. While gain output ratio (GOR), being the ratio of MD water produced to the amount of water produced by a single evaporation stage, varied from 6.5 to 7.0 over the run time, no trends in performance to predict longer term behavior were observed from this trial. STEC and GOR values can generally be found now in published pilot trial experiences, but SEEC is more difficult to find. Future publications should report the STEC and SEEC to highlight to MD adopters the energy use experiences for their given application example.

Ruiz-Aguirre et al. (Ruiz-Aguirre et al. 2018) tested two Aquastill AGMD modules, one being 7.2 m^2 and another being 24 m^2. Using actual seawater feed and solar collectors as a heat source, they both contained different internal features where the larger active area module contained longer channels (5.0 m for 24 m^2 vs. 1.5 m for 7.2 m^2) and thus the study focused on how internal module differences lead to optimal performance. It was found that the longer module uniquely enabled maximizing the productivity without losing thermal efficiency, while the shortest module needed a compromise between both. Specifically, the long module gave a flux of 1.35 L/m^2/h with a STEC requirement of 106.6 kWh/m^3 but the shorter module with maximized productivity and thermal efficiency led to a flux of 4.20 L/m^2/h and STEC of 295.6 kWh/m^3. This work highlights, on a commercially available design, how the features of the module need to be customized for the context of operation. Flux is linked to total membrane area requirement (and in turn plant capital cost) and clearly a 3–4-fold difference in flux leading to a likewise reduction in total membrane area would be a substantial saving in terms of both membrane capital cost and footprint. However, it has clearly come at the penalty of a threefold increase in energy demand, so the cost of that additional heat needs to be weighed up against the membrane plant capital cost saving.

Meaningful documentation of findings from longer terms tests (lasting at least 1 week) are harder to find in the open literature. One study conducted over two years saw a pilot system running for a total of 2400 h using solar collectors as the source of thermal energy (Guillén-Burrieza et al. 2012), albeit with synthetic feedwater. This system featured two different precommercial MD desalination modules from Keppel Seghers with a total membrane area of 9 m^2 each. The first was a compact prototype and gave a thermal efficiency of 1805 kWh/m^3 while the other was a multistage system requiring a lower STEC of 294 kWh/m^3. Several findings were documented from this experience, including to show that the MD modules worked well as heat exchangers and lost little heat to their surroundings. In their situation, both modules were designed quite differently, that is, with and without internal heat recovery. The internal heat recovery design performed best with highest fluxes and STEC of 5.09 L/m^2/h and 294 kWh/m^3, respectively. Some details were given on the membrane behavior from the model seawater solution tested over the 10 total months of testing both modules. Product water quality was found to be consistently good, where electrical conductivity values only exceeded 4 μS/cm at the start of experiments and was explained to have occurred from crystal growth through the membrane and wetting. But overall no negative effects from fouling over the longest module test time (7 months) were observed and no cleaning of the membranes or system was reported. This may be because of the low water recoveries used (always <5%), thus not leading to any noticeable concentration rise of the dissolved solids. With this aside, this is evidence that MD systems and current membranes can operate for practically long periods of time. However, low recoveries do not suit all situations, and often cleaning is a requirement for membranes and associated equipment. So experiences with operating long times with higher recoveries and membrane cleaning also need to be demonstrated.

16.2.3 Performance Optimization

Pilot scale units were also trialled with a main objective to find optimal performance using solar thermal energy located at the PSA in Spain using a model seawater solution (Ruiz-Aguirre et al. 2017a). Using Response Surface Methodology (RSM), the key operational measures of permeate flux and STEC were optimized (maximum flux and minimum STEC) on a Solar Spring Oryx 150 permeate gap MD (PGMD) module with 10 m^2 of active surface area. It was found that a flux as high as 2.66 L/m^2/h could be achieved with the minimum STEC of 255.8 kWh/m^3. Further, the analysis concluded that evaporator inlet temperature (temperature of the warmed seawater fed to the AGMD's membrane channel) had the strongest influence on flux and STEC, in comparison with feed flow rate which is another MD operational setting. Condenser inlet temperature (temperature of the cool seawater fed into the AGMD module's condenser channel) on the other hand influenced only the membrane flux leading to the conclusion that for optimal thermal efficiency, additional cooling is not needed if the feed water for desalination is 30 °C or less.

The RSM approach was again carried out on the same Solar Spring Module with Artificial Neural Networks (ANN) to understand MD performance in terms of flux and thermal efficiency with varying parameters similar to the prior work (Ruiz-Aguirre et al. 2017a): inlet temperature, evaporator inlet temperature, and feed flow rate, but with the addition of incoming feed salt concentration (Gil et al. 2018c). Specifically, feed concentrations of 70 g/L and 105 g/L were chosen to represent the performance when desalting RO brines. With a fixed temperature of the heated seawater to the AGMD membrane channel, for applications that require higher flux (water production), it is better to operate with lower temperatures of the seawater fed to the AMGD condenser channel and larger module cycle flow rates. But this changes if higher thermal efficiency (lowest STEC) is required, where lower cycle flow rates and higher condenser channel feed temperatures are ideal for 70 g/L salinity, whereas for 105 g/L this changed to operating at larger module cycle flow rates (but still with higher condenser temperatures).

While optimization of performance is clearly needed to gain best flux and minimal STEC values, evaluation is challenging for MD since there are many influencing conditions like available temperature, operation temperatures, flow rates, and module design. In each case, these conditions have a competing purpose to also allow best performance found by researchers when treating the desired solution to minimize fouling. Complicating factors also include the constraints of available heat (varies between highly constrained and abundant (Dow et al. 2016; Guan et al. 2015). Therefore, more optimization studies are needed, but they now must be conceived, carried out, and produce results in specific applications of need, as optimizing without targeting a specific application will lead to too many input variables and in turn a highly conditional and complex array of optimisations.

16.3 INDUSTRIAL WATER TREATMENT

Conceptually, application of MD to treat industrial waters is very promising, but is the most challenging due to numerous types of wastewaters carrying a myriad of probable membrane foulants. Trials showing experiences on industrial wastewater are less common than trials on seawater. Project goals can be very different, where extracting fresh water from the sea often has the sole objective to provide a fresh water supply, while treating industrial wastewater is often driven by a means to deal with a waste issue. The value of the captured fresh water varies considerably and its ability to be reused depends on the water being treated, along with the general availability of fresh water at the location. While numerous bench scale tests have been carried out on a number of industrial wastewaters, this section outlines the experiences specifically on pilot plants where issues such as fouling of actual wastewaters (complex and variable in composition), heat and process integration, long-term performance, and economics are uniquely being directly addressed.

16.3.1 Municipal Waste Water

Woldemariam et al. (Woldemariam et al. 2016) reported experiences from trialling a Xzero AB AGMD unit at pilot scale for the purification of effluent from a municipal wastewater treatment plant, presented in Figure 16.6. The heat supply proposed in this work came from district heating. A specific proposition of this trial was the MD system's ability to reject pharmaceutical residues, which was supported as it was found that MD rejected the pharmaceuticals from the wastewater to below measurable quantities. From a performance perspective, they reported STEC values of up to 875 kWh/m^3 without heat recovery, and as low as 105 kWh/m^3 with heat recovery. The energy analysis was carried out using data collected over a 40-day period (300 h) of intermittent operation (with overnight shutdowns). Pharmaceutical removal testing was run for three trials within a 10-month period, lasting about a week each. This was

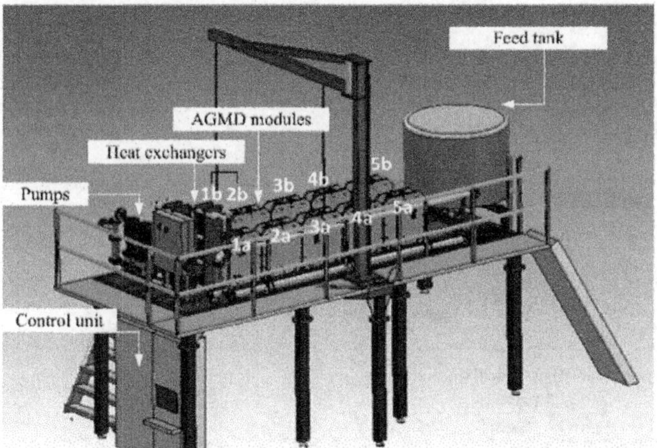

FIGURE 16.6 Illustration of the pilot system using Xzero AB MD modules to treat municipal waste water (Woldemariam et al. 2016). Original color image available from source reference. Copyright (2016) Elsevier.

followed by a concentrating test lasting another three weeks (also with overnight shutdowns).

In this trial, the economic basis was an important feature of the investigation. Figure 16.7 shows the economic breakdown of the different components from the MD pilot plant. An interesting finding here is the relative cost of the modules over other major components, particularly heat exchangers. Overall, the MD modules account for about a third of the capital cost breakdown. This is no surprise considering the modules are the main feature of the equipment and are relatively new technology compared to the other equipment being more routine such as pumps, tanks, and heat exchangers.

The scenarios explored in their study included three cases which revealed a story of benefit and cost relevant to various practical interests. One case had advantages of least heat demand, but had disadvantages of low flux/high module numbers. In another case with highest flux leading to least number of modules and associated capital cost, but having the disadvantage of no heat recovery with the associated highest unit cost of water purification. A final case by using heat recovery low unit costs of water was achieved, but led to low fluxes and in turn more membrane area (and capital cost). This economic study is valid considering it used real cost data including a commercial MD module setup in a valid scenario. Up until now, MD economic studies have been based on assumed data in ideal scenarios which have so far helped to understand practical drivers. However, now more published studies like the Woldemariam wastewater application which includes details of economic breakdowns of real MD plants (much more detailed than single line economic

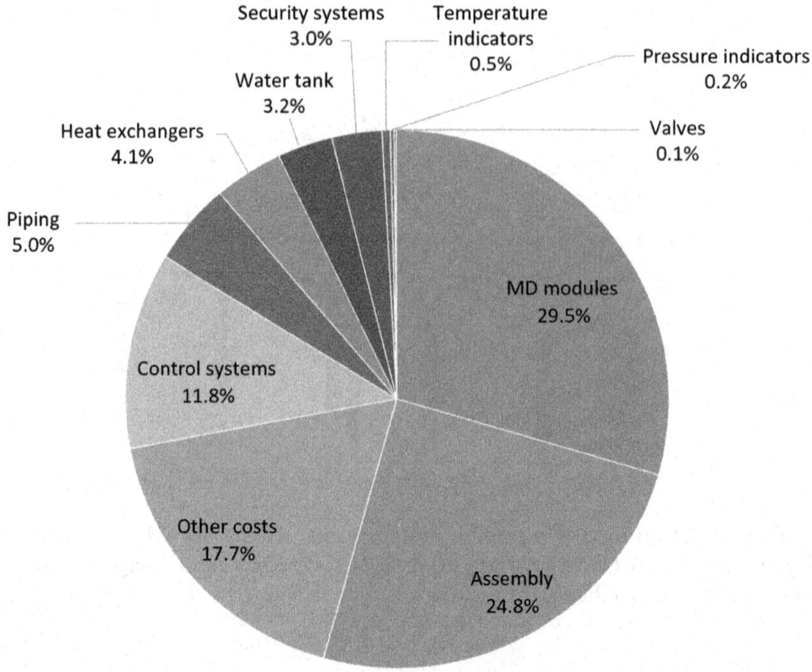

FIGURE 16.7 Capital cost breakdown of the various components in a real MD reference plant using Xzero AB AGMD modules (Woldemariam et al. 2016). Copyright (2016) Elsevier.

values) in valid applications are critically needed as references to build confidence in the technology by the industry.

16.3.2 Dairy Processing

MD has been considered for application in dairy processing (Hausmann et al. 2011, 2013a, 2013b, 2014). However more recently, a pilot test was conducted for application to treat waste from cheese pickling (rich in NaCl and dairy compounds) on a 14.4 m^2 AGMD module from Aquastill (Eykens et al. 2018). A photograph of the pilot setup showing the Aquastill AGMD module is shown in Figure 16.8. With the given specific need to address costly disposal of saline wastes, a cost analysis for the MD solution could be presented. For a requirement of 650 m^3/year of distillate, this would require four 24m^2 AGMD modules. The cost was highly sensitive to the cost of the heat as has been observed in previous studies on MD economics applied to seawater desalination scenarios (Kesieme et al. 2013). Without utilisation of waste heat, the lowest cost estimated in the 2018 publication was 35 Euros/m^3 (\$39 USD/m^3) for the 24 m^2 module using 34 kW of thermal energy, whereas

FIGURE 16.8 Pilot plant to treat dairy waste using an Aquastill AGMD module (Eykens et al. 2018). Original color image available from source reference. Copyright (2018) Elsevier.

if all thermal energy is supplied by waste heat, the lowest cost is 12 Euros/m^3 (\$13.5 USD/m^3). However, the cost to discharge the brine was estimated at 50–150 Euro/m^3 (\$56–170 USD/m^3), while the salts in the brine have a value of 20 Euros/m^3 (\$23 USD/m^3). Based on the savings made by reducing the waste volume, even without using waste heat, it was concluded that MD is economically justified for application in managing waste from saline cheese pickle. Overall, application to dairy industry is promising, particularly in managing saline wastes.

16.3.3 Ion Exchange Regeneration Waste from a Power Station

A pilot plant was installed on site at a power station to treat the wastewater produced from ion exchange regeneration (Dow et al. 2016). The trial presented a unique challenge whereby the goal was to harness only waste thermal energy and not utilise high grade thermal energy that would normally be used to produce electric power. This led to the constraint whereby the thermal energy exhausted from the plant was abundant in quantity yet limited in temperature to 40°C, which in turn became the hot feed temperature to the MD pilot. Figure 16.9 shows the waste heat collection system, where tapping the heat from the boiler circulation pumps as shown were the most effective heat supply.

A custom built multilayer DCMD module with 0.67m^2 of active surface area was operated over three months, and was cooled using the same estuarine water that was used to cool the power station's condensers. Long-term operation achieved a water recovery of 92.8%, but the membrane analysed at the end of the trial was fouled from supporting equipment corrosion byproducts

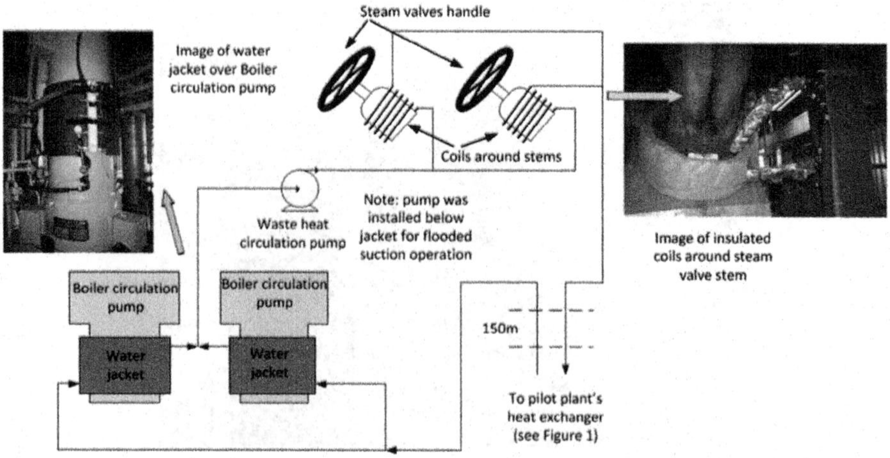

FIGURE 16.9 Schematic representation of the waste heat recovery system used in a pilot trial MD system to treat ion exchange regeneration wastewater from a power station (Dow et al. 2016). Original color image available from source reference. Copyright (2016) Elsevier.

(e.g. heat exchangers). As shown in Figure 16.10, both the feed and permeate cycles accumulated fouling over time. For the permeate side fouling, a breach in the heat exchanger allowing saline river water contaminants to reach the membrane was found due to corrosion which was considered the cause of the fouling. Like the trial at PSA (Guillen-Burrieza et al. 2014), such practical experiences from issues related to the supporting equipment are found over these longer terms studies. Findings from this pilot study gave real evidence of how the DCMD system could manage the variable temperature property of the waste heat, which translated to variable flux output of the MD plant. Energy assessment with no internal heat recycling (GOR ≤1) found that the

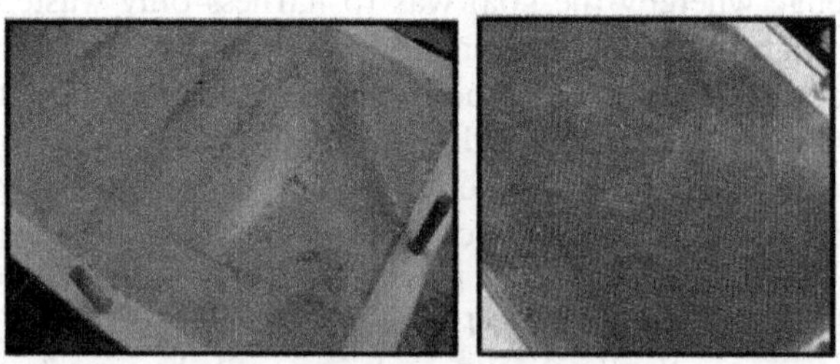

FIGURE 16.10 Appearance of membrane feed side (left) and permeate side (right) of membrane after treating ion exchange regeneration wastewater for 3 months and no cleaning (Dow et al. 2016). Original color image available from source reference. Copyright (2016) Elsevier.

abundant low grade thermal energy of 500 MW exhausted by the power station operating at peak capacity could theoretically produce up to 8,000 kL/day of clean distillate for reuse.

16.3.4 Textile Industry Wastewater

The custom-made DCMD module with no heat recovery system trialled at the power station presented above (Dow et al. 2016) was improved in design and scaled up for a trial conducted on textile wastewater (Dow et al. 2017). As shown in Figure 16.11a, the $6.4m^2$ membrane area module was installed on site at a textile mill with heat supplied from the site's steam heat service condensate return and cooled by dry (air) coolers while operated for 65 days (operating on weekdays only). The recovery of the water reached 91.6% by the end of the trial and no wetting was observed. The feed water was assessed as part of the study, being either effluent directly from the site, or after biological treatment (conventional activated sludge process). Preliminary bench testing with the novel use of a foam fractionator pre-treatment greatly minimised the membrane wetting from the surfactant rich samples taken directly from the site, however the risk of wetting was still high and instead the biologically treated wastewater was used in the trial as it showed no wetting issues. The foam fractionator shown in Figure 16.11b was still used during the trial as it also assisted to remove surface-active compounds including proteins likely to be present in the biologically treated effluent. An important aspect of this trial

FIGURE 16.11 Pilot plant setup to treat textile wastewater for 3 months, using a custom made multilayer DCMD module (a). The image on the right (b) shows the foam fractionator in start-up mode. During steady-state operation, the foam collapsed and was removed via the drain. Original color image available from source reference. Reproduced from (Dow et al. 2017) by permission of The Royal Society of Chemistry.

was the use of cleaning, where in this case the membrane fouling was predominantly organic material and caustic cleaning was found to effectively restore membrane flux (returned to 81% of initial productivity after a series of caustic cleans). No issues were found during shutdown of the plant during weekends such as membrane integrity loss as was concluded as a requirement from pilot tests previously (Guillen-Burrieza et al. 2014), suggesting appropriate shutdown methods were used in this trial on textile wastewater. As with most industry trials, integration assessments are required. It was found that the site waste heat was enough to treat the textile wastewater by MD for a potential zero (or near zero) liquid discharge situation, but required selectively isolating saline waste stream from the various site effluents and using RO to pre-concentrate the saline waste (~4,000 mg/L). Treatment of the combined saline wastes by RO before MD was assessed as the better overall solution as the stream was still relatively low in salinity and RO is a low cost, low energy option in this low concentration starting range.

16.3.5 Coal Bed Methane Water

Water from coal-bed methane bodies (or coal seam gas, CSG) is saline and presents issues in its handling when extracting gas. MD was applied to the RO concentrate from a CSG wastewater using a 7.2m^2 AGMD system provided by Aquastill (Duong et al. 2015). The trial lasting 140 h showed that over time the distillate flux production decreased, and the STEC increased (175 kWh/m^3 to 250 kWh/m^3) as a result of membrane fouling. This study also provided the SEEC, identified earlier in this chapter as being harder to find in published studies, as being 1.1 kWh/m^3. In considering the integration into the CSG application, solar thermal energy was proposed and it was determined that 1 ha of flat plate solar collectors would be required to treat 118 m^3/day of CSG brine using the AGMD system. In addition to the context assessment in a practical application, the brief trial raised important issues such as the change in STEC as a result of fouling from an actual wastewater.

16.4 NOVEL INTEGRATIONS WITH OTHER TECHNOLOGIES

This section is devoted to presenting cases where MD has been integrated with other technologies and demonstrated at pilot scale. Specifically, this relates to major equipment units that extend the function of the MD, and does not include pre-treatment required for effective MD operation which has been described in the earlier sections on sea/groundwater desalination and industrial wastewater treatment. While this section is devoted to pilot scale studies which provide critical knowledge on challenges associated with larger scales and sustained simultaneous operation of the hybrid process, it is important to point out that novel concepts for integration with other processes are being developed at bench scale, such as with forward osmosis for further concentrating RO brine (Martinetti et al. 2009), with photo-Fenton to eliminate bacterial spores (Ruiz-Aguirre et al. 2017b) integration with solar ponds (Suárez et al. 2015) and conversion of heat to electricity by use of reverse electrodialysis (RED) (Micari et al. 2019). In surveying the literature on integrated (hybrid) processes with MD, it was found to date that only integration with solar panels and membrane bioreactors have reached the stage of pilot testing.

16.4.1 Solar Panels

As MD develops to commercial reality, the process needs to work with technologies also being developed for future industries, which includes solar panels. Specifically for MD being a thermal technology, solar thermal collectors are utilised. The key issue is the intermittency of solar energy on the operation of seawater MD was explored by Gil et al (Gil et al. 2018b). They applied a complete feedback control architecture for an experimental solar powered MD facility using a 24 m^2 Aquastill AGMD system. The control system was shown to fix a suitable operating temperature for the MD thus assisting with the operation.

Guillén-Burrieza et al (Guillén-Burrieza et al. 2011) conducted pilot trials over 4 months on a Scarab AB unit with 2.8m^2 of membrane area using up to 35 g/L NaCl solution to explore performance features of the pilot system when connected to a static solar collector field as the thermal energy supply where no heat storage was used. It was found the MD operation was able to accommodate for the transient nature of solar thermal energy. Further, the pilot trial experience showed fresh water production and thermal energy consumption was different to laboratory trials. An interesting finding from this study was the salinity was higher than expected in the permeate, suspected to be due to membrane wetting or damage, or an issue with the module or equipment. This occurred despite using model NaCl solution, which highlights the potential for permeate quality integrity loss associated with physical aspects of operation and not entirely from chemical aspects associated with fouling and wetting.

The issue of irregular thermal energy supply from solar energy was explored by (Gil et al. 2018a). A model was developed to optimise thermal efficiency (via reduced STEC), distillate (flux) production maximisation and cost savings, with optimisation of solar energy use by Nonlinear Model Predictive Control (NMPC) scheme. The system schematic is shown in Figure 16.12, which shows the key components of the solar field, MD system but also a thermal storage system. Operation was explored using an Aquastill AGMD module with 24m^2 of active surface area. An example finding on a clear day is presented in Figure 16.13. Patterns of irradiance and in turn temperatures highlight how the MD system must be ready to accommodate for the variability of the incoming solar energy. While various operational parameters are shown, an important measure is the distillate flux production (Figure 16.13d) which reached 30 L/h corresponding to a flux of 1.25 L/m^2/h. Tests were also presented in the case of a cloudy day, where the system managed the MD operation, in particular noting that the MD operation continued after the solar field turned off due to the residual tank temperature enabling the module to operate over 60°C. The main conclusion from this study was that the proposed control system is able to improve the product fresh water production and make cost savings in the case of a 20ha tomato greenhouse with future work planned to explore night operation.

16.4.2 Pilot Integration to Membrane Bioreactors (MBR)

Morrow et al piloted MD together with an osmotic membrane bioreactor (OMBR-MD) to treat wastewater from an activated sludge water treatment process (Morrow et al. 2018). The experimental process is shown in Figure 16.14. The system with MD was operated for 50 h, where average MD flux on the 0.74 m^2 pilot unit was 0.8 $L/m^2/h$. The integrated system led to a NH_4^+ removal of 90.2% and COD removal of 98.4%. This setup is a promising process, and indeed operating as a pilot scale system has presented meaningful results to potential MD technology adopters.

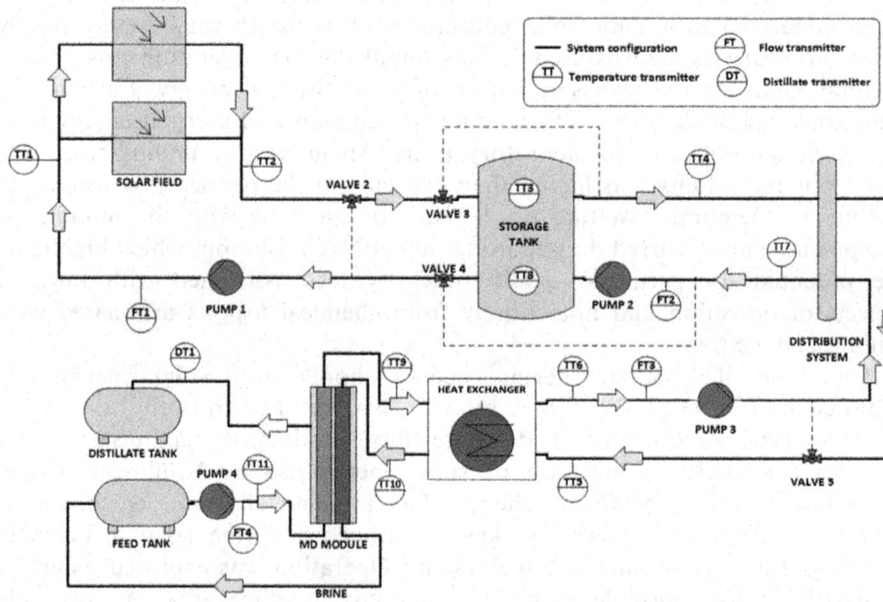

FIGURE 16.12 Flow diagram of the MD pilot plant using an Aquastill AGMD module including solar collectors and heat storage for a solar driven seawater desalination application (Gil et al. 2018a). Original color image available from source reference. Copyright (2018) Elsevier.

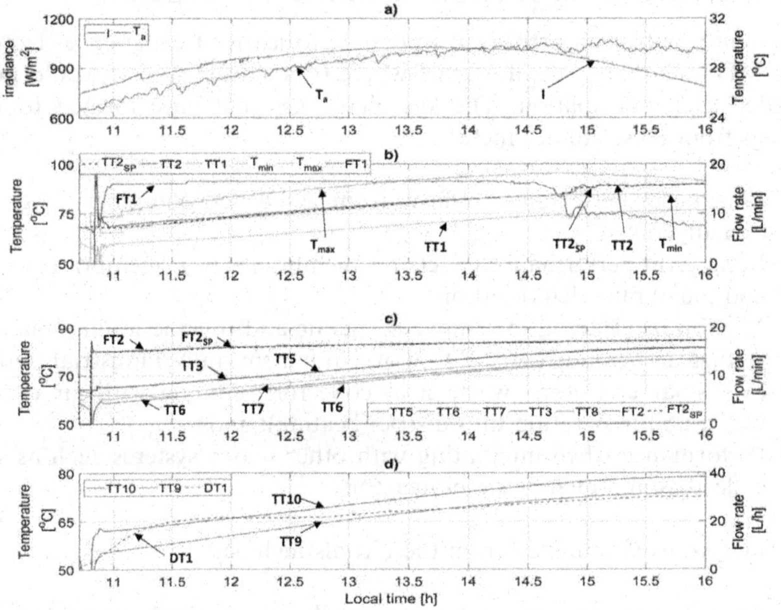

FIGURE 16.13 Performance results of the solar thermal MD plant for the example of a clear sky day, showing (a) meteorological conditions, (b) solar field variables, (c) tank and distribution system variables and (d) MD module variables (Gil et al. 2018a). Original color image available from source reference. Copyright (2018) Elsevier.

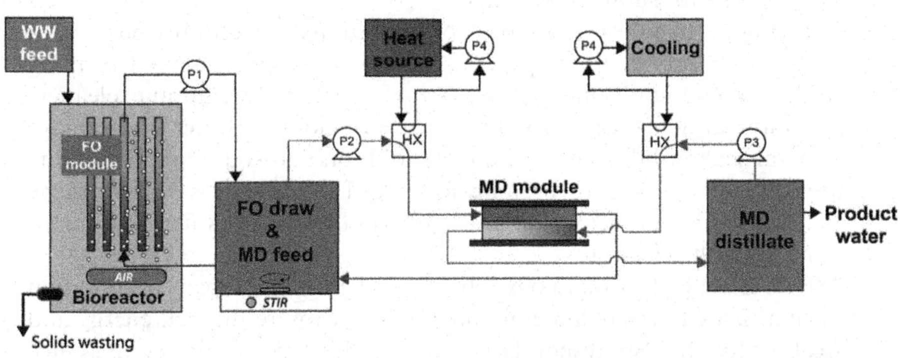

FIGURE 16.14 Schematic of the integrated osmotic membrane bioreactor -MD system in direct configuration. Forward osmosis and MD processes work simultaneously from one tank while the draw solution from forward osmosis is the feed to the MD (Morrow et al. 2018). Original color image available from source reference. Copyright (2018) Elsevier.

16.5 LESSONS LEARNT FROM DEMONSTRATING MD

In currently published articles, it's possible to learn of experiences from operating MD technology at pilot scale where bench scale studies would not have provided such information. The key challenges that have formed topics and findings from these studies include:

- Long term performance effects from associated equipment (e.g. heat exchangers);
- Long term performance effects on membranes (e.g. membrane fouling and membrane deterioration);
- Membrane cleaning to remove inorganic and organic fouling material;
- Actual integration of MD pilot equipment into industrial settings (e.g. solar collectors, waste heat collectors, cooling with air or river water, and integration into district heating); and
- Performance when integrating with other major systems such as solar collectors and membrane bioreactors.

The key lessons illuminated from these trials include:

- Associated equipment can unexpectedly cause formation of corrosion byproducts or even leak, leading to membrane fouling from the byproducts or leaked cooling water. This could be avoided by engineering protections into future plants, however such occasions are real and not always readily foreseeable, and thus the risk needs to be managed into the engineering design;
- Currently commercial membranes perform very well in both sea/groundwater and industry wastewater applications showing that with current commercial membranes, MD can be a practical success. However, for tackling certain fouling issues, non-commercial modified membranes are showing benefits;
- Cleaning of membranes as part of operational procedures has so far been successful, where employing judicious practice to avoid forming wetting compounds has shown use of common membrane cleaners like mineral acids, organics acid or even deionised water can remove mineral scale, while caustic (NaOH) solution and sodium hypochlorite are effective to remove organic fouling. The success is measured by the ability for the membrane to substantially return its flux and maintain high salt rejection;
- MD is a versatile technology, effectively being a compact evaporation system that can easily integrate into sites to capture thermal energy and treat water. It also responds to variable thermal energy sources like waste heat and solar energy by varying its performance with no apparent negative impacts on the equipment or membrane if managed correctly. Because of this, it has enabled applications at pilot scale to not only 'plug in' to existing industrial process schemes, but to explore novel

integrations such as with solar collectors/heat storage, and even with membrane bioreactors; and

- MD costing shows reasonable capital cost breakdowns from reference plants, and economic assessments for specific industrial settings has shown economic viability.

16.6 CURRENT GAPS AND FUTURE NEEDS – GOING BEYOND PILOT TRIALS

The currently available results of experiences from pilot trials shows promise and even ability for MD to be considered for commercial use in industry, particularly at smaller scales (e.g. several modules). However, it may still be regarded as too risky compared to more mature options and there is still much more that needs to be done to improve confidence in its use. Commercialising any technology into water treatment is challenging due to the conservative nature of the water industry to assure reliable water quality for a given price that allows operators to honour certain environmental and health obligations. Also, the prospect of replacing current equipment at the end of its life, which can last decades, are limited. However, there may be opportunities where treatment objectives have currently no cost effective solution, or a demand for a new plant comes up where available technologies to be installed will be scoped widely.

Suppliers are available to provide MD solutions, and this chapter has presented real trials on their equipment for the practical value that made the trial happen. However, to consider a technology that is relatively new to the industry presents a practical challenge for its adoption. Risks of fouling and wetting are present, but work has shown that these can be overcome with appropriate pre-treatment or avoiding risky wastewaters until suitable membranes become available. Effective trials have been conducted on available membranes, so while the membrane supply is not regarded as an impediment, custom designed membranes with special chemistries to minimise fouling are yet to become readily available from suppliers.

Another issue for MD developers from research and industry sectors is the aforementioned lack of information found documenting the experiences in implementing MD technology at pilot scale. While this chapter has attempted to highlight the progress found in the open literature, it was already known at the time of writing that various MD pilot trials are underway around the world, especially for industrial water treatment, which have not yet made results publicly available. However, it is reassuring that these experiences are likely to be published and when available, another review will need to be conducted to capture all their experiences. This will greatly assist in presenting progress in the application of MD technology which is developing in interesting ways especially with novel process integrations to resolve current industry issues around water security and waste reduction.

Based on this assessment of recently conducted trials that have been presented in literature, the following future work is suggested:

- Continue to conduct pilot trials in applications of need using commercially orientated module designs;
- Publish findings in the open literature (not just in conference proceedings) to make a permanent record and more widely available source for the experience;
- In assessing the results, present important parameters such as flux, membrane area, plant footprint, equipment photos, STEC and more importantly the SEEC;
- Undertake repeated cleaning cycles and record performance under standard conditions;
- Carry out an autopsy of the membrane at the end of the trial period;
- Run trials at reasonable scales (>1 m^2 membrane area) on commercially oriented modules for long periods of time, being more than a week. Trials lasting several months and hopefully years, should be undertaken;
- Analyse and present findings of the experiences in operating the associated equipment (heat exchangers, pumps, instruments, etc.). Experiences include corrosion, long term scaling and means to manage the operation of the equipment; and
- Run a trial for several years to show currently unknown parameters such as membrane lifetime (replacement frequency) and reliable operating cost information.

16.7 CONCLUSION

This chapter has presented a review of the recent experiences gained from researchers worldwide to trial MD technology in applications of direct industry relevance. Recent published findings so far have provided critical information on long-term performance of equipment and membranes, optimal heat and flux performance, effective membrane cleaning approaches, integration into complete solutions, long-term data on application to treat industrial wastewater and combination with other major systems. This has been a major positive step to giving reference plant knowledge that gives confidence to installers that MD can be part of a solution to provide clean drinking water and help resolve current waste issues by the industry. So, while much effort has been undertaken already on bench scale systems and more recently pilot trials, moving the technology into industry would be greatly benefited from efforts focussed on presenting the outcomes of future planned pilot demonstrations operating for sufficient periods of time.

ACKNOWLEDGEMENTS

The authors would like to acknowledge the funding from the Australian Research Council Discovery Project (DP17012391) which partially supported the writing of this chapter.

REFERENCES

Andrés-Mañas, J.A., Ruiz-Aguirre, A., Acién, F.G. and Zaragoza, G. (2018) Assessment of a pilot system for seawater desalination based on vacuum multi-effect membrane distillation with enhanced heat recovery. Desalination 443, 110–121.

Camacho, L., Dumée, L., Zhang, J., Li, J.-D., Duke, M., Gomez, J. and Gray, S. (2013) Advances in membrane distillation for water desalination and purification applications. Water 5(1), 94–196.

Cassard, H.M. and Park, H.G. (2018) How to select the optimal membrane distillation system for industrial applications. Journal of Membrane Science 565, 402–410.

Dow, N., Duke, M., Zhang, J., O'Rielly, T., Li, J.-D., Gray, S., Ostarcevic, E. and Atherton, P. (2010). Demonstration of solar driven membrane distillation in remote Victoria, p. 036. Australian Water Association, Brisbane, Queensland, Australia.

Dow, N., Gray, S., Li, J.-D., Zhang, J., Ostarcevic, E., Liubinas, A., Atherton, P., Roeszler, G., Gibbs, A. and Duke, M. (2016) Pilot trial of membrane distillation driven by low grade waste heat: Membrane fouling and energy assessment. Desalination 391, 30–42.

Dow, N., Villalobos Garcia, J., Niadoo, L., Milne, N., Zhang, J., Gray, S. and Duke, M. (2017) Demonstration of membrane distillation on textile waste water: Assessment of long term performance, membrane cleaning and waste heat integration. Environmental Science: Water Research & Technology 3, 433–449.

Duong, H.C., Chivas, A.R., Nelemans, B., Duke, M., Gray, S., Cath, T.Y. and Nghiem, L.D. (2015) Treatment of RO brine from CSG produced water by spiral-wound air gap membrane distillation – A pilot study. Desalination 366 (0), 121–129.

Duong, H.C., Cooper, P., Nelemans, B., Cath, T.Y. and Nghiem, L.D. (2016) Evaluating energy consumption of air gap membrane distillation for seawater desalination at pilot scale level. Separation and Purification Technology 166, 55–62.

Eykens, L., De Sitter, K., Boeckaert, C., Boeckx, J. and Borgmans, G. (2018) Recovery of brines from cheese making using membrane distillation at lab and pilot scale. Journal of Food Engineering 219, 52–61.

Eykens, L., Hitsov, I., De Sitter, K., Dotremont, C., Pinoy, L. and Van der Bruggen, B. (2017) Direct contact and air gap membrane distillation: Differences and similarities between lab and pilot scale. Desalination 422, 91–100.

García-Fernández, L., Khayet, M. and García-Payo, M.C. (2015) Pervaporation, vapour permeation and membrane distillation. Basile, A., Figoli, A. and Khayet, M. (eds), pp. 317–359. Woodhead Publishing, Oxford.

Gil, J.D., Roca, L., Ruiz-Aguirre, A., Zaragoza, G. and Berenguel, M. (2018a) Optimal operation of a Solar Membrane Distillation pilot plant via Nonlinear Model Predictive Control. Computers & Chemical Engineering 109, 151–165.

Gil, J.D., Roca, L., Zaragoza, G. and Berenguel, M. (2018b) A feedback control system with reference governor for a solar membrane distillation pilot facility. Renewable Energy 120, 536–549.

Gil, J.D., Ruiz-Aguirre, A., Roca, L., Zaragoza, G. and Berenguel, M. (2018c) Prediction models to analyse the performance of a commercial-scale membrane distillation unit for desalting brines from RO plants. Desalination 445, 15–28.

Guan, G., Yang, X., Wang, R. and Fane, A.G. (2015) Evaluation of heat utilization in membrane distillation desalination system integrated with heat recovery. Desalination 366, 80–93.

Guillén-Burrieza, E., Blanco, J., Zaragoza, G., Alarcón, D.-C., Palenzuela, P., Ibarra, M. and Gernjak, W. (2011) Experimental analysis of an air gap membrane distillation solar desalination pilot system. Journal of Membrane Science 379(1), 386–396.

Guillen-Burrieza, E., Ruiz-Aguirre, A., Zaragoza, G. and Arafat, H.A. (2014) Membrane fouling and cleaning in long term plant-scale membrane distillation operations. Journal of Membrane Science 468, 360–372.

Guillén-Burrieza, E., Zaragoza, G., Miralles-Cuevas, S. and Blanco, J. (2012) Experimental evaluation of two pilot-scale membrane distillation modules used for solar desalination. Journal of Membrane Science 409–410, 264–275.

Hausmann, A., Sanciolo, P., Vasiljevic, T., Kulozik, U. and Duke, M. (2014) Performance assessment of membrane distillation for skim milk and whey processing. Journal of Dairy Science 97(1), 56–71.

Hausmann, A., Sanciolo, P., Vasiljevic, T., Ponnampalam, E., Quispe-Chavez, N., Weeks, M. and Duke, M. (2011) Direct contact membrane distillation of dairy process streams. Membranes 1(1), 48–58.

Hausmann, A., Sanciolo, P., Vasiljevic, T., Weeks, M., Schroën, K., Gray, S. and Duke, M. (2013a) Fouling mechanisms of dairy streams during membrane distillation. Journal of Membrane Science 441(0), 102–111.

Hausmann, A., Sanciolo, P., Vasiljevic, T., Weeks, M., Schroën, K., Gray, S. and Duke, M. (2013b) Fouling of dairy components on hydrophobic polytetrafluoroethylene (PTFE) membranes for membrane distillation. Journal of Membrane Science 442(0), 149–159.

Kesieme, U.K., Milne, N., Aral, H., Cheng, C.Y. and Duke, M. (2013) Economic analysis of desalination technologies in the context of carbon pricing, and opportunities for membrane distillation. Desalination 323(0), 66–74.

Lu, K.-J., Zuo, J. and Chung, T.-S. (2016) Tri-bore PVDF hollow fibers with a super-hydrophobic coating for membrane distillation. Journal of Membrane Science 514, 165–175.

Mansour, S., Giwa, A. and Hasan, S.W. (2018) Novel graphene nanoplatelets-coated polyethylene membrane for the treatment of reject brine by pilot-scale direct contact membrane distillation: An optimization study. Desalination 441, 9–20.

Martinetti, C.R., Childress, A.E. and Cath, T.Y. (2009) High recovery of concentrated RO brines using forward osmosis and membrane distillation. Journal of Membrane Science 331(1–2), 31–39.

Micari, M., Cipollina, A., Giacalone, F., Kosmadakis, G., Papapetrou, M., Zaragoza, G., Micale, G. and Tamburini, A. (2019) Towards the first proof of the concept of a reverse electrodialysis – Membrane distillation heat engine. Desalination 453, 77–88.

Morrow, C.P., Furtaw, N.M., Murphy, J.R., Achilli, A., Marchand, E.A., Hiibel, S.R. and Childress, A.E. (2018) Integrating an aerobic/anoxic osmotic membrane bioreactor with membrane distillation for potable reuse. Desalination 432, 46–54.

Ruiz-Aguirre, A., Andrés-Mañas, J.A., Fernández-Sevilla, J.M. and Zaragoza, G. (2017a) Modeling and optimization of a commercial permeate gap spiral wound membrane distillation module for seawater desalination. Desalination 419, 160–168.

Ruiz-Aguirre, A., Andrés-Mañas, J.A., Fernández-Sevilla, J.M. and Zaragoza, G. (2018) Experimental characterization and optimization of multi-channel spiral wound air gap membrane distillation modules for seawater desalination. Separation and Purification Technology 205, 212–222.

Ruiz-Aguirre, A., Polo-López, M.I., Fernández-Ibáñez, P. and Zaragoza, G. (2017b) Integration of membrane distillation with solar photo-Fenton for purification of water contaminated with Bacillus sp. and Clostridium sp. spores. Science of The Total Environment 595, 110–118.

Song, L., Ma, Z., Liao, X., Kosaraju, P.B., Irish, J.R. and Sirkar, K.K. (2008) Pilot plant studies of novel membranes and devices for direct contact membrane distillation-based desalination. Journal of Membrane Science 323(2), 257–270.

Suárez, F., Ruskowitz, J.A., Tyler, S.W. and Childress, A.E. (2015) Renewable water: Direct contact membrane distillation coupled with solar ponds. Applied Energy 158, 532–539.

Woldemariam, D., Kullab, A., Fortkamp, U., Magner, J., Royen, H. and Martin, A. (2016) Membrane distillation pilot plant trials with pharmaceutical residues and energy demand analysis. Chemical Engineering Journal 306, 471–483.

Zhang, P., Knötig, P., Gray, S. and Duke, M. (2015) Scale reduction and cleaning techniques during direct contact membrane distillation of seawater reverse osmosis brine. Desalination 374, 20–30.

Zuo, G. and Wang, R. (2013) Novel membrane surface modification to enhance anti-oil fouling property for membrane distillation application. Journal of Membrane Science 447, 26–35.

Index

Page numbers in italics indicate figures, and those in bold indicate tables.

N

O

P

Q

R

S